T0184468

Machine Learning Applications in Electronic Design Automation

Haoxing Ren • Jiang Hu

Editors

Machine Learning Applications in Electronic Design Automation

 Springer

Editors
Haoxing Ren
NVIDIA
Austin, TX, USA

Jiang Hu
Texas A&M University
College Station, TX, USA

ISBN 978-3-031-13076-2   ISBN 978-3-031-13074-8   (eBook)
https://doi.org/10.1007/978-3-031-13074-8

© The Editor(s) (if applicable) and The Author(s), under exclusive license to Springer Nature Switzerland
AG 2022
This work is subject to copyright. All rights are solely and exclusively licensed by the Publisher, whether
the whole or part of the material is concerned, specifically the rights of translation, reprinting, reuse
of illustrations, recitation, broadcasting, reproduction on microfilms or in any other physical way, and
transmission or information storage and retrieval, electronic adaptation, computer software, or by similar
or dissimilar methodology now known or hereafter developed.
The use of general descriptive names, registered names, trademarks, service marks, etc. in this publication
does not imply, even in the absence of a specific statement, that such names are exempt from the relevant
protective laws and regulations and therefore free for general use.
The publisher, the authors, and the editors are safe to assume that the advice and information in this book
are believed to be true and accurate at the date of publication. Neither the publisher nor the authors or
the editors give a warranty, expressed or implied, with respect to the material contained herein or for any
errors or omissions that may have been made. The publisher remains neutral with regard to jurisdictional
claims in published maps and institutional affiliations.

This Springer imprint is published by the registered company Springer Nature Switzerland AG
The registered company address is: Gewerbestrasse 11, 6330 Cham, Switzerland

"To the loving memory of my mother: Zhong Yunlu."
Haoxing Ren

"To my wife Min Zhao."
Jiang Hu

Preface

Electronic design automation (EDA) is a software technology that attempts to let computers undertake chip design tasks so that we can handle complexities beyond manual design capabilities. Although conventional EDA techniques have led to huge design productivity improvement, they face the fundamental limit that most EDA problems are NP hard and therefore have no polynomial-time algorithms for optimal solutions. As chip complexity increases to dozens of billions of transistors, such limitation becomes even more pronounced and there is a compelling need to have innovative changes.

Luckily, machine learning (ML) technology has recently made unprecedented progress and created phenomenal impacts to numerous computing applications. Deep learning (DL) models such as convolutional neural networks (CNNs) solved the image recognition problem; generative adversarial networks (GAN) generate impressive real-looking images; reinforcement learning (RL) agent beats the best human GO players; and so forth. The tremendous success motivates the study of ML techniques for advancing the state of the art of EDA technology. Indeed, this is what has been happening in the past several years. In almost all major EDA conferences and journals, ML EDA papers grew from sporadic appearances to mainstream in a few years. Such growing interest and research activities led to the book *Machine Learning in VLSI Computer-Aided Design* edited by Elfadel, Boning, and Li in 2019 as well as two survey papers published in 2021—one in the ACM Transactions on Design Automation of Electronic Systems and the other in the IEEE Transactions on Computer-Aided Design. Although the book by Elfadel, Boning, and Li provides in-depth review on various topics including lithography, physical design, yield prediction, post-silicon performance analysis, reliability and failure analysis, power and thermal analysis, analog design, logic synthesis, verification, and neuromorphic design, it could not include many newer techniques published in 2019 or later, which are substantially important. In this regard, our book offers the latest view for this fast progressing field. The two survey papers provide a very comprehensive coverage; however, their reviews are generally brief due to limited space. As such, readers may still need to search and find the original papers for understanding important details. Our goal is to provide a one-stop solution that balances depth and breadth.

This book is organized into three parts. Part I, including Chaps. 1–6, is focused on machine learning-based design prediction techniques. Chapter 1 introduces the problem of design QoR (quality of results) prediction in general and highlights many important issues of this application such as synergy with optimization, infrastructure, and pitfalls. Chapter 2 describes techniques on how to predict routability in terms of design rule violations in the placement stage. Chapter 3 introduces techniques for predicting net length and timing at the logic synthesis stage and how to predict crosstalk effect in placement stage without any trial routing. Chapter 4 describes power and switching activity prediction using graph neural network (GNN) models. Chapter 5 covers IR drop, EM hotspot, and thermal prediction. Chapter 6 introduces application of ML in testability prediction.

Part II (Chaps. 7–12) is centered around machine learning-based design optimization techniques. Chapter 7 describes the application of reinforcement learning (RL) in logic synthesis, such as synthesis flow optimization and cut choices in technology mapping. Chapter 8 shows an industrial approach of deep reinforcement learning for macro placement. Chapter 9 proposes a novel technique that leverages ML infrastructure for fast cell placement optimization. In Chap. 10, GNN and RL techniques are described for optimizations in 3D and 2.5D designs. Chapter 11 introduces RL applications in routing. Chapter 12 covers a Bayesian optimization approach for analog circuit sizing.

Part III (Chaps. 13–18) emphasizes the application of ML techniques on various specific domains in EDA. Chapter 13 discusses various ML approaches for design space exploration. Chapter 14 discusses the application of ML in functional verification, which is a key bottleneck in many chip design processes. Chapter 15 introduces deep learning techniques in design for manufacturability including lithography modeling, litho-hotspot detection, and mask optimization. Chapter 16 is focused on ML applications in FPGA designs, including QoR prediction in high-level synthesis and design space exploration. Chapter 17 introduces ML techniques for automatic analog layout generation including placement and routing. Finally, Chapter 18 introduces ML applications in system level performance and power models.

This book is targeted to EDA researchers, graduate students, and practitioners with basic background knowledge of EDA and machine learning. Therefore, this book does not dedicate specific chapters on machine learning background. Moreover, tutorials on machine learning can be easily found in plenty of literature and online resources. Meanwhile, to be self-contained, some chapters cover their related machine learning background.

Although we strive to reach a compromise between breadth and depth, this book is by no means an encyclopedia of machine learning EDA. It puts an emphasis on currently popular ML approaches such as deep learning and reinforcement learning, and provides little coverage on older techniques such as linear regression and support vector machines. Also, ML for EDA is a fast-changing field. As we are preparing this book, new techniques are continuously published and cannot be included immediately. Nevertheless, we hope this book will serve as an easy approachable stepping stone for people to enter the research and development in this field.

Contents

About the Editors

Haoxing Ren (Mark) was born in Nanchang, China, in 1976. He received two BS degrees in electrical engineering and finance, and an MS degree in electrical engineering from Shanghai Jiao Tong University, China, in 1996 and 1999, respectively; an MS in computer engineering from Rensselaer Polytechnic Institute in 2000; and a PhD in computer engineering from the University of Texas at Austin in 2006. From 2000 to 2015, he worked at IBM Microelectronics and Thomas J. Watson Research Center (after 2006) developing physical design and logic synthesis tools and methodology for IBM microprocessor and ASIC designs. He received several IBM technical achievement awards including the IBM Corporate Award for his work on improving microprocessor design productivity. After his 15-year tenure at IBM, he had a brief stint as a technical executive at a chip design start-up developing server-class CPUs based on IBM OpenPOWER technology. In 2016, Mark joined NVIDIA Research, where he currently leads the Design Automation research group, whose mission is to improve the quality and productivity of chip design through machine learning and GPU-accelerated tools. He has published many papers in the field of design automation including several book chapters in logic synthesis and physical design. He has also received best paper awards at International Symposium on Physical Design (ISPD) in 2013, Design Automation Conference (DAC) in 2019, and IEEE Transactions on Computer-Aided Design of Integrated Circuits and Systems in 2021.

Jiang Hu received his BS in optical engineering from Zhejiang University (China) in 1990, MS in physics in 1997, and PhD in electrical engineering from the University of Minnesota in 2001. He worked with IBM Microelectronics from January 2001 to June 2002.

In 2002, he joined the electrical engineering faculty at Texas A&M University. His research interests include design automation of VLSI circuits and systems, computer architecture, hardware security, and machine learning applications. His honors include receiving a best paper award at the ACM/IEEE Design Automation Conference in 2001, an IBM Invention Achievement Award in 2003, a best paper award at the IEEE/ACM International Conference on Computer-Aided Design in 2011, a best paper award at the IEEE International Conference on Vehicular Electronics and Safety in 2018, and a best paper award at the IEEE/ACM International Symposium on Microarchitecture in 2021. He has served as technical program committee member for DAC, ICCAD, ISPD, ISQED, ICCD, DATE, ISCAS, ASP-DAC, and ISLPED. Jiang Hu is the general chair for the 2012 ACM International Symposium on Physical Design. He has served as an associate editor for IEEE Transactions on CAD and the ACM Transactions on Design Automation of Electronic Systems. He received the Humboldt Research Fellowship in 2012. He was named an IEEE Fellow in 2016.

Part I
Machine Learning-Based Design Prediction Techniques

Chapter 1
ML for Design QoR Prediction

Andrew B. Kahng and Zhiang Wang

1.1 Introduction

Semiconductor technology scaling is challenged on many fronts that include pitch reduction, patterning flexibility, wafer processing cost, interconnect resistance, and manufacturing variability. The difficulty of continuing Moore's law lateral scaling beyond the foundry 5 nm/3 nm nodes has been widely lamented. Scaling boosters (buried interconnects, backside power delivery, supervias), next device architectures (VGAA FETs, CFETs), ever-improving design technology co-optimizations, and use of the vertical dimension (heterogeneous multi-die integration, monolithic 3D VLSI) all offer potential extensions of the industry's scaling trajectory. In addition, various "rebooting computing" paradigms –quantum, approximate, stochastic, adiabatic, neuromorphic, etc.– are being actively explored.

No matter how future extensions of semiconductor scaling materialize, the industry already faces a crisis: design of new products in advanced technology nodes costs too much. Indeed, the 2001 *International Technology Roadmap for Semiconductors* [23] noted that "cost of design is the greatest threat to continuation of the semiconductor roadmap". Cost pressures rise when incremental technology and product benefits fall. The industry transition from 40nm to 28nm brought as little as 20% power, performance, or area (PPA) benefit. Today, going from foundry 7nm to 5nm, without aggressive deployment of "scaling boosters" the PPA benefit is significantly less, and products may well realize only one or possibly two of these PPA wins. As a result, there is now a tremendous focus on improvement of IC design quality of results (QoR) within available design resources, such as licenses, engineers, and schedule.

A. B. Kahng (✉) · Z. Wang
University of California San Diego, La Jolla, CA, USA
e-mail: abk@eng.ucsd.edu; zhw033@ucsd.edu

© The Author(s), under exclusive license to Springer Nature Switzerland AG 2022
H. Ren, J. Hu (eds.), *Machine Learning Applications in Electronic Design Automation*, https://doi.org/10.1007/978-3-031-13074-8_1

Improving product design outcomes without changing the underlying design enablement (process, device, interconnect, IPs, tooling) is a key form of *design-based equivalent scaling* [29] that extends the usefulness of leading-edge technology to more designers and new products. A powerful lever for this is the use of machine learning (ML) techniques, both inside and "around" electronic design automation (EDA) tools.

Generally speaking, machine learning for QoR improvement tries to answer the following types of questions:

- **Prediction**: For example, will *RouteOpt* finish with clean sign-off (e.g., less than 1000 DRVs) by tomorrow night?
- **Classification**: For example, out of 50 candidate floorplans and budgeting solutions, which three should go forward into trial SP&R?
- **Estimation**: For example, how many hold buffers will the tool eventually add into a given post-CTS layout?
- **Guidance**: For example, what synthesis, placement, and routing (SP&R) tool setup/script will obtain the best QoR within the next 36 h?

All of these questions require modeling and prediction of downstream flow steps and their QoR outcomes. The predictive models (e.g., of wirelength, power, area, timing, etc.) become objectives or guides for optimizations, via a "modeling stack" that reaches up to system, architecture, and even project and enterprise levels. Furthermore, design QoR prediction improves design efficiency and quality. For example, physical design (PD) engineers typically perform multiple runs in parallel with different tool configurations –up to limits of compute and license resources– to perform power-performance-area (PPA) exploration. Each run consumes significant schedule and resources. An early-stop mechanism guided by design QoR prediction can reduce wasted resources and allow them to be redirected to more promising runs. The more efficient search of the design solution space improves design outcomes within the given design schedule.

This chapter will introduce machine learning-aided approaches for design quality of results (QoR) prediction. The remainder of this chapter is organized as follows. Section 1.2 presents various challenges for QoR prediction. Section 1.3 reviews several machine learning techniques that have been widely used in QoR prediction. Sections 1.4 and 1.5 illustrate design QoR prediction in two contexts –timing estimation and design space exploration– with specific examples. Section 1.6 concludes with a view of future directions for ML-based design QoR prediction.

1.2 Challenges of Design QoR Prediction

Design QoR is the result of a long (from days to months) design process involving multiple complex tools, each of which incorporates numerous optimization heuristics while offering the designer thousands of command-option and parameter combinations. At the same time, tool versions, tool flows, process nodes, libraries and IPs, product goals, and the product designs themselves are always changing. As a consequence, design QoR prediction has several fundamental challenges.

1.2.1 Limited Number of Samples

Training a high-quality QoR predictor usually requires rich and representative data. However, in many contexts the total number of samples available in a given dataset is limited by a lack of relevant testcases. Further, long runtimes of synthesis, placement, and routing (SP&R) tools can be a bottleneck in collecting enough samples. For example, a DNN accelerator [13] may easily have 5–10 million placeable instances: the SP&R runtime for a design of such complexity will be on the order of a week, which means that data collection for training may require several months. Besides, a given dataset can have very few or no positive samples. For example, in the static timing analysis (STA) domain, samples might consist of combinational paths (i.e., signal paths from launch flip-flop to sequentially adjacent capture flip-flop), and positive samples might consist of negative-slack paths as reported by STA tools. Then, the dataset for STA-critical paths prediction may be imbalanced because most paths are noncritical.

Various approaches can be utilized to relieve the negative impact of limited data samples. These include:

- **Sampling.** When the number of samples is very limited, sampling techniques such as Latin hypercube sampling (LHS) can be used to make the dataset more representative. Sampling is also at the heart of derivative-free or black-box optimization methods, which can also be used to predict QoR outcomes.
- **Feature engineering.** For example, in the above STA-critical paths prediction, a set of design-related features can be developed for describing a path. If an effective set of features is used, an accurate prediction model can be built rather easily. Otherwise, it can be very difficult to learn a good model.

Other potentially relevant ML techniques range from data augmentation to transfer and low-shot learning. However, the challenge of limited sample size will always exist, given the enormous space of potential chip plans and tool configurations, and the limited compute resources available for development of QoR prediction models.

1.2.2 Chaotic Behaviors of EDA Tools

Physical design tools and flows today are inherently unpredictable. A root cause is that many complex heuristics have been accreted upon previous complex heuristics. Thus, tools have become noisy or even chaotic [25, 32]–meaning that a small change to input conditions causes a very large change in outcome– particularly when they are pushed to their QoR limits. Figure 1.1 demonstrates the chaotic behavior of EDA

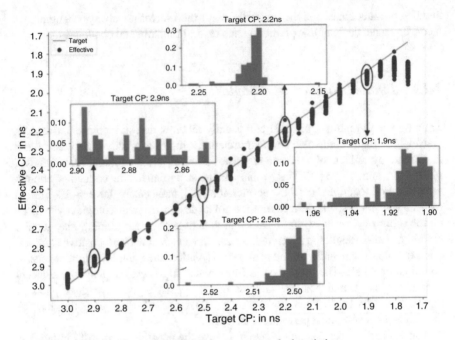

Fig. 1.1 Distributions of actual clock period versus target clock period

tools. The figure shows the mapping between target clock period (target CP) and actual clock period (effective CP). The x-axis has 26 target clock periods, spaced 50 ps apart. For each target clock period, 101 results are generated by stepping the target clock period by 0.01 ps in the $[-0.5\,ps, +0.5\,ps]$ interval. We can see that the actual clock period can change quite a lot even if the target clock period changes by as little as 0.01 ps, especially when the target clock period corresponds to a target frequency near the maximum that is achievable.

Unpredictability of design implementation leads to unpredictability of the design schedule. Since product companies must strictly meet design and tapeout schedules, the design target (PPA) must be guardbanded, impacting both product quality and profitability. Relevant considerations thus include the following. First, current heuristics and tools are chaotic when designers demand best-quality results. Second, when designers want predictable results, they must aim low. Third, the complexity of ML models for QoR prediction can increase significantly with data samples at the limits of tools. Fourth, special attention should be paid to avoid overfitting when improving the accuracy of ML models. Last, more samples at the limits of tools are needed for ML models to capture the chaotic behaviors of EDA tools, which typically implies that techniques such as importance sampling must be deployed.

1.2.3 Actionable Predictions

Design QoR prediction must be useful and actionable. The most basic form of this challenge: a prediction that 5 GHz clock frequency is achievable within a 2 W power budget is not useful unless accompanied by a "certificate", i.e., a flow script that actually results in this performance and power efficiency. A deeper form of this challenge arises from the fact that every major design phase, such as physical implementation, is actually a stage-by-stage sequential process, wherein the outcome of the current stage strongly depends on the outcomes from previous stages. Therefore, acting on a prediction can sometimes change what is being predicted. Thus: **Be careful what you ask for.**

Chan et al. [9] illustrate this phenomenon in the macro placement context. In Fig. 1.2, the left-side figures show original solutions from automatic macro placement by two P&R tools. Then, for the same testcase, some of the macros (dark-blue rectangles around the periphery in the right-side figures) are *preplaced* in their original locations, simulating the "perfect prediction" of their placements. We see that the positions of remaining macros (orange rectangles in the right-side figures) end up being quite different from the original solutions (corresponding orange rectangles in the left-side figures). In other words, when "correct" predictions of

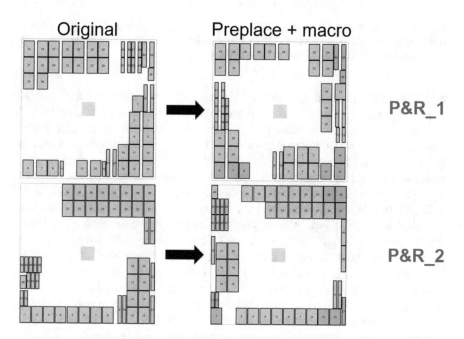

Fig. 1.2 Visualization of macro placement solutions when a subset of macros are preplaced for the *SweRV_wrapper* design [44]. The center light-blue square is a small macro placement blockage, and dark-blue rectangles around the periphery denote fixed macros based on the original solutions [9]

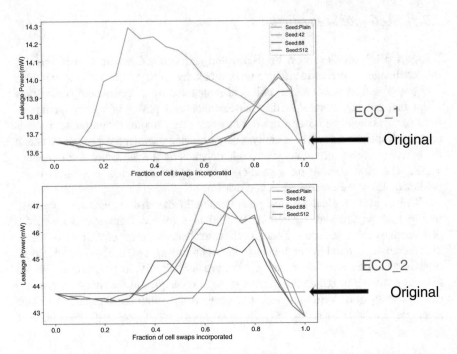

Fig. 1.3 Visualization of ECO solutions when a *subset* of cell swaps from the original solution produced by each of two commercial ECO tools is provided as a "hint" to leakage reduction. Seed: Plain means that the order of cell swaps is not changed

macro locations are used, this changes the placement outcomes of other macros. Applying a (perfect, partial) prediction can change the outcome that is predicted.

Figure 1.3 gives another example of this phenomenon, in the engineering change order (ECO) power optimization context. During ECO leakage power recovery, an ECO tool will try to swap LVT cells with HVT cells (or high drive strength cells with low drive strength cells) to recover leakage power without affecting timing sign-off. Here, we first record all the cell swaps reported by a given ECO tool, in order. We then randomly shuffle all the cell swaps with different random seeds and feed a subset of cell swaps, with varying size, as a starting point for the ECO tool. We can see that for both ECO tools, the leakage recovery outcome is no longer the same as originally achieved. Notably, when given a "large hint" of around 70–80% of its original swap-cell list, each tool performs worse than if it were given no hint at all. This means that if a predictor can predict 70–80% of the ECO tool's original swap-cell list and backend designers adopt the prediction (i.e., executing the corresponding cell swaps before launching the ECO tool), the post-ECO QoR may not be what they expect. Therefore, be careful what you ask for.

1.2.4 Infrastructure Needs

To enable ML-based design QoR prediction, considerable new infrastructure is required. For example, standards for ML model encapsulation, model application, IP-preserving model sharing, etc. will likely be required before any training data, data generation tasks, or ML models can be shared across multiple organizations. Overall, infrastructure requirements are too numerous to list. However, four specific requirements are the following. First, design owners, foundries and EDA providers should be comfortable that their IP (design function, technology parameters, protected syntax, etc.) is sufficiently protected. This might be achieved by standard anonymization (privacy preservation) and obfuscation mechanisms. Second, ML researchers, foundries, and EDA tool users should be comfortable that their use of ML to improve IC design enablement and flows does not risk IP infringement claims. For example, consensus is needed as to whether it is permissible to share a learning-based model of a delay calculator's "error versus SPICE." Third, academic researchers should be attracted with a critical mass of ML modeling challenges, supporting data, and incentives (e.g., a "Kaggle [27] for machine learning in IC design"). Fourth, in the longer term, the design, EDA, and research communities must share responsibility for deployment of a standard ML platform for EDA and IC design modeling that spans design metrics collection, tool and flow model generation, design-adaptive tool and flow configuration, and prediction of tool and flow outcomes.

The "standard ML platform for EDA and IC design modeling" requirement recalls the METRICS [14, 31] initiative, proposed nearly 20 years ago as a standard platform and industry infrastructure for measurement of the IC design process. A new METRICS2.1 [26] version, developed by the OpenROAD Project [10] in collaboration with the IEEE CEDA Design Automation Technical Committee, has been recently released. Figure 1.4 gives an overview of the METRICS2.1 infrastructure.

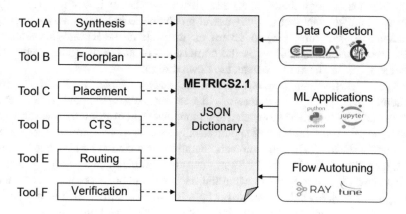

Fig. 1.4 Overview of METRICS2.1 infrastructure

METRICS2.1 supports the analysis of how flow parameter settings affect QoR outcomes, as well as the building of machine learning applications to predict tool and flow outcomes. A typical EDA design flow invokes multiple engines, with each engine (e.g., global placement) having multiple parameters to guide the heuristic optimization that it performs. The parameter settings for a given engine will affect not only the results produced by that engine but also the results of the entire flow. METRICS2.1 provides consistent reporting of metrics across flow stages, enabling the collection of run metrics from large-scale designs of experiments that vary tool and flow parameters in a controlled manner. Moreover, because METRICS2.1 captures all parameter settings and commit versions used in a given run, collected data is inherently reproducible. The collected data serves many purposes, e.g., (i) showing evolution of PPA metrics throughout the flow, (ii) enabling recovery of parameter settings that led to particular outcomes, and (iii) giving insights into trends and interrelationships between values of multiple parameters. Typical experiment types and purposes include (i) running one design on one platform to study the impact of parameter settings for that design on that platform, (ii) running multiple designs on the same platform to model and predict outcomes for unseen designs on that platform, and (iii) running multiple designs on multiple platforms to build predictive models.

1.2.5   The Bar for Design QoR Prediction

We close this section by discussing the context and the threshold for ML-based design QoR prediction methods to impact real-world design practice.

Design QoR prediction has been well-recognized as a critical need ever since the "design productivity gap" was highlighted by SEMATECH in 1993. Indeed, METRICS [14] cited "the basic problem of unpredictable design success" as a key motivation, while contemporaneous discussions of emerging hierarchical planning methodology cited "the need for forward prediction in a convergent design process" and "the fact that the only useful prediction technologies we know of are constructive predictors" [28]. A review of prospects in the RTL-to-GDS domain [19] noted that predictability "is the cornerstone of any convergence-improving approach: It is the heart of scalable, top-down design."

Figure 1.5 (left) is reproduced from the design chapter in the 2009 ITRS Roadmap [41]. The figure's message is that earlier knobs in the flow (system level, architecture, RTL design) grow relatively more powerful over time. However, while design space exploration (DSE) should ideally explore more powerful knobs more thoroughly (ideal "DSE"), attention and iterations still tend to be biased toward the RTL downflow (today's "DSE"). This is because optimizing and exploring in early stages have only limited value when the backend cannot be predicted accurately enough, and high-level decisions do not correlate to what can actually be closed and signed off. This again points to prediction of design QoR as a fundamental need.

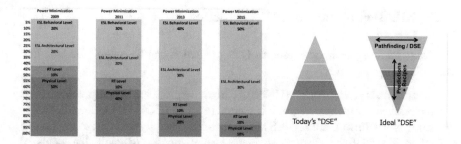

Figure DESN5   Evolving Role of Design Phases in Overall System Power Minimization

Fig. 1.5 Left: growing impact of higher-level design stages on system-level power optimization with advancing technology and system complexity [24]. Right: today's design space exploration (DSE) cannot accurately explore higher-level design stages, whereas an Ideal DSE would spend effort where it can pay off the most. Drawn from [30]

ML-based design QoR prediction *coexists* and *competes* with other approaches that share the same goals. [19] noted that EDA has always worked with three basic categories of prediction approaches. (i) *Statistical* prediction methods (e.g., based on wireload models or rent parameter values) are characterized as being efficient but unable to deliver the required accuracy. In particular, design quality metrics of interest, such as maximum clock frequency or noise margin, are maximum or extreme (as opposed to average or ensemble) statistics and hence very difficult to predict. Arguably, today's ML-based prediction methods can be seen as the next generation of statistical methods. (ii) *Constructive* prediction methods esti- mate future outcomes by "quick-and-dirty, under-the-hood" means, such as quick placement or trial routing. Even 20 years ago, such methods "track[ed] previous estimates of logic, placement, and routing congestion", and "use[d] the estimates to formulate next solutions" [19], much like today's data- and experience-driven ML approaches. However, because constructive prediction methods essentially run a given optimization heuristic in order to predict what that heuristic will do, the constructive approach does not substantially improve scalability of the design process. (iii) *Assume and enforce* methods involve making some concrete assumption about a property of the design –e.g., the signal propagation delay on a specific timing arc– and then ensuring that the assumption holds true in the final design outcome. The approach is often traced back to Grodstein et al. [16]. Enforcement of an assumption takes some amount of unpredictability out of the picture; in this sense, assume-and-enforce can be grouped with many other design methodology concepts, such as staggered repeaters, noise-free signaling fabrics, registering of block I/Os, etc. By improving predictability, these methodology concepts reduce the urgency of developing new prediction methods, This, too, is a challenge for ML-based design QoR prediction.

1.3 ML Techniques in QoR Prediction

Machine learning techniques used in QoR prediction can be roughly divided into the following categories.

- **Graph neural networks (GNNs).** Circuit netlists are usually modeled as hypergraphs or graphs; hence, it is natural to apply GNNs for QoR prediction.
- **Long short-term memory (LSTM) networks.** The physical design implementation is actually a stage-by-stage sequential process, wherein the status of the current stage highly depends on the outcomes from previous stages. As a variant of recurrent neural networks (RNN), the LSTM network can capture such time series information by memorizing the important information from previous stages.
- **Reinforcement learning (RL).** The algorithms implemented inside EDA tools have many exposed parameter settings that users can tune to achieve desired power-performance-area (PPA) outcomes. Whereas physical design engineers might execute massively parallel tool runs with different tool configurations in a quest to achieve decent QoR, this is both timing-consuming and resource-inefficient. Reinforcement learning can train an autonomous agent that iteratively learns to tune parameter settings with "no human in the loop."
- **Other models.** Traditional ML techniques, such as (i) linear regression, (ii) feedforward neural network, (iii) gradient boosted decision trees, and (iv) random forest (RF), can also be leveraged for QoR prediction.

We briefly introduce the abovementioned ML techniques in the following subsections.

1.3.1 Graph Neural Networks

In a graph neural network (GNN) [36, 39] with an input graph $G(V, E)$, each node $v \in V$ is associated with a vector of features $h_v^{(0)} = (x_1, x_2, \ldots, x_d)$. For each node v, $N(v) = \{u | \exists e = \{u, v\} \in E\}$ denotes the neighborhood of v. A trained GNN takes $G(V, E)$ as input and generates the node embedding $ENC(v) = (y_1, y_2, \ldots, y_m)$ for each node. Then, the graph embedding $ENC(G)$ for the entire graph can be obtained by performing global pooling on the node embeddings, i.e.,

$$ENC(G) = global\_pooling(ENC(v) \mid v \in V) \tag{1.1}$$

Examples of the *global_pooling* function are *max*, *min*, and *mean*. The node embedding of each node can be used for node-level applications such as node classification, and the graph embedding of the entire graph can be used for graph-level applications, such as graph classification.

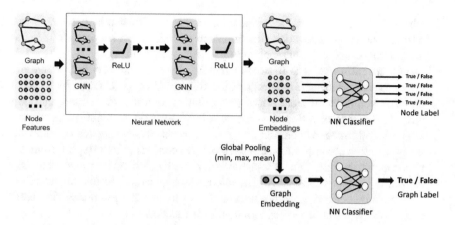

Fig. 1.6 An example architecture of graph neural networks (GNNs)

Figure 1.6 shows an example architecture of graph neural networks (GNNs) for node classification and graph classification.

The key idea in the GNN approach is to generate the node embedding through a neighbor message-passing mechanism that serves to aggregate information from the local neighborhood. Based on the choice of neighbor message-passing mechanism, GNNs can be further classified into graph convolutional neural networks (GCNs) [34], GraphSAGE [17], and graph attention networks (GATs) [47].

- **Graph convolutional neural networks (GCNs)** mirror the same basics of convolutional neural networks (CNNs). Given a graph $G(V, E)$ and a feature vector $h_v^{(0)} = (x_1, x_2, \ldots, x_d)$ for each node $v \in V$ as input, each GCN layer can be written as a nonlinear function

$$h_v^{(l+1)} = \sigma \left(W^{(l)} \cdot h_v^{(l)} \right) \tag{1.2}$$

  where $W^{(l)}$ is the trainable weight matrix for the lth neural network layer and $\sigma(\cdot)$ is a nonlinear activation function such as a rectified linear unit (ReLU).

- **GraphSAGE** is a widely used framework for graph classification tasks in the ML literature. GraphSAGE first performs mean, min, or max-pooling neighborhood aggregation and then updates the node representation by applying a linear projection on top of the concatenation of the current node representation and its neighborhood aggregation. Given a graph $G(V, E)$ and a feature vector $h_v^{(0)} = (x_1, x_2, \ldots, x_d)$ for each node $v \in V$ as input, each GraphSAGE layer can be written as a nonlinear function

$$h_v^{(l+1)} = \sigma \left(W^{(l)} \cdot concat \left(h_v^{(l)}, aggregate \left(\{h_u^{(l)}, \forall u \in \mathcal{N}(v)\} \right) \right) \right) \tag{1.3}$$

where $W^{(l)}$ is the trainable weight matrix for the lth neural network layer, $\sigma(\cdot)$ is a nonlinear activation function such as ReLU, and $aggregate(\cdot)$ is the neighborhood aggregation function (e.g., mean, min, or max).

- **Graph attention neural networks (GATs)** leverage masked self-attentional layers to address the shortcomings of the GCNs and GraphSAGE, which assign uniform weight to different nodes in a neighborhood. GATs enable specifying different weights to different nodes in a neighborhood, without requiring any kind of costly matrix operation (such as inversion) or dependence on knowing the graph structure upfront. A single GAT layer consists of two steps: *weighting*, which applies a shared linear transformation (parameterized by a *trained weight matrix*, W) to every node, and *aggregation*, which generates the output feature for each node by collecting feature information of its neighboring nodes. Formally, each GAT layer can be written as a nonlinear function

$$h_v^{(l+1)} = \sigma\left(\sum_{u \in \mathcal{N}(v)} \alpha_{vu} W^{(l)} h_u^{(l)}\right) \tag{1.4}$$

where $W^{(l)}$ is the trainable weight matrix for the lth neural network layer, $\sigma(\cdot)$ is a nonlinear activation function such as ReLU, and α_{vu} is the attention coefficient from node u to node v. The attention coefficient α_{vu} is given by

$$\alpha_{vu} = \frac{exp(a(W^{(l)}h_v, W^{(l)}h_u))}{\sum_{k \in \mathcal{N}(v)} a(W^{(l)}h_v, W^{(l)}h_k)} \tag{1.5}$$

where a is a shared attentional mechanism on the nodes, which can be a single-layer feedforward neural network.

1.3.2 Long Short-Term Memory (LSTM) Networks

Long short-term memory (LSTM) networks [20] are a special class of recurrent neural networks (RNN), which can effectively memorize information from previous steps, thus addressing the challenge of long-term dependencies. The capability to learn long-term dependencies makes LSTM networks well-suited for prediction based on sequential data. In contrast to other neural network types such as GNNs, LSTM networks consist of a chain of repeating neural network blocks. Figure 1.7 shows a prototypical architecture of LSTM networks.

The magic of LSTM networks lies at the repeating block of neural networks ("A" in Fig. 1.7). As shown in Fig. 1.7, an LSTM network has two types of states: (i) hidden states, also known as short-term memory, for example, h_{t-1} and h_t are the hidden states of previous and current timestep, respectively; and (ii) cell states, also known as long-term memory, for example, C_{t-1} and C_t are the cell states of previous

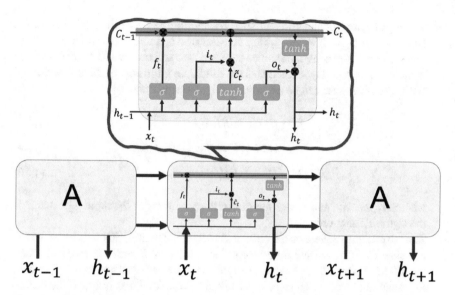

Fig. 1.7 Prototypical architecture of LSTM networks, adapted from [46]. x_{t-1}, x_t and x_{t+1} are the input of previous, current, and next timestep, respectively. C_{t-1} and C_t are the cell states of previous and current timestep, respectively. h_{t-1}, h_t, and h_{t+1} are the hidden states of previous, current, and next timestep, respectively. Here, h_{t-1}, h_t, and h_{t+1} can serve as inputs to other neural networks for generating the final prediction

and current timestep, respectively. Note that the cell state carries information along through all the timesteps, which is due to the introduction of the top horizontal line running from left to right enclosed in the highlighted region. With some linear operations along this line, the cell state C allows information to flow through the entire LSTM network; this enables the LSTM network to remember important information from inputs at previous timesteps, thereby obtaining long-term memory. Three gates in each LSTM block serve to update the cell state, as follows.

- The **forget gate** takes as input both the hidden state h_{t-1} at the previous timestep and the current input x_t then generates a vector f_t to determine what information should be forgotten. All the values in f_t are between zero and one, which represent the degree of "forget," i.e., zero means to totally forget and one means to keep the original information as it is. Formally, the forget gate can be described using the equations

$$f_t = \sigma(W_f \times [h_{t-1}, x_t] + b_f) \tag{1.6}$$

$$C_t = f_t \cdot C_{t-1} \tag{1.7}$$

where W_f and b_f are the trainable parameters of the forget gate and σ is the sigmoid function.

- The **input gate** updates the cell state by adding the new information from the input x_t at current timestep, which consists of a sigmoid layer and a tanh layer. The sigmoid layer generates values between zero and one to decide which values to update. The tanh layer calculates the candidate information added into the cell state. The input gate can be described using the equations

$$i_t = \sigma(W_i \times [h_{t-1}, x_t] + b_i) \tag{1.8}$$

$$\tilde{C}_t = tanh(W_c \times [h_{t-1}, x_t] + b_c) \tag{1.9}$$

$$C_t = C_t + i_t \cdot \tilde{C}_t \tag{1.10}$$

where W_i, b_i, W_c, and b_c are the trainable parameters of the input gate and σ is the sigmoid function.
- The **output gate** generates the hidden state h_t at the current timestep based on the cell state C_t at the current timestep; it also consists of a sigmoid layer and a tanh layer. The sigmoid layer generates values between zero and one to decide which information to output. The tanh layer calculates the candidate output information. The output gate can be described using the equations

$$o_t = \sigma(W_o \times [h_{t-1}, x_t] + b_o) \tag{1.11}$$

$$h_t = o_t \cdot tanh(C_t) \tag{1.12}$$

where W_o and b_o are the trainable parameters of the output gate and σ is the sigmoid function.

1.3.3 Reinforcement Learning

Reinforcement learning (RL) trains an agent by trial and error in an interactive environment using feedback from its own actions and experience. RL can be formalized as a Markov decision process (MDP) [18], which consists of:

- A set of states S
- A set of actions $\mathcal{A}$
- State transition $\mathcal{T}(s_{t+1}|s_t, a_t)$, which maps a state-action pair (s_t, a_t) at timestep t onto a distribution of states s_{t+1} at timestep $t + 1$
- A reward function $\mathcal{R}(s_t, a_t, s_{t+1})$

A key concept in reinforcement learning is the policy π, which maps the agent's states to actions. At each timestep t in an episode (the path from initial state to a terminal state), the agent observes some state $s_t \in S$ and picks an action $a_t \in \mathcal{A}$ based on policy π. The environment then returns the next state $s_{t+1} \sim \mathcal{T}(s_{t+1}|s_t, a_t)$ and an associated scalar reward $r_t = \mathcal{R}(s_t, a_t, s_{t+1})$. The goal of RL is to learn an optimal policy π^*, such that the agent can maximize the accumulated reward (also known as *return*) in an episode:

$$G_t = \sum_t^T \gamma^t r_t \tag{1.13}$$

Here, T is the timestep when the agent enters the terminal state ($T = \infty$ if there is no terminal state defined), and $\gamma \in [0, 1)$ is the discount factor which enables the agent to pay more attention to short-term rewards.

In order to determine the optimal policy π^*, the value function $V(s)$ and action-value function $Q(s, a)$ are widely used in RL algorithms. The value function $V^\pi(s_t)$ describes the expected return starting from state s_t and thereafter following policy π:

$$V^\pi(s_t) = \mathbb{E}_\pi \left[\sum_{k=0}^\infty \gamma^k r_{t+k} | s = s_t \right] \tag{1.14}$$

The action-value function $Q(s_t, a_t)$ describes the expected return after taking action a_t in state s_t and thereafter following policy π:

$$Q^\pi(s_t, a_t) = \mathbb{E}_\pi \left[\sum_{k=0}^\infty \gamma^k r_{t+k} | s = s_t, a = a_t \right] \tag{1.15}$$

If we define the optimal value function $V^*(s)$ and the optimal action-value function $Q^*(s, a)$ in terms of optimal policy π^* as

$$V^*(s) = V^{\pi^*}(s) = max_\pi V^\pi(s) \tag{1.16}$$

$$Q^*(s, a) = Q^{\pi^*}(s, a) = max_\pi Q^\pi(s, a) \tag{1.17}$$

then the relationship between $V^*(s)$ and $Q^*(s, a)$ can be described as

$$V^*(s) = \max_a Q^*(s, a) \tag{1.18}$$

Given the definition of value function and action-value function, the Bellman equation reveals the behavior of the optimal policy π^*:

$$V^*(s_t) = \mathbb{E}_{\pi^*} \left[r_t + \gamma V^*(s_{t+1}) \right] \tag{1.19}$$

$$Q^*(s_t, a_t) = \mathbb{E}_{\pi^*} \left[r_t + \gamma max_{a_{t+1}} Q^*(s_{t+1}, a_{t+1}) \right] \tag{1.20}$$

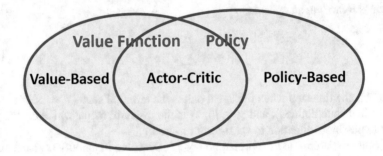

Fig. 1.8 The relationship between three types of RL methods

which allows us to find an optimal policy via dynamic programming. More specifically, based on the approach used to obtain the optimal policy π, there are three types of RL approaches, as follows (see Fig. 1.8).

- **Value-based** algorithms first learn the optimal action-value function $Q^*(s, a)$ and then obtain the optimal policy π^* implicitly by choosing the action a that promises the maximum $Q^*(s, a)$ of state s, i.e.,

$$\pi^*(s) = arg \max_a Q^*(s, a) \tag{1.21}$$

Here, the action-value function $Q_w(s, a)$ can be represented by any of several function approximators, such as lookup table, neural network, etc. Among function approximators, the neural network is most commonly used and is trained to approximate the optimal action-value function $Q^*(s, a)$ by exploiting the Bellman equation:

$$\Delta w_t = (r_t + \gamma Q_w(s_{t+1}, a_{t+1}) - Q_w(s_t, a_t))\nabla_w Q_w(s_t, a_t) \tag{1.22}$$

Here, w is the set of trainable parameters of the action-value neural network. The most well-known algorithms are the deep Q-learning algorithm [40] and its enhancements.
- **Policy-based** algorithms directly learn a parameterized policy $\pi_\theta(a_t|s_t)$ mapping states into actions. Policy-based algorithms update the policy neural network $\pi_\theta(a_t|s_t)$ directly through stochastic gradient ascent in the direction of the value:

$$\Delta\theta_t = G_t\nabla_\theta log\pi_\theta(a_t|s_t) \tag{1.23}$$

Here, θ is the set of trainable parameters of the policy neural network, and G_t is the accumulated reward of the current episode defined in Eq. 1.13. The most well-known algorithms are the policy gradient algorithm [43] and its enhancements.
- **Actor-critic** algorithms combine value-based and policy-based methods. In the actor-critic approach, two networks (the actor network and the critic network)

work together: the critic network estimates the action-value function $Q_w(s, a)$ in the same manner as in value-based algorithms (Eq. 1.22), while the actor network updates the parameterized policy $\pi_\theta(a|s)$ in the direction suggested by the critic network, according to

$$\Delta\theta_t = Q_w(s_t, a_t)\nabla_\theta log\pi_\theta(a_t|s_t) \qquad (1.24)$$

Here, w and θ are the set of trainable parameters of the critic network and actor network, respectively. The most well-known algorithms are the *advantage actor-critic (A2C)* algorithm [41] and its enhancements.

1.3.4 Other Models

Other classical ML techniques can also be leveraged for QoR prediction. The most frequently encountered techniques among these are the following.

- **Linear regression** models the output variable y as a linear combination of input features $\{x_1, x_2, \ldots, x_n\}$, which can be represented as $y = \sum_{i=1}^{n} w_i \times x_i$. Here, $\{w_1, w_2, \ldots, w_n\}$ are the trainable parameters, which are to be determined to minimize the loss function during training process.
- **Feedforward neural network** is the simplest type of artificial neural network, which consists of an input layer, followed by a series of hidden layers and an output layer. Each hidden layer consists of a set of neurons, each of which performs nonlinear transformation on the outputs from the previous layer, i.e., $y_i^{(l)} = f(\sum_{i=1}^{n} w_{ij} \times y_j^{(l-1)})$. Here, $y_i^{(l)}$ is the output of the ith neuron of the lth hidden layer, $y_j^{(l-1)}$ is the output of the jth neuron of the $(l-1)^{st}$ hidden layer, f is the nonlinear activation function, and w_{ij} are the trainable parameters. We can see that the feedforward neural network can be used to model nonlinear relationships.
- **Gradient boosted decision trees** combine multiple decision trees, known as weak predictors, into a strong predictor. New trees are added sequentially to minimize the loss function during the learning process. *XGBoost* [50] is a commonly used library for implementation of the gradient boosted decision trees algorithm.
- **Random forest** [5] also uses decision trees but trains each tree independently using random samples of data. The final prediction is generated based on voting over a set of trees or from the average of the predictions generated by the individual trees.

The main advantage of these classical ML techniques lies in their simplicity. But they may be too simple for complicated QoR prediction tasks.

1.4   Timing Estimation

Power, performance or timing, and area are the three basic types of design QoR. During the design process, there is usually a trade-off between power, timing, and area according to the constraints set by the designers. However, a chip must meet timing constraints in order to operate correctly at the specified clock frequency. Otherwise, particularly in the application-specific standard product context where there is no "binning," the chip can be thrown away. Given this, timing is one of the most important design QoR criteria. Among all the timing metrics, worst negative slack (WNS) and total negative slack (TNS) are still the most commonly used. WNS of the circuit is the largest violation of the timing constraint –i.e., the most negative value of required time minus arrival time– seen among all the timing paths in the circuit. For setup timing constraints, negative slack means that the sum of path delay and other delays exceeds the target clock period. TNS is the sum of all the negative slacks (typically, taken over all timing endpoints) in the design. In this section, we illustrate ML-based post-route TNS prediction in the early stages of the design flow by reviewing in detail the work of Lu et al [37]. Other exemplary works include the following: (i) Cao et al. [7] apply convolutional neural network (CNN)-based feature engineering and ensemble modeling to predict circuit path delays across multiple voltages and process corners; (ii) Barboza et al. [4] predict pre-routing timing with the aid of linear regression, feedforward neural network, and random forest; and (iii) Chan et al. [8] implement boosting with support vector machine (SVM) regression to predict possible timing failures at the floorplan stage.

1.4.1   Problem Formulation

To achieve the best-possible final timing outcomes, design teams often perform power, timing, and area explorations using parallel physical design (PD) runs with different tool configurations. This is a very high-stress process in light of tight tapeout schedules and limited tool license and computing resources. Importantly, when seeking the absolute limits of what can be achieved in design implementation, most of the runs are "doomed" to fail, i.e., not able to meet the given timing constraints. If the doomed runs can be identified in the early stages of the design flow, then the chip design turnaround time (TAT) can be improved and the computing resources can be utilized in a more efficient manner. Lu et al. [37] propose an end-to-end machine learning (ML) framework to predict doomed runs in early stages of the design flow, i.e., *given a netlist n with a clock period cp and a cell density d, prediction in the early stages of the design flow whether a physical design implementation will successfully satisfy the timing constraint in terms of post-route TNS.*

1.4.2   Estimation Flow

The basic idea is to model the physical design implementation as a sequential process and perform post-route TNS prediction starting from the early stages of the design flow. In this way, an ongoing run that is doomed in terms of post-route TNS can be terminated early. Figure 1.9 shows an overview of the doomed runs prediction flow. Three intermediate physical stages from the placement and CTS process, i.e., detailed placement, PlaceOpt, and CTSOpt, are selected to perform sequential modeling. For each intermediate stage, graph neural networks (GNNs) are used to perform per-stage TNS prediction by taking the netlist graph as input. In addition, a long short-term memory (LSTM) network is used to perform The final TNS prediction by taking the GNN-based embeddings from the three modeling stages as sequential inputs. With all TNS predictions available, a doomed run will be terminated immediately. The power of this GNN-based LSTM prediction flow comes from the following aspects.

- The netlists are originally represented as graphs, where the edge connectivity captures the connection topology of instances and the node features capture the instance characteristics. The graph neural network can perform meaningful graph embedding by fully exploiting the edge connectivity and node features of the graph.
- The graph neural network can be trained as a universal graph encoder, which is able to embed the netlists that come in different sizes and from various stages of the design flow into meaningful representations of the same dimension.

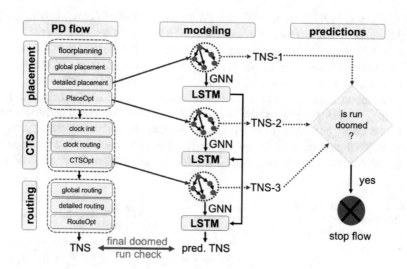

Fig. 1.9 Overview of the doomed runs prediction flow, adapted from [37]

- The physical design implementation is actually a stage-by-stage sequential process, where the outcome of the current stage highly depends on the outcomes from previous stages. The LSTM network can capture such sequential information by taking the GNN-based embeddings of the netlists from across different PD stages.

1.4.3 Feature Engineering

The inputs to the GNN-based LSTM prediction flow are graphs with node features from the three different stages of detailed placement, PlaceOpt, and CTSOpt. At each physical design stage, the netlist is transformed into a graph using a fully connected clique model, i.e., connecting each pair of instances in each net. As mentioned in Sect. 1.3.1, the key idea in the GNN approach is to generate node embeddings by aggregating information from the local neighborhood. Therefore, the fully connected clique model is chosen to create a rich neighborhood for each node. The node features are the timing and power information extracted from timing reports, power reports, and technology files, as summarized in Table 1.1. Note that all the node features are manually selected based on domain expertise.

1.4.4 Machine Learning Engines

The GNN-based LSTM framework is mainly composed of two components: graph neural networks (GNNs) for netlist encoding and a long short-term memory (LSTM) network for sequential flow modeling. The detailed architecture of the GNN-based LSTM framework is shown in Fig. 1.10. The left side presents the sequential flow modeling, where the GNN-based embeddings of the netlists from different physical implementation stages are taken as sequential inputs to the LSTM network. The right side presents the single-stage modeling, where the GNN-based embedding of the netlist of the current stage is fed into a dedicated feedforward neural network to obtain TNS prediction at the current stage. Here, the *mean* function is used as the global pooling function (see Eq. 1.1 for details) to transform node embeddings

Table 1.1 Node features for the GNN-based LSTM doomed runs prediction flow. Drawn from [37]

Name	Description
Worst slack	Worst slack of cell
Worst output slew	Maximum transition of output pin
Worst input slew	Minimum transition of input pin
Driving net power	Switching power of driving net
Internal power	Cell internal power
Leakage power	Cell leakage power

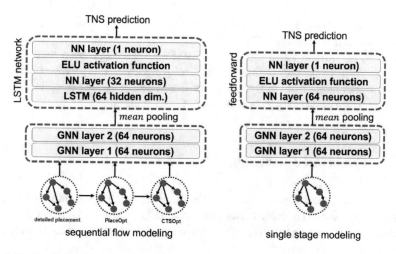

Fig. 1.10 Detailed architecture of the GNN-based LSTM framework. Drawn from [37]

into a graph embedding of the entire netlist. And, the exponential linear unit (ELU) activation function is defined as

$$ELU(x) = \begin{cases} x & \text{if } x > 0 \\ \exp(x) - 1 & \text{if } x <= 0 \end{cases} \tag{1.25}$$

Note that the GNN-based LSTM framework is still a supervised framework. Additionally, all three physical implementation stages share the same GNN module, thereby improving the generalizability of the trained framework.

1.5 Design Space Exploration

Design space exploration (DSE) is an essential technology for improvement of design QoR. For a given design and physical implementation flow, there may be three types of parameters: (i) architectural parameters, such as the bitwidth of a CPU core; (ii) design constraints, such as clock period and die size of the chip; and (iii) knobs of EDA tools, such as the timing-driven and congestion-driven options of the placement engine. The simplest approach of design space exploration is running massive physical design (PD) runs with different configurations, as noted in Sect. 1.4. However, ML-driven automated design space exploration is attracting more and more attention in recent years. In this section, we illustrate the design space exploration on knobs of EDA tools with Agnesina et al.'s work [1, 2]. Other exemplary works include the following: (i) Neto et al. [42] utilize a deep neural network (DNN) to automatically decide which optimizer should handle different portions of the netlist during logic synthesis (see also Hosny et al. [21] and Ziegler et al. [51]); (ii) Wang et al. [48] adopt graph convolutional neural networks (GCN) and reinforcement learning (RL) to achieve automatic transistor sizing; and (iii) Hu et al. [22] use Monte Carlo tree search (MCTS), a technique from reinforcement learning (RL), to find an optimized application-specific NoC architecture. Other related works range from analytical model-based design space exploration [33] to heuristic-based physical synthesis tool parameters autotuning [15].

1.5.1   Problem Formulation

Recently, researchers have sought to automate power-performance-area (PPA) optimization with the aid of reinforcement learning (RL). Agnesina et al. [2] propose a method for formulating the PPA optimization problem as an RL problem, where the SP&R tools and flows are modeled as a black-box function PPA_{tool}, and the input netlists N and knobs or parameters P of SP&R tools and flows are the independent variables of PPA_{tool}. Formally, the problem is defined as follows: *Given a netlist n, find the optimal parameter setting* $p^* = argmax_{p \in P} PPA_{tool}(n, p)$, *where* P *is the space of parameters of SP&R tools and flows.* The black-box function PPA_{tool} can only be evaluated at a query point (n, p), where queries are very expensive.

In a related work, Agnesina et al. [1] also illustrate the above idea within the context of 2D VLSI placement. More specifically, the RL formulation for the placement parameter optimization problem is defined as follows: *Given a netlist n, find* $p^* = argmin_{p \in P} HPWL(p)$, *where* P *is the set of parameter combinations of the placement engine and* $HPWL$ *is the post-placement half-perimeter wirelength of the returned placement solution.*

1.5.2   Estimation Flow

As described in Sect. 1.3.3, in order to train an RL agent that tunes the parameter settings of the placement engine autonomously, it is necessary to define the following four key elements:

- **States**: The set of netlists N and all possible parameter settings (P) of the placement engine. Table 1.2 shows the targeted parameters of the placement engine, which consists of 6×10^9 possible parameter settings. A single state s consists of a unique netlist n and a specific parameter setting p.
- **Actions**: The set of actions that the agent can use to update the parameter setting, i.e., the value of p. Note that the RL agent cannot change the netlist n in an episode. Given that there are different types of parameters (Boolean, enumerate, integer), placement stages (global, detailed), and *effort* (true, false), the actions are defined for groups of parameters having the same properties. For example, the action *flip* applies for Boolean parameters, and the actions *up* and

Table 1.2 Targeted parameters of the placement engine. *Stage* indicates when the parameter is applied, e.g., at global placement or detailed placement; *effort* indicates whether the parameter is related to the effort level, e.g., high, medium, or low. Reproduced from [1]

Name	Description	Type	Stage	Effort	# val
Eco max distance	Maximum distance allowed during placement legalization	Integer	Detailed	False	[0, 100]
Legalization gap	Minimum site gap between instances	Integer	Detailed	False	[0, 100]
Max density	Controls the maximum density of local bins	Integer	Global	False	[0, 100]
Eco priority	Instance priority for refine place	Enum	Detailed	False	3
Activity power-driven	Level of effort for activity power-driven placer	Enum	Detailed	True	3
Wirelength opt	Optimizes wirelength by swapping cells	Enum	Detailed	True	3
Blockage channel	Creates placement blockages in narrow channels between macros	Enum	Global	False	3
Timing effort	Level of effort for timing-driven placer	Enum	Global	True	2
Clock power-driven	Level of effort for clock power-driven placer	Enum	Global	True	3
Congestion effort	The effort level for relieving congestion	Enum	Global	True	3
Clock gate aware	Specifies that placement is aware of clock gate cells in the design	Bool	Global	False	2
Uniform density	Enables even cell distribution	Bool	Global	False	2

Table 1.3 The set of actions, reproduced from [1]

Action id	Description
1	Flip Boolean parameters
2	Up integer parameters
3	Down integer parameters
4	Up effort parameters
5	Down effort parameters
6	Up parameters related to detailed placement
7	Down parameters related to detailed placement
8	Up parameters related to global placement (do not touch the Boolean parameters)
9	Down parameters related to global placement (do not touch the Boolean parameters)
10	Invert-mix timing vs. congestion vs. WL efforts
11	Do nothing

down applies for integer parameters. Further, an action that does not change the current parameter setting is also added; this can serve as a trigger to reset the environment. In summary, there are 11 different actions $\mathcal{A}$ presented in Table 1.3.

- **State transition**: Given a state (s_t) and an action (a_t), the next state (s_{t+1}) is the same netlist with updated parameter settings.
- **Reward**: The reward is defined as the negative of post-placement wirelength $(-HPWL)$ given by the placement engine. The reward increases if the action improves the parameter setting in terms of minimizing post-placement wirelength. To train an agent across various netlists which may differ significantly in terms of post-placement wirelength, the reward is normalized by a baseline post-placement wirelength, i.e.,

$$R_t = \frac{HPWL_{Human\_Baseline} - HPWL_t}{HPWL_{Human\_Baseline}} \tag{1.26}$$

where $HPWL_{Human\_Baseline}$ is an expected baseline post-placement wirelength for the netlist, which can be obtained by running the placement engine manually once.

With these definitions, the RL agent for placement parameter optimization can learn from interacting with its *environment* (placement engine) over a number of episodes. As shown in Fig. 1.11, at each timestep t of an episode, the agent receives a state s_t and selects an action a_t from a set of possible actions $\mathcal{A}$ according to its policy π, where π maps states to actions. In return, the agent receives a reward signal R_t and transitions to the next state s_{t+1}. The episode ends when the agent reaches a terminal state, after which the next episode starts. Note that in an episode, the netlist n will not change during the entire episode, i.e., only the parameter setting in state s will be modified at each timestep. Moreover, the placement engine (*environment*) will run the entire placement process (including both global placement and detailed placement) for the netlist under the parameter setting in state s at each timestep, thereby obtaining the reward based on the post-placement wirelength. Also for each

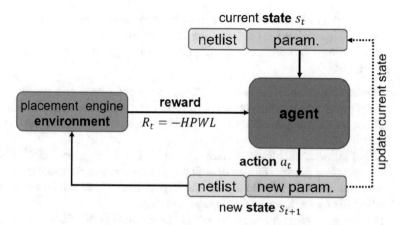

Fig. 1.11 Overview of the RL placement parameter optimization flow, reproduced from [1]

episode, the agent starts with a given netlist and its human parameter set and then terminates after 16 timesteps.

1.5.3 Feature Engineering

Recall that a single state s consists of a unique netlist n and a specific parameter setting p of the placement engine. The representation of netlists should be able to transfer the knowledge across the netlists, which may vary significantly in terms of post-placement wirelength. Therefore, the trained RL agent can generalize its tuning process to unseen netlists. Here, the netlist representation consists of meta knowledge (number of cells, floorplan area, etc.), graph topological features, and graph embedding obtained by a graph neural network. All the netlist features are summarized as follows:

- **Metadata features** include #cells, #nets, #cell pins, #IOs, #nets with fanouts between five and ten, #nets with fanouts greater than ten, #FFs, total cell area (um^2), #hard macros, and macro area (um^2).
- **Topological graph features** include graph features, which capture complex topological characteristics (e.g., connections and spectral) of the netlist. Before extracting these graph topological features, the original netlist n is converted into a graph $G(V, E)$ using a fully connected clique model, i.e, connecting each pair of cells in each net. Salient details include the following: (i) global signals such as reset, clock, or VDD/VSS are ignored; (ii) multiple connections between two cells are merged into one; and (iii) self-loops are eliminated. From the obtained graph representation G, graph topological features highly related to the objectives of the placement engine are extracted. For example, connectivity features such as the number of connected components, size of maximal clique,

and rich club coefficient are important to capture congestion information. The maximum logical level related to long timing paths reflects the difficulty of meeting timing constraints. In summary, the following graph topological features are extracted:

- *Number of connected components*: A connected component of a graph G is a subgraph S of G in which there is a path between any two nodes. The number of connected components can be calculated in linear time using depth-first search [45].
- *Size of the maximal clique*: A clique of a graph G is a subgraph S of G, which is a complete graph. The size of a clique S is the number of nodes in S. The maximal clique is the clique with the largest size, which can be calculated efficiently by the Born-Kerbosch algorithm [6].
- *Chromatic number*: A *proper coloring* is an assignment of colors to the nodes of a graph G such that no two adjacent nodes have the same color. The chromatic number of a graph is the minimum number of colors in a proper coloring of that graph, which can be computed using the method proposed in [12].
- *Maximum logic level*: The logic level of two flip-flops is the maximum number of stages in any combinational path between them. The maximum logic level is the maximum logic level over all pairs of flip-flops.
- *Rich club coefficient (RCC)*: Let $G_k = (V_k, E_k)$ be the filtered graph of G with only the nodes of degree larger than k. Then, $RCC_k = \frac{2|E_k|}{|V_k|(|V_k|-1)}$ [11]. The $\{RCC_k\}$ captures the connectivity between high-degree nodes.
- *Clustering coefficient (CC)*: The clustering coefficient [49] measures the cliquishness of nodes of neighborhoods, which is defined as

$$CC = \frac{1}{|V|} \sum_{i \in V} \frac{|e_{jk} : v_j, v_k \in \text{Neighbors}(i), e_{jk} \in E|}{\text{degree}(i)(\text{degree}(i) - 1)} \tag{1.27}$$

- *Spectral characteristics*: Two eigenvalues of the Laplacian matrix of the graph G are calculated: the second smallest eigenvalue (Fiedler value) and the largest eigenvalue (spectral radius).

- **Features from graph neural network (GNN)** are the graph embedding $(ENC(G))$ obtained using an unsupervised *GraphSAGE* (see Sect. 1.3.1 for a detailed explanation of *GraphSAGE*). As depicted in Fig. 1.12, the GNN takes simple node features including degree, fanout, Area, and encoded cell type as inputs and generates node embedding $(ENC(v))$ with a size of 32 for each node. Then, the graph embedding $(ENC(G))$ is obtained by performing *mean* pooling or aggregation on the node embeddings as follows:

$$ENC(G) = mean(ENC(v)|v \in V) \tag{1.28}$$

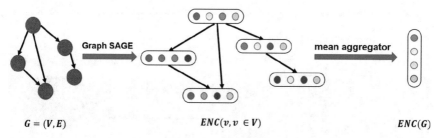

Fig. 1.12 Graph embedding using an unsupervised *GraphSAGE*. First, 32 features ($ENC(v)$) for each node in the graph are extracted through the *GraphSAGE* networks. Then, 32 features ($ENC(G)$) for the entire graph are obtained through calculating the mean among all nodes for each feature. Drawn from [1]

1.5.4 Machine Learning Engines

The *advantage actor-critic (A2C)* framework (see Sect. 1.3.3 for a detailed explanation of A2C) is adopted to train the RL agent for placement parameter optimization. The agent is represented as a deep neural network shown in Fig. 1.13. The concatenation of placement parameters p with features extracted from the input netlist n is first passed through two feedforward fully connected (FC) layers with *tanh* activation function followed by an FC linear layer. This is followed by a long short-term memory (LSTM) module with layer normalization [3] (see Sect. 1.3.2 for a detailed explanation of LSTM networks). The introduction of the LSTM module is motivated by the fact that traditional optimization methods (e.g., simulated annealing) are based on recurrent approaches. Moreover, a sequence-to-one global attention mechanism [38] is added to help the LSTM module focus on important parts of the recursion. Finally, the hidden state of the LSTM module is fed into the policy (actor) network and the value (critic) network, which each include two FC layers with *tanh* activation function. The policy network and the value network respectively use softmax layer and linear layer as the output layer. The specification of each neuron layer is shown in Table 1.4.

Aside from network architectures of the RL agent, there are other two main issues for training a placement parameter optimization agent: (i) *latency of tool runs*, because it takes minutes to hours to perform one placement, and (ii) *sparsity of data*, because there is no database of the possible millions of netlists, placed designs, or layouts. In order to relieve both issues, a parallel version of A2C [35] is implemented. As depicted in Fig. 1.14, an agent learns from the experiences of multiple actors interacting in parallel with their own copy of the environment. The learning updates are applied synchronously through a deterministic implementation that waits for each actor to finish its segment of experience (according to the current policy provided by the step model) before performing a single batch update to the weights of the network.

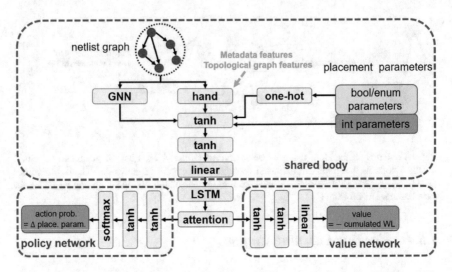

Fig. 1.13 Overall network architecture of the RL agent. The combination of an LSTM module with an attention mechanism enables the learning of complex recurrent optimization algorithms. Drawn from [1]

Table 1.4 Specification of each neuron layer. Drawn from [1]

Part	Input	Hidden	Output
Shared body	79	(64, 32) (tanh)	16 (linear)
LSTM (6 unroll)	16	16	16 × 6
Attention	16 × 6	W_a, W_c	16
Policy	16	(32, 32) (tanh)	11 (softmax)
Value	16	(32, 16) (tanh)	1 (linear)

Fig. 1.14 Synchronous parallel learner. The global network sends actions to the actors through the step model. Each actor gathers experiences from their own environment. Drawn from [1]

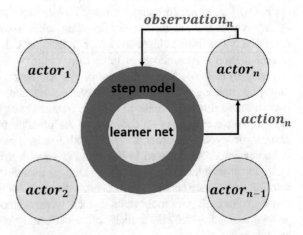

1.6 Summary

In this chapter, we have introduced motivations, challenges, and basic techniques that underlie ML-based approaches for design QoR prediction.

ML-based design QoR prediction is a powerful lever for design-based equivalent scaling that can amplify the benefits of semiconductor technology realized by designers and new products. However, achieving general-purpose, high-quality QoR prediction is challenged by the limited number of samples, chaotic behavior of EDA tools and flows, the difficulty of making actionable predictions in a given flow context, and missing infrastructure elements. The long-standing availability of alternative (e.g., constructive or assume-and-enforce) design QoR prediction methods that have been well-honed over decades brings added challenges to the development and deployment of ML-based QoR prediction.

With advances and improved understanding of ML techniques such as GNN, LSTM, and RL, researchers have made rapid progress in the ML-based QoR prediction arena. This chapter has summarized several key techniques and described in detail example methodologies for design QoR prediction in the timing estimation and design space exploration contexts.

References

1. Agnesina, A., Chang, K., Lim, S.K.: VLSI placement parameter optimization using deep reinforcement learning. In: Proceedings of the IEEE/ACM International Conference on Computer Aided Design (ICCAD), pp. 1–9 (2020)
2. Agnesina, A., Pentapati, S., Lim, S.K.: A general framework for VLSI tool parameter optimization with deep reinforcement learning. In: Proceedings of the International Conference on Neural Information Processing Systems (NeurIPS) (2021)
3. Ba, J.L., Kiros, J.R., Hinton, G.E.: Layer normalization (2016).Preprint. arXiv:1607.06450
4. Barboza, E.C., Shukla, N., Chen, Y., Hu, J.: Machine learning-based pre-routing timing prediction with reduced pessimism. In: Proceedings of the ACM/IEEE Design Automation Conference (DAC), pp. 1–6 (2019)
5. Breiman, L.: Random forests. In: Machine Learning, pp. 5–32 (2001)
6. Bron, C., Kerbosch, J.: Algorithm 457: finding all cliques of an undirected graph. Commun. ACM **16**(9), 575–577 (1973)
7. Cao, P., Bao, W., Wang, K., Yang, T.: A timing prediction framework for wide voltage design with data augmentation strategy. In: Proceedings of the IEEE/ACM Asia and South Pacific Design Automation Conference (ASP-DAC), pp. 291–296 (2021)
8. Chan, W.-T.J., Chung, K.-Y., Kahng, A.B., MacDonald, N.D., Nath, S.: Learning-based prediction of embedded memory timing failures during initial floorplan design. In: Proceedings of the IEEE/ACM Asia and South Pacific Design Automation Conference (ASP-DAC), pp. 178–185 (2016)
9. Chan, T.-B., Kahng, A.B., Woo, M.: Revisiting inherent noise floors for interconnect prediction. In: Proceedings of the ACM/IEEE International Workshop on System-Level Interconnect Problems and Pathfinding, pp. 1–7 (2020)
10. Chen, J., Jiang, I.H.-R., Jung, J., Kahng, A.B., Kim, S., et al.: DATC RDF-2021: design flow and beyond. In: Proceedings of the ACM/IEEE International Conference on Computer-Aided Design (ICCAD), pp. 1–6 (2021)
11. Colizza, V., Flammini, A., Serrano, M.A., Vespignani, A.: Detecting rich-club ordering in complex networks. Nat. Phys. **2**, 110–115 (2006)

12. Coudert, O.: Exact coloring of real-life graphs is easy. In: Proceedings of the Design Automation Conference (DAC), pp. 121–126 (1997)
13. Esmaeilzadeh, H., Ghodrati, S., Gu, J., Guo, S., Kahng, A.B., et al.: VeriGOOD-ML: an open-source flow for automated ML hardware synthesis. In: Proceedings of the ACM/IEEE International Conference on Computer-Aided Design (ICCAD), pp. 1–7 (2021)
14. Fenstermaker, S., George, D., Kahng, A.B., Mantik, S., Thielges, B.: METRICS: a system architecture for design process optimization. In: Proceedings of the Design Automation Conference (DAC), pp. 705–710 (2000)
15. Geng, H., Chen, T., Sun, Q., Yu, B.: Techniques for CAD tool parameter auto-tuning in physical synthesis: a survey (invited paper). In: Proceedings of the IEEE/ACM Asia and South Pacific Design Automation Conference (ASP-DAC), pp. 635-640 (2022)
16. Grodstein, J., Lehman, E., Harkness, H., Grundmann, B., Watanabe, Y.: A delay model for logic synthesis of continuously-sized networks. In: Proceedings of the IEEE International Conference on Computer Aided Design, pp. 458-462 (1995)
17. Hamilton, W., Ying, Z., Leskovec, J.: Inductive representation learning on large graphs. In: Proceedings of the International Conference and Workshop on Neural Information Processing Systems (NeurIPS), pp. 1024–1034 (2017)
18. He, Z., Zhang, L., Liao, P., Ma, Y., Yu, B.: Reinforcement learning driven physical synthesis (invited paper). In: Proceedings of the IEEE International Conference on Solid-State & Integrated Circuit Technology (ICSICT), pp. 1–4 (2020)
19. Hill, D., Kahng, A.B.: Guest editors' introduction: RTL to GDSII – from foilware to standard practice. IEEE Desig. Test Comput. $21(1)$, 9–12 (2004)
20. Hochreiter, S., Schmidhuber, J.: Long short-term memory. Neural Comput. $9(8)$, 1735–1780 (1997)
21. Hosny, A., Hashemi, S., Shalan, M., Reda, S.: DRiLLS: deep reinforcement learning for logic synthesis. In: Proceedings of the IEEE/ACM Asia and South Pacific Design Automation Conference (ASP-DAC), pp. 581–586 (2020)
22. Hu, Y., Mettler, M., Mueller-Gritschneder, D., Wild, T., Herkersdorf, A., Schlichtmann, U.: Machine learning approaches for efficient design space exploration of application-specific NoCs. ACM Trans. Desig. Auto. Electron. Syst. $25(5)$, 1–27 (2020)
23. International technology roadmap for semiconductors. http://www.itrs2.net/itrs-reports.html
24. International technology roadmap for semiconductors 2009 design chapter. https://www.dropbox.com/sh/ia1jkem3v708hx1/AAB1fo1HrYIKClJNk0dB7YrCa?dl=0&preview=Design.pdf
25. Jeong, K., Kahng, A.B.: Methodology from chaos in IC implementation. In: Proceedings of the International Symposium on Quality Electronic Design, pp. 885–892 (2010)
26. Jung, J., Kahng, A.B., Kim, S., Varadarajan, R.: METRICS2.1 and flow tuning in the IEEE CEDA robust design flow and OpenROAD. In: Proceedings of the ACM/IEEE International Conference on Computer-Aided Design (ICCAD), pp. 1–9 (2021)
27. Kaggle: your machine learning and data science community. https://www.kaggle.com
28. Kahng, A.B.: Classical floorplanning harmful? In: Proceedings of the International Symposium on Physical Design (ISPD), pp. 207–213 (2000)
29. Kahng, A.B.: The ITRS design technology and system drivers roadmap: process and status. In: Proceedings of the Design Automation Conference (DAC), pp. 1–6 (2013)
30. Kahng, A.B.: Open-source EDA: if we build it, who will come? In: Proceedings of the IFIP/IEEE International Conference on Very Large Scale Integration (VLSI-SoC), pp. 1–6 (2020)
31. Kahng, A.B., Mantik, S.: A system for automatic recording and prediction of design quality metrics. In: Proceedings of the IEEE International Symposium on Quality Electronic Design, pp. 81–86 (2001)
32. Kahng, A.B., Mantik, S.: Measurement of inherent noise in EDA tools. In: Proceedings of the International Symposium on Quality Electronic Design, pp. 206–211 (2002)
33. Kahng, A.B., Li, B., Peh, L.-S., Samadi, K.: ORION 2.0: a fast and accurate NoC power and area model for early-stage design space exploration. In: Proceedings of the Design, Automation & Test in Europe Conference & Exhibition (DATE), pp. 423-428 (2009)

34. Kipf, T.N., Welling, M.: Semi-supervised classification with graph convolutional networks (2016). Preprint. arXiv:1609.02907
35. Lasse, E., Hubert, S., Remi, M., Karen, S., Volodymir, M., et al.: IMPALA: scalable distributed deep-RL with importance weighted actor-learner architectures. In: Proceedings of the International Conference on Machine Learning (ICML), pp. 1406–1415 (2018)
36. Lopera, D.S., Servadei, L., Kiprit, G.N., Hazra, S., Wille, R., Ecker, W.: A survey of graph neural networks for electronic design automation. In: Proceedings of the ACM/IEEE Workshop on Machine Learning for CAD (MLCAD), pp. 1–6 (2021)
37. Lu, Y.-C., Nath, S., Khandelwal, V., Lim, S.K.: Doomed run prediction in physical design by exploiting sequential flow and graph learning. In: Proceedings of the IEEE/ACM International Conference on Computer Aided Design (ICCAD), pp. 1–9 (2021)
38. Luong, T., Pham, H., Manning, C.D.: Effective approaches to attention-based neural machine translation. In: Proceedings of the Conference on Empirical Methods in Natural Language Processing, pp. 1412–1421 (2015)
39. Ma, Y., He, Z., Li, W., Zhang, L., Yu, B.: Understanding graphs in EDA: from shallow to deep learning. In: Proceedings of the International Symposium on Physical Design (ISPD), pp. 119–126 (2020)
40. Mnih, V., Kavukcuoglu, K., Silver, D., Graves, A., Antonoglou, I., Wierstra, D., et al.: Playing atari with deep reinforcement learning (2013). Preprint. arXiv 1312.5602
41. Mnih, V., Badia, A.P., Mirza, M., Graves, A., Lillicrap, T., Harley, T., Silver, D., Kavukcuoglu, K.: Asynchronous methods for deep reinforcement learning. In: Proceedings of the International Conference on Machine Learning (PMLR), pp. 1928–1937 (2016)
42. Neto, W.L., Austin, M., Temple, S., Amaru, L., Tang, X., Gaillardon, P.-E.: LSOracle: a logic synthesis framework driven by artificial intelligence: invited paper. In: Proceedings of the IEEE/ACM International Conference on Computer-Aided Design (ICCAD), pp. 1–6 (2019)
43. Sutton, R.S., McAllester, D., Singh, S., Mansour, Y.: Policy gradient methods for reinforcement learning with function approximation. In: Proceedings of the International Conference on Neural Information Processing Systems (NeurIPS), pp. 1057–1063 (1999)
44. SweRV RISC-V CoreTM 1.1 from western digital. https://github.com/westerndigitalcorporation/swerv_eh1
45. Tarjan, R.: Depth-first search and linear graph algorithms. In: Proceedings of the Annual Symposium on Switching and Automata Theory (SWAT), pp. 114–121 (1971)
46. Understanding LSTM networks. https://colah.github.io/posts/2015-08-Understanding-LSTMs/
47. Veličković, P., Cucurull, G., Casanova, A., Romero, A., Liò, P., Bengio, Y.: Graph attention networks (2017). Preprint. arXiv:1710.10903
48. Wang, H., Wang, K., Yang, J., Shen, L., Sun, N., Lee, H.-S., Han, S.: GCN-RL circuit designer: transferable transistor sizing with graph neural networks and reinforcement learning. In: Proceedings of the ACM/IEEE Design Automation Conference (DAC), pp. 1–6 (2020)
49. Watts, D.J., Strogatz, S.H.: Collective dynamics of 'small-world' networks. Nature 393(4), 440–442 (1998)
50. XGBoost. https://xgboost.readthedocs.io/en/stable/python/python_intro.html
51. Ziegler, M.M., Liu, H.-Y., Gristede, G., Owens, B., Nigaglioni, R., Carloni, L.P.: A synthesis-parameter tuning system for autonomous design-space exploration. In: Proceedings of the Design, Automation & Test in Europe Conference & Exhibition (DATE), pp. 1148–1151 (2016)

Chapter 2
Deep Learning for Routability

**Zhiyao Xie, Jingyu Pan, Chen-Chia Chang, Rongjian Liang,
Erick Carvajal Barboza, and Yiran Chen**

2.1 Introduction

With the advance of semiconductor technology, an increasing number of complicated design rules need to be followed in VLSI design, and a chip can only be
taped-out after passing the design rule checking (DRC). This basic requirement is
often difficult to be satisfied in modern chip design, especially when routability is
not adequately considered in early design stages. In light of this fact, routability
is widely recognized as a main objective in placement, and routability prediction
has received considerable attention in both academic research and industrial tool
development. Routability prediction at early design stages can help designers and
tools perform preventive measures so that design rule violations can be avoided in a
proactive manner.

In some industrial design flows, fast trial global routing is often employed for
overall routability prediction at the placement stage. However, this is still too slow
from the routability prediction point of view, as it needs to be invoked many times
in the placement process. In comparison, even the full-fledged global routing is not
accurate enough to identify precise locations of DRC hotspots, due to complicated
design rules imposed upon design layout for manufacturing. Overall, existing

Z. Xie (✉)
Hong Kong University of Science and Technology, Clear Water Bay, Hong Kong
e-mail: eezhiyao@ust.hk

J. Pan · C.-C. Chang · Y. Chen
Duke University, Durham, NC, USA
e-mail: jingyu.pan@duke.edu; chenchia.chang@duke.edu; yiran.chen@duke.edu

R. Liang · E. C. Barboza
Texas A&M University, College Station, TX, USA
e-mail: liangrj14@tamu.edu; ecarvajal@tamu.edu

© The Author(s), under exclusive license to Springer Nature Switzerland AG 2022
H. Ren, J. Hu (eds.), *Machine Learning Applications in Electronic Design
Automation*, https://doi.org/10.1007/978-3-031-13074-8_2

routing-based solutions are neither fast enough for overall routability forecast nor accurate enough for pinpointing DRC hotspots.

In recent years, machine learning (ML), especially deep learning (DL), has been demonstrated as a powerful technique in many fields. In VLSI design, many DL-based routability estimators have achieved promising results. Unlike traditional algorithms, these data-driven techniques avoid constructing solutions from scratch and achieve orders-of-magnitude acceleration by directly learning complex correlations from prior data. Compared with traditional ML methods, DL-based methods are superior in capturing the global information from a larger region of the circuit layout. In this chapter, we will cover DL-based routability estimators in detail, from the background to the latest research efforts, with more focus on feature engineering and DL model architecture design. In addition, we will also cover existing explorations on the efficient deployment of these routability estimators in the physical design flow to benefit the final chip quality.

2.2 Background on DL for Routability

2.2.1 Routability Prediction Background

2.2.1.1 Design Rule Checking (DRC) Violations

In practice, layout routability may be evaluated with different metrics. The most accurate, ultimate, and widely adopted measurement of routability is design rule violation (DRV). DRC verifies whether a specific layout meets the constraints derived according to manufacturing process requirements. Such checking of design rules is an essential part of physical design flows. Although design rules cannot guarantee success for manufacturing a design, their violations definitely make the manufacturing more difficult and increase the failure rate.

Design rules are provided by process engineers and/or fabrication facilities. Each process technology has its own set of rules, commonly defined in a DRC rule deck file. The number and complexity of DRC rules increase as the transistor feature size shrinks at advanced technology nodes. As a result, routability prediction for different technology nodes may require essentially different methods and ML models. Some basic and common types of DRC rules include minimum width, minimum spacing, minimum area, wide metal jog, misaligned via wire, special notch spacing, end of line spacing, etc. Most design rules can be broadly categorized into three types:

- Size rules. They define the minimum length or width of components/shapes in a layout.
- Separation rules. They define the minimum distance between two adjacent objects at each layer.

- Overlap/enclosure rules. They define the minimum amount of overlap/coverage between two connected shapes in two adjacent layers.

Background Information
DRC violation (DRV) can be accurately evaluated based on rule decks after routing, using physical verification sign-off tools. Widely used commercial tools include Siemens EDA (formerly Mentor Graphics) Calibre®, Cadence Pegasus™ and PVS, Synopsys IC Validator™, etc. Before the sign-off stage, some digital layout tools like Cadence Innovus™ and Synopsys IC Compiler™ II also provide quick design rule checking after routing. These tools can mark the precise locations with DRV and return the total number of DRVs in the whole layout.

2.2.1.2 Routing Congestion and Pin Accessibility

Although DRC violation is the ultimate optimization goal, it is not the only metric to measure routability. In practice, routability is also approximated with the routing congestion at early design stages like global routing. For each region on the layout, congestion measures the gap between the demand of routing resources and the supply. This measurement is based on the routing model. In a basic routing model, the entire layout is tessellated into an array of global routing cells (gcells), also referred to as grids or bins. They are further gridded using horizontal and vertical gridlines, referred to as routing tracks, along which wires can be created. Each gcell can only accommodate only a finite number of routing tracks, and the number of available tracks is referred to as its routing *supply* or *capacity*. In comparison, the routing *demand* is contributed by each wire crossing the gcell, requiring one routing track in either horizontal or vertical direction. The congestion of each gcell is measured by the number of excess tracks in routing *demands* over the *supply*, referred to as track overflow. Such overflow for horizontal and vertical tracks is calculated separately and is set to zero if supply is larger or equal to demand. Routers also report the overall percent of gcells with overflow in an entire layout, as an indicator of the overall congestion. The existence of routing congestion often results in detoured wires, poor layer assignment, or even incomplete routes containing opens and shorts. It is viewed as one major contributor to DRC violations.

Compared with DRC violation, routing congestion is much easier to estimate based on placement and routing solutions. In some industrial tools, routing congestion can be measured during global routing (GR), or even at early global routing (eGR), also named trial routing (TR). Thus, it is commonly used as an early measurement of layout routability, and many traditional routability improvement techniques are based on it. However, many studies [5] have demonstrated the mis-correlation between routing congestion and DRC violations, especially at advanced technology nodes when the complexity of design rules increases. This also makes routability improvement increasingly difficult.

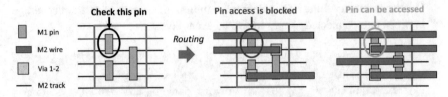

Fig. 2.1 Pin accessibility example: the detailed routing around pins in a standard cell [9]

At advanced technology nodes like sub-20nm, as design rules keep increasing, pin accessibility becomes another key contributor to DRC violations. As Fig. 2.1 demonstrates, the problem arises from the difficulty to route to standard cell pins on low-level metal layers, even when the routing congestion is low. At an advanced node, even a standard cell designed for easy pin access may not be easily routable if surrounded by other cells, which restrict wire access to its pins. As indicated in previous works [30], the pin access problem is partially contributed by high pin densities in a local region, but they are not strictly correlated.

2.2.1.3 Relevant Physical Design Steps

Now, we inspect the routability problem in a design flow, which typically starts with a design in register-transfer level (RTL). After logic synthesis tools convert design RTL into a gate-level netlist, physical design is performed, where all design components, including macros, cells, and wires, are instantiated with concrete geometric representations on metal layers. Routability problems and the majority of routability estimations arise at this stage. A typical physical design flow consists of several major steps, including floorplanning, power planning, placement, clock tree synthesis (CTS), routing, and physical verification. Each physical design step may involve multiple sub-steps. For example, placement includes global placement (GP) and detailed placement (DP). Routing includes global routing (GR) and detailed routing (DR). In addition, optimizations are performed after major steps including placement, CTS, and routing.

Background Information

All routing blockages and different types of wires, including power grids, clock wires, and signal nets, all share the common area resource in a layout. Thus, decisions at many layout steps directly affect routability, and there exist multiple trade-offs between routability and other design objectives. During floorplanning, a higher utilization rate for smaller area directly leads to less routing spaces and worse routability. During power planning, if more routing resources are devoted to power grids, the design suffers less from IR drop

(continued)

violations but results in worse routability. During CTS, a clock tree can achieve better skew and clock tree capacitance with wire sizing but requires more space and often leads to additional DRC violations. Lastly, if routability is not considered and optimized during floorplanning and placement, the layout may achieve better wirelength, timing, and power, at the cost of more DRC violations. In summary, improving routability is not a stand-alone design problem. Better routability not only helps secure DRC clean at physical verification but also allows trade-offs for improvement on other design objectives according to specific design goals.

2.2.1.4 Routability Prediction

As mentioned, DRC violations can only be precisely measured after routing finishes, when the room for fixing DRV has become very limited. Besides fixing DRV manually, one option is to perform engineering changing order (ECO), which tries to complete unrouted and partially routed nets while maintaining existing wires as much as possible. Another fixing method is to delete part of existing wires and reroute them. But it is difficult for these minor modifications at the post-routing stage to fix all violations for layouts with poor routability. As a result, designers have to trace back to earlier stages, change their layout solution accordingly, and start a new design iteration. It can take many iterations to reach a DRV-clean layout, leading to a very long turnaround time.

To improve layout routability, design rule violations should be avoided with preventive measures in a proactive manner. This heavily relies on early routability prediction methods. However, accurate early routability prediction is difficult since the behavior of placement and routing engines in modern EDA tools is highly complex and rather unpredictable. One possible solution is to develop some fast trial routing algorithms, but it is hard to achieve ideal accuracy and speed at the same time. Another promising research direction nowadays is to learn from prior data by developing data-driven routability estimators with machine learning (ML) algorithms. A main strength of ML methods is the automatic extraction of complex correlations between separated design steps based on prior knowledge. Once the ML model has been trained, it can produce routability predictions in a very short time, without constructing solutions from scratch.

For routability prediction, researchers have tried different prediction granularities according to their application scenarios. Figure 2.2 shows two common routability prediction scenarios with different granularities. Some coarse-grained predictions only evaluate the overall routability of an entire layout. Such routability is usually measured with the total number of DRC violations, also named DRV count. Generally, it is easier to achieve DRV clean for layouts with less DRVs. Another similar metric is the total number of nets with DRC violations, also named violated

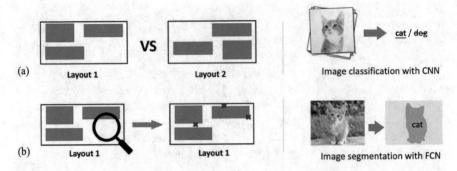

Fig. 2.2 Two common routability prediction scenarios. (**a**) Use coarse-grained prediction on the overall routability of an entire layout to identify more routable layouts among candidates. (**b**) Use fine-grained prediction on detailed DRV locations to guide mitigation techniques

net count. Since the same net may cause multiple DRC violations, using violated net count instead of DRV count avoids counting the same problem repeatedly. Such a coarse-grained prediction evaluates the whole layout and enables the identification of more routable layout solutions among many candidates. In comparison, fine-grained routability prediction tries to pinpoint the detailed locations with DRC violations. This guides layout modifications at early stages to proactively prevent DRC violations. Based on these predictions, many different applications of routability estimators have been proposed.

2.2.2 DL Techniques in Routability Prediction

The most commonly adopted DL techniques in routability prediction are from computer vision, including convolutional neural networks (CNN) and fully convolutional network (FCN) [17]. The basic idea is to process the layout input like processing an image for image classification, segmentation, and generation. There are many variations for each DL technique. For CNN models in routability prediction, popular models include ResNet [10], DilatedNet [28], and DeepLabv3 [8]. For FCN models, popular models include the vanilla FCN model [17] and the U-Net model [23]. In addition, some prior works train their FCN-based model with a conditional generative adversarial network (cGAN) [19] framework.

2.2.2.1 CNN Methods

Compared with traditional machine learning algorithms, CNN learns more abstract patterns from images. A typical structure of CNN is composed of convolutional (Conv) layers, pooling (Pool) layers, and fully connected (FC) layers. The final output is a single vector of class scores, whose length equals the number of classes. The ResNet [10] model is a classical variant of CNN model proposed in 2015. It takes advantage of scalable residual blocks to allow skipping layers. As a result, it well solved the gradient vanishing problem when the depth of CNN increases. Besides the standard convolution, another widely adopted convolution is named atrous or dilated convolution (DC) [28]. It introduces another parameter to convolutional layers called the dilation rate, which defines a spacing between the values in a kernel. A 3×3 kernel with a dilation rate of 2 will have the same field of view as a 5×5 kernel, while only using 9 parameters. Compared with basic convolution, it can effectively enlarge receptive fields of filters.

2.2.2.2   FCN Methods

Compared with traditional CNN targeting image classification, the FCN, a CNN variant without FC layers, is firstly proposed to perform end-to-end semantic segmentation. Given an arbitrary input size, it can output an image with its size equal to the input. Many FCNs [17] adopt an encoder-decoder framework. In the encoder, by downsampling operators, the depth and spatial dimensions of the feature map gradually get deeper and smaller. In the decoder, the width and height of the feature map are gradually recovered to those of the input by upsampling operators. Transposed-convolutional (Trans) layers are usually added at the decoder part to upsample feature maps and control the size of the final output. Such architecture is widely used in many computer vision problems, like crowd counting and biomedical image segmentation. Besides eliminating FC layers, many FCNs have multiple shortcuts, which concatenate feature maps in the front directly to feature maps near the end. A popular example is U-Net [23] for medical image segmentation. As a result, both longer and shorter paths exist between the input layer and the final output layer. Such multipath architecture reserves both shallow and deep embedding information.

2.2.2.3   GAN Methods

Generative adversarial networks (GANs) are used in unsupervised tasks. A GAN consists of a generator G and a discriminator D. The discriminator D distinguishes between samples generated from the generator and samples from the training dataset. The generator G generates a mapping of input samples that cannot be distinguished by the discriminator D. In the conditional GAN (cGAN) [19], ordinary GAN is extended to a conditional model by conditioning both the generator and discriminator with some extra information, which could be any kind of auxiliary information, such as class labels or data from other modalities [19].

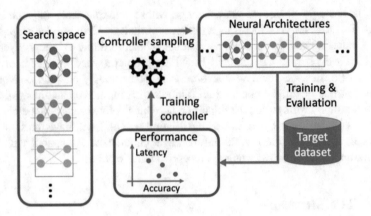

Fig. 2.3 A basic NAS framework

2.2.2.4 NAS Methods

A recent study in routability prediction explores the automated development of estimator structures with very little designer expertise or human effort. It utilizes the popular neural architecture search (NAS) technique [21]. NAS automatically conducts architecture engineering to find effective neural network models for specific tasks without (or with minimum) human interventions. It has demonstrated great potential in applications like image classification, object detection, and semantic segmentation. Figure 2.3 shows a basic NAS framework. It consists of three key ingredients: *search space, evaluation strategy*, and *search strategy*. The *search space* defines a family of candidate architectures that can be explored in NAS. The *evaluation strategy* determines the metric to estimate the design quality (e.g., accuracy) of a candidate architecture and provides feedback to the search process. The *search strategy* is the method to explore the search space and guide the search process toward the choice of a promising ML model.

2.2.3   Why DL for Routability

Data-driven routability estimators are initially constructed with traditional ML models, including support vector machine (SVM)-based estimator [4, 5], multivariate adaptive regression spline (MARS)-based estimator [4, 31], and artificial neural network (ANN)-based estimator [25, 26]. These ML models typically only process a limited number of input features. For fine-grained routability prediction on DRV locations, traditional ML methods are applied to make decisions based on a small cropped region with limited features from the layout. Such a small input region strongly limits the *receptive field* (the field visible to the model) of these traditional ML methods. As a result, they cannot capture the *global information* from a larger

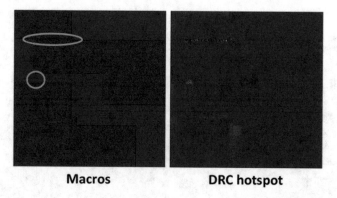

Macros   **DRC hotspot**

Fig. 2.4 Macros and DRC hotspot distribution. Orange circles indicate regions with high DRVs

region of the layout. For example, the nets spanning a large region have a large impact on the routability. Another important global impact is the tendency for DRV hotspots to aggregate in the space between adjacent macros. This is illustrated in Fig. 2.4. A cropped region is usually too small to capture the impact from neighboring macros. In addition, routability can also be affected by clock wires and power grids, which may also be captured by a large receptive field.

DL models are naturally good fits to capture the global information from large regions. In computer vision, various CNN/FCN models have been designed to identify the objects or semantic of images by capturing the pattern of a whole image. Many routability estimators [6, 7, 11, 12, 15, 18, 20, 27, 29] utilize these DL techniques to process placement or routing solutions like images. Also, DL models have strong abilities to process the interactions among different features/channels, like the RGB colors in different channels. For routability prediction, researchers can extract many relevant features for routability based on their own intuitions and then feed these features into DL models as different channels.

Notice that it is actually not necessary for DL-based models to take the whole layout as input. If developers choose to take cropped regions as input, DL models can still support a much larger cropped region as input compared with traditional ML models.

Here, we provide a simple case study on fine-grained DRV location prediction, assuming F different features are defined for each method. The whole layout is tessellated into $w \times h$ grids. More details of such setup will be introduced in the next section on methodologies.

- For traditional ML-based methods [31] without neighboring information, to make predictions on each 1×1 grid, the F features are measured only

(continued)

on this grid; thus, the model input is a vector with length F. The receptive field is 1×1 grids.

- For traditional ML-based method [5] with both small (1×1 grid) and large (3×3 grids) measurement windows, there are F features collected with the small window and another F features with the large window. The model input will be a vector with length $2F$. The receptive field is 3×3 grids.
- For traditional ML-based method [25] with features on neighboring grids calculated separately, for a cropped region with 3×3 grids, the model input is a vector with length $9F$. The receptive field is 3×3 grids.
- For DL-based methods applied on the whole layout, the raw input is in the shape of $w \times h \times F$. Their receptive fields are equal to or smaller than $w \times h$ grids and can be calculated [2] based on actual model structures.

This case study verifies our previous claim that DL-based estimators can process more inputs and support a much larger receptive field.

2.3 DL for Routability Prediction Methodologies

We will introduce the general flow for routability prediction with DL methods in this section. It includes four major steps: data generation and augmentation, feature engineering, DL model architecture design, and model training and inference. The prediction flow itself can be stand-alone, without being part of placement or routing algorithms. The majority of previous routability prediction methods are applied after the floorplanning stage, including post-global placement, post-detailed placement, and post-global routing. When applied at later stages, more layout design decisions have been made; thus, more features are available and the prediction can be more accurate. As a trade-off, there is less room for DRV mitigation at later stages. Before floorplanning, in contrast, no layout information is available, and the design has to be processed as a graph with graph neural network (GNN) based on the gate-level netlist, instead of an image. Very few works [14] target such pre-layout stage due to the difficulty. Except for this rare case, the general flow for routability prediction based on layouts is introduced step-by-step in this section.

2.3.1 Data Preparation and Augmentation

The model construction process starts with data generation. Designers collect representative circuit designs and then go through the physical design process with EDA tools to generate training data. In this process, relevant raw data is dumped out for feature and label extraction and preprocessing.

Before starting the data generation process, designers need to decide whether the model targets *cross-design*, according to their application scenario. A cross-design ML estimator means the estimator directly applies to new designs that are not in the training set. Usually, researchers require the new design to be different from training designs at the netlist level. Thus, the following examples are not viewed as cross-design: (1) models trained and tested with different layout implementation of the same netlist and (2) models trained and tested on multiple designs, which appear in both training and testing sets. Most representative routability estimators are cross-design, which means they do not need model retraining or fine-tuning for new designs. In this case, the only prediction cost is the short model inference time, which is the main advantage of performing predictions with ML models. However, it is more challenging for cross-design models to achieve high accuracy on all new designs never seen by the model, especially for new designs that are largely different from the training data. Notice that, although cross-designs support training and testing on different designs, almost all estimators require the same technology library used for training and testing designs. For different technology nodes, the complex correlation between features and routability can be quite different, preventing the model from learning a more generalizable pattern.

To construct a typical cross-design ML model, designers start with collecting representative circuit designs and constructing a training dataset based on them. Many prior works perform their experiments on different private industrial benchmarks. This makes the direct comparison among different solutions very difficult. One popular public benchmark is from the ISPD'15 detailed-routing-driven placement contest [3]. Based on these circuit designs, the physical design process can be finished with commercial layout tools, which are viewed as the source of ground truth. All raw data related to features and labels are collected and dumped out at corresponding design stages.

To construct high-performance DL-based estimators, a sufficient amount of training data is necessary. Take those widely used datasets in computer vision as an example, CIFAR-10 and MNIST contain 50 thousand and 60 thousand training images, respectively. The larger-scale dataset ImageNet contains more than 14 million images. In comparison, it is challenging to generate as much training data in routability prediction. The challenge from limited data is twofold. First, it is difficult to collect a large number of different circuit designs for training data generation, especially for researchers in academia. Second, even if many circuit designs can be collected, generating a large amount of labeled data can be highly time-consuming. The generation of each label requires going through the standard physical design flow once. As a result, routability estimators are usually trained with a limited amount of training data. Currently, there is no consensus on the minimum amount of data to use for training, considering the largely different scenarios in different works. But as a rule of thumb, it is recommended that at least hundreds of layouts from several different designs should be generated for training cross-design estimators.

Although we face challenges from limited training data, there are several operations that can enlarge the size of the training dataset. For the same circuit design, multiple netlists may be generated by providing different parameters to the logic synthesis tool. Similarly, for the same netlist, multiple layouts can be generated. For designs with macros, changing the locations of macros can very efficiently generate many essentially different layout solutions. In addition, based on the same layout solution, researchers can perform data augmentation on-the-fly during training, which is well supported by current deep learning techniques. This on-the-fly augmentation avoids enlarging the storage requirement of the training dataset. The most common augmentation techniques for routability problem include horizontal flipping, vertical flipping, and rotation by 180°. Some researchers suggest more aggressive augmentation techniques like rotation by 90°, 270°, random cropping and padding, random cropping and scaling, etc. However, these augmentation techniques tend to create representations of unrealistic layouts and thus are not very reasonable in this routability prediction problem and should be used very carefully.

Based on the generated raw data, input features X and labels Y are generated in the subsequent data preprocessing stage. Just like the basic routing model in EDA tools, for routability prediction, the whole layout is firstly tessellated into a two-dimensional matrix of equal-sized grids/tiles, with w grids in the width and h grids in the height. A visualization of the label extraction process is shown in Fig. 2.5. Notice that this step is not directly related to the gcell or the routing algorithm, and it is up to the designers to decide the grid size in tessellation. There is a trade-off between the granularity and the computation cost. A finer-grained tessellation with a smaller grid size allows more detailed identification of DRV hotspot locations, but at the expense of a higher computation cost during preprocessing, training, and inference.

After the layout tessellation, each feature or the label of the layout is extracted as a two-dimensional density distribution map. Details of extracted features will be introduced in the next subsection. To calculate this, for each grid on the tessellated layout, we measure the feature value in this grid as a real number. Then, each feature of the whole layout with $w \times h$ grids is in $\mathbb{R}^{w \times h}$. Assuming altogether there are F different input features, they can be calculated independently and then stacked together as one input tensor $X \in \mathbb{R}^{w \times h \times F}$. Similarly, for the label collection, assuming we are performing DRV hotspot detection, the label can be extracted in the same way and denoted as $Y \in \mathbb{R}^{w \times h}$. Sometimes designers are only interested

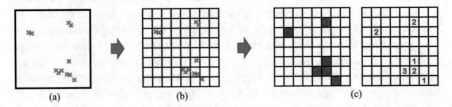

Fig. 2.5 Visualization on the feature and label extraction. (**a**) The layout with DRC violation distributions as labels. (**b**) The tessellation of the layout with $w \times h$ grids. (**c**) The label on DRC violation distribution extracted as $Y \in \mathbb{R}^{w \times h}$ for regression (right) or $Y' \in \{0, 1\}^{w \times h}$ for binary classification (left)

in whether there is any violation at each grid, so this task degrades to a binary classification problem for each grid, and the label is $Y' \in \{0, 1\}^{w \times h}$. This process is also shown in Fig. 2.5. For DRV count prediction, the label is summation of all numbers of violations, $y = \text{sum}(Y) = \sum_{j=1}^{w} \sum_{i=1}^{h} Y[i][j] \in \mathbb{R}$.

2.3.2 Feature Engineering

Feature engineering plays a key role in DL for routability prediction methodologies since it determines the upper limit of the performance of ML methods. If certain key information is missing in input features, it is almost impossible for the ML model to learn the corresponding correlation and make accurate predictions. Notice that available features depend on the stage we apply the model. For example, if the model targets to be applied before placement, then only macro locations can be included as features, while cell locations are unknown yet. Figure 2.6 shows the basic physical design flow with available features and labels at each design stage. The selection of features heavily relies on designers' expertise in both physical design and deep learning, and the solution is specific to this routability prediction problem. After years of exploration, there are multiple fundamental features that are recognized by most EDA researchers and engineers. They can be roughly categorized into several types. We introduce these features in detail below.

2.3.2.1 Blockage

A routing blockage defines a region where routing is not allowed on specific layers. This is often considered as a hard constraint and directly affects the final design routability. A higher density of blockages naturally leads to poor layout routability. The blockage information includes locations of macros, cells, and pins on the layout. To extract more information for the ML engine, different types of blockages can be captured separately into different two-dimensional density distribution maps. Below are some of the most commonly adopted blockage-related features:

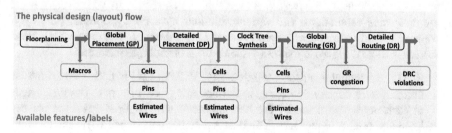

Fig. 2.6 A basic physical design flow with available features and labels at each design stage

- Macro density. After floorplanning, the locations of all macros are fixed. The regions occupied by macros are included in features.
- Cell density. Cell locations are mostly fixed after detailed placement. The density distribution of cells is included in features.
- Pin density. Besides macros and cells as blockages, the density distribution of all pins from both macros and cells is included in features.

Besides these widely adopted features, other features that may be useful include:

- Decomposed cell density. In some works, to capture more information, the cell density of different types of cells is extracted separately as different features. For example, densities of flip-flop cells, clock tree cells, and fixed cells that cannot be moved can be additional features besides the density of all cells.
- Decomposed pin density. Similar to cell density, pin density can also be decomposed into multiple features, including macro pins at each metal layer, the pins of flip-flop cells, and the pins of cells on the clock tree.
- Horizontal/vertical track capacity maps. As mentioned in the background, routing congestion depends on the gap between the *demand* of routing tracks and the *supply*. As a result, the density map of routing track *supply* can also be a useful input feature. Track capacity is measured separately in two directions: horizontal and vertical. Thus, we have separate features for the two directions.
- Detailed pin configurations. As mentioned, pin accessibility problems can cause DRC violations. In some works [15] targeting advanced technology nodes, the detailed pin patterns/configurations are captured specifically with "high-resolution" as input features.

2.3.2.2 Wire Density

In each region, a higher wire density will lead to higher demand on routing tracks and thus higher routing track overflow, more routing congestions, and likely more DRC violations. To capture this effect, wire density information is an essential input to routability estimators. Compared with blockage information, which is directly available after locations of components are fixed by placement, the wire distribution on a layout cannot be known explicitly until routing finishes, when it is too late to perform routability prediction. As a result, the extraction of wire density itself requires some estimations, and various solutions have been proposed. Below are some commonly adopted wire density-related input features. The most widely used feature is RUDY. Figure 2.7 shows visualizations of multiple features.

- RUDY. Rectangular uniform wire density (RUDY) [24] is derived by the total uniform wire density spreading in the bounding box of a net. The wirelength of each net is estimated by the half-perimeter wirelength (HPWL) of the net bounding box. A higher RUDY indicates a higher wire density. At each location (x, y) on the layout, the RUDY contributed by the ith net is:

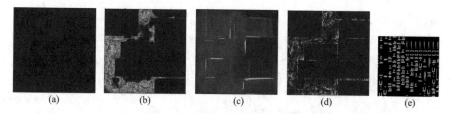

Fig. 2.7 Visualization of some common extracted features. (**a**) Pin density. (**b**) Macro density. (**c**) Long-range RUDY. (**d**) RUDY pins. (**e**) Detailed pin configurations in high resolution (from part of the layout)

$$\text{RUDY}^i(x, y) = \begin{cases} \text{HPWL}(i)/\text{area}(i) & \text{if } (x, y) \text{ is in the bounding box of } i\text{th net} \\ 0 & \text{Otherwise} \end{cases}$$

where area(i) is the bounding box area of the ith net. The final RUDY at location (x, y) is the summation of contributions from all nets in the design.

$$\text{RUDY}(x, y) = \sum_i \text{RUDY}^i(x, y)$$

RUDY is originally proposed as a pre-routing congestion estimator [24] and is first adopted as an input feature in the work of [27].

- Short/long-range RUDY map. The original RUDY feature can be decomposed into long-range and short-range RUDY. Long-range RUDY is from nets covering a distance longer than a threshold. Similarly, short-range RUDY is for nets shorter than this threshold. According to the work of [27], there is a stronger correlation between long-range RUDY and DRV than short-range ones.
- Horizontal/vertical net density map. Similar to long-range RUDY, these two features estimate how many nets are expected to go through each grid horizontally and vertically [7]. The probability of the grid being routed through by a net will be evenly distributed to each grid in a column or a row in the net bounding box. Thus, in the horizontal (vertical) direction, all grids covered by the same net will receive the same density from this net, which equals one divided by the number of covered grids in this row (column).
- RUDY pins. It is similar to pin density, while the contribution of each pin to the density map now equals the long-range RUDY of the net connected with it.

In addition to these methods, some works adopt a heuristic named fly line, which is a line that connects pins. It is also referred to as flight line in some previous works [6]. There are different types of fly lines that may reflect wire congestions. Figure 2.8 demonstrates examples of these wire density features, and some explanations are given below:

- Pair-wise fly lines. For each net, the pair-wise fly lines are connected from each pin to all the other pins of the same net.

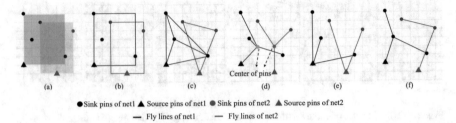

Center of pins

(a)   (b)   (c)   (d)   (e)   (f)

● Sink pins of net1 ▲ Source pins of net1 ● Sink pins of net2 ▲ Source pins of net2
— Fly lines of net1   — Fly lines of net2

Fig. 2.8 Wire density feature examples. (**a**) RUDY. (**b**) Bounding box. (**c**) Pair-wise fly lines. (**d**) Star fly lines. (**e**) Source-sink fly lines. (**f**) MST fly lines

- Star fly lines. For each net, the star fly line connects each pin to the center of the pins in the net.
- Source-sink fly lines. In timing optimization, tools tend to connect sinks with the source through the shortest path. To capture this effect, source-sink fly line connects the source pin with all sink pins in the same net.
- MST fly lines. The three aforementioned fly line features tend to overestimate the routing demand. In traditional routing, the minimum spanning tree (MST) is an effective algorithm to guide the router. Thus, the fly line connects all the edges in its MST.

2.3.2.3 Routing Congestion

As introduced, the routing congestion of a layout is a good, although not perfect, indicator of routability and DRC violations. Thus, estimations on congestion are very useful input features. Besides being a feature, it is also used as the prediction label in many prior works. There are two types of congestion reports that can be generated by commercial layout tools, as shown below:

- Trial routing (TR) congestion. After detailed placement, the trial global routing, also denoted as trial routing, can be performed. It produces an estimation of routing congestion, named TR congestion.
- Global routing (GR) congestion. Compared with trial routing, the full-fledged global routing can generate a more detailed congestion map, denoted as GR congestion.

2.3.2.4 Pin Accessibility

Besides congestion, pin accessibility is another main cause of DRC violations, especially at the advanced technology nodes. As a result, DRVs may correlate with pin shapes and the proximity relationship among pins. Compared with other input features, such fine-grained pin patterns at the advanced node are difficult to be directly quantitatively measured when using large grids, with each grid containing multiple pins. To solve this, one option is to use high-resolution "images" showing the pin patterns/configurations as input features [15]. However, such a high resolution may lead to a higher computation cost in the routability estimator.

2.3.2.5 Routability Label

Besides all features mentioned above, we also discuss different types of labels used in routability prediction in this subsection.

- All DRC violations. Most prior works directly predict all DRC violations in the layout. Some works use accurate DRC violations from sign-off validation EDA tools, while others adopt fast DRV estimations from digital layout tools after detailed routing finishes.
- Part of DRC violations. Some prior works choose to only predict certain types of DRC violations (like low metal layer short violations) in order to improve the layout flow for better routability.
- Routing congestion. Considering the complexity and difficulty in DRV prediction, many works choose to only predict the routing congestion for better accuracy, especially for models targeting to be applied at earlier design stages.

2.3.3 DL Model Architecture Design

As mentioned, the most widely used DL models for routability prediction are CNN- and FCN-based methods, targeting coarse-grained and fine-grained routability prediction tasks, respectively. We use the term "FCN based" broadly to include all DL models with both downsampling and upsampling structures. In this part, we will introduce the model architecture design. It starts with the common operators and connections, which are the basic building blocks of estimators. After that, we introduce representative prior works as case studies to demonstrate the design of different routability estimators.

2.3.3.1 Common Operators and Connections

We start with the detailed introduction of operators used in CNN/FCN methods for routability prediction. The fundamental operators include convolutional layers, pooling layers, and fully connected (FC) layers. CNN performs downsampling before the FC layers in order to enlarge its input receptive field and capture global information of the whole circuit layout. The downsampling in CNN-based routability estimators is usually achieved by either a maximum pooling layer or a convolutional layer with stride equal to 2. The stride is a parameter of the

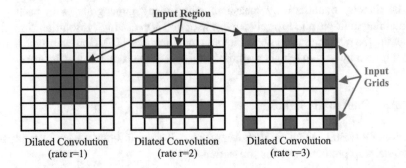

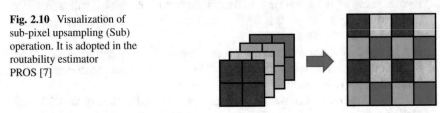

Fig. 2.9 The input regions of atrous or dilated convolution with rate $r = 1, 2, 3$ for a 3×3 kernel

Fig. 2.10 Visualization of sub-pixel upsampling (Sub) operation. It is adopted in the routability estimator PROS [7]

neural network's filter that modifies the amount of movement over the input. Many researchers prefer the latter option for downsampling since it allows one more convolutional operation. After the downsampling, the output is flattened and processed by multiple FC layers, which generate one vector or scalar as the prediction. To evaluate the overall routability, CNN-based estimators perform binary classification or regression.

In addition to the standard convolutional layer, the atrous convolution, also known as dilated convolution, is constructed by inserting zeros between each value in the kernel. An atrous convolution with dilation rate r would insert $r - 1$ zeros between adjacent values in the kernel. A visualization of atrous is shown in Fig. 2.9. When rate $r = 1$, it degrades to a regular convolution operator. This operator is designed to further increase the receptive field at the cost of higher memory usage. It is commonly seen in image segmentations and starts to be adopted for routability prediction in recent years [7].

Compared with CNN, FCN performs both downsampling and upsampling, or named encoding and decoding, in order to achieve a two-dimensional output as the fine-grained prediction on DRC violation distribution. The downsampling process is performed with the front half of the FCN structure, which is usually very similar to the CNN model with convolutional and pooling layers. After downsampling, the upsampling is commonly performed with transposed-convolutional layers in the latter half of the FCN structure. In addition to transposed convolution, the other common choice of upsampling is the sub-pixel upsampling block (Sub). It rearranges the elements of $W \times H \times r^2 C$ tensor to form a $rW \times rH \times C$ tensor with pixel shuffling. Figure 2.10 provides a simple example, where $W = 2, H = 2, r = 2, C = 1$, converting a tensor in $2 \times 2 \times 4$ to a tensor in $4 \times 4 \times 1$. This operation adds no extra trainable parameters and takes less computation cost. Notice that standard convolutions are also applied during the upsampling process, and the width and height of the final output are commonly designed to be the same as the input.

These widely adopted operations need to be connected to build the routability estimator. Besides the standard connections between every two neighboring layers, the *shortcut* or *skip connection* is commonly adopted in CNN-/FCN-based routability estimators. Such shortcuts can be roughly categorized into short-range and long-range shortcuts. The short-range shortcut can be achieved with structures like the residual block (resBlock) in ResNet [10]. It adds the input value of this residual block to the output. The addition is element-wise and thus requires exactly the same dimension between the two ends of the shortcut. The size of a residual block typically contains only two convolution layers, limiting the range of such shortcut. The long-range shortcut can be based on structures like U-Net. It applies to FCN-based estimators, which concatenate or add feature maps in the front (before downsampling and upsampling) directly to feature maps near the end (after downsampling and upsampling). In this way, it supports a combination of both shallow and deep feature maps, and the successive layers can learn from the feature mixture. If it performs feature map concatenation instead of addition, it only requires the same width and height between the two ends of shortcut, without limiting the dimension in channels.

These are the basic building blocks of routability estimators; then, we introduce representative prior works as case studies below. The visualizations of these estimator structures are shown in Fig. 2.11.

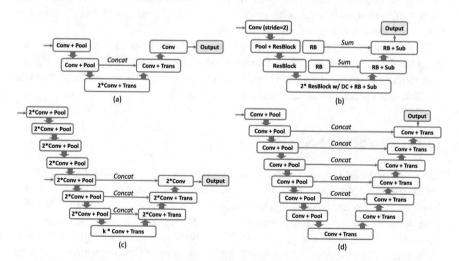

Fig. 2.11 Model structures for fine-grained DRC hotspot location detection as case studies. (**a**) RouteNet [27] structure. (**b**) PROS [7] structure. DC stands for dilated convolution, RB stands for refinement block, and Sub stands for sub-pixel upsampling block. (**c**) J-Net [15] structure. (**d**) Painting [29] generative framework structure

2.3.3.2   Case Study: RouteNet [27]

The work RouteNet [27] proposed two different models, one for the overall routability prediction and the other to pinpoint the DRC hotspots. A CNN-based model is adopted to predict the overall routability in RouteNet. It directly adopts the 18-layer ResNet but replaces the output layer to produce a single-scalar prediction, indicating four different levels of routability. In addition, this 18-layer ResNet is already pretrained on the ImageNet. Thus, the training process can be viewed as transfer learning with data about routability. This model is applied after global placement.

In comparison, to pinpoint the DRC hotspots, RouteNet proposes an FCN-based model with six convolutional layers, two pooling layers, two trans-convolutional layers, and one long-range shortcut structure. According to its ablation study, using fewer convolutions, removing the shortcut, or removing pooling/trans-convolution will all lead to accuracy degradation. This model is applied after global routing.

2.3.3.3   Case Study: PROS [7]

The work of PROS [7] proposed a more complex FCN-based model to pinpoint GR congestion locations before routing. Compared with DRC hotspot detection, this is also a fine-grained prediction task but easier to obtain a higher accuracy. In the downsampling (encoding) part of its model, similar to RouteNet [27], it directly adopts a whole pretrained ResNet network and then performs transfer learning. To avoid too much reduction in feature map size while maintaining the receptive field, some standard convolutions with stride 1 or 2 are replaced by dilated convolution (DC) with stride set to 1 and rate set to 2 or 4.

In the upsampling (decoding) part of its model, it adopts the sub-pixel upsampling blocks (Sub) to upsample and recover the fine-grained predictions. In addition, it introduces refinement blocks (RB), which are quite similar to residual blocks in ResNet, to perform convolutions on the feature maps.

2.3.3.4   Case Study: J-Net [15]

The work of J-Net [15] proposed an FCN-based model to pinpoint DRC hotspots before routing. Compared with previous works, it is targeted to the advanced sub-10nm process nodes, where pin accessibility becomes a major contributor to DRC violations. The main model structure is also FCN based with three long-range shortcuts. To deal with the pin accessibility problem, it adopts "high-resolution" pin configuration images as extra inputs and designs extra downsampling structures to extract pin accessibility information from these pin configuration inputs.

2.3.3.5 Case Study: Painting [29]

The work of Painting [29] adopted a training framework based on cGAN for routability prediction on FPGA. The generative network structure is an FCN-based model with five long-range shortcuts. This FCN-based model also generates predictions on routing congestion locations. But instead of directly optimizing this FCN-based generative network like the prior works, it is trained under the cGAN framework together with a CNN-based discriminator network. The generative network is trained to produce predictions similar to labels such that it confuses the discriminator network. Currently, the difference in performance between this cGAN-based training method and common FCN-based methods still remains rather unclear. For routability predictions on FPGA, we also observe some research efforts [1] adopting such cGAN-based framework and training methods.

2.3.3.6 Case Study: Automated Model Development [6]

Most aforementioned routability models are very carefully designed and achieve good results on their own benchmarks. However, these very well-designed methods usually take a long development time and high engineering efforts. A recent work proposes to avoid this cost by automating the model development step by neural architecture search (NAS) methods [6]. It supports a large search space allowing various types of operations and highly flexible connections. Candidate operations include standard convolutions, atrous convolutions, and mixed depth-wise convolutions. In the search space, a model is represented by a graph. Specifically, vertices represent operations, and edges indicate the directed connections of operations. Its search space is shown in Fig. 2.12. As for the search strategy, it samples edges from multiple completely ordered graphs with adjustable probabilities defined on each component of the graph. After each sampling, the sampled model is trained and evaluated on the validation set. Then, the evaluation result is used to update

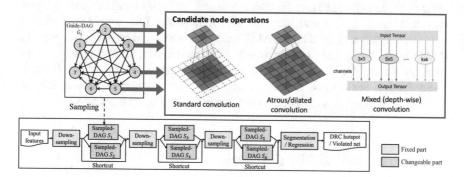

Fig. 2.12 The graph-based search space for routability estimators, in the work of [6]

the sampling probability. The basic idea is if the performance is good, then the probability of corresponding structures being sampled again becomes higher.

Compared with human-developed routability estimators, the automated-developed estimators construct much more parallel branches and flexible interactions. This inherits from the topology of the graph-based search space. Also, the framework supports many different convolutional operators, increasing the diversity of feature representations. When comparing generated models between the coarse-grained and fine-grained predictions, the generated model for the fine-grained DRC hotspot detection is significantly more complex [6]. It indicates that the essential difference between these two routability prediction tasks is captured and reflected by the two discovered models. In summary, automated model development relieves developers from the long model development cycle and may become the trend of designing routability estimators in the future. However, notice that this method only automates the model development step, while the data generation and feature engineering still rely on human engineers. We believe there is still a long way toward fully automated routability estimator development.

2.3.4   DL Model Training and Inference

After the feature and label collection and preprocessing finishes and the model architecture is determined, the DL model training and testing is a relative simple task. In the training process, regularization like L2 norm is commonly adopted to avoid overfitting. The batch normalization should be applied. For optimizers, the common choices are stochastic gradient descent (SGD) and Adam [13]. Some researchers tend to believe parameter tuning with Adam optimizer is easier. The hyperparameters such as learning rate and regularization strength are adjusted according to the model and the training dataset.

The training and inference are usually accelerated by one GPU card. The training process usually finishes in one day, and the inference of one layout usually only takes seconds. But notice that the feature extraction overhead for each testing layout may take much longer time than the model inference time itself. Such feature extraction overhead may be reduced if the model is well integrated into a physical design tool. During training, for a large FCN model with a large input size, the memory size of GPU may limit the maximum batch size, which affects batch normalization quality and the model accuracy. This memory limit may require multi-GPU training or adopting the cropping of layouts as the input.

2.4   DL for Routability Deployment

Besides the prediction of design routability, another equally important topic is how to apply these developed estimators in real design flows to actually benefit design routability, which is designers' ultimate goal. However, this topic is less explored, and researchers haven't reached a consensus on a unified best solution to deploy routability estimators. In the following, we will introduce some representative research explorations on this topic.

2.4.1 Direct Feedback to Engineers

The most straightforward way to apply routability estimators is to directly provide feedback to human engineers. After finishing the placement of a design, engineers can apply the model to predict the routing congestion or the DRC violation that will emerge after the routing stage. Based on this prediction, engineers can select more routable layout solutions or proactively improve the layout solution to achieve better routability.

2.4.2 Macro Location Optimization

In the macro placement process, there is no effective cost metric to accurately evaluate the output layout quality at this early stage, since its final performance depends on the subsequent highly complex cell placement and routing stages. The oversimplified early estimations on the HPWL and area cannot precisely reflect the layout qualities after routing. Therefore, DL-based routability estimators are good fit to improve the quality of macro placement. The work of [11] integrates the DL model into the simulated annealing (SA)-based macro placement algorithm. After deriving a placement result from perturbation of SA, it utilizes the DL model to predict the number of design rule violations to see whether this solution will be accepted or not. With the integration of a DL model, it achieves solutions with fewer DRVs.

2.4.3 White Space-Driven Model-Guided Detailed Placement

Routability estimators can be applied to mitigate DRC violations proactively in the placement flow. This is achieved by spreading the white space of the regions with potential DRC hotspots [5]. After predicting the DRV hotspot locations, the mitigation method collects the white space in local regions around those estimated hotspots. Then, it redistributes white space among overlapped local windows and keeps incrementally moving cells to redistribute white space. After this re-legalization method, the output layout achieves fewer DRVs after routing.

2.4.4 Pin Accessibility-Driven Model-Guided Detailed Placement

Based on a pin accessibility-focused routability estimator, a model-guided detailed placement algorithm [30] can guide the detailed placer to avoid generating DRV-prone pin patterns. First, the estimator is used to generate a set of cell spacing rules, which applies to any designs using the same cell library. The rules define the minimum spacing between every pair of cells. These spacing rules are integrated into a detailed placer for optimization, minimizing the total amount of inserted placement blockages in a cell row considering cell orientations [30]. With the pre-inserted placement blockages, all pin patterns with predicted bad pin accessibility will be removed in the following legalization step. In this way, this model-guided detailed placement can effectively reduce pin accessibility-induced DRVs in the subsequent routing step.

2.4.5 Integration in Routing Flow

Besides improving the placement step, another perspective is to use routability estimators to facilitate global routing [7]. In global routing, based on congestion predictions, the routing cost of gcell congestion can be modified. For the gcells that are predicted congested, the router can shrink the available tracks so that the router will reduce the number of wires passing through these potential congestion areas. Also, the router can increase the wire/via cost of these congestion gcells when routing nets spanning large regions. Nets spanning large regions can find a way to detour and avoid potentially congested regions. With the help of a DL model, the router can reduce DRV count and routing congestions.

2.4.6 Explicit Routability Optimization During Global Placement

Some recent works [16] choose to explicitly optimize routability by directly integrating the routing congestion prediction into the global placement objective function as a new penalty term. In this way, the placer can utilize gradients to adjust all movable cells toward better routability during the optimization of such placement objective function. This may be the most direct and explicit way to optimize placement solutions with respect to a given routability estimator.

2.4.7 Visualization of Routing Utilization

In addition to most applications on optimizing layout solutions, some works [29] apply routability estimators in visualization. They visualize the routing utilization on-the-fly during FPGA placement and generate such real-time forecast results in GIFs or videos.

2.4.8 Optimization with Reinforcement Learning (RL)

Last, besides the deployment of DL-based routability estimators, the work of [22] indicates that the order of nets to be routed can significantly impact the routing quality. It proposes an RL-based algorithm to learn the ordering policy that minimizes the DRC violations from the net features.

2.5 Summary

In this chapter, we introduce deep learning-based methods for routability estimations. After introducing background on both routability and relevant deep learning algorithms, we emphasize the importance of global information and model receptive field, which motivates the adoption of DL models for routability predictions. After that, we introduce the methodology in detail, covering topics including data preparation and augmentation, feature engineering, model architecture design, and model training and inference. Finally, we introduce existing explorations in the application and deployment of these routability estimators in the physical design flow.

Although there have been many existing research efforts showing promising results in this direction, we believe DL-based routability estimations still face several challenges. First, the accuracy of DL estimators may degrade when applied to certain new designs, making them unreliable in practice. In addition, it is usually challenging to get access to adequate training data with sufficient diversity in VLSI design, which limits the development of high-quality and generalized DL estimators. In addition, the model size of recent DL estimators keeps increasing. While it may not be a problem as a stand-alone tool, this may prevent an efficient integration of the DL model into existing chip design flows. Considering these challenges, we believe DL-based routability estimators should target higher accuracy, better generalization, less inference cost, and improved interoperability in design flow in the future.

References

1. Alawieh, M.B., Li, W., Lin, Y., Singhal, L., Iyer, M.A., Pan, D.Z.: High-definition routing congestion prediction for large-scale FPGAs. In: Asia and South Pacific Design Automation Conference (ASP-DAC) (2020)
2. Araujo, A., Norris, W., Sim, J.: Computing receptive fields of convolutional neural networks. Distill (2019) doi:10.23915/distill.00021, https://distill.pub/2019/computing-receptive-fields
3. Bustany, I.S., Chinnery, D., Shinnerl, J.R., Yutsis, V.: ISPD 2015 benchmarks with fence regions and routing blockages for detailed-routing-driven placement. In: International Symposium on Physical Design (ISPD) (2015)
4. Chan, W.T.J., Du, Y., Kahng, A.B., Nath, S., Samadi, K.: BEOL stack-aware routability prediction from placement using data mining techniques. In: International Conference on Computer Design (ICCD) (2016)
5. Chan, W.T.J., Ho, P.H., Kahng, A.B., Saxena, P.: Routability optimization for industrial designs at sub-14nm process nodes using machine learning. In: International Symposium on Physical Design (ISPD) (2017)
6. Chang, C.C., Pan, J., Zhang, T., Xie, Z., Hu, J., Qi, W., Lin, C.W., Liang, R., Mitra, J., Fallon, E., Chen, Y.: Automatic routability predictor development using neural architecture search. In: International Conference on Computer-Aided Design (ICCAD) (2021)
7. Chen, J., Kuang, J., Zhao, G., Huang, D.J.H., Young, E.F.: PROS: A plug-in for routability optimization applied in the state-of-the-art commercial EDA tool using deep learning. In: International Conference On Computer Aided Design (ICCAD) (2020)
8. Chen, L.C., Papandreou, G., Schroff, F., Adam, H.: Rethinking atrous convolution for semantic image segmentation (2017). Preprint. arXiv:1706.05587
9. Ding, Y., Chu, C., Mak, W.K.: Pin accessibility-driven detailed placement refinement. In: International Symposium on Physical Design (ISPD) (2017)
10. He, K., Zhang, X., Ren, S., Sun, J.: Deep residual learning for image recognition. In: IEEE Conference on Computer Vision and Pattern Recognition (CVPR) (2016)
11. Huang, Y.H., Xie, Z., Fang, G.Q., Yu, T.C., Ren, H., Fang, S.Y., Chen, Y., Hu, J.: Routability-driven macro placement with embedded cnn-based prediction model. In: Design, Automation & Test in Europe Conference & Exhibition (DATE) (2019)
12. Hung, W.T., Huang, J.Y., Chou, Y.C., Tsai, C.H., Chao, M.: Transforming global routing report into DRC violation map with convolutional neural network. In: International Symposium on Physical Design (ISPD) (2020)
13. Kingma, D.P., Ba, J.: Adam: a method for stochastic optimization (2014). Preprint. arXiv:1412.6980
14. Kirby, R., Godil, S., Roy, R., Catanzaro, B.: CongestionNet: routing congestion prediction using deep graph neural networks. In: International Conference on Very Large Scale Integration (VLSI-SoC) (2019)
15. Liang, R., Xiang, H., Pandey, D., Reddy, L., Ramji, S., Nam, G.J., Hu, J.: DRC hotspot prediction at sub-10nm process nodes using customized convolutional network. In: International Symposium on Physical Design (ISPD) (2020)
16. Liu, S., Sun, Q., Liao, P., Lin, Y., Yu, B.: Global placement with deep learning-enabled explicit routability optimization. In: Design, Automation & Test in Europe Conference & Exhibition (DATE) (2021)
17. Long, J., Shelhamer, E., Darrell, T.: Fully convolutional networks for semantic segmentation. In: IEEE Conference on Computer Vision and Pattern Recognition (CVPR) (2015)
18. Maarouff, D., Shamli, A., Martin, T., Grewal, G., Areibi, S.: A deep-learning framework for predicting congestion during FPGA placement. In: International Conference on Field-Programmable Logic and Applications (FPL) (2020)
19. Mirza, M., Osindero, S.: Conditional generative adversarial nets (2014). Preprint. arXiv:1411.1784

20. Pan, J., Chang, C.C., Xie, Z., Li, A., Tang, M., Zhang, T., Hu, J., Chen, Y.: Towards collaborative intelligence: routability estimation based on decentralized private data. In: Design Automation Conference (DAC) (2022)
21. Pham, H., Guan, M., Zoph, B., Le, Q., Dean, J.: Efficient neural architecture search via parameters sharing. In: International Conference on Machine Learning (ICML) (2018)
22. Qu, T., Lin, Y., Lu, Z., Su, Y., Wei, Y.: Asynchronous reinforcement learning framework for net order exploration in detailed routing. In: Design, Automation & Test in Europe Conference & Exhibition (DATE) (2021)
23. Ronneberger, O., Fischer, P., Brox, T.: U-Net: Convolutional networks for biomedical image segmentation. In: International Conference on Medical Image Computing and Computer-Assisted Intervention (MICCAI) (2015)
24. Spindler, P., Johannes, F.M.: Fast and accurate routing demand estimation for efficient routability-driven placement. In: Design, Automation & Test in Europe Conference & Exhibition (DATE) (2007)
25. Tabrizi, A.F., Rakai, L., Darav, N.K., Bustany, I., Behjat, L., Xu, S., Kennings, A.: A machine learning framework to identify detailed routing short violations from a placed netlist. In: Design Automation Conference (DAC) (2018)
26. Tabrizi, A.F., Darav, N.K., Rakai, L., Bustany, I., Kennings, A., Behjat, L.: Eh? predictor: a deep learning framework to identify detailed routing short violations from a placed netlist. IEEE Trans. Comput. Aided Des. Integr. Circuits Syst. (TCAD) (2019)
27. Xie, Z., Huang, Y.H., Fang, G.Q., Ren, H., Fang, S.Y., Chen, Y., Hu, J.: RouteNet: routability prediction for mixed-size designs using convolutional neural network. In: International Conference on Computer-Aided Design (ICCAD) (2018)
28. Yu, F., Koltun, V.: Multi-scale context aggregation by dilated convolutions (2015). Preprint. arXiv:1511.07122
29. Yu, C., Zhang, Z.: Painting on placement: Forecasting routing congestion using conditional generative adversarial nets. In: Design Automation Conference (DAC) (2019)
30. Yu, T.C., Fang, S.Y., Chiu, H.S., Hu, K.S., Tai, P.H.Y., Shen, C.C.F., Sheng, H.: Pin accessibility prediction and optimization with deep learning-based pin pattern recognition. In: Design Automation Conference (DAC) (2019)
31. Zhou, Q., Wang, X., Qi, Z., Chen, Z., Zhou, Q., Cai, Y.: An accurate detailed routing routability prediction model in placement. In: Asia Symposium on Quality Electronic Design (ASQED) (2015)

Chapter 3
Net-Based Machine Learning-Aided Approaches for Timing and Crosstalk Prediction

Rongjian Liang, Zhiyao Xie, Erick Carvajal Barboza, and Jiang Hu

3.1 Introduction

In digital circuit design, timing is a primary design objective that needs to be considered since the very early design stages. Accurate timing prediction is very challenging at early stages due to the absence of information determined by later stages in the design flow. For example, locations of cells and the exact routing topology are critical for timing analysis, but they are not available until cell placement stage and detailed routing stage, respectively, have been executed. However, early design stages have relatively ample room for changes that can fix timing problems in a proactive manner. As a design proceeds to later stages, the design flexibility diminishes although timing estimation becomes more and more accurate. In addition, even at the sign-off stage where all detailed layout parameters are determined, the license cost of EDA tools and the runtime overhead to consider the complex signal integrity (SI) effects hinder the acquisition of accurate timing reports.

In conventional VLSI design flow, two tactics are employed to address the uncertainty in timing analysis. One is to take an overly pessimistic estimation to ensure that no timing violations will occur after routing. However, such pessimism causes overdesign that wastes power, area, and optimization time. The other is to iterate back to early stages, i.e., cell placement or even logic synthesis stage, when the desired timing-power trade-off cannot be achieved at sign-off. However,

R. Liang (✉) · E. C. Barboza · J. Hu
Texas A&M University, College Station, TX, USA
e-mail: liangrj14@tamu.edu; ecarvajal@tamu.edu; jianghu@tamu.edu

Z. Xie
Hong Kong University of Science and Technology, Clear Water Bay, Hong Kong
e-mail: eezhiyao@ust.hk

© The Author(s), under exclusive license to Springer Nature Switzerland AG 2022
H. Ren, J. Hu (eds.), *Machine Learning Applications in Electronic Design Automation*, https://doi.org/10.1007/978-3-031-13074-8_3

an additional iteration may not guarantee success, and multiple iterations would grossly increase design turnaround time.

Crosstalk is one of the major threats to SI and timing closure in ASIC design. Through capacitive coupling, a signal switching of one net causes crosstalk noise and incremental delay at its neighboring nets [29]. Noise amplitude can even reach up to 30% of V_{DD} [25], which may exceed the threshold voltage of transistors and lead to glitches, imposing a risk of logic errors and unwanted switching power consumption. Moreover, the coupling capacitance itself serves as extra load, increasing both signal delay and internal power dissipation; the incremental delay due to coupling capacitance along a timing path can reach 300 ps [26], comparable to clock periods of modern high-performance processors.

Similar to timing analysis, an accurate estimation of crosstalk effects is only possible after detailed routing, where few rooms remain to fix all the crosstalk-induced design problems, even if we identify every crosstalk issue [29]. In this regard, many research efforts have been undertaken to predict and mitigate crosstalk problems at earlier design stages, e.g., placement [20, 26]. However, a majority of crosstalk-driven placement works resort to global routing or trial routing for obtaining an approximate estimate of routing landscape and thereby capacitive coupling of nets. An evident drawback here is that global or trial routing is time-consuming and thus induces huge runtime costs.

In recent years, machine learning (ML) techniques have been adopted to improve the predictability of timing and crosstalk effects at different stages of the chip design flow. As for the preplacement stages, ML solutions have been developed for the estimate of the overall wirelength of a netlist [19], lengths of a few selected paths [12], and the net length as well as delay for each individual net [33]. Note that wirelength/net length is an important optimization objective in VLSI design since interconnect is a dominating factor for performance and power in advanced technology nodes. Timing and crosstalk effect estimation at the cell placement stage have been studied in [2] and [17], respectively. ML techniques have also been leveraged to calibrate non-SI timing to SI timing [14], non-SI to non-SI, or SI to SI between different timers [9]. Unlike conventional methods either being very inaccurate or very runtime expensive, ML-aided solutions can be fast yet accurate, reaching a new balance point between runtime and accuracy. Estimation results can essentially benefit design automation by providing early and high-fidelity feedback and guiding proactive actions in early stages to improve design outcomes.

A majority of the aforementioned methods utilize net-based models. Here, net-based models include those modeling the timing or crosstalk properties of individual cell, net, or logic stage. The timing performance of a design can be inferred by propagating cell delays and net delays along the netlist. Crosstalk occurs between physically adjacent nets. Hence, it is natural to build net-based models. Net-based prediction results can be easily integrated into existing EDA flow, since various timing and crosstalk optimization techniques target cells or nets, e.g., gate sizing [6], buffer insertion [11], and layer assignment [30]. Compared with directly modeling path delays or properties of a netlist, net-based estimation provides more detailed information to guide optimization techniques.

In this chapter, we will cover net-based machine learning-aided approaches for timing and crosstalk prediction in detail. A comprehensive review is presented in Sect. 3.2, followed by four representative case studies introduced in Sect. 3.3 to 3.6. Finally, conclusions are drawn in Sect. 3.7.

3.2 Backgrounds on Machine Learning-Aided Timing and Crosstalk Estimation

3.2.1 Timing Prediction Background

Timing analysis determines if the timing constraints imposed on a digital circuit are met. According to whether input vectors are applied, there are two types of timing analysis, i.e., static timing analysis (STA) and dynamic timing analysis. STA checks static timing requirements without applying any input vectors, while dynamic timing analysis applies input vectors and verifies whether output vectors are correct. Compared to dynamic timing analysis, STA is faster as it does not need to simulate multiple input vectors, and it is more thorough since it determines the worst-case timing. All the timing prediction methods introduced in this chapter belong to STA.

We use the example in Fig. 3.1 to illustrate the basic concepts in timing analysis. A timing arc represents the timing relationship between two pins or input/output ports. Timing arcs can be roughly divided into two categories, i.e., net arcs and cell arcs. For example, in Fig. 3.1, the timing arc between Pin c in Net A and Pin y in Net C is a cell arc, while the arc between Pin y and Pin e in Net C is a net arc. The arc from an input pin of a net driver to a sink pin of this net describes the timing for a logic stage. Some previous works develop separate models for net arcs and cell arcs [33], while others build logic stage-based models [2, 14]. For simplicity, a few pre-routing STA methods [2] do not differentiate among rising delays and falling delays. A data path starts from an input port of a design or a clock pin of a sequential cell (e.g., flip-flop/latch/register) and ends at a data input pin of a sequential cell or an output port of the design. For example, in Fig. 3.1, the path from the input pin of Flip-flop F1 to Pin g of Flip-flop F3 is a data path. A clock path starts from the clock input port of the design to a clock pin of a sequential cell. Path delays can be calculated by propagating cell delays and net delays along timing paths. Signal arrival time is the time in which signal arrivals at the pins/ports. Slack is defined as the difference between signal arrival time and the required arrival time. Positive slacks mean that timing constraints are met, while negative slacks imply violations of timing constraints.

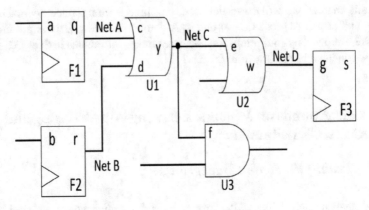

Fig. 3.1 A circuit example [2]

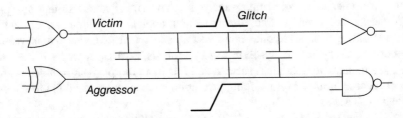

Fig. 3.2 Illustration of crosstalk noise

3.2.2 Crosstalk Prediction Background

Crosstalk refers to the undesirable electrical interaction caused by the capacitive cross-coupling between physically neighboring nets [24]. In advanced technology nodes, metal wires tend to be tall and thin and routed close to each other, leading to increased coupling capacitance between neighboring nets. As shown in Fig. 3.2, due to the coupling capacitance, the rising edge on the aggressor net causes a noise bump or glitch on the victim net, which should be constant at logic 0 or 1. In addition, crosstalk can also lead to signal delays by changing the times at which signal transitions occur, as shown in Fig. 3.3.

3.2.3 Relevant Design Steps

The netlist of a design is generated at logic synthesis stage. It is possible to build net-based models for timing and crosstalk estimation at every major design step since logic synthesis. There is a trade-off between accuracy of timing/crosstalk prediction and design flexibility across relevant design stages. An ideal timing closure and crosstalk avoidance flow can be as follows:

Fig. 3.3 Illustration of transition slowdown or speedup induced by crosstalk [24]

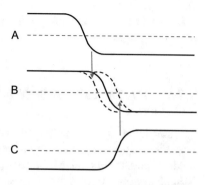

1. During logic synthesis and placement stage, significant timing violations and crosstalk risks are identified and resolved by leveraging the greater design flexibility. Typical optimization techniques at logic synthesis and placement stage include buffer insertion [11], gate sizing [6], and logic restructuring [21].
2. Most of the remaining timing violations and crosstalk risks are eliminated in global routing via optimization techniques, such as buffer insertion, layer assignment [30], and area routing [27].
3. In detailed routing, complete timing closure and crosstalk avoidance is achieved with the precise timing and crosstalk evaluation.

In conventional design flows, there are mainly two kinds of timing models that are often employed for circuit logic and physical synthesis. One is the sign-off model that evaluates if a circuit design satisfies timing specifications. It estimates gate delay using lookup tables [18] or current source model, where slew rate is considered, and wire delays use high-order models [23]. Additionally, sign-off timing analysis tools consider many complicated details such as rising/falling switching, crosstalk, false paths, and simultaneous switching. These models are accurate but very slow. The other is a relatively fast model that is often invoked within synthesis and optimizations. In this case, a gate is modeled as an RC switch and wires are modeled as RC trees, and delay is computed using the Elmore method [7]. Gate and wire delays are collected through addition/subtraction and max/min operations in PERT traversals [4] to obtain signal arrival time, required arrival time, and slack at each node in a circuit. However, such a model is not accurate, as it does not consider potential wire detours due to congestion avoidance and layer assignment impact. Moreover, higher-order interconnect model has no ground to carry out accurate delay computation without wire parasitic information. The aforementioned reasons have motivated recent research in ML-based approaches for fast yet precise timing prediction.

A majority of previous crosstalk avoidance methods target at routing stages [22, 27, 29, 34], the placement stage solutions [20, 26] are still far from being practical largely due to the dependence on trial/global routing, which can easily take more than a half hour for a modern design. In addition, timing analysis tools that consider complicated crosstalk effects at sign-off timing analysis are usually very slow. ML-aided methods for crosstalk prediction have shown great potential in addressing these challenges.

3.2.4 ML Techniques in Net-Based Prediction

Machine learning techniques utilized in net-based prediction can be roughly divided into three categories, i.e., graph neural networks (GNNs), decision tree-based models, and traditional ML techniques. Netlists in fact are graphs; thus, it is natural to apply GNNs for net-based prediction. In decision tree-based models, knowledge is represented by a set of binary decision-making. Such high-level interpretation of knowledge allows decision tree-based models to easily incorporate features from different sources. Traditional ML techniques, such as linear models, multilayer artificial neural networks (ANNs), and support vector machines (SVMs), are also leveraged for timing and crosstalk prediction. We briefly introduce the aforementioned ML techniques in the upcoming paragraphs.

Graph Neural Networks GNN [31] models are composed of multiple sequential convolution layers, as shown in Fig. 3.4. Each layer generates a new embedding for every node based on the previous embeddings. For node n_k with node features O_k, we denote its embedding at the tth layer as $h_k^{(t)}$. Its initial embedding is the node features $h_k^{(0)} = O_k$. In each layer t, GNNs calculate the updated embedding $h_k^{(t)}$ based on the previous embedding of the node itself $h_k^{(t-1)}$ and its neighbors $h_b^{(t-1)} | n_b \in \mathcal{N}(n_k)$.

We show one layer of graph convolutional network (GCN) [16], GSage [8], and graph attention network (GAT) [28] below. Notice that there exist other expressions of these models. The two-dimensional learnable weight at layer t is $W^{(t)}$. In GAT, there is an extra one-dimensional weight $\theta^{(t)}$. The operation [||] concatenates two vectors into one longer vector. Functions σ and g are sigmoid and Leaky ReLU activation function, respectively.

GCN (with self-loops):

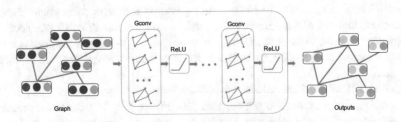

Fig. 3.4 Illustration of the graph neural networks (adapted from [31])

$$h_k^{(t)} = \sigma \left(\sum_{n_\beta \in \mathcal{N}(n_k) \cup \{n_k\}} a_{k\beta} W^{(t)} h_\beta^{(t-1)} \right)$$

$$\text{where } a_{k\beta} = \frac{1}{\sqrt{deg(k)+1}\sqrt{deg(\beta)+1}} \in \mathbb{R}$$

GSage:

$$h_k^{(t)} = \sigma \left(W^{(t)} \left[h_k^{(t-1)} \| \frac{1}{deg(k)} \sum_{n_b \in \mathcal{N}(n_k)} h_b^{(t-1)} \right] \right)$$

GAT:

$$h_k^{(t)} = \sigma \left(\sum_{n_\beta \in \mathcal{N}(n_k) \cup \{n_k\}} a_{k\beta} W^{(t)} h_\beta^{(t-1)} \right)$$

$$\text{where } a_{k\beta} = softmax_\beta(r_{k\beta}) \quad \text{over } n_k \text{ and its neighbors,}$$

$$r_{k\beta} = g \left(\theta^{(t)\top} \left[W^{(t)} h_\beta^{(t-1)} \| W^{(t)} h_k^{(t-1)} \right] \right) \in \mathbb{R}$$

Here, we briefly discuss the difference between these methods. GCN scales the contribution of neighbors by a predetermined coefficient $a_{k\beta}$, depending on the node degree. GSage does not scale neighbors by any factor. In contrast, GAT uses learnable weights W, θ to firstly decide node n_β's contribution $r_{k\beta}$ and then normalize the coefficient $r_{k\beta}$ across n_k and its neighbors through a softmax operation. Such a learnable $a_{k\beta}$ leads to a more flexible model. For all these GNN methods, the last layer's output embedding $h_k^{(t)}$ is connected to a multilayer ANN.

Decision Tree-Based Models Knowledge is represented by a set of binary decision-making in decision tree models. An advantage of these models is that they can ensemble knowledge from different sources due to its high-level interpretation of knowledge and problems. It is especially important for timing and crosstalk prediction since layout, electrical, and logic parameters are all need to be considered. Strong learners can be obtained by ensembling simple decision trees. Random forest [3] and gradient boosted decision trees (GBDT) [5] (shown in Fig. 3.5) are two popular ensemble models. In random forest models, decision trees are used as parallel learners, and each tree is fit to a set of bootstrapping samples taken from the original dataset. Bootstrapping means randomly selecting samples from the original dataset with replacement. The final prediction result is obtained by averaging the results of decision trees. In GBDT models, each decision tree is fit

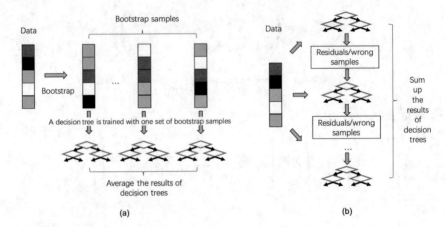

Fig. 3.5 Illustration of the random forest model and the GBDT model. (**a**) Random forest. (**b**) Gradient boosted decision trees

to the residuals from previous ones, and the final result is obtained by summing up the results of all trees.

Traditional ML Techniques Linear models, such as the linear regression and the logistic regression, are also utilized for timing and crosstalk estimation. The main advantage of the linear model is its simplicity. However, it might be too simple for complicated learning tasks.

An ANN consists of multiple fully connected layers and activation layers (e.g., sigmoid and ReLu). A key strength is its ability to capture nonlinear attributes in data [10]. Another advantage is that off-the-shelf neural network engines allow easy customization to the loss function. The main drawback of neural network models is the lack of interpretability.

SVMs find hyperplanes in a high-dimensional feature space that distinctly separates data points from different classes. Key advantages of SVMs are their robustness against noisy data and their effectiveness in high-dimensional spaces.

3.2.5   *Why ML for Timing and Crosstalk Prediction*

A significant challenge for timing and crosstalk prediction is that they are determined by the complicated joint effects of layout, electrical, and logic parameters. Thanks to its strong data-driven learning capability, ML is a natural good fit for such complex modeling tasks. Many previous ML-based solutions for timing and crosstalk estimation put an emphasis on feature engineering. The extracted features can be roughly divided into four categories, i.e., layout features, electrical features, logic structure features, and timing reports generated by EDA tools. The extracted features are then fed into ML engines to deliver timing and crosstalk prediction outcomes. The timing reports used as input are generated at early design stages or without considering complicated crosstalk effects. They incorporate information that is captured by EDA tools, but there exists a gap between these reports and the timing outcome at later stages or at a more accurate analysis mode. Other features enable ML-based estimators to further reduce the gap.

We introduce in detail four representative net-based ML-aided solutions for timing and crosstalk prediction in the following sections. The first one targets at the preplacement net length and timing prediction. The second and the third one focus on pre-routing timing and crosstalk effect prediction, respectively. The last one calibrates the non-SI timing to SI timing at sign-off. These four case studies cover several design steps from logic synthesis to sign-off. Since the available information varies at different steps, these case studies utilize different input features. We emphasize the problem formulation, prediction flow, feature engineering, and machine learning engines in the introduction of these case studies.

3.3 Preplacement Net Length and Timing Prediction

3.3.1 Problem Formulation

The work by Xie et al. [33] targets at the preplacement net length and timing prediction. The net length refers to the half perimeter wirelength (HPWL) of the bounding box of the net after placement. It is a key proxy metric for optimizing timing and power. The timing report generated by an industrial timer after placement is used as the ground truth for timing prediction. ML-based preplacement net length and timing prediction contribute to accurate evaluation of timing and power performance of synthesis solutions.

3.3.2 Prediction Flow

Figure 3.6 shows the overall preplacement flow for both individual net size and timing predictions. It is applied before layout and predicts post-placement design objectives. Prediction results can benefit optimization and evaluation for both synthesis and placement. For the net length estimation, a fast version named Net^{2f} and an accuracy-oriented version named Net^{2a} are developed. As Fig. 3.6 shows, both versions extract features directly from the netlist, while Net^{2a} further captures global information by performing clustering on the circuit netlist.

The timing estimator is constructed and applied to directly predict the delay of each individual timing arc, including cell arcs and net arcs. Besides features used by net size prediction, the preplacement timing report from commercial EDA

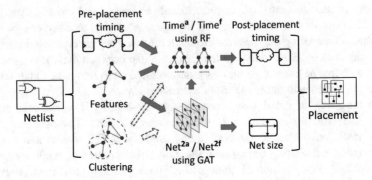

Fig. 3.6 The net size and timing prediction flow [33]

tools is also used as the input. The timing estimators also utilize the information from net size predictions as important input features. Considering the different properties between cell arcs and net arcs, two separate timing prediction models are constructed. Then, based on the inference result, the slack of each circuit node is obtained by traversing the graph with predicted delay values.

3.3.3   Feature Engineering

3.3.3.1   Features for Net Length Prediction

Both global and local topology information are incorporated in [33] for net length and timing prediction. Graph distance between two nodes is evaluated by the number of hops along the shortest path between them. Local information includes the information about the estimated net itself or from its one- to two-hop neighboring nets. In contrast, global information means the pattern behind the topology of the whole netlist or the information from nets far away from the net to be estimated.

The *local* information utilized for net length estimation is shown as follows:

- Physical features: the net's driver area, the sum of areas of all the cells of a net
- Logic structure features: the fan-out size (number of sinks) of the net, the fan-in size (number of input pins of the net's driver cell), the summation and the standard deviation of all neighboring nets' fan-in and fan-out

To capture *global* information, an efficient multilevel partitioning method hMETIS [15] to divide one netlist into multiple clusters/partitions is utilized. The partition method minimizes the overall cut between all clusters and thus provides a global perspective. A few novel global features are extracted based on the clustering results. The most important intuition behind the global features extraction is that,

Table 3.1 Preplacement features for timing prediction [33]

For each cell arc
Preplacement delay of the arc itself
Source pin information: capacitance, slew, slack
All net-size-relevant features of the following net
Predicted size of previous net
For each net arc
Source pin information: max capacitance, slew, slack
Sink pin information: capacitance
All net-size-relevant features of the net
Predicted size of the following net

for a high-quality placement solution, on average, the cells assigned to different clusters tend to be placed far away from each other.

3.3.3.2 Features for Timing Prediction

Table 3.1 summarizes selected features for cell arcs and net arcs. All these features in Table 3.1 are from the three main sources, as summarized below:

- All relevant slew, delay, and slack information from the preplacement timing report.
- Electrical and logic structure information of all relevant nets and cells. It includes the global information captured by performing clustering on the netlist.
- The prediction outcome of Net^{2f}/Net^{2a}.

3.3.4 Machine Learning Engines

3.3.4.1 Machine Learning Engine for Net Length Prediction

To apply graph-based methods, each netlist is converted to one directed graph, as shown in Fig. 3.7. Different from most GNN-based EDA tasks, net length prediction focuses on nets rather than cells. Thus, each net is represented by a node. For each net n_k, it is connected with its fan-ins and fan-outs through their common cells by edges in both directions. The common cell shared by both nets on that edge is called its *edge cell*. For example, in Fig. 3.7b, net n_3 is connected with nets n_4 and n_5 through its sinks c_G and c_H; it is connected with nets n_1 and n_2 through its driver c_D. The edges through edge cell c_G is denoted as $n_3 \rightarrow n_5$ and $n_5 \rightarrow n_3$. The edge cell c_G can also be referred to as c_{35} or c_{53}. Edges in different directions are differentiated by assigning different edge features to $n_3 \rightarrow n_5$ and $n_5 \rightarrow n_3$. After the directed graph is generated, a customized GAT model is applied to the graph to deliver the net length estimation.

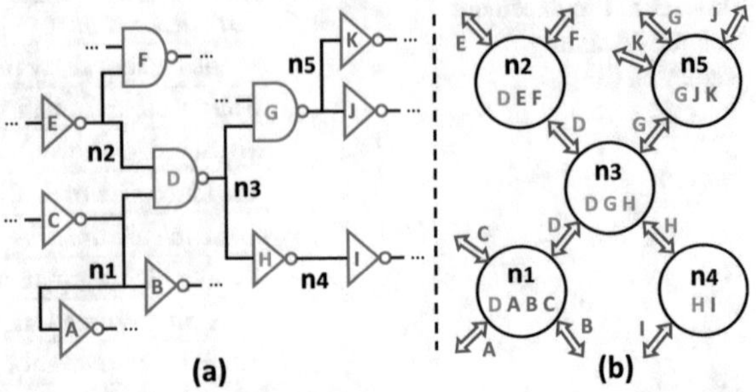

Fig. 3.7 (a) Part of a netlist.
(b) The corresponding
graph [33]

(a)   (b)

3.3.4.2 Machine Learning Engine for Preplacement Timing Prediction

Based on extracted features of the two different types of timing arcs, one cell-arc
model and one net-arc model are developed based on the random forest algorithm.
Instead of directly predicting the ground truth post-placement delay of each arc,
the model by Xie, et al. is actually trained to predict the difference between
preplacement and the ground truth post-placement timing. Then, the final predicted
delay is the summation of both preplacement delay and the prediction of the
incremental delay. This strategy helps the model to directly capture wire-load-
induced delay based on the preplacement timing report.

3.4 Pre-Routing Timing Prediction

3.4.1 Problem Formulation

To handle timing uncertainty due to the lack of routing information, designers tend
to make very pessimistic predictions, which causes overdesign that wastes chip
resources or design effort. To reduce such pessimism, Barboza et al. [2] study
the problem of calibrating pre-routing timing to post-routing timing based on ML
techniques.

3.4.2 Prediction Flow

Figure 3.8 shows the pre-routing timing prediction flow proposed in [2]. An ML
model for predicting the slew of the individual logic stage is first constructed and
trained. Besides the pre-routing timing report and the extracted features, the output
of the slew model is used to train the model for the prediction of logic stage
delay. Then, these models are applied for inference of logic stage delays in PERT
traversals [4] of circuit graph in order to obtain arrival time, required arrival time,
and slack of each circuit node. The logic stage delay model does not differentiate

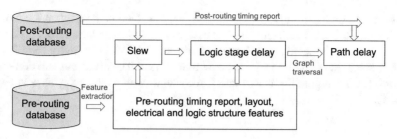

Fig. 3.8 The pre-routing timing prediction flow

among multiple input pins of a driver gate. The different signal arrival times at different input pins are considered during PERT traversal instead of net delay model.

3.4.3 Feature Engineering

Features form input to ML models and their selection is of critical importance for the effectiveness of model application. Net-based delay and slew models share the same features, which are elaborated as follows:

- **Driver and sink capacitance:** Driver output capacitance is generally proportional to its driving strength. Total sink capacitance presents load to the driver. Both of them are the determining factors for net delay and slew. For a net with large capacitive load or on critical paths, buffers may be inserted after placement. The effect of buffer insertion is contained in training data.
- **Distance between the driver and the target sink:** The model predicts delay and slew of one sink at a time, and this sink is called target sink. The horizontal and vertical distances from the driver of the target sink is generally proportional to the corresponding wire delay, especially when buffers are inserted [1].
- **Max driver input slew:** It is obtained using its own ML model, as mentioned above. Here, slew rate is defined to be the signal transition time. Hence, a small slew means sharp signal transition. Both net delay and sink slew are affected by driver input slew. As different types of logic gates may have different number of input pins, we use the maximum slew among all input pins of net driver as feature. This is to accommodate that a machine learning model normally requires fixed input size.
- **Context sink locations:** When a model is applied to estimate the delay/slew of the target sink of a net, the other sinks are called context sinks. For example, consider Net C in Fig. 3.1. When the delay to sink e is estimated, sink f serves as a context sink. Besides contributing to total load capacitance, the locations of context sinks affect routing, buffering, and thus delay/slew at the target sink. Since the number of context sinks varies from one net to another while machine learning model requires input of fixed size, the characteristics of context sink

locations are captured with statistical signatures. One is the median location of all context sinks, which tells roughly how far the context sinks are from the driver. The other is the standard deviations of context sink locations in x-y coordinates. The standard deviation indicates how much the sinks spread out and correlates with the corresponding interconnect tree size as well as the delay/slew at target sink.

3.4.4 Machine Learning Engines

A couple of machine learning engines are constructed and compared in [2], including a linear regression model with L1 regularization, an ANN model, and a random forest model. Experimental results demonstrate that the random forest model achieves the best performance among all the machine learning engines. It is reported that the random forest-based pre-routing timing estimation solution reaches accuracy near post-routing sign-off analysis.

3.5 Pre-Routing Crosstalk Prediction

3.5.1 Problem Formulation

The work by Liang et al. [17] targets three crosstalk classification tasks at placement stage, i.e., identifying

1. The nets likely to have large coupling capacitance
2. The nets likely to have large crosstalk-induced noise
3. The nets likely to have long incremental delay due to crosstalk

Figure 3.9 shows the Venn diagram of the above three sets of crosstalk-critical nets. We can find that these sets have overlaps, but they are not identical.

Fig. 3.9 The Venn diagram of three crosstalk-critical net categories [17]. The value on each region shows the proportion of the nets belonging to the region, normalized against the total number of crosstalk-critical nets

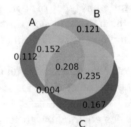

- A: Nets with large coupling capacitance;
- B: Nets with large crosstalk-induced noise;
- C: Nets with long incremental delay.

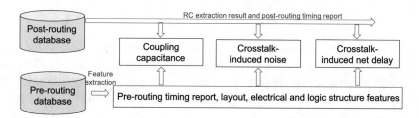

Fig. 3.10 Crosstalk modeling flow

3.5.2   Prediction Flow

The crosstalk modeling flow is shown in Fig. 3.10. Input features and ground truth information are extracted from the placement and the post-routing databases, respectively. By training and evaluating the prediction performance of candidate ML models, e.g., XGBoost, the most effective feature sets and the best models for three crosstalk classification problems are determined, which can be used for fast identification of problematic nets in new placement instances.

The raw input to the crosstalk prediction engine includes placed DEF file, standard cell libraries, and the STA results generated after placement. The placed DEF contains the circuit netlist and the locations of cells and input/output ports after placement. The standard cell library files are used to get the physical, electrical, and logical properties of the cells in the circuit. The STA results generated after placement give timing information, such as cell delays, wire delays, and transition times. Net-based features are extracted from these files and then fed into ML models. The ground truth information is extracted from the parasitic information file, and the timing report in SI mode is generated after detailed routing.

3.5.3   Feature Engineering

3.5.3.1   Probabilistic Congestion Estimation

Routing congestion strongly correlates with crosstalk, since coupling capacitance tends to occur in congested areas. A probabilistic technique for congestion analysis –RUDY (elaborated in Chapter 2)– is employed in [17] due to its great runtime advantage over other techniques and its good correlation with the post-routing solution. Nets can be divided into long-range nets and short-range nets according to their HPWL. In [32], it is shown that routing congestion has a stronger correlation with long-range nets than with shorter ones. In this regard, a longRangeRUDY feature is also extracted by considering only long-range nets when computing the total wire volume.

3.5.3.2   Net Physical Information

Net physical information is necessary for crosstalk estimation, because different routing topology of a net exposes it to different aggressors. Also, it leads to different electrical characteristics of interconnects, which have impacts on crosstalk noise and incremental delay. A few net-topology-related features are extracted as follows:

- **HPWL** of the net's bounding box
- **area** of the net's bounding box
- **fan-out** of the driver cell, i.e., the number of sinks of the net
- **max-ss-distance**: the maximal distance between the driver cell and the sink cells

3.5.3.3   Product of the Wirelength and Congestion

As illustrated in [20], the total coupling capacitance of a net is proportional to the product of its wirelength and the unit coupling capacitance. The product of the HPWL and the RUDY/long-range-RUDY of a net is utilized as an indicator of its coupling capacitance.

- **HPWL-RUDY/HPWL-longRangeRUDY**: the product of HPWL and RUDY/ longRangeRUDY of a net

3.5.3.4   Electrical and Logic Features

The electrical properties of cells play an important role in crosstalk. For example, the crosstalk noise is affected by the driver cell's resistance [26]. Also, the logic type of the driver cell may affect the switching activity of the net and consequently affect the noise and incremental delay. However, given the complicated timing models used by modern cell libraries, it is difficult to capture all crosstalk-related properties of a cell. To address this problem, a logic-based encoding is proposed in [17], along with the output capacitance, to represent a cell. First, library cells are consolidated into groups according to logic types. Each group contains cells with the same logic but can have different fan-in counts (e.g., a two-input NAND and a three-input NAND belong to the same group) and different sizes. A variant of the one-hot encoding is utilized to encode the gate groups. Specifically, a vector of length N_g, the total number of gate groups, with only one nonzero entry is used to describe which group the cell belongs to. Unlike assigning 1 to the nonzero entry in the one-hot encoding, the cell's fan-in is assigned to the nonzero entry. An additional feature, output capacitance, is used to capture the size of the gate. A few other electrical and logical structure features are also extracted.

- l_0 to l_{N_g}: the logic-based encoding of the driver cell
- **sourceCap**: the output capacitance of the driver cell
- **sinkCap**: the sum of the input capacitance of sink cells
- **fan-in**: the fan-in of the driver cell

3.5.3.5 Timing Information

The pre-routing timing report gives a rough estimation of wire delays and slews, which is informative for crosstalk prediction. For example, smaller slew often means stronger driving strength, consequently more resistant to aggregator's effect in terms of incremental delay.

- **wireDelay**: the longest wire delay from the driver to the sinks. Note it is a rough prediction from the pre-routing STA
- **outputSlew** of the driver cell from the pre-routing STA

3.5.3.6 Neighboring Net Information

Crosstalk noise and incremental delay depend not only on coupling capacitance but also on the coupling location [26] (near the driver or sink cells) and the aggressors' driving strength. The coupling location is estimated according to the relative location of the net's bounding box and its neighbors' bounding boxes.

- **#Neighboring nets**: the number of neighboring nets
- **#Neighboring long-range nets** : the number of neighboring long-range nets
- **mean-, std-, max-overlapArea**: the average/standard deviation/maximum of overlap areas between the net's bounding box and neighboring nets' bounding boxes
- **mean-dist-source-overlap**: the average distance between the driver cell and the geometric centers of overlap regions
- **weighted-dist**: the average distance between the driver cell and the geometric centers of overlap regions, weighted by the area of each overlap region
- **dist-source-maxOverlap**: the distance between the driver cell and the geometric center of the largest overlap region

The sourceCap and the outputSlew features are used to represent the driving strength of a net. If a net is surrounded by nets with strong driving strength, then its aggressors are likely to have strong driving strength. The following features are employed to capture neighboring nets' driving strength:

- **mean-, std-sourceCap**: the average/standard deviation of sourceCap of neighboring nets
- **sourceCap-maxOverlap**: the sourceCap of the neighboring net that has the largest overlap with the target net
- **mean-, std-outputSlew**: the average/standard deviation of outputSlew of neighboring nets
- **outputSlew-maxOverlap**: the outputSlew of the neighboring net that has the largest overlap with the target net

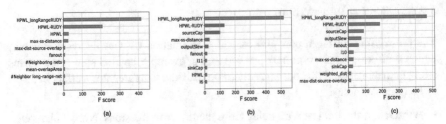

Fig. 3.11 Top 10 important features: (**a**) in coupling capacitance, (**b**) in crosstalk noise, and (**c**) in incremental delay estimations [17]

3.5.4 Machine Learning Engines

Several popular ML techniques, i.e., logistic regression, ANN, random forest, XGBoost, and GNN models, were constructed to model the mappings from the extracted features to coupling capacitance, crosstalk-induced noise, and incremental delay. For each technique, three independent models were trained and fine-tuned for the three classification tasks. Experimental results show that the XGBoost method achieves the best performance among all the ML engines.

After training an XGBoost-based model, users can check which features are most important in building the decision trees. Importance can be defined from various aspects. One commonly used metric is "Gain", which is the improvement in accuracy brought by a feature. Figure 3.11 shows the top 10 important features in coupling capacitance, crosstalk-induced noise, and incremental delay predictions, in terms of "Gain." It can be seen that the layout features play an important role in the three crosstalk estimation tasks because crosstalk heavily depends on layout. As for crosstalk-induced noise and incremental delay prediction, it can be found that electrical features (e.g., sourceCap and sinkCap), logical features (e.g., logic-based encoding for the driver cell: l_6, l_{10}, and l_{11}), the timing information (e.g., outputSlew), and the neighboring net information (e.g., the weighted-dist and the max-dist-source-overlap) also have great importance.

3.6 Interconnect Coupling Delay and Transition Effect Prediction at Sign-Off

3.6.1 Problem Formulation

The runtime and license costs of SI-enabled timing analysis are typically much larger than those in non-SI mode. The work by Kahng et al. [14] investigates the problem of calibrating sign-off non-SI timing to SI timing with ML techniques. To be specific, the incremental slew, incremental delay due to SI, and SI-aware path

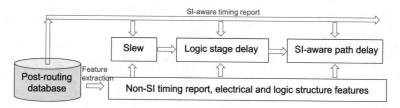

Fig. 3.12 Flow of calibrating non-SI timing to SI timing

delay are estimated given the reports of a sign-off timer that performs only non-SI analysis.

3.6.2 Prediction Flow

Figure 3.12 depicts the flow of calibrating non-SI timing to SI timing. Given a post-routing instance, the non-SI timing report, the electrical and logic structure parameters are extracted as input features. The model for predicting incremental slew due to SI is first constructed and trained, whose outcome is used as input for the training of the SI-induced incremental delay estimator. The predicted delay is utilized for predicting the SI-aware path delay.

3.6.3 Feature Engineering

The following features are extracted for the prediction of SI-aware timing at sign-off:

- Non-SI timing report: slew in non-SI mode, min/max rise/fall delta arrival times between worst aggressor and victim, toggle rate of a victim net, path delay in non-SI mode
- Electrical features: resistance of an arc, coupling capacitance of an arc, ratio of coupling capacitance to the total capacitance, logical effort of the driver cell
- Logic structure: ratio of arc's stage to the total number of stages

It is interesting to note that the work by Kahng et al. does not include layout parameters as input features, since layout information is reflected in parameters such as coupling capacitance, total capacitance, and wire resistance.

Fig. 3.13 The ensemble
model for SI-aware timing
estimation (adapted
from [14])

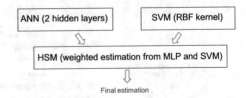

3.6.4 Machine Learning Engines

An ensemble model that integrates an ANN and a SVM is leveraged in [14]. The hybrid surrogate modeling (HSM) [13] is utilized to combine the predicted values from the ANN and SVM models and obtain the final estimation, as shown in Fig. 3.13.

3.7 Summary

In this chapter, we introduce net-based ML-aided approaches for timing and crosstalk prediction. Timing and signal integrity are the fundamental objectives in digital circuit design. Conventional timing and crosstalk effect prediction methods are usually either too slow or very inaccurate. Recent ML-aided approaches have demonstrated great potential in providing fast yet accurate prediction. Most of these works build net-based models. After introducing the background on timing and crosstalk modeling as well as relevant ML techniques, we present four representative net-based ML-aided solutions, emphasizing the problem formulation, prediction flow, feature engineering, and ML engines. The key difference between these case studies is in the feature engineering part, since the case studies target different design steps and/or different design properties. There are many other related problems to be studied. A significant one is integrating the ML-based estimators into EDA flows to investigate their impacts on the timing and SI closure.

References

1. Alpert, C.J., Hu, J., Sapatnekar, S.S., Sze, C.: Accurate estimation of global buffer delay within a floorplan. IEEE Trans. Comput. Aided Des. Integr. Circuits Syst. **25**(6), 1140–1145 (2006)
2. Barboza, E.C., Shukla, N., Chen, Y., Hu, J.: Machine learning-based pre-routing timing prediction with reduced pessimism. In: Proceedings of Design Automation Conference (DAC), pp. 1–6 (2019)
3. Breiman, L.: Random forests. Mach. Learn. **45**(1), 5–32 (2001)
4. Chang, H., Sapatnekar, S.S.: Statistical timing analysis considering spatial correlations using a single PERT-like traversal. In: Proceedings of International Conference On Computer Aided Design (ICCAD), pp. 621–625 (2003)

5. Chen, T., Guestrin, C.: XGBoost: a scalable tree boosting system. In: Proceedings of the International Conference on Knowledge Discovery and Data Mining (DMKD), pp. 785–794 (2016)
6. Coudert, O.: Gate sizing for constrained delay/power/area optimization. IEEE Trans. VLSI Syst. **5**(4), 465–472 (1997)
7. Elmore, W.C.: The transient response of damped linear networks with particular regard to wideband amplifiers. J. Appl. Phys. **19**(1), 55–63 (1948)
8. Hamilton, W., Ying, Z., Leskovec, J.: Inductive representation learning on large graphs. In: Proceedings of the International Conference and Workshop on Neural Information Processing Systems (NIPS), pp. 1024–1034 (2017)
9. Han, S.S., Kahng, A.B., Nath, S., Vydyanathan, A.S.: A deep learning methodology to proliferate golden signoff timing. In: Proceedings of Design, Automation & Test in Europe Conference & Exhibition (DATE), pp. 1–6 (2014)
10. Hornik, K., Stinchcombe, M., White, H.: Multilayer feedforward networks are universal approximators. Neural Netw. **2**(5), 359–366 (1989)
11. Hu, J., Sapatnekar, S.S.: Simultaneous buffer insertion and non-Hanan optimization for VLSI interconnect under a higher order awe model. In: Proceedings of the International Symposium on Physical Design (ISPD), pp. 133–138 (1999)
12. Hyun, D., Fan, Y., Shin, Y.: Accurate wirelength prediction for placement-aware synthesis through machine learning. In: Proceedings of Design, Automation & Test in Europe Conference & Exhibition (DATE), pp. 324–327 (2019)
13. Kahng, A.B., Lin, B., Nath, S.: Enhanced metamodeling techniques for high-dimensional IC design estimation problems. In: Proceedings of Design, Automation & Test in Europe Conference & Exhibition (DATE), pp. 1861–1866 (2013)
14. Kahng, A.B., Luo, M., Nath, S.: SI for free: machine learning of interconnect coupling delay and transition effects. In: Proceedings of the International Workshop on System Level Interconnect Prediction (SLIP), pp. 1–8 (2015)
15. Karypis, G., Aggarwal, R., Kumar, V., Shekhar, S.: Multilevel hypergraph partitioning: applications in VLSI domain. IEEE Trans. VLSI Syst. **7**(1), 69–79 (1999)
16. Kipf, T.N., Welling, M.: Semi-supervised classification with graph convolutional networks (2016). Preprint. arXiv:1609.02907
17. Liang, R., Xie, Z., Jung, J., Chauha, V., Chen, Y., Hu, J., Xiang, H., Nam, G.J.: Routing-free crosstalk prediction. In: Proceedigs of the International Conference on Computer Aided Design (ICCAD), pp. 1–9 (2020)
18. Lin, S., Lillis, J.P.: Interconnect Analysis and Synthesis. John Wiley & Sons Inc., Hoboken (1999)
19. Liu, Q., Ma, J., Zhang, Q.: Neural network based pre-placement wirelength estimation. In: Proceedings of the International Conference on Field-Programmable Technology (FPT), pp. 16–22 (2012)
20. Lou, J., Chen, W.: Crosstalk-aware placement. IEEE Des. Test Comput. **21**(1), 24–32 (2004)
21. Lou, J., Chen, W., Pedram, M.: Concurrent logic restructuring and placement for timing closure. In: Proceedings of International Conference on Computer-Aided Design (ICCAD), pp. 31–35 (1999)
22. Parakh, P.N., Brown, R.B.: Crosstalk constrained global route embedding. In: Proceedings of the International Symposium on Physical Design (ISPD), pp. 201–206 (1999)
23. Pillage, L.T., Rohrer, R.A.: Asymptotic waveform evaluation for timing analysis. IEEE Trans. Comput. Aided Des. Integr. Circuits Syst. **9**(4), 352–366 (1990)
24. PrimeTime SI: Crosstalk delay and noise. www.synopsys.com/support/training/signoff/primetimesi-fcd.html (Synopsys)
25. Rahmat, K., Neves, J., Lee, J.F.: Methods for calculating coupling noise in early design: a comparative analysis. In: Proceedings of the International Conference on Computer Aided Design (ICCAD), pp. 76–81 (1998)
26. Ren, H., Pan, D., Villarubia, P.G.: True crosstalk aware incremental placement with noise map. In: Proceedings of the International Conference on Computer Aided Design (ICCAD), pp. 402–409 (2004)

27. Tseng, H.P., Scheffer, L., Sechen, C.: Timing- and crosstalk-driven area routing. IEEE Trans. Comput. Aided Des. Integr. Circuits Syst. **20**(4), 528–544 (2001)
28. Veličković, P., Cucurull, G., Casanova, A., Romero, A., Lio, P., Bengio, Y.: Graph attention networks (2017). Preprint. arXiv:1710.10903
29. Vittal, A., Marek-Sadowska, M.: Crosstalk reduction for VLSI. IEEE Trans. Comput. Aided Des. Integr. Circuits Syst. **16**(3), 290–298 (1997)
30. Wu, D., Hu, J., Mahapatra, R., Zhao, M.: Layer assignment for crosstalk risk minimization. In: Proceedings of Asia and South Pacific Design Automation Conference (ASP-DAC), pp. 159–162 (2004)
31. Wu, Z., Pan, S., Chen, F., Long, G., Zhang, C., Philip, S.Y.: A comprehensive survey on graph neural networks. IEEE Trans. Neural Netw. Learn. Syst. **32**(1), 4–24 (2020)
32. Xie, Z., Huang, Y.H., Fang, G.Q., Ren, H., Fang, S.Y., Chen, Y., Hu, J.: RouteNet: routability prediction for mixed-size designs using convolutional neural network. In: Proceedings of the International Conference on Computer-Aided Design (ICCAD), pp. 1–8 (2018)
33. Xie, Z., Liang, R., Xu, X., Hu, J., Duan, Y., Chen, Y.: Net2: a graph attention network method customized for pre-placement net length estimation. In: Proceedings of Asia and South Pacific Design Automation Conference (ASP-DAC), pp. 671–677 (2021)
34. Zhou, H., Wong, M.D.F.: Global routing with crosstalk constraints. IEEE Trans. Comput. Aided Desig. Integr. Circuits Syst. **18**(11), 1683–1688 (1999)

Chapter 4
Deep Learning for Power and Switching Activity Estimation

Yanqing Zhang

4.1 Introduction

Today's computing systems are power-constrained. From datacenters to mobile devices, limits on thermal or electrical power impact achievable performance. Therefore, it is important for designers to be aware of how much power their designed components will consume so as not to surpass these limits during deployment. Fast and accurate power estimation is a critical part of all aspects of digital VLSI development flows today.

Various design stages employ tools and algorithms that model or calculate the power consumption according to their corresponding level of abstraction. For example, architectural power analysis requires estimating average power over hundreds to millions of cycles [1] to assess if power requirements are met during certain actual deployment workload scenarios. Power integrity sign-off requires accurate power analysis on physically annotated gate-level netlists. Dynamic power optimizations during logic synthesis, clock gate insertion, or place-and-route steps are necessary to meet power targets. RTL, high level synthesis (HLS), and architectural level power estimation are used for efficient design turnaround time when assessing the effects on unit or total design power when design parameter tuning (such as size of cache/SRAMs, number of components, bit-width, DVFS, technology scaling, etc.) and adding/changing features is done.

Currently, one of the two most common implementations of power estimation is power calculation from gate-level simulation. Power calculation involves performing gate-level simulation by inputting a gate-level netlist and input testbench vectors into a gate-level simulator. The results of the simulation provide a detailed profile of

Y. Zhang (✉)
NVIDIA, Santa Clara, CA, USA
e-mail: yanqingz@nvidia.com

© The Author(s), under exclusive license to Springer Nature Switzerland AG 2022
H. Ren, J. Hu (eds.), *Machine Learning Applications in Electronic Design
Automation*, https://doi.org/10.1007/978-3-031-13074-8_4

the switching activity exhibited by each gate during the testbench. Then, the gate-level netlist and simulation results are fed, as input, to power analysis tools such as Synopsys PrimePower [2] to generate cycle-level power traces or provide the average power consumption number over a certain power window. Power analysis tools perform power calculation by essentially doing a matrix multiply of the switching activity of each gate (also known as toggle counts or toggle rates) and a compiled lookup table (LUT) of the power consumption that each switching activity incurs. The values of the LUT for the per switch "cost-of-action" is often determined by gate characterization measurements (commonly in the form of a LIBERTY file), extracted RC parasitics (commonly in SPEF format), and design/frequency constraints (commonly in SDC format).

The other common implementation of power estimation is through switching activity estimators (SAEs). In most power analysis use cases, annotating toggle counts/rates with vectors captured from real workloads is highly preferred to vectorless methods because of their accuracy. A vectorless power analysis is when no real testbench/workload input vectors are given during gate-level simulation. Instead, a flat probability of switching is assigned as input to each timing startpoint (input ports and sequential elements outputs), and the toggle rates of each ensuing Boolean logic gate in the pipeline can be statistically calculated from the assigned probabilities. The statistical engine that calculates toggle rates of the Boolean logic gates is known as a switching activity estimator (SAE). Average toggle rates for unit inputs and registers can be gathered over a window of interest from RTL simulations that would precede any gate-level simulation.

Typical workflows for performing power estimation using power calculation and SAEs are depicted in Figs. 4.1 and 4.2, respectively. The figures also summarize the main trade-offs of the two common approaches. Accurate vector-based power analysis is desired, but it also requires running gate-level simulations. Gate-level simulations can be very slow, typically 10–1000 cycles/s, depending on activity factor and size of design, leading to long turnaround times (hours to days). Slow

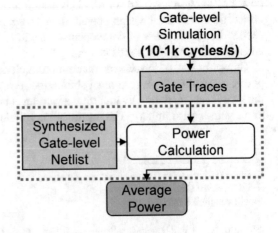

Fig. 4.1 A typical workflow for implementing power estimation using gate-level simulation and a power calculation tool. Gate-level simulation speeds may be as low as 10 cycles/s

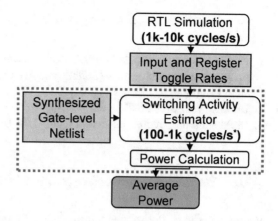

Fig. 4.2 A typical workflow for implementing power estimation using a higher level-of-design-abstract simulation (such as RTL simulation) and a switching activity estimator

simulations negatively impact design productivity for power analysis and are impractical for use in dynamic power optimizations, such as automatic clock or data gating or gate resizing during place-and-route. As a result, in such cases, it is more typical to use an SAE. Although SAEs are fast, they are inaccurate, due to issues such as signal correlation or reconvergence. Simple versions of SAEs can quickly provide a power estimation number by propagating toggle rate activity through the netlist instead of the entire per-cycle waveforms, but at a significant cost of accuracy due to their statistical nature.

In this chapter, we describe various conventional methods of power estimation, before portraying the advantages that deep learning models can provide for this task. We will see that modeling using deep learning techniques aim to find a "sweet spot" balance between speed and accuracy for power estimation. While deep learning models retain faster speed compared to slow gate-level simulation, we will find they also provide better accuracy than conventional methods due to their ability to model highly nonlinear functions. Toward the end of this chapter, two case studies focus on post-synthesis (to post-layout), gate-level power estimation using deep learning models. The focus on post-synthesis (to post-layout) power estimation is due to its overarching usage, being able to provide a detailed power profile for the design on a per-cycle or per power window basis (average power estimation).

4.2 Background on Modeling Methods for Switching Activity Estimators

A natural research question that arises from the dilemma between accurate but slow gate-level simulation and a fast but inaccurate simple probability-based SAE is as follows: Is there a method of power estimation that can achieve both fast speeds and a high level of accuracy? To this end, there has been an abundance of research work, most of which has settled on three main approaches to build power estimation models for various levels of design abstraction. The three main approaches have been:

1. Build a more sophisticated statistical model/equation that incorporates numerous factors that contribute to switching activity that simpler equations don't consider [3–5]
2. Build a "cost-of-action" model [1, 6–9]
3. Build a learning/regression-based model [10–12, 14–16]

4.2.1   Statistical Approaches to Switching Activity Estimators

Figure 4.3 shows the problem formulation of switching activity estimation and its application to power analysis. Exact solutions require gate-level simulations to propagate traces from register outputs and unit inputs (typically captured from RTL simulation or FPGA emulation) through a combinational logic netlist. It is common in many cases that only average power is needed. In such cases, it is inefficient to use traditional event-driven simulators when only the final average toggle rates (α) on gate outputs are needed for the dynamic power calculation.

Another approach, shown in Fig. 4.4(top), is to move away from simulation altogether and instead propagate average toggle rates from inputs to outputs per gate, based on a derived equation (a simplistic, basic form of one such equation is depicted on the right side of Fig. 4.4) from static probabilities computed from the Boolean logic expression of each gate [3–5]. A similar approach is often used in commercial power analysis tools. Many commercial tools deploy an internal switching activity estimator to aid in the power analysis flow for situations, where toggle rates from gate-level simulation are not available. To determine the toggle rates of the Boolean logic gates, tools first determine random waveform patterns on the timing startpoints that match the annotated toggle rates on those startpoints (over

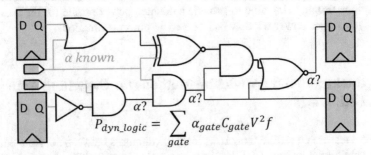

Fig. 4.3 Depiction of statistical switching activity estimation problem. The goal is to derive all the unknown combinational logic toggle rate αs from previously known toggle rate αs of input ports and sequential elements outputs (timing startpoints) and the Boolean logic function of each combinational gate

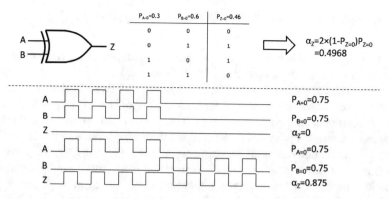

Fig. 4.4 (Top) Example of using a simple statistical SAE derived from logic gate truth table to calculate the toggle rate on the output of the gate. (Bottom) An example showing the shortcomings of this approach—because of multiple input switching conditions and correlated inputs, it is not sufficient to derive output toggle rates only from input toggle rate/value probabilities

a time window of multiple cycles determined by the internal SAE). Then, these random waveform patterns are propagated throughout the logic using zero-delay simulation, from which the logic gates' toggle rates can be calculated. Doing so rids the SAE of any inaccuracies caused by logic reconvergence correlation and multiple input pin switching scenarios, but has drawbacks such as the following: It does not address input value correlations, and this method still requires some gate-level simulation. Similarly mentioned in [3], statistical approaches to switching activity estimation are fast but can be highly inaccurate, since derived equations often do not consider reconvergence logic correlations, input pin value correlations, and multiple input pin switching scenarios. An example scenario for the shortcomings from multiple input pin switching situations and input pin value correlations is depicted in Fig. 4.4(Bottom). Some work has improved the accuracy of the statistical equation propagation approach by adding analytical terms to the derived equation that addresses one or more of the above issues, such as [4] focusing on multiple input pin switching scenarios and [5] focusing on logic reconvergence. Finally, in [3], a "hybrid" approach of using both a statistical equation and simulation is taken. In [3], a custom algorithm tags gates in the design that have reconverging inputs and simulates those gates by recording their Boolean logic expression with regard to primary inputs instead of propagated toggle rates. This rids the SAE of logic reconvergence and input value correlation issues, but one drawback with this approach is the massive memory requirement needed for recording the analytical logic expressions, which scales with logic depth and number of primary inputs. Also, some gate-level simulation is still required for those tagged gates and their fan-in cones.

4.2.2 "Cost-of-Action"-Based Power Estimation Models

One way to get around having to do gate-level simulation on higher levels of abstraction (such as the architectural level) is to build a model based on characterization. The idea is to characterize, to the granularity feasible given by resource and time constraints, the power of subcomponents of the top-level design when those subcomponents perform an action. The triggers and count of those actions are known through higher level simulations than gate-level simulation (such as instruction execution-based architectural simulation). Once the cost of power per action and the action count is known, a modeled total power can be derived by summing the costs across all subcomponents, either on an average power over window or per-cycle basis (similar to the equation at the bottom of Fig. 4.3, but on a "macro" level). For architectural "cost-of-action" power analysis models, typical action triggers include cache miss/hit rate, type of instruction executed on core, or if data was moved. While "cost-of-action"-based models are much more convenient to perform architectural analysis on and provides a good model to reason through architectural decisions related to power, they suffer when estimating power for the same testbenches for sign-off and/or power verification quality accuracy. This is mainly because it is difficult to capture all the nuanced scenarios the subcomponent may exhibit during actual testbench execution during characterization.

For example, [1] builds a fine-grained "cost-of-action" model for various deep learning (DL) accelerator design based on data movement characteristics. The authors carefully break down the actions into compute and memory energy overheads as it is stressed the characteristics of these two types of actions are quite distinct in the DL accelerator field. [7] is a well-known computer architect "cost-of-action" model. McPat [7] decomposes entire processors into low-level components and then focuses on extrapolating regular structures before summation to get the total power. [8] deploys a "middle ground" approach between "cost-of-action" modeling and full simulation. In [8], "smart sampling" of the testbench cycles is done so that only critical cycles of the entire testbench is fully simulated. These critical cycles either have a high rate of reoccurrence for the duration of the testbench and can thus be representative of numerous cycles of the testbench or have high toggle coverage of the underlying gates in the design, thus representing the rare, nuanced high power spike outlier cycles of the testbench. In this way, a per-cycle "cost-of-action" model is built from only simulating a select set of cycles out of the entire testbench. Finally, [9], built for design space exploration of DL accelerators, takes a fine-grained approach to "cost-of-action" model building. All subcomponents in the top-level design can be decomposed into an intermediate representation that maps to known components, whose power characterization had thoroughly been done through micro-benchmarks.

4.2.3 Learning/Regression-Based Power Estimation Models

As mentioned previously, an alternative is to analyze power above gate level. However, "cost-of-action" models break down when they fail to capture the nuances of different power profiles that the "actions" being modeled may exhibit. For accurate power characterization, many low-level details of the circuit need to be modeled to exhibit these nuances, including the standard cell parameters, sizing of the gates, and clock gating status of the registers, which gate-level simulation covers.

One method in model building that can get around fully running slow gate-level simulations, but still possibly retain some of the nuances of power consumption of the design on a cycle-by-cycle or average-over-power-window basis, is to build a learning-based model. In the past, these efforts typically make use of measured constants or simple curve fitting techniques such as linear regression to characterize the power of a given circuit, improving the speed of power analysis while retaining some low-level detail, but still at the expense of overall estimation accuracy.

The overall method is to train the regression models with ground truth gate-level simulation and gate-level power analysis results. Parameters in the regression model then undergo curve fitting to build a power profile across the different ranges those parameters exhibit during different testbenches. For example, earlier works in RTL power estimation use simple regression models, such as linear regression and regression trees, to characterize small circuit blocks [10]. Average power and cycle-by-cycle power of the whole design can be obtained by summing up the outputs from multiple models, thus taking a hybrid learning-based model and "cost-of-action" model approach. Sunwoo et al. [11] uses linear models to characterize larger modules, where heavy feature engineering (it is ensured that the characteristics of the testbenches used during training and test/inference are similar) and feature selection (only the "top-N" deemed-important registers in the design become parameters in the model) are applied to reduce the complexity of power models. A more recent work [12] uses a single linear model to characterize the entire designs. A feature selection technique based on the singular value decomposition (SVD) is applied to reduce model complexity so that the regression model can be compact. Xie et al. [13] uses deep learning to do feature selection before constructing the regression model. All of [11–13] can provide cycle-by-cycle power estimates. Van den Steen et al. [14] builds a linear regression model on the architectural level for the entire out-of-order processor design units by extracting performance and application metrics common to all processors, such as instruction type, branch predictor states, and cache miss/hit rate. Lee et al. [15] tries to match SystemC activity with gate-level activity to perform power analysis at the SystemC coding stage. They also choose a linear model, albeit with much human intervention, to map SystemC variables' activity to the corresponding gate-level activity, as well as to map SystemC cycles to gate-level simulation cycles. Finally, [16] uses a simple multilayer perceptron (MLP) to model the specific behavior of memory SRAM macros.

In general, most existing learning-based techniques either model small circuit sub-modules or try to use simple regression models to characterize the whole design. Sub-module-level modeling cannot accurately reflect the power consumption of intermediate logic, and simple regression models, such as linear models, are not a good fit for large, complex designs. It is also very difficult for simple analytical or linear regression models to capture the complex, nonlinear relationship between register toggles, total switching capacitance, different logic gate functions, and final power.

4.3 Deep Learning Models for Power Estimation

While past learning-based modeling approaches try to find a good balance between speed and accuracy for power estimation, the simplicity of those regression models don't meet the needs for large designs, which commonly have over 10k 100k registers and 1Ms of combinational gates. Simple regression models don't have enough capacity to capture the highly nonlinear relationships between input port/register toggles and combinational logic, which include complex situations, such as logic reconvergence, input correlations, and multiple pin switching scenarios. The advent of deep learning models and deep learning applications has led to many model architectures that have parameter capacities upward in the millions range. This enables deep learning model architectures to learn highly complex input/output relationships. Given the observed complexity of power estimation, it follows we may also benefit from studying the efficacy of deep learning models for power estimation.

Multilayer perceptron (MLP) contains only fully connected network layers and is more computer-efficient than other deep learning models such as convolutional neural networks (CNN). However, the parameter count of MLP grows quickly with respect to the feature size of the design, resulting in over-fitting and training convergence issues, which may not be suitable for power estimation applications. Depending on the model design, deep learning power estimation model features will scale with design size or input waveform length/count. CNNs have shown impressive performance in image classification tasks. For power estimation purposes, since the power of a certain logic cone is only correlated with a small set of registers, the convolutional windows of CNNs can gather useful information, if input "images" can be created from information from the design netlist and the input "images" are constructed in such a way that registers that feed into common logic cones are "spatially" close to each other in the constructed image. CNNs are most effective when there are spatial relationships in their 2D inputs. While design netlists and timing startpoint waveforms do not constitute an "image" in the traditional sense, there exist certain embedding techniques that can translate this information into a "pseudo-image" format based on the connectivity of the netlist and characteristics of the waveforms. The following section (PRIMAL) in this chapter details such an approach, where register-to-"pixel" mapping reflects the connectivity or physical placement of the registers. In addition, thanks to the structure of convolutional layers, CNN is a more scalable choice than MLP for large designs, since the parameter count of a CNN does not increase significantly as the input image size (related to design size) grows.

Graph neural networks (GNNs) are a powerful neural network architecture for machine learning on graphs [17], with many applications in social networking [18] or scene labeling [19]. GNNs operate by assigning node and edge features on a graph and then sharing the features with neighbor nodes through message passing. GNN architectures are amenable to the power estimation modeling problem, as well as many netlist-centric VLSI modeling problems, because of the observation that netlists are graphs [20]. One popular type of GNN is the graph convolutional network (GCN). GCNs perform message passing through three steps of message sending, message reduction, and node transformation [18]. While there are many variations of GCNs and GNNs, the GCN model architecture sets up a common theme for the versatility and uniqueness of graph learning: Because the embeddings on each node in the graph receive information/features from its neighbors before undergoing convolution (or other forms of transformations), the model learns from not only the input features but also the local connectivity of the graph. This is helpful for power estimation models, since combinational logic toggles are related to both netlist connectivity and input waveform characteristics. The following section (GRANNITE) in this chapter describes in detail a GNN approach to power estimation modeling, where it is set up that learned parameters attain their values from input feature data as well as structure of the input netlist graph. In conclusion, using GNNs in the area of machine learning for EDAs is a growing field, because a netlist is a graph, which provides good inductive bias for GNN models [20–24].

4.4 A Case Study on Using Deep Learning Models for Per Design Power Estimation

This section gives an in-depth description of PRIMAL [25], a learning-based framework that enables fast and accurate power estimation. PRIMAL trains ML models with design verification testbenches for characterizing the power of reusable circuit building blocks. The trained models can then be used to generate detailed power profiles of the same blocks under different workloads. Several established ML models are evaluated for this task, including traditional ML models such as ridge regression, gradient tree boosting, as well as DL models such as multilayer perceptron (MLP), and convolutional neural network (CNN). For average power estimation, the results of [25] show ML-based techniques can achieve an average error of less than 1% across a diverse set of realistic benchmarks. Speed-wise, this technique outperforms a commercial RTL power estimation tool (whose purpose is similarly also to forego full gate-level simulation and estimate power) by 15×.

Fig. 4.5 PRIMAL trains ML-based power models for reusable IPs. Using the trained models, detailed power traces are obtained by running ML model inference on previously available simulation traces from RTL

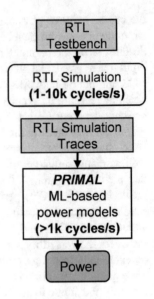

For cycle-by-cycle power estimation, PRIMAL is on average 50× faster than a commercial gate-level power analysis tool, with an average error of less than 5%.

Figure 4.5 illustrates the inference flow of PRIMAL, which only requires inputs from RTL simulation to rapidly generate accurate power estimates (>1k cycles per second). By greatly reducing the required number of gate-level simulation cycles, PRIMAL allows designers to perform power-directed design space exploration in a much more productive manner.

The following sections will deep dive into these topics related to PRIMAL:

- The ML-based methodology for PRIMAL which performs rapid power estimation with RTL simulation traces. The trained ML models can provide accurate, cycle-by-cycle power inference for user workloads even when they differ significantly from those used for training.
- Experiments using several different types of established ML models for power estimation are presented. Discussion on trade-offs between accuracy, training time, and inference speed is given. The study suggests that nonlinear models, especially CNNs, can effectively learn power-related design characteristics for large circuits. Results demonstrate that PRIMAL is at least 50× faster on average than a commercial tool for cycle-accurate power estimation with a small error. Notably, the CNN-based approach is 35× faster than the commercial tool with a 5.2% error when used on a RISC-V core. PRIMAL also achieves a 15× speedup over a commercial RTL power analysis tool for average power estimation.
- Feature engineering techniques to construct "image" representations from register traces is explored. The constructed features are used by the CNN models.

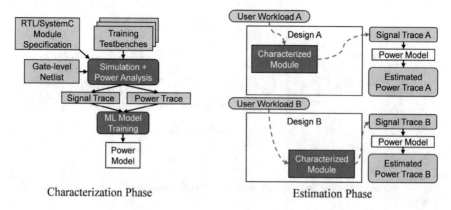

Fig. 4.6 Two phases of the PRIMAL workflow—power models are trained once per module and then used across different workloads and in different designs that instantiate the module

4.4.1 PRIMAL Methodology

Figure 4.6 shows the two phases of the PRIMAL workflow. The characterization phase (Fig. 4.6(left)) requires an RTL HDL description of the module, the gate-level netlist, and a set of unit-level testbenches for training. RTL simulation traces and the power traces generated by gate-level power analysis are used to train the ML models. The characterization process only needs to be performed once per IP block. The trained power models can then be used to estimate power for different workloads as illustrated in Fig. 4.6(right).

It is important to note that the training testbenches may be very different from the actual user workloads. For example, designers can use functional verification test-benches to train the power models, which then generalize to realistic workloads. By using state-of-the-art ML models, PRIMAL can accommodate diverse workloads and model large, complex circuit blocks. The ML models are trained for cycle-by-cycle power estimation to provide detailed power profiles and enable more effective design optimization.

4.4.2 List of PRIMAL ML Models for Experimentation

PRIMAL explores a set of established ML models for power estimation. The classical ridge linear regression model is used as a baseline. PRIMAL also experiments with gradient tree boosting, a promising nonlinear regression technique [26]. For linear models and gradient tree boosting models, principal component analysis (PCA) [27] is applied to the input data to reduce model complexity and overfitting. The efficacy of deep learning models (MLP and CNNs), which are capable of approximating more complex nonlinear functions, is explored.

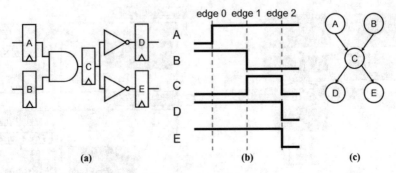

Fig. 4.7 Simple circuit example (**a**) with waveform of register outputs (**b**) and register connection graph (**c**)

4.4.2.1 Feature Construction Techniques in PRIMAL

This subsection describes the feature construction procedure using the circuit in Fig. 4.7a as an example. Figure 4.7b shows the register waveform, where each "edge" in the figure corresponds to a clock rising edge. Register switching activities in the simulation traces are used as input features, because register switching activities are representative of the circuit's state transitions. In addition, there is a one-to-one correspondence between registers in RTL and gate-level netlist (assuming no register retiming optimizations are done during gate synthesis). Because cycle-accurate power traces from gate-level simulation are used as ground truth, the ML models are essentially learning the complex relationship between the switching power for all gate-level cells and register switching activities.

4.4.2.2 Feature Encoding for Cycle-by-Cycle Power Estimation

For cycle-by-cycle power estimation, RTL register and I/O signal switching activities are the input features without any manual feature selection. Switching activities of both internal registers and I/O signals are required to capture complete circuit state transitions. These features can be easily collected from RTL simulation. Because PRIMAL targets cycle-by-cycle power estimation, each cycle in the simulation trace is constructed as an independent "sample."

A concise encoding, which PRIMAL refers to as switching encoding, represents each register switching event as a 1 and non-switching event as a 0. For an RTL module with n registers, each cycle in the RTL simulation trace is represented as a $1 \times n$ vector. Figure 4.8a shows the corresponding encoding for the waveform in Fig. 4.7b. Each vector in Fig. 4.8a represents one cycle in the waveform. PRIMAL uses this one-dimensional (1D) switching encoding for all but the CNN models. The same feature encoding is used in [12].

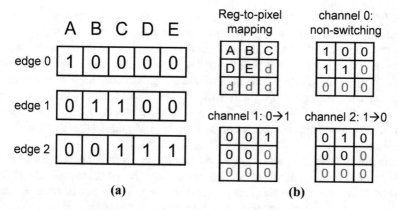

Fig. 4.8 Basic feature encoding methods—(**a**) 1D switching encoding and (**b**) default 2D encoding for edge 1 in Fig. 4.7b

In order to leverage well-studied two-dimensional (2D) CNN models, PRIMAL also provides an input feature encoding of a three-channel 2D "image" representation for every cycle in the register trace. For an RTL module with n registers, PRIMAL uses a $\sqrt{n} \times \sqrt{n} \times 3$ image to encode one cycle in the RTL simulation trace. One-hot encoding is used in the channel dimension to represent the switching activities of each register: non-switching is represented as [1, 0, 0], switching from zero to one is represented as [0, 1, 0], and switching from one to zero is represented as [0, 0, 1]. PRIMAL refers to this encoding as default 2D encoding. Figure 4.8b shows how to encode edge 1 of the waveform in Fig. 4.7b. If the total number of pixels in the image is greater than n, padding pixels are added to the image, shown as d's in Fig. 4.8b. These padding pixels do not represent any register in the module, and they have zero values in all three channels. Every other pixel corresponds to one register in the module. For this default 2D encoding, the registers are mapped by their sequence in the training traces. For example, since in Fig. 4.7b the order of registers is A, B, C, D, and E, in each channel the top-left pixel in Fig. 4.8b corresponds to A, the top-right pixel is mapped to C, and the center pixel refers to E.

4.4.2.3 Mapping Registers and Signals to Pixels

Since the gate-level netlist of the design is available during the characterization phase, it is possible to use the outputs of logic synthesis tools to map RTL registers to netlist nodes. Because only register and I/O switching activities are used, all combinational components can be ignored, and only register connection graphs need to be extracted when processing the gate-level netlist. The graph for the example circuit in Fig. 4.7a is shown in Fig. 4.7c. Each node in the graph corresponds to one register in the design, and two nodes are connected if their corresponding registers are connected by some combinational datapath (timing start-end pair).

PRIMAL proposes two graph-based methods for generating register-to-pixel mappings, which introduce local structures into the images according to the structural similarities between nodes. Notice that the proposed graph-based mapping methods only change the register mapping in the width and height dimensions of the image: The channel-wise one-hot encoding for every register is preserved. Each register's contribution to each pixel is proportional to the overlapping area of the register's occupied region and the pixel. In other words, with the graph-based encoding methods the pixel values are nonnegative real numbers rather than binary numbers.

The first method is based on graph partitioning, in which the graph is recursively divided into two partitions of similar sizes, and the partitions are mapped to corresponding regions in the image (see Fig. 4.9a). The area allocated for each partition is computed according to the number of nodes in the partition. The second method is based on node embedding. Node embedding techniques map each node in the graph to a point in a vector space, where similar nodes are mapped close to each other in the vector space. PRIMAL flow for embedding-based register mapping is shown in Fig. 4.9b. PRIMAL uses node2vec [28] for node embedding and then applies PCA and t-SNE [29] to project the vector representations to 2D space. The resulting 2D vector representations are scaled according to the image size and indicate the mapping locations of the registers.

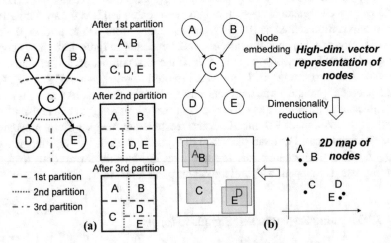

Fig. 4.9 Graph-based register mapping schemes—(**a**) Register mapping based on graph partitioning. The register connection graph is recursively partitioned into two parts. Each partition also divides the map into two nonoverlapping parts. (**b**) Register mapping based on node embedding. Node embedding maps each graph node as a point in high-dimensional space, and then dimensionality reduction techniques project the high-dimensional representations onto the 2D space. In the generated mapping, each register occupies a unit square whose area is equivalent to one pixel

4.5 PRIMAL Experiments

PRIMAL is implemented in Python 3.6, leveraging networkx [30], metis [31], and node2vec package. MLP and CNN models are implemented using Keras [32]. Other ML models are realized in scikit-learn [33] and XGBoost [34]. Experiments are done on a server with an Intel Xeon E5-2630 v4 CPU and 128GB RAM. Neural network training and inference is done on a NVIDIA 1080Ti GPU. Logic synthesis is done with Synopsys Design Compiler, targeting a 16nm FinFET standard cell library. The RTL register traces and gate-level power traces are obtained from Synopsys VCS and PTPX, respectively. Gate-level power analysis is performed on another server with an Intel Xeon CPU and 64GB RAM using a maximum of 30 threads.

Table 4.1 lists the benchmarks used to evaluate PRIMAL. Benchmarks include a number of fixed- and floating-point arithmetic units from [35]. Two more complex designs are also tested—a NoC router used in a CNN accelerator and a RISC-V processor core. The RISC-V core is an RV64IMAC implementation of the open-source Rocket Chip Generator [36] similar to the SmallCore instance. Different portions of random stimulus traces are used for training and test sets for the arithmetic units. For the NoC router and the RISC-V core, functional verification testbenches are selected for training and realistic workloads for test. The NoC router experiment uses actual traces of mesh network traffic from a CNN accelerator SoC. In the RISCV experiment, dhrystone, median, multiply, qsort, towers, and vvadd form the set of test workloads.

4.5.1 Power Estimation Results of PRIMAL

Figure 4.10 summarizes the results for gate-level power estimation from RTL level simulation traces. Table 4.2 shows the approximate training time for each type of model. RTL register traces are the raw input and the feature construction techniques described in "*Feature Construction Techniques in PRIMAL*" are applied. Two percent of the training data is used as a validation set for hyper-parameter tuning of the ML models. They are also used for early stopping when training the deep neural networks.

All models except CNNs use the 1D switching encoding, while CNNs use the 2D image encoding methods. PCA is applied for ridge regression and gradient tree boosting to reduce the size of input features to 256. MLP models have three layers for the arithmetic units and four layers for the NoC router and the RISC-V core. An open-source implementation of ShuffleNet V2 [37] is used for CNN-based power estimation because of its parameter-efficient architecture and fast inference speed. The v0.5 configuration in [37] is used for the arithmetic units, while the v1.5 configuration is used for the NoC outer and RISC-V core. The CNN models are trained from scratch. CNN default, CNN partition, and CNN embedding in Fig. 4.10

Table 4.1 Benchmark information. PRIMAL is evaluated with a diverse set of benchmark designs. For NoC router and RISC-V core, the test sets are realistic workloads, which are potentially different from the corresponding training set

Design	Description	Register + I/O signal count	Gate count	Gate sim throughput (cycles/s)	Training set (#cycles)	Test set (# cycles)
qadd_pipe	32-bit fixed point adder	160	838	1250	Random stimulus (480 k)	Random stimulus (120 k)
qmult_pipe {1,2,3}	32-bit fixed point multiplier with {1,2,3} pipe stages	{384, 405, 438}	{1721, 1718, 1749}	{144.9, 135.1, 156.3}	Random stimulus (480 k)	Random stimulus (120 k)
float_adder	32-bit floating point adder	381	1239	714.3	Random stimulus (480 k)	Random stimulus (120 k)
float_mult	32-bit floating point multiplier	372	2274	454.5	Random stimulus (480 k)	Random stimulus (120 k)
NoC router	Router for CNN accelerator	5651	15,076	44.7	Unit-level testbenches (910 k)	Convolution tests (244 k)
RISC-V Core	RISC-V Rocket Core	24,531	80,206	45	RISC-V ISA tests (2.2 M)	RISC-V benchmark (1.7 M)

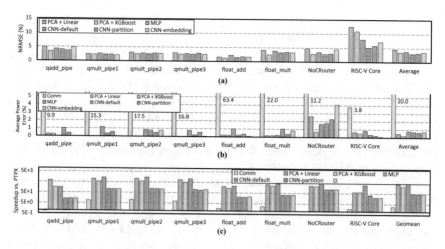

Fig. 4.10 Performance of different machine learning models on test sets—The ML models used by PRIMAL achieve high accuracy for both cycle-by-cycle and average power estimation, while offering significant speedup against both gate-level simulation-based power analysis and a commercial RTL power analysis tool (Comm). PRIMAL is also significantly more accurate than Comm in average power estimation. (**a**) Cycle-by-cycle estimation error. (**b**) Average power estimation error. (**c**) Speedup vs Commercial gate simulation-based power analysis

Table 4.2 Training time of different ML models

Design	PCA	Ridge regression	XGBoost	MLP	CNN
Arithmetic units	~10 min	~1 min	~15 min	~25 min	~3 h
NoC router	~7 h	~15 min	~1 h	~1.5 h	~10 h
RISC-V Core	~20 h	~30 min	~1.5 h	~7 h	~20 h

refer to the default 2D encoding, graph-partition-based register mapping, and node-embedding-based register mapping methods introduced in *Feature Construction Techniques in PRIMAL*, respectively.

4.5.2   Results Analysis

Figure 4.10c presents the speedup of the commercial RTL power analysis tool and the PRIMAL techniques against a commercial gate-level simulation-based power analysis tool. Notice that for PRIMAL, the reported speedup is for model inference only, which is the typical use case. While the commercial RTL power analysis tool is only ~3× faster than gate-level simulation power analysis on average, all ML models achieve much higher estimation speed. Even the most compute-intensive CNN models provide ~50× average speedup against gate-level simulation-based power analysis. Linear model, XGBoost, and MLP have an additional 8×, 5×, and 10× speedup compared with CNNs, respectively. Note that the linear and gradient tree boosting models are executed on CPU, while MLP and CNN inference is performed on a single GPU. As a result, if more efficient implementations of the ML models and more compute resources are available, higher speedup can be expected with a modest hardware cost. For small designs, linear model and gradient tree boosting are almost always more favorable choices, since the neural network models do not provide significant accuracy improvement but require much more compute and training effort. For complex designs such as the RISC-V core, CNN provides the best accuracy with $\sim$35$\times$ speedup, while other models are faster but less accurate. Viewing Table 4.2, we see that training time also becomes a considerable consideration for larger designs.

PRIMAL is a learning-based framework that enables fast and accurate power estimation for ASIC designs. Using state-of-the-art ML/DL models, PRIMAL can be applied to complex hardware such as a RISC-V core, and the trained power models can generalize to workloads that are dissimilar to the training testbenches. The ML-based techniques achieve less than 5% and 1% average error for cycle-by-cycle and average power estimation, respectively. The error is less for average power estimation as some of the per-cycle inaccuracy is amortized over a power window. Compared with gate-level simulation-based power analysis, PRIMAL provides at least 50$\times$ speedup across the selection of benchmarks. This means deep learning modeling provides a good accuracy/speed trade-off point for power estimation. The main shortcoming is that PRIMAL must be retrained for each new design encountered, and the training time itself is long, especially for larger, more complex designs (Table 4.2). Depending on the size of the design, various ML models may provide the best training time overhead/accuracy/speedup/inference time trade-off.

4.6 A Case Study on Using Graph Neural Networks for Generalizable Power Estimation

Although the trained ML models in PRIMAL can infer average power (the common use case) for a **new** workload on the same design, a new ML model must be trained for each new design encountered. This is problematic for fast power analysis of changing netlists or new designs, since training can take hours or days.

This section introduces GRANNITE [22], a GPU-accelerated novel graph neural network (GNN) model for fast, accurate, and transferable (or generalizable) vector-based average power estimation. During training, GRANNITE learns how to propagate average toggle rates through combinational logic: A netlist is represented as a graph, register states and unit inputs from RTL simulation are used as model input features, and combinational gate toggle rates are used as labels. The trained GNN model can then infer average toggle rates on a new workload of interest or **new netlists** from RTL simulation results in a few seconds. Compared to traditional power analysis using gate-level simulations, GRANNITE achieves >18.7$\times$ speedup with an error of only <5.5% across a diverse set of benchmark circuits. Compared to a GPU-accelerated conventional simplistic probabilistic switching activity estimation (SAE) approach, GRANNITE achieves much better accuracy (on average 25.9% lower error) at similar runtimes.

4.6.1 GRANNITE Introduction

GRANNITE (which stands for Graph neural network inference for transferable power estimation) is a supervised learning-based SAE for average power inference that foregoes the need for gate-level simulation. During training (Fig. 4.11), GRANNITE takes gate-level netlists and corresponding known input port and register (timing startpoints) toggle rates over a power window from simulation as input features. Ground truth toggle rates per logic gate from gate-level simulation are taken as labels to train against. The trained model can then be used as a learned SAE (Fig. 4.12), inferring logic gate toggle rates from new input toggle rate features over a new window of interest for the same designs, or new designs. The inferred toggle rates can then be easily translated into industry-standard formats

Fig. 4.11 GRANNITE training flow

Fig. 4.12 GRANNITE inference flow. Throughput is based on a modest power window of 1000 cycles

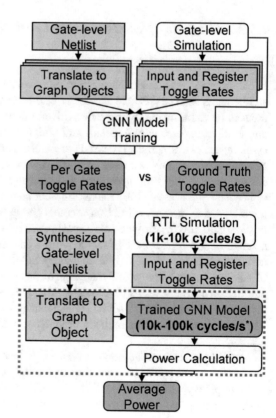

such as the Switching Activity Interchange Format (SAIF) for average power analysis by commercial tools over the window of interest. Compared to previous machine learning (ML)-based power estimation approaches [13, 25], GRANNITE is more transferable (or generalizable), since it is able to infer power on new gate-level netlists without requiring retraining. This is accomplished by using a novel graph neural network (GNN) model architecture [17, 18] for fast, accurate, and transferable SAE. By achieving an equivalent throughput of >10 k cycles/second (assuming a modest window size of 1000 cycles) and skipping gate-level simulation, GRANNITE greatly improves productivity and turnaround time for average power analysis use cases. The GPU-accelerated implementation of GRANNITE achieves $>18.7\times$ speedup and $<5.5\%$ error on average compared to the traditional gate-level simulation approach. As a comparison to previously proposed SAE methods, a GPU-accelerated implementation of the simplistic probabilistic SAE using a similar graph message passing framework to GRANNITE is also presented. It achieves $40\times-1125\times$ speedup with similar accuracy (31% average error) compared to a commercial power analysis tool implementing a similar simplistic probabilistic SAE.

4.6.2   The Role of GPUs in Gate-Level Simulation and Power Estimation

In recent years, researchers have proposed novel simulation methods using parallel architectures such as GPUs to accelerate gate-level simulation. These methods have focused on either using hybrid event-driven (gates are only simulated if a change on the input is detected/scheduled) and oblivious (all gates are simulated at every cycle regardless of activity) methods or parallelizing simulation across cycles and gates to achieve more speedup [38–40]. However, Amdahl's law effects or GPU device memory can limit speedups, and typically some amount of manual tuning is needed. Holst et al. [39] makes improvements to the memory issue using a novel dynamic memory allocation scheme, and they achieve a throughput of $\sim$300 million gate$\times$cycles$\times$toggle rate (events) per second. This would correspond to a range of 1.5–15 k cycles/s for 100 k-1mil gate-sized designs (assuming a high activity toggle rate of 0.2). GPU-enabled parallel simulation can bring the efficacy of gate-level simulation within the range of faster RTL simulation! So, the massive parallelism available on GPUs should continue to inspire new improvements to gate-level simulation speeds. GRANNITE also makes use of GPUs for training speedups. The baseline comparison simplistic probabilistic SAE is also mapped to GPUs for speedup.

In GRANNITE, the authors compare a probabilistic SAE engine in a widely used commercial power analysis tool to a GPU-accelerated PyTorch/DGL framework implementation and find that the GPU implementation achieves similar or better accuracy at $>40\times$ speedup. For the remainder of this section, the GPU-accelerated PyTorch/DGL implementation is used as the baseline comparison for the GRANNITE ML-based SAE, and it will be referred to as the baseline implementation in later subsections.

4.6.3 GRANNITE Implementation

This section describes the GRANNITE GNN model architecture and implementation, written in PyTorch [41] with the Deep Graph Library (DGL) package [42]. The model is a variation of a popular type of GNN, the graph convolutional network (GCN) [18]. While GCNs perform neighbor message passing on all nodes in parallel, this variation does so in a levelized, sequential manner. The levelization is the natural logic stage levelization of the netlist graph. The final resulting node features (or embeddings) then become the output of the graph network layer. Figure 4.13 gives a depiction of the GRANNITE GNN layer message passing mechanisms. Due to the message passing mechanisms, this means GRANNITE learns model parameters from input feature data as well as structure of the input graph. Since a logic netlist is represented as a graph, during training, it is expected that the message passing steps will learn to propagate toggle rates through the netlist from one logic level to the next. In this way, GNNs can learn an approximate solution to SAE based on the netlist features, graph structure, and labeled training data.

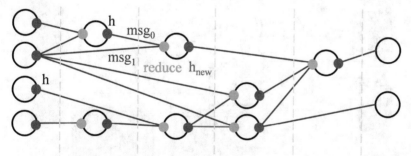

Step 1: Message sending Step 2: Message Reduction   Step 3: Node Transformation
msg = f(edge_features, h) reduce=f(msg$_0$,msg$_1$,msg$_2$...) h$_{new}$=f(reduce,h,node_features)

Fig. 4.13 Depiction of graph neural network layer mechanisms in GRANNITE. Same level nodes are processed in parallel, while different level nodes are processed in sequence. Step 1 sends a message across each edge to the next node, applying a function of local edge features and predecessor node's propagating features. Step 2 applies a function to conglomerate all incoming messages into one reduced message. Step 3 applies a function of the reduced message, local node features, and previous node features to attain the new propagating node features

4.6.3.1   Toggle Rate Features

Toggle rates of input ports and register outputs (timing startpoints) are used as input to the GRANNITE model, gathered from gate-level simulation during training (ground truth labels) or RTL simulation during inference (the inference inputs). These toggle rate input features are the "source" of the switching activity of the yet unknown toggles in the combinational logic and are readily available prior to gate-level simulation (during inference). The features are encoded into an array format with four dimensions representing the signals' [*chance to stay low, stay high, switch to low, or switch to high*] over the duration of the training power window.

4.6.3.2   Graph Object Creation

The other input to GRANNITE is the graph representation of the gate-level netlist, translated via a custom Python script into a DGL graph object, shown in Fig. 4.14. Gates are mapped to graph nodes and output-pin-to-net-to-input-pin connections into graph edges. The translation process automatically splits multiple output gates, such as full adders, into two separate nodes. The translation preserves both graph connectivity information and local node and edge features that contain characteristics of each gate and net, shown in Table 4.3. All node and edge features can be derived from the truth table of each logic cell instance. The purpose of

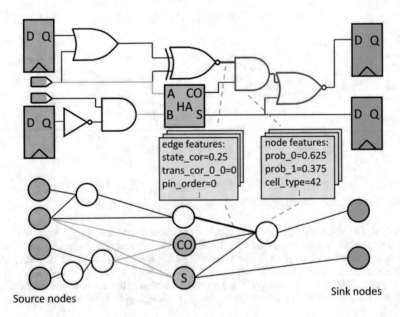

Fig. 4.14 Translating gate-level netlists to graph objects. Multiple output gates are split into multiple nodes. The process also records node and edge features

Table 4.3 Summary of local node/edge features

Type	Description (count)	NAND2/A pin example value
Node	Intrinsic state probabilities (2)	prob_0 = 0.25
Node	Intrinsitic transition probability (1)	prob_sw = 0.1875
Node	Boolean tag if gate is inverting logic (1)	inv = 1
Edge	Pin state to output state correlation (1)	state_cor = 0.5
Edge	Pin transition to output pin transition correlations (16)	trans_cor_0_to_1 = 1.0

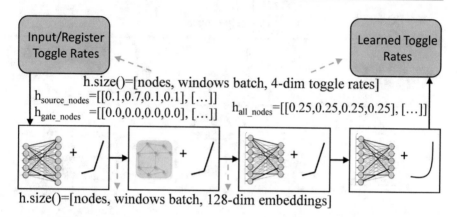

Fig. 4.15 Diagram of GRANNITE architecture. Arrays have three dimensions as multiple power windows can be enabled to be batched during training/inference

translating truth tables into features is so that the GNN model can hopefully differentiate between different types of logic from the different values of these node/edge features and help correctly learn to deduce the underlying toggle rate propagation mechanisms. Pin state to output state correlation is the "logical weight" of each input pin in the Boolean logic function. For example, the values for two of the three input pins (edge connections in the graph) in an AOI21 gate will be 0.25, while one of the input pins has value 0.5. The meaning of *pin transition to output pin transition correlation* is interpreted as "if this input pin exhibits behavior of [*stay low, stay high, switch high to low, switch low to high*], what are the chances, assuming random inputs on all input pins, that the output pin exhibits behavior of [*stay low, stay high, switch high to low, switch low to high*]?".

4.6.3.3 GRANNITE Architecture

Figure 4.15 describes the overall architecture of GRANNITE. The GNN model consists of 1 fully connected (FC) layer followed by one GNN layer and concludes with two FC layers. The first FC layer maps the low (four) dimension input toggle rate features to a higher dimension space (GRANNITE chose 128 dimensions). The dimensions represent different learned switching activity embeddings. In essence,

Implementation	GRANNITE	Baseline

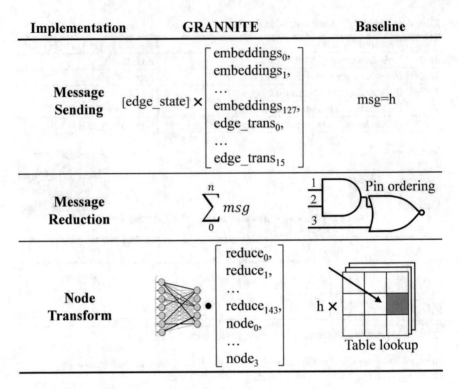

Fig. 4.16 GNN layer message passing definitions

the function of the GNN model is to learn the complex, nonlinear relationship between input toggle rates, logic, netlist structure, and output toggle rates. Figure 4.16 defines the GNN message passing mechanisms.

First, message sending concatenates the predecessor nodes' embeddings with local edge transition features before matrix multiplying with local edge state features. Second, message reduction sums all incoming messages. Last, the node transform function concatenates the reduced message with local node features before passing through an FC layer inside the GNN layer. Thus, the calculated embeddings on each node contain both information from predecessor nodes and local node and edge features. Messages are passed from first gate/node stage to last in a levelized manner using the *prop_nodes* function from DGL. The result of the GNN layer sees embeddings on every logic node/gate in the graph, and the last two FC layers map the high-dimension embeddings back into low (four) dimension output toggle rate features. Since the desired model is highly nonlinear, GRANNITE provides nonlinearity in the network by adding LeakyReLu activations on the first three layers and a softmax activation on the last layer, as the four-dimensional toggle rate features necessarily sum to 1. The loss function during training is mean square error (MSE) loss between the ground truth per logic gate toggle rate and the sum of the third and fourth element of the output layer tensor (which represents the [*chance to switch low, chance to switch high*] behavior, or in other words, the predicted toggle rate).

During training, it's expected that the learned embeddings will contain both predecessor and local information and that the GNN will learn the correct output toggle rates based on both local logic functions and reconvergence correlation caused by predecessors.

4.7 GRANNITE Results

To evaluate GRANNITE, GNN model training and inference experiments are conducted on an NVIDIA Tesla V100 GPU with 16 GB device memory. GRANNITE and the baseline probabilistic SAE both run on the same logic netlists, synthesized from RTL to a 16nm FinFET standard cell library using a commercial logic synthesis tool.

For training and testing datasets, 26 benchmark circuits listed in Table 4.4 are used, which contain small- to medium-sized units with a wide range of average toggle rates (0.01–0.30). Open-source RTL is used for fixed- and floating-point arithmetic units [35]. The open-source network-on-chip router is written in SystemC and synthesized to RTL by a high level synthesis (HLS) tool [43]. The RISC-V core is an RV64IMAC implementation of the open-source RocketChip generator [36], similar to the SmallCore instance. Twenty randomly chosen datapath IP blocks supplied by a commercial logic synthesis tool (∼1k gates each) are also chosen to increase the diversity of standard cell gate types during training.

Multiple GNN models are trained on 25 of the 26 total circuits, leaving 1 circuit outside of the training data set to be the test data set. In this way, the transferability/generalization of the proposed GNN architecture can be verified on all 26 circuits (in other words, 26 different GNN models are trained for each circuit to become the 1 circuit in the test set). In addition to this round-robin "$n - 1$" training/test regime, one more GNN model is trained on all 26 circuits, and tested on different power windows than the windows in the training data, thereby verifying

Table 4.4 Benchmark circuits for GRANNITE

Design	Description	Gate count	Stimulus (40 k cycles)
qadd_pipe	32-bit fixed point adder	774	Random
qmult_pipe	32-bit fixed point multiplier	1410	Random
fadd	32-bit floating point adder	961	Random
fmult	32-bit floating point multiplier	2005	Random
NoC router	Wormhole router with virtual channels	10,330	Operation mode tests
RISC-V Core	RISC-V Rocket Core (small core)	56,243	Dhrystone benchmark
Datapath units	Shifters, encoders, muxes, leading 0/1 detectors, ...	∼1000 each	Random

accuracy on new workloads running on designs in the training set. The input toggle rate features for training are constructed by calculating toggle rates across a sliding power window of 750 cycles out of a total of 40 k simulated cycles for each design. Models are trained for 10 epochs (due to slow GNN model training speed). Due to the limited data, batch size of 1 is chosen (i.e., 1 power window trained per backprop calculation). All training set circuits are trained in parallel by using DGL's dgl.batch() API. After training, inference is performed on new power windows of 1000 cycles on the new test design, or new power windows on the training designs. The output toggle rates are then translated into a SAIF file and average power is computed by a commercial power analysis tool. The same test power windows are used in the baseline probabilistic SAE implementation.

Table 4.5 shows the accuracy achieved by GRANNITE compared to the baseline. For brevity, only the mean relative error metrics for the 20 unnamed datapath units are listed. Only the combinational average power consumption is reported as sequential power is known (from the register traces).

Compared to the baseline probabilistic SAE, GRANNITE achieves much better overall accuracy (5.3% vs. 31.4%) and achieves <10% error for all the benchmarks but the NoC router in the testing (transferable power estimation) results. In all benchmarks, GRANNITE's error decreases from "testing" to "validation" (here, "validation" meaning the test circuit was seen during training, but the test power window was not), suggesting that expanding training data will improve accuracy. Since NoC router is a larger benchmark, it's expected that there is increased likelihood that the training data did not see some feature patterns present in NoC router, which would explain the slightly worse error for the NoC router design.

Table 4.6 shows the speedup achieved when using GRANNITE over the traditional flow of using gate-level simulation. The quoted throughput is based on running inference at 1000 cycle power windows, which is well within the range for average power estimation purposes. The GPU-accelerated baseline SAE throughput

Table 4.5 Average power results and error comparison. % error in parentheses vs. ground truth gate-level simulations. "Testing" refers to when both design and power window are excluded from the training set. "Validation" refers to running inference on a new power window when the design is included in the training set

Design	Ground truth average power (mW)	Baseline predicted power (mW)	GRANNITE inference power, testing (mW)	GRANNITE inference power, validation (mW)
qadd_pipe	0.553	0.550 (0.5%)	0.564 (2.0%)	0.563 (1.9%)
qmult_pipe	2.018	2.006 (0.6%)	2.156 (6.8%)	2.084 (3.3%)
fadd	0.068	0.103 (51.5%)	0.071 (4.0%)	0.068 (0.3%)
fmult	0.219	0.563 (157.1%)	0.215 (1.9%)	0.221 (1.0%)
NoC router	1.036	1.088 (5.0%)	0.897 (13.4%)	0.918 (11.4%)
RISC-V	0.923	0.904 (2.0%)	0.997 (8.0%)	0.873 (5.4%)
datapath_units		N/A (3.3%)	N/A (1.0%)	N/A (0.9%)
Average		(31.4%)	(5.3%)	(3.4%)

Table 4.6 Speed comparison for GRANNITE. Throughput based on 1000 cycle power windows at max batch size. Speedup ($\times$) in parentheses. GRANNITE has an additional benefit of allowing batching of power windows to be calculated in parallel

Design	Batch size = 1 inference latency (s)	Max windows batch size	Per-cycle simulation throughput (kHz)	GNN inference throughput (kHz)
qadd_pipe	0.293	2200	2.5	5050.9 (2044.9X)
qmult_pipe	0.401	1400	1.9	1679.5 (888.3X)
fadd	0.304	1200	8.8	3024.5 (343.7X)
fmult	0.342	950	6.2	1259.9 (202.5X)
NoC router	0.435	175	1.4	194.7 (141.0X)
RISC-V	0.703	30	2.4	44.4 (18.7X)

Table 4.7 *Baseline* vs. commercial tool comparison. Throughput is calculated with 1000 cycle power windows

Design	Commercial estimator error (%)	Baseline error (%)	Commercial estimator throughput (kHz)	Baseline estimator throughput (kHz)	Speedup (X)
qadd_pipe	36.2	0.5	2	355.2	177.6
qmult_pipe	52.1	0.6	1	40.6	40.6
fadd	48.5	51.5	2	2251	1125.5
fmult	137.4	157.1	1.2	646.1	538.4
NoC router	19.2	5	0.3	146.4	585.6
RISC-V	2.8	2.8	0.2	22.4	112.0

is reported in Table 4.7 and is similar to GRANNITE. Since both are implemented in PyTorch/DGL, many power windows can be easily batched in parallel during one run of inference. The minimum speedup at max batch size is 18.7$\times$, greatly outperforming the traditional approach. Of note is that speedup can be even greater with larger power window size, for example, >187$\times$ for 10k cycle windows.

4.7.1 Analysis

GRANNITE is a novel GNN model that achieves fast, accurate, and transferable/-generalizable average power estimation by inferring output toggle rates on logic gate cells and foregoing the need for slower gate-level simulation. GRANNITE achieves less than 5.5% average error and at worst 13.4% error for **new benchmark circuits** and new power windows during inference mode, which is a vast improvement over the 31% average error given by a baseline probabilistic estimator implementation. In addition, it achieves >18.7$\times$ speedup when compared to traditional per-cycle gate-level simulation. This approach can help in alleviating some signal correlation inaccuracies, which is an issue commonly plaguing non -imulation, switching activity estimation ideas.

Compared to PRIMAL, GRANNITE has both less speedup and more error. The advantage GRANNITE displays is in foregoing the need for training a different model for each individual design. Depending on the use case, GRANNITE may be more desirable for this reason. This case study further shows that such advanced state-of-the-art ML model approaches provide a good balance between speed and accuracy for the problem of power estimation.

4.8 Conclusion

This chapter has introduced the concepts of achieving balance between speed and accuracy for efficient switching activity estimation, which in turn translate to efficient power estimation in VLSI design flows. The chapter puts forth that conventional gate-level simulation is slow and conventional switching activity estimation is inaccurate. Thus, modeling is the best bet to achieve balance between speed and accuracy. The chapter gives an overview of the modeling methods used for switching activity estimation: sophisticated statistical equations, "cost-of-action" models, and learning/regression-based models. The advent of deep learning models provides another opportunity to improve upon power modeling, and the chapter summarizes the benefits of deep learning for power modeling, before focusing on two specific examples to show their strengths: using CNNs for per-cycle power estimation and GNNs for average power estimation.

References

1. Yang, T.-J., Chen, Y.-H., Emer, J., Sze, V.: A method to estimate the energy consumption of deep neural networks. In: 2017 51st Asilomar Conference on Signals, Systems, and Computers, pp. 1916–1920. IEEE, Piscataway (2017)
2. "PrimePower" https://www.synopsys.com/implementation-and-signoff/signoff/primepower.html
3. Nourani, M., Nazarian, S., Afzali-Kusha, A.: A parallel algorithm for power estimation at gate level. In: MWSCAS, pp. I–511 (2002)
4. Mehta, H., Borah, M., Owens, R.M., Irwin, M.J.: Accurate estimation of combinational circuit activity. In: Proceedings of the 32nd annual ACM/IEEE Design Automation Conference (DAC '95)
5. Najm, F.N.: Transition density: a new measure of activity in digital circuits. IEEE Trans. Comput. Aided Desig. Integr. Circuits Syst. 12(2), 310–323 (1993). https://doi.org/10.1109/43.205010
6. Kurian, G., Neuman, S.M., Bezerra, G., Giovinazzo, A., Devadas, S., Miller, J.E.: Power modeling and other new features in the graphite simulator. In: 2014 IEEE International Symposium on Performance Analysis of Systems and Software (ISPASS), pp. 132–134 (2014). https://doi.org/10.1109/ISPASS.2014.6844471
7. Li, S., Ahn, J.H., Strong, R.D., Brockman, J.B., Tullsen, D.M., Jouppi, N.P.: McPAT: an integrated power, area, and timing modeling framework for multicore and manycore architectures. In: 2009 42nd Annual IEEE/ACM International Symposium on Microarchitecture (MICRO), pp. 469–480 (2009)

8. Kim, D., et al.: Strober: fast and accurate sample-based energy simulation for arbitrary RTL. In: 2016 ACM/IEEE 43rd Annual International Symposium on Computer Architecture (ISCA), pp. 128–139 (2016). https://doi.org/10.1109/ISCA.2016.21
9. Shao, Y.S., Reagen, B., Wei, G., Brooks, D.: The aladdin approach to accelerator design and modeling. IEEE Micro. 35(3), 58–70 (2015). https://doi.org/10.1109/MM.2015.50
10. Ravi, S., Raghunathan, A., Chakradhar, S.: Efficient RTL power estimation for large designs. In: Proceedings of the 16th International Conference on VLSI Design, pp. 431–439 (2003). https://doi.org/10.1109/ICVD.2003.1183173
11. Sunwoo, D., Wu, G.Y., Patil, N.A., Chiou, D.: PrEsto: an FPGA-accelerated power estimation methodology for complex systems. In: 2010 International Conference on Field Programmable Logic and Applications, pp. 310–317 (2010). https://doi.org/10.1109/FPL.2010.69
12. Yang, J., Ma, L., Zhao, K., Cai, Y., Ngai, T.-F.: Early stage real-time SoC power estimation using RTL instrumentation. In: The 20th Asia and South Pacific Design Automation Conference, pp. 779–784 (2015). https://doi.org/10.1109/ASPDAC.2015.7059105
13. Xie, Z., Xu, X., Walker, M., Knebel, J., Palaniswamy, K., Hebert, N., Hu, J., Yang, H., Chen, Y., Das, S.: APOLLO: an automated power modeling framework for runtime power introspection in high-volume commercial microprocessors. In: The 54th Annual IEEE/ACM International Symposium on Microarchitecture
14. Van den Steen, S., et al.: Analytical processor performance and power modeling using micro-architecture independent characteristics. IEEE Trans. Comput. 65(12), 3537–3551 (2016). https://doi.org/10.1109/TC.2016.2547387
15. Lee, D., John, L.K., Gerstlauer, A.: Dynamic power and performance back-annotation for fast and accurate functional hardware simulation. In: 2015 Design, Automation & Test in Europe Conference & Exhibition (DATE), pp. 1126–1131 (2015)
16. Stockman, M., et al.: A novel approach to memory power estimation using machine learning. In: 2010 International Conference on Energy Aware Computing, pp. 1–3 (2010). https://doi.org/10.1109/ICEAC.2010.5702284
17. Zhou, J., et al.: Graph Neural Networks: A Review of Methods andApplications (2018). CoRR, vol. abs/1812.08434
18. Kipf, T.N., Welling, M.: Semi-Supervised Classification with Graph Convolutional Networks (2016). CoRR, vol. abs/1609.02907
19. Shuai, B., Zuo, Z., Wang, G., Wang, B.: DAG-Recurrent Neural Networks For Scene Labeling (2015). CoRR, vol. abs/1509.00552
20. Wang, H., et al.: GCN-RL circuit designer: transferable transistor sizing with graph neural networks and reinforcement learning. In: 2020 57th ACM/IEEE Design Automation Conference (DAC), pp. 1–6 (2020). https://doi.org/10.1109/DAC18072.2020.9218757
21. Ma, Y., Ren, H., Khailany, B., Sikka, H., Luo, L., Natarajan, K., Yu, B.: High performance graph convolutional networks with applications in testability analysis. In: Proceedings of the 56th Annual Design Automation Conference 2019 (DAC '19), Article 18, pp. 1–6. Association for Computing Machinery, New York (2019). https://doi.org/10.1145/3316781.3317
22. Zhang, Y., Ren, H., Khailany, B.: GRANNITE: graph neural network inference for transferable power estimation. In: Proceedings of the 57th ACM/EDAC/IEEE Design Automation Conference (DAC '20), Article 60, pp. 1–6. IEEE Press, Piscataway (2020)
23. Kunal, K., Poojary, J., Dhar, T., Madhusudan, M., Harjani, R., Sapatnekar, S.S.: A general approach for identifying hierarchical symmetry constraints for analog circuit layout. In: Proceedings of the 39th International Conference on Computer-Aided Design (ICCAD '20), Article 120, pp. 1–8. Association for Computing Machinery, New York (2020). https://doi.org/10.1145/3400302.3415685
24. Lu, Y.-C., Nath, S., Kiran Pentapati, S.S., Lim, S.K.: A fast learning-driven signoff power optimization framework. In: 2020 IEEE/ACM International Conference On Computer Aided Design (ICCAD), pp. 1–9 (2020)
25. Zhou, Y., Ren, H., Zhang, Y., Keller, B., Khailany, B., Zhang, Z.: PRIMAL: power inference using machine learning. In: Proceedings of the 56th Annual Design Automation Conference 2019 (DAC '19), Article 39, pp. 1–6. Association for Computing Machinery, New York (2019). https://doi.org/10.1145/3316781.3317884

26. Mason, L., et al.: Boosting Algorithms as Gradient Descent. In: Proceedings of the Advances in Neural Information Processing Systems (2000)
27. Jolliffe, I.: Principal component analysis. In: International Encyclopedia of Statistical Science. Springer, Berlin (2011)
28. Grover, A., Leskovec, J.: node2vec: scalable feature learning for networks. In: International Conference on Knowledge Discovery and Data Mining (2016)
29. Maaten, L.v.d., Hinton, G.: Visualizing data using t-SNE. J. Mach. Learn. Res. (2008)
30. Hagberg, A., et al.: Exploring network structure, dynamics, and function using networkX. Technical report. Los Alamos National Lab. (LANL), Los Alamos (2008)
31. Karypis, G., Kumar, V.: A fast and high quality multilevel scheme for partitioning irregular graphs. J. Scient. Comput. (1998)
32. Keras: The Python Deep Learning library (2018). https://keras.io/
33. Pedregosa, F., et al.: Scikit-Learn: machine learning in python. J. Mach. Learn. Res. (2011)
34. Chen, T., Guestrin, C.: XGBoost: a scalable tree boosting system. In: Proceedings of the International Conference on Knowledge Discovery and Data Mining (2016)
35. OpenCores.org, Fixed Point Math Library for Verilog :: Manual (2018). https://opencores.org/project/verilog_fixed_point_math_library/manual
36. Asanović, K., et al.: The Rocket Chip Generator. Technical Report. UCB/EECS-2016-17. Department of Electrical Engineering and Computer Sciences, University of California, Berkeley (2016)
37. Ma, N., et al.: Shufflenet v2: practical guidelines for efficient CNN architecture design (2018). Preprint. arXiv:1807.11164
38. Chatterjee, D., DeOrio, A., Bertacco, V.: Event-driven gate-level simulation with GP-GPUs. In: DAC, pp. 557–562 (2009)
39. Holst, S., Imhof, M.E., Wunderlich, H.-J.: High-throughput logic timing simulation on GPGPUs. TODAES **20**, 1–22 (2015)
40. Zhu, Y., Wang, B., Deng, Y.: Massively parallel logic simulation with GPUs. Trans. Design Automat. Electron. Syst. **16**(3), 29:1–29:20 (2011)
41. Paszke, A., et al.: Automatic differentiation in pytorch. In: NIPSW (2017)
42. Wang, M., et al.: Deep graph library: towards efficient and scalable deep learning on graphs. In: ICLR Workshop on Representation Learning on Graphs and Manifolds (2019)
43. Khailany, B., et al.: A modular digital VLSI flow for high-productivity SoC design. In: DAC, pp. 72:1–72:6 (2018)

Chapter 5
Deep Learning for Analyzing Power Delivery Networks and Thermal Networks

Vidya A. Chhabria and Sachin S. Sapatnekar

5.1 Introduction

As a consequence of aggressive technology scaling, on-chip power density has been on an increasing trend over time. This has led to major challenges in designing integrated circuits (ICs) in advanced technology nodes. To design high-performance ICs, it is essential to build tools that help overcome twin aspects of high power densities: (1) successfully delivering power to the gates through reliable design power delivery networks (PDNs) that can meet supply level constraints and (2) efficiently removing the heat generated due to power dissipation through adequate heat-removal paths. From an analysis perspective, two critical and time-consuming simulations are performed several times during the design cycle:

- *PDN analysis*, which diagnoses the goodness and reliability of the PDN by checking (i) whether the voltage drops from the power pads to the gates are within specified limits and (ii) whether the wire current densities satisfy reliability constraints related to electromigration (EM)
- *Thermal analysis*, which checks the feasibility of a placement/floorplan solution by computing on-chip temperature to check for temperature hotspots

The core underlying computations required for PDN analysis (for both IR drop and EM) and thermal analysis are fundamentally similar, solving partial differential equations (PDEs) of similar form.

Static and transient thermal analysis The on-chip temperature distribution is governed by the parabolic PDE:

V. A. Chhabria (✉) · S. S. Sapatnekar
University of Minnesota, Minneapolis, MN, USA
e-mail: chhab011@umn.edu; sachin@umn.edu

© The Author(s), under exclusive license to Springer Nature Switzerland AG 2022
H. Ren, J. Hu (eds.), *Machine Learning Applications in Electronic Design
Automation*, https://doi.org/10.1007/978-3-031-13074-8_5

$$k_t \nabla^2 T + g(\mathbf{r}, t) = \rho c_p \frac{\partial T(\mathbf{r}, t)}{\partial t} \tag{5.1}$$

Here, $\mathbf{r}$ is the spatial coordinate of the point at which temperature is being analyzed, t is time (in seconds), g is the power density per unit volume (in W/m^3), c_p is the heat capacity of the chip material (in J/kg K), and ρ is the density of the chip material (in kg/m^3). Therefore, finding an on-chip temperature profile involves solving $T(r, t)$ given a power density distribution $g(r, t)$ of the chip.

Traditional techniques for solving (5.1) in either a steady state (i.e., with $\partial T(\mathbf{r}, t)/\partial t = 0$) or transient state are based on the finite difference method (FDM) or finite element method (FEM): both methods discretize the differential operator or the temperature field across space and time. In a steady state, the solution to the above PDE amounts to solving a system of linear equations of the form $G\mathbf{T} = \mathbf{P}$ [60, 64] where G is a $N \times N$ conductance matrix representing connected conductances on the grid, $\mathbf{T}$ is a $N \times 1$ vector of unknown temperatures, and $\mathbf{P}$ is a vector of the generated power density values for each element. While G is sparse, in industry-sized designs N is in the order of tens of millions.

Static and Transient IR Drop Analysis The on-chip PDN is responsible for transmitting voltages and currents to each cell in the design. However, due to the parasitics in the PDN, a voltage drop is induced between the power pads and the cells in the design. Large voltage drops in the PDN can hurt chip performance and, in the worst case, its functionality. Consequently, it is essential to introduce checks that verify that worst-case IR drop values in the PDN are within specified limits. Simulating the PDN to calculate the IR drop in a PDN requires solving a differential equation obtained through modified nodal analysis:

$$G\mathbf{V}(t) + C \frac{\partial \mathbf{V}(t)}{\partial t} = \mathbf{J}(t) \tag{5.2}$$

where G is the conduction matrix, C is the diagonal capacitance matrix, and $\mathbf{V}(t)$ and $\mathbf{J}(t)$ are vectors of voltages and currents at specific instances of time. In a steady state, this reduces to $G\mathbf{V} = \mathbf{J}$, similar to its thermal counterpart.

Motivation for DL One of the major challenges with these analyses is the overhead of extremely large runtimes. The underlying computational engines that form the crux of both analyses are similar: both simulate networks of conductances and current/voltage sources by solving a large system of linear equations [60, 64] with millions to billions of variables. In modern industry designs, a single full-chip temperature or IR drop simulation can take several hours.

Although research on analyzing large networks that represent PDN or thermal networks has delivered significant gains in efficiency through the use of multigrid methods [34, 37], multiscale methods [1], hierarchical techniques [36, 62], or frequency domain approaches [65], these methods remain computationally prohibitive for full-chip analysis, especially if they are to be invoked repeatedly within the inner loop of an optimization scheme, as in [51–53, 57]. Most PDN analysis methods work with general power grid topologies with no specific structure, resulting in high analysis costs. Accelerating these analyses opens the door to optimizations that iteratively invoke these engines under the hood.

The analysis of these large networks is equally important for optimization, which may involve repeated expensive calls to an analysis engine. Several techniques that aim to build optimized PDNs have been proposed in the past few decades, based on heuristics or formal optimization formulations. The works [53, 57] proposed linear and nonlinear programming-based optimization solutions, respectively; a heuristic for signal congestion-aware PDN synthesis was developed in [52], and an algorithm that uses successive chip partitioning, together with local PDN refinements were described in [51].

ML techniques can help overcome this computationally expensive challenge by the following: (i) building fast and accurate ML-inference based analyzers or (ii) predicting correct-by-construction solutions that bypass the use of slow PDN analyzers. This chapter provides an overview of ML techniques for PDN and thermal analysis, PDN optimization, and PDN benchmark generation.

5.2   Deep Learning for PDN Analysis

There have been several works [14, 15, 18, 24, 39, 58] that use ML techniques to overcome runtime challenges in PDN analysis. The works in [22, 24, 39] address incremental IR drop analysis and are not intended for full-chip estimation. One class of ML techniques in [56, 58] divides the chip into regions (*tiles*), where CNNs operate on each tile and its near neighbors. However, selecting an appropriate tile and window size is nontrivial: small windows could violate the principle of locality [19], causing inaccuracies, while large windows could result in large models with significant runtimes for training and inference. Alternative approaches in [14, 15, 18] bypass window size selection by providing the entire power map as input, allowing a U-Net [47] to learn the window size for accurate estimation of IR drop and accurate EM hotspot classification.

There has been little prior work that uses ML for EM hotspot classification in PDNs. The work in [28] uses a GAN to predict transient stress profiles for multisegment interconnect tree topologies by using images of the tree topologies and their current densities as input. Since the current density is an input, this method does not overcome the runtime challenge inherent in PDN analysis. The work in [14] overcomes the limitations of [28] by performing EM hotspot classification using a single inference on the entire chip instead of smaller regions. It encapsulates the estimation of current densities on a per-segment basis into a one-time training step and uses the on-chip power, PDN density distributions, power pad locations, and temperature to predict EM-prone segments during inference directly.

The rest of this section focuses on the use of CNNs for IR drop estimation [58] and U-Nets for PDN analysis (IR drop and EM hotspot classification) [14, 15, 18].

5.2.1   CNNs for IR Drop Estimation

PowerNet [58] leverages a CNN to capture 2D spatial distributions of power to predict vectorless dynamic IR drop. The work tessellates a chip into an array of tiles or regions and predicts the mean IR drop of all instances in the tile.

5.2.1.1   PowerNet Input Feature Representation

PowerNet extracts raw features from a standard design flow, such as the instance location, power, toggle rates r_{tog}, and timing windows, which are then processed to generate power maps using a spatial and temporal decomposition technique. The technique creates a spatial power map based on estimates of the power dissipation for each tile. Since the IR drop in each tile is not just simply proportional to its tile power but also depends on its neighborhood power, both spatial and temporal power distributions are used as features in PowerNet.

Five types of power maps are used as features: (i) internal power, p_i; (ii) switching power, p_s; (iii) toggle-rate-scaled power, given by $p_{sca} = r_{tog} \times (p_i + p_s) + p_l$, where p_l is leakage power; (iv) total power, $p_{all} = p_i + p_l + p_s$; and (v) power at time instance t, p_t obtained from a timing decomposition method. The first four power distributions are static, and the last is a vector of power maps for each time step.

5.2.1.2   PowerNet Architecture

Figure 5.1 shows the architecture of PowerNet, which uses a CNN to predict the worst transient IR drop across the clock period. The clock period is split into N time steps, and the power at each time step j is represented by $p_t[j]$. All N time-decomposed power maps $p_t[j]$ are processed separately by the same CNN model, together with all other static power maps p_{sca}, p_{all}, p_i, and p_s. These are all inputs to a CNN with N outputs, o_j, and the maximum output is denoted as $o_{max} = \text{Max}(o_j \mid j \in [1, N])$ is the predicted worst IR drop for the analyzed tile. This is then repeated for all tiles on the chip. This maximum structure highlights the instant time leading to the peak IR drop and guides the CNN to predict the worst IR drop for the clock period. PowerNet uses a tile size of $5\,\mu\text{m} \times 5\,\mu\text{m}$ and the power information in a 31×31 tiled neighborhood (window) that forms the features of the CNN.

The PowerNet CNN has four convolutional layers, two max pool layers, and two fully connected layers, as shown in Figure 5.1. During cross-validation, the

Fig. 5.1 PowerNet using a maximum CNN architecture based on [58]

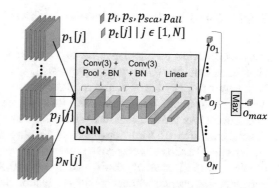

Table 5.1 Area under ROC comparing PowerNet against other ML models

Design	$5\,\mu m \times 5\,\mu m$ tile size			$1\,\mu m \times 1\,\mu m$ tile size		
	XGBoost	CNN	PowerNet	XGBoost	CNN	PowerNet
D1	0.80	0.82	0.95	0.79	0.81	0.92
D2	0.93	0.92	0.97	0.92	0.89	0.95
D3	0.84	0.91	0.95	0.83	0.85	0.91
D4	0.80	0.86	0.97	0.75	0.82	0.93

CNN structure and hyperparameters (e.g., N) are tuned. Batch normalization (BN) is applied to accelerate model convergence, and the Adam [33] method is used for optimization. The L1 loss function, the mean absolute error between the prediction and label, is used for training.

5.2.1.3 Evaluation of PowerNet

PowerNet reports evaluations on four industry designs implemented in a sub-10nm technology node and compared against two other ML models: (i) an XGBoost model similar to the work in [22] and a (ii) CNN without the maximum structure. Even though PowerNet predicts the IR drop value of a tile directly, it is evaluated by its ability to classify a tile as a hotspot or not. Therefore, Table 5.1 compares the area under the receiver operating characteristics (AUROC) curve of PowerNet against the XGBoost and CNN models. PowerNet is found to have a larger AUROC and, therefore, more accurate than the XGBoost and CNN models.

The PowerNet-generated IR drop heatmaps are compared against a commercial tool in [58]. The comparison shows that PowerNet can capture most IR drop hotspots. In addition, it is stated that when the tile size is increased from $1\,\mu m \times 1\,\mu m$ to $5\,\mu m \times 5\,\mu m$, the accuracy of PowerNet increases. This behavior could be attributed to the fact that when the size of the tile increases, the CNN operates over a larger window, providing it with more global information. In general, this problem of finding the tile size and window size is nontrivial.

5.2.2 Encoder-Decoder Networks for PDN Analysis

We now overview an alternative set of techniques, using U-Nets [47] for IR drop prediction, that have recently been proposed [14, 15, 18], which have been shown to be superior to CNN-based methods. These works translate the PDN analysis tasks into image-to-image translation tasks for static analysis or sequence-to-sequence translation tasks for transient analysis. Fully convolutional (FC) encoder-decoder-based generative (EDGe) networks are employed for rapid and accurate PDN analysis. FC EDGe networks have proven to be very successful with image-related problems with 2D spatially distributed data [4, 40, 47, 49] when compared to other networks that operate without spatial correlation awareness [22, 24]. Based on this translation and the use of EDGe networks, the work in [14] proposes two networks: (i) IREDGe, a static and transient IR drop estimator, and (ii) EMEDGe, an EM hotspot classifier.

Figure 5.2 shows a structure of an EDGe network. Its encoder/downsampling path captures global features of the 2D distributions of power dissipation and produces a low-dimensional state space. The decoder/upsampling path transforms the state space into the required detailed outputs (IR drop contours or EM hotspots). The EDGe network is well-suited for PDN analyses because:

(a) The convolutional nature of the encoder captures the dependence of PDN analysis on the *spatial distributions of power*. Unlike CNNs, EDGe networks contain a decoder that acts as a generator to convert the extracted power and PDN density features into accurate IR drop contours and EM hotspots.

(b) The trained EDGe network model for static analysis is *design-independent*: it only stores the weights of the convolutional kernel. The same filter can be applied to *any* chip of any size for a given technology and packaging solution. The selection of the network topology (convolution filter size, number of convolution layers) is related to the expected sizes of the hotspots rather

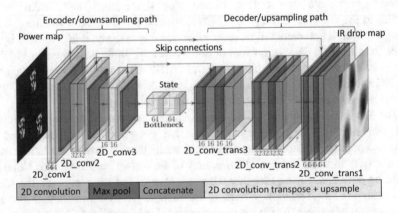

Fig. 5.2 U-Net-based PDN analysis

than the size of the chip: these sizes are similar for a given application domain, technology, and package.

(c) Unlike methods like [28, 58], where finding the right tile and window size for accurate analysis is challenging, *the choice of window size is treated as an ML hyperparameter tuning problem* to decide the amount of spatial input information.

5.2.2.1   PDN Analysis as an Image-to-Image Translation Task

PDN analysis can be mapped to an image-to-image translation task, where all inputs and outputs are represented as images (Fig. 5.3). The three input images are:

(i) *Power Map*: The layout database provides the locations of all standard cells and blocks in the layout and the power database (Fig. 5.3a,b), obtained after analyzing the design for power using a power analysis tool [7], and provides the power per instance or block in the design. This information is used to create a 2D representation of the current distributions as a current map, as shown in Fig. 5.11a, where each pixel is the sum of power values of all instances in the region represented by the pixel.

(ii) *Effective Distance-to-Power-Pad*: Fig. 5.3f shows a typical "checkerboard" power pad layout for flip-chip packages [2, 59]. The effective distance of each

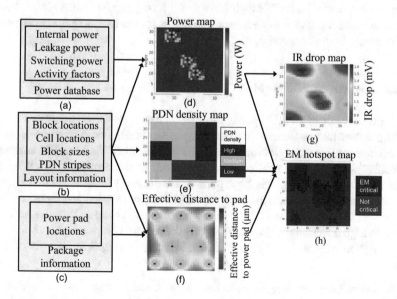

Fig. 5.3 Mapping PDN analysis problems into ML-based image-to-image translation tasks: (**a**) power database, (**b**) physical information, (**c**) package information, (**d**) power map, (**e**) PDN density map, (**f**) effective distance to pad, (**g**) IR drop map, and (**h**) EM hotspot map

instance, d_e, to N_p power pads is defined as $d_e^{-1} = d_1^{-1} + d_2^{-1} + \ldots + d_{N_p}^{-1}$, where d_i is the distance of the ith power pad from the instance and d_e is a proxy for the resistance to the pads.

(iii) *PDN Density Map*: The spatial distribution of the average PDN pitch across the chip provides the PDN density map. For example, for PDN styles similar to [13, 51], where the chip uses region-wise uniform PDNs, the average PDN density in each region across all metal layers provides the PDN density for the region (Fig. 5.3e). Intuitively, d_e and the PDN density map together reflect the equivalent resistance of all paths between the instance and the pads.

A single power map is provided as an input for static IR drop prediction across the chip (Fig. 5.3g). In contrast, a sequence of power maps over time is provided for analyzing the worst-case dynamic IR drop over the time period [14, 18]. The output IR drop map is an image representation of the IR drop across the chip. Each image pixel represents the worst IR drop over all PDN nodes on the lowermost layer in that region, which is connected to the switching devices. Static IREDGe uses a U-Net-based model, and transient IREDGe uses 3D U-Net [18].

On-chip temperature dictates EM limits that are essential to capture for EM hotspot classification. Therefore, in addition to the three inputs to IREDGe, EMEDGe takes the on-chip temperature map as input. EMEDGe classifies each region as either EM-prone or EM-safe, and its output is an EM hotspot map, where a pixel is classified as EM-prone if any PDN segment in its region is EM-prone. A U-Net topology is also used to predict the EM hotspot locations. EMEDGe is metal layer-specific, i.e., a PDN with five metal layers requires five U-Net models to predict EM hotspots for these layers. The inputs for the five models are identical, but the labels are layer-specific.

5.2.2.2 U-Nets for PDN Analysis

CNNs successfully extract 2D spatial information for image classification and image labeling tasks with low-dimensional outputs (class or label). For PDN analysis tasks, the required outputs are high-dimensional distributions of IR drop and EM hotspots, respectively, where the dimensionality corresponds to the number of pixels of the image and the number of pixels is proportional to the size of the chip. This calls for a generator network that can translate the extracted low-dimensional input features (e.g., power, PDN, effective distance to pad features, etc.) from a CNN-like encoder back into high-dimensional data representing the required output.

The EDGe network for static IR and EM analysis (Fig. 5.2) consists of two networks:

(a) *Encoder/Downsampling Network* Like a CNN, the network utilizes a sequence of 2D convolution and max pooling layer pairs that extract critical features from the high-dimensional input feature set. The convolution operation performs a weighted sum on a sliding window across the image [21], and the max pooling layer reduces the dimension of the input data by extracting the maximum value from a sliding window across the input image. In Fig. 5.2, the feature dimension is halved at each stage by each layer pair, and after several such operations, an encoded, low-dimensional, compressed representation of the input data is obtained. For this reason, the encoder is also called the downsampling path. Intuitively, downsampling helps understand the *what* (e.g., "Does the image contain power or EM or IR hotspots?") in the input image but tends to be imprecise with the *where* information (e.g., the precise locations of the hotspots). The decoder stages recover the latter.

(b) *Decoder/Upsampling Network* Intuitively, the generative decoder is responsible for retrieving the *where* information that was lost during downsampling. This network distinguishes an EDGe network from its CNN counterpart. The decoder is implemented using the transpose convolution [21] and upsampling layers. Upsampling layers are functionally the opposite of a pooling layer and increase the dimension of the input data matrix by replicating rows and columns.

Use of Skip Connections The outputs of the PDN analysis are strongly correlated to the input power—a region with high power on the chip could potentially have an IR drop or EM hotspots in its vicinity. U-Nets [47] utilize *skips* connections between the downsampling and upsampling paths (Fig. 5.2), which take information from one layer and incorporate it using a *concatenation* layer at a deeper stage, skipping intermediate layers, and append it to the embedding along the z-dimension.

Skip connections combine the local embeddings from the downsampling path with the global power information from the upsampling path, allowing the underlying input features to shuttle to the layers closer to the output directly. This helps recover the fine-grained (*where*) details that are lost in the encoder network during upsampling in the decoder for detailed IR drop contours and EM hotspots.

5.2.2.3 3D U-Nets for IR Drop Sequence-to-Sequence Translation

U-Nets consist of 2D convolutional layers that perform the convolution operation on the input across *all* the input channels of the input features. Due to the averaging nature of the 2D convolution operation, the resulting embedding loses fine-grained detailed information in each channel. While this averaging effect helps capture spatial variations of power for the static IR drop analysis problems, it fails to capture detailed, fine-grained information, such as the on-chip switching activity that is critical to the transient IR drop problem. Therefore, for the transient IR drop problem, a 3D U-Net is used, inspired by [18], which uses 3D convolutional layers in the encoder path instead of 2D convolutional layers. The 3D convolutional layer restricts the number of input channels on which the convolutional layer operates to a small local window based on the specified filter size. Thus, the 3D convolutional layer works on a few time-steps of power instead of all time-steps at once. This prevents the loss of fine-grained switching information due to the averaging effect across all channels.

5.2.2.4 Regression-Like Layer for Instance-Level IR Drop Prediction

While [14, 58] predict IR drop of a region or tile represented by a pixel, [18] predicts IR drop at the instance-level for a fine-grained accuracy. This is enabled by using a regression-like layer at the end of the decoder path that uses instance-level input features and multiplies them with the predicted coefficients (β_i) by the U-Net-like structure. The coefficient predicted for every tile is then multiplied with per-instance features. This capability is particularly useful for instance-level IR drop predictions in IR-aware STA and IR drop mitigation applications. This architecture provides two other key advantages over prior ML-based solutions:

(a) *Improved transferability* compared to the prior art, as the instance-based features, helps capture fine-grained variations that are otherwise lost due to the averaging nature of U-Net convolutions, as well as tile-level features that capture the bigger picture. Instead of learning the IR drop values directly as in [15, 58], the U-Net-like structure learns the relationship (β_i) between the features and the IR drop values.

(b) *Model interpretability* as the predicted coefficients provides information on the weights associated with each feature. The coefficients correspond to feature sensitivity and allow a designer to assess the root causes of an IR drop violation.

5.2.2.5 Encoder-Secoder Network Training

IREDGe and EMEDGe are trained to learn IR drop contours and EM hotspots, respectively. The training process consists of "golden" data generation and ML model training, as explained below.

Data Generation For both static and transient PDN analysis, an open-source modified nodal analysis-based IR drop solver PDNSim [12] is used to generate "golden" datapoints.[1] The data is processed separately for training IREDGe and EMEDGe.

IR Drop Analysis The simulation generates the IR drop map, which is represented at the same resolution as the input power map, where each pixel represents the maximum IR drop of all PDN nodes in the region covered by the pixel.

For transient IR drop analysis, the U-Net is trained to predict the worst IR drop for an overlapping sequence of ten time-steps. Although a 3D convolution layer works on a small local window of neighboring time-steps, an overlapping sequence of time-steps for training and inference is still used. The overlapping sequence provides the model with past and future switching activity in the neighborhood to

---

[1] The open-source version of PDNSim is for static IR drop analysis only. Therefore, a modified version of the source code to perform dynamic IR drop analysis.

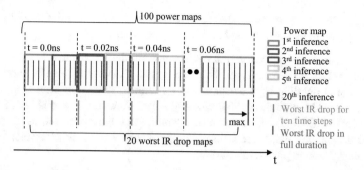

Fig. 5.4 Transient IREDGe power map sequence used for training and inference. The U-Net-based model with 3D convolutions predicts IR drop for each inference

accurately predict the worst IR drop for the current sequence. The number of time steps and the number of overlapping time steps considered during training are ML hyperparameters tuned for best accuracy.

Figure 5.4 illustrates how transient IREDGe uses the sequence of power maps over 100 time-steps. The figure shows that the worst IR drop is predicted for every ten time-steps in the sequence of power maps but uses an overlapping five-time-step sequence of power maps. Therefore, to predict the IR drop for a $t = 2$ns duration, 20 inferences are performed at a 0.1ns time step granularity. This finer granularity with multiple inferences, where each inference predicts the worst IR drop for a fraction of the clock period, provides better accuracy than feeding all the 100 time-steps to a single inference.

EM Hotspot Classification During training, the current densities in each segment are inferred from [12] and compared against the technology-specific EM limits to flag each segment in the PDN as EM-critical or not. As in IR drop analysis, the per-segment information is converted into a pixel-level EM hotspot map, annotating a pixel as EM-critical if any PDN segment in its region has an EM-critical segment.

Unlike IR drop analysis, where a single U-Net model is used, the EM problem uses separate models to predict EM hotspots for each layer in the PDN. For example, for a PDN that uses five layers, five U-Nets are trained, one for each layer. In the testcases used in [14], the metal layers M1 and M2 have 30–50% of their PDN segments as EM-critical, M5 has less than 7% of EM-critical segments, and M8 and M9 have zero EM-critical segments in the dataset. The small number of EM-critical segments in M5 makes it challenging to capture hotspots in this layer due to a severe data imbalance. This is addressed by using aggressive thresholds across all layers such that the fraction of the number of EM hotspots in M5 increases from 7 to 15%, i.e., 15% of M5 is *EM-prone* while 7% is EM-critical. While using an aggressive threshold does increase the number of false positives on the original threshold, it ensures EMEDGe does not miss any EM-prone regions. The flagged regions can be checked with a more accurate detailed analysis on a much smaller scale than the original problem (a few thousand nodes compared to millions).

Model Training The dataset is split for training, validation, and test sets for each of the models trained for each layer. The training dataset is normalized by subtracting the mean and dividing by the standard deviation and is used to train the network using an ADAM optimizer [33], where the loss function is a pixel-wise mean square error (MSE). The convolutional operation in the encoder and the transpose convolution in the decoder are followed by ReLU activation to add nonlinearity and L2 regularization to prevent overfitting.

While the general architecture between all the U-Net-based models is similar, there are some differences. The key differences lie in the convolution layer type (2D vs. 3D), the number of layers, and filter sizes, because the sizes of the hotspots (IR drop, EM, or temperature) are different. These differences between these models are listed in Table 5.2. The table also lists the number of trainable parameters in each model. It is important to note that the number of parameters in the model is independent of the size of the chip but scales with the size of the hotspot, which is generally similar for a given application domain, technology, and packaging choice. A change in hotspot size or resolution demands a change in the number of layers and receptive field sizes of the models to accurately capture IR drop, temperature, and EM hotspots.

5.2.2.6 Evaluation of EDGe Networks for PDN Analysis

A region-wise error metric, IR_{err}, is defined as the absolute difference between the IR of the ground-truth and the predicted image from IREDGe. The percentage mean

Table 5.2 IREDGe and EMEDGe ML model parameters

Layer hyperparameters		Static ThermEDGe	Transient ThermEDGe	Static IREDGe	Transient IREDGe	EMEDGe (all models)
conv1	Filter size	5×5	5×5	3×3	$3 \times 3 \times 3$	3×3
conv_trans1	# filters	64	64	64	32	32
conv2	Filter size	3×3	3×3	3×3	$3 \times 3 \times 3$	3×3
conv_trans1	# filters	32	32	32	32	32
conv3	Filter size	3×3	–	3×3	$3 \times 3 \times 3$	3×3
conv_trans3	# filters	16	–	16	64	64
conv4	Filter size	–	–	–	$3 \times 3 \times 3$	3×3
conv_trans4	# filters	–	–	–	128	128
ConvLSTM	filter size	–	7×7	–	–	–
	# filters	–	16	–	–	–
Max pool layer filters		2×2	2×2	2×2	2×2	2×2
#Trainable parameters		132,769	235,521	133,921	458,337	261,089

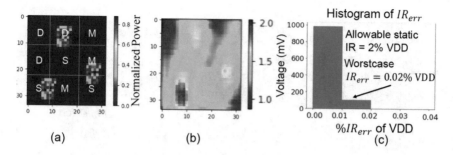

Fig. 5.5 Static IREDGe result: (**a**) power and PDN density (D is dense template, M is a medium density template, and S is a sparse template), (**b**) predicted IR drop maps, and (**c**) histogram of IR_{err}

and the maximum error are listed as a fraction of VDD $= 0.7$ V for IR drop analysis. To evaluate EMEDGe, standard metrics such as accuracy, F1 scores, and area under the ROC curve (AUROC) are used.

Static IREDGe Accuracy Across five testcases evaluated, IREDGe has a maximum average IR_{err} of 0.008% and a maximum IR_{err} 0.05% of VDD, well below the static IR drop limits of 1–2.5% of VDD.

Figure 5.5 shows the result of static IREDGe in detail. The input power maps and PDN density maps are shown in Fig. 5.5a. Six power bumps are used in a checkerboard pattern. Figure 5.5b shows the predicted IR drop contours for the input, and Fig. 5.5c shows the histogram of IR_{err}, where the worst IR_{err} is under 0.02% of VDD.

Size-Independence As stated earlier, the EDGe model can predict IR drop/temperature contours of a chip of a different size than the training set. Across five testcases of size 68×32, IREDGe has a maximum average IR_{err} of 0.01% and worst-case error of 0.068%, using a model trained on 34×32 power maps.

Transient IREDGe Accuracy Transient IREDGe-generated IR drop contours are compared against the ground-truth solver for five different testcases. The average and maximum IR_{err} are 0.08% and 0.43% of VDD, respectively, well below industry-standard transient IR drop limits of 5% of VDD.

A detailed testcase, with five power bumps in a checkerboard pattern, the inputs, including the power map (at $t = 1$ ns) and the PDN density map, is shown in Fig. 5.6a. The ground-truth solution over the entire 2ns period is shown in Fig. 5.6b and is compared against the IREDGe solution. The histogram in Fig. 5.6c represents the IR_{err} as a fraction of VDD for all pixels and shows a worst-case IR_{err} of 0.35% of VDD.

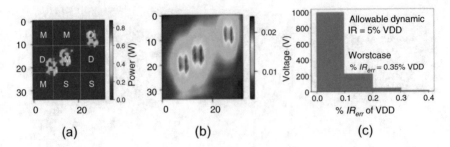

Fig. 5.6 Transient IREDGe data for T21: input (**a**) power map at time $t = 1$ ns and PDN density map (D is dense template, M is a medium density template, and S is a sparse template), (**b**) predicted worst-case dynamic IR drop map for time period 0 to 2 ns, and (**c**) histogram of IR_{err}

IREDGe Compared with PowerNet A version of PowerNet [58] is implemented in [14] for evaluation. For a fair comparison, IREDGe is trained under a fixed PDN density, and fixed power pad locations are used to train PowerNet.

Qualitatively, IREDGe has three advantages over PowerNet:

(i) *Tile and Window Size Selection:* PowerNet is sensitive to the tile size and window. It is stated in [58] that when the size of the tile is increased from $1\,\mu m \times 1\,\mu m$ to $5\,\mu m \times 5\,\mu m$ and the size of the resulting window is increased to represent a 31×31 window of $25\,\mu m^2$ tiles instead of $1\,\mu m^2$ tiles, the accuracy of PowerNet improves. This behavior is expected in IR analysis, where the accuracy increases as more global information is available, until a certain radius after which the principle of locality holds [19]. IREDGe bypasses this tile-size selection problem entirely by providing the entire power map as input to IREDGe and allowing the network to learn the window size needed for accurate IR estimation.

(ii) *Runtimes and Accuracy:* At iso-accuracy, for a fixed testcase with a tile size of $25\,\mu m \times 25\,\mu m$ and a 100 time-step simulation, transient IREDGe is $9.7\times$ faster than PowerNet for a chip of area 0.68mm^2 and $15.3\times$ faster for a chip of area 1.36mm^2, as it requires a single inference in space and 20 inferences in time. In contrast, PowerNet requires $34 \times 32 \times 100$ inferences across space and time.

(iii) *Pixelated IR Drop Maps:* Since PowerNet uses a CNN to predict IR drop on a tile-by-tile basis ($5\,\mu m \times 5\,\mu m$), the resulting IR drop image is pixelated, and the predicted region value does not correlate well with neighboring regions.

EMEDGe Evaluation Since EMEDGe operates on a region level, where all the PDN segments in the region are represented by a single pixel, the fine-grained information at the PDN-segment level is lost. To fairly evaluate EMEDGe, this inaccuracy is accounted for by comparing against the ground-truth for each PDN segment. The EMEDGe-predicted EM-prone regions are converted into segment-level detail by assigning all segments in the region to the same class (EM-prone or not) as the pixel.

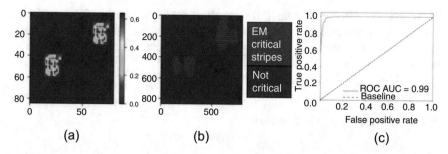

Fig. 5.7 EMEDGe evaluation on for layer M1: input (**a**) input chip power map, (**b**) predicted EM-prone PDN segments, and (**c**) ROC curve for on a per-PDN-segment basis

Figure 5.7 compares the EM-critical segments reported by the ground-truth solver and the EM-prone segments reported by EMEDGe for M1 and an input power map shown in Fig. 5.7a. Visually, the ground -truth EM hotspots are near-identical to the EMEDGe-predicted hotspots in Fig. 5.7b. Since the EM hotspot analysis is formulated as a classical binary classification problem, the receiver operating characteristic (ROC) curve, which demonstrates the performance of the model for different thresholds, is shown. The area under the ROC (AUROC) curve denotes the accuracy of the model. For the testcase in Fig. 5.7, the M1 layer prediction has AUROC of 0.99, as shown in Fig. 5.7c, demonstrating good classification accuracy. Note that even though the training is performed with an aggressive threshold, the results are evaluated against PDK-specified original thresholds. In addition to AUROC, other metrics, such as F1 score, and the number of true and false positives and negatives are summarized. Across the five testcases considered, the average F1 scores for the M1, M2, and M5 models are 0.90, 0.86, and 0.81, respectively. These numbers show that the models are able to classify the minority class correctly. Due to the aggressive threshold, the model predicts no false positives.

Inference Runtime Analysis The inference times for the EDGe networks for these testcases on an NVIDIA GeForce RTX 2080Ti GPU are a few milliseconds, significantly faster than the ground-truth analysis tools that require tens of minutes. The one-time cost of training the EDGe networks is quickly amortized over multiple uses within a design cycle and over multiple designs.

5.3   Deep Learning for Thermal Analysis

For thermal analysis, existing ML-based solutions primarily focus on coarser system-level thermal modeling [29, 30, 48, 61]. The work in [56] predicts temperature at a finer granularity but uses coarse temperature estimates as an input. The methods in [29, 48] use post-silicon performance metrics as inputs to predict the full-chip temperature and are not suitable for predicting temperature during the design process. Similar to static IR drop analysis described in Sects. 5.2.2, [15] uses encoder-decoder networks to develop TherMEDGe, a full-chip static and dynamic thermal analyzer that uses power maps as input to generate temperature contours as output.

5.3.1 Problem Formulation

The static thermal analysis problem is converted into an image-to-image translation task, where the input is a single power map and output is a single IR drop map. The dynamic thermal analysis task is converted into a sequence-to-sequence translation task, where the input is a sequence of power maps and the output a is sequence of IR drop maps at different time steps. ThermEDGe leverages U-Nets [47] for the static analysis task and long short-term memory models (LSTMs) [25] with an encoder-decoder-like architecture for the image-to-image and sequence-to-sequence translation tasks. The computationally expensive FDM analysis is encapsulated into a one-time cost of training ThermEDGe. For a trained network, a rapid inference predicts temperature values of the chip for any given input power map.

5.3.2 Model Architecture for Thermal Analysis

The architecture of static ThermEDGe is the same as that of static IREDGe and is defined in Fig. 5.2 and Table 5.2. The reasons for selecting a U-Net for this task are identical as the reasons for using U-Nets for static IREDGe as justified in Sect. 5.2.2. However, unlike IREDGe that uses 3D convolutional networks for transient analysis, transient ThermEDGe uses LSTM-based models due to the long thermal time constants. The structure of transient ThermEDGe is shown in Fig. 5.8. The core architecture is an EDGe network, similar to the static analysis problem described in Sect. 5.2.2, except that the network uses additional LSTM cells to account for the time-varying component. The figure demonstrates the time-unrolled LSTM, where input power frames are passed to the network one frame at a time. The LSTM cell accounts for the history of the power maps to generate the output temperature frames. In this case, because the temporal variations are much slower than for PDNs, an LSTM is adequate.

As in Fig. 5.8, the encoder consists of convolution and max pooling layers to downsample and extract critical local and global spatial information, and the decoder consists of upsampling and transpose convolution layers to upsample the encoded output. However, in addition, transient ThermEDGe has LSTM layers in both the encoder and decoder. While the basic LSTM cell uses fully connected layers within each gate, the work in [14] uses a variation of an LSTM cell called a convolutional LSTM (ConvLSTM) [50]. In this cell, the fully connected layers in each gate are replaced by convolution layers that capture spatial information. Thus,

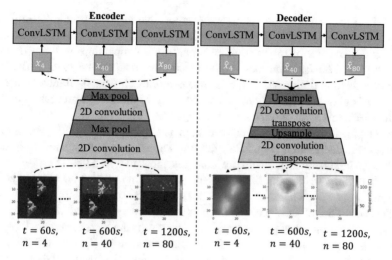

Fig. 5.8 LSTM-based network for transient ThermEDGe

the LSTM-based EDGe network obtains a spatiotemporal view that enables accurate inference.

5.3.3   Model Training and Data Generation

For both the static and thermal analysis problems, the golden data is obtained from Ansys-Icepak [3] simulations. A 50-datapoint set is created for the static thermal analysis problem, where a single datapoint consists of a power map and static temperature maps. For the transient thermal analysis case, the training data consists of 150 datapoints with time-varying workloads as features and the time-varying temperatures as labels. For each testcase, 45 time-step simulations are generated that range from 0 to 3000 s, with irregular time intervals from the thermal simulator. The LSTM-based network is trained using constant time steps of 15 s, which enables easy integration with existing LSTM architectures, which have an implicit assumption of uniformly distributed time steps, without requiring features to account for the time.

5.3.4   Evaluation of ThermEDGe

The work in [14] performs a comparison between temperature maps generated by a commercial tool temperature and by ThermEDGe-generated temperature map for five testcases. The runtime of static ThermEDGe, for each of the five testcases of size 34×32, is reported as 1.1 ms. Across the five testcases, static ThermEDGe has an average T_{err} of $0.63\,°C$ and a maximum T_{err} of $2.93\,°C$. These numbers are a small fraction of the maximum ground-truth temperature of these testcases ($85–150\,°C$). The fast runtimes imply that the method can be used in the inner loop of a thermal optimizer, e.g., to evaluate various chip configurations under the same packaging solution (typically chosen early in the design process). For such applications, this level of error is very acceptable.

A graphical view of the predicted map is depicted in Fig. 5.9. For a given input power distribution (Fig. 5.9a), the true temperature is compared against the ThermEDGe-generated temperature contours plots in Fig. 5.9b. The discrepancy is visually seen to be small. Numerically, the histogram in Fig. 5.9c shows the distribution of T_{err}, where the worst-case T_{err} is 2.63% of the temperature corner.

Transient ThermEDGe predicts the output 200-frame temperature sequence at a 15 s interval for the input power sequence. The results are summarized in Table 5.3. Across the five testcases, the prediction has an average T_{err} of 0.52% and a maximum T_{err} of 6.80% as shown. The maximum T_{err} in the testcases occur during transients that do not have long-lasting effects (e.g., on IC reliability). The magnitude of the maximum T_{err} at sustained peak temperatures is much lower and is similar to the average T_{err}. The inference runtime to generate a 200-frame temperature contour sequence takes is 10ms. In light of millisecond runtimes, the errors are negligible.

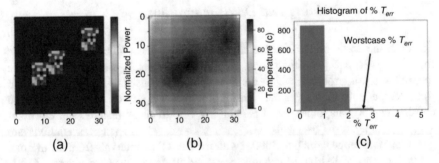

Fig. 5.9 Static ThermEDGe: (**a**) power distribution, (**b**) predicted temperature, and (**c**) histogram of T_{err}

Table 5.3 Transient ThermEDGe results across five testcases

Transient ThermEDGe		
Test	Avg. T_{err} (C)	Max T_{err} (C)
T6	0.51 (0.49%)	5.59 (5.32%)
T7	0.58 (0.55%)	6.17 (5.88%)
T8	0.57 (0.54%)	5.83 (5.55%)
T9	0.52 (0.50%)	6.32 (6.02%)
T10	0.56 (0.53%)	7.14 (6.80%)

5.4   Deep Learning for PDN Synthesis

PDN analysis techniques can diagnose problems and are complemented by auto-mated synthesis methods that meet voltage and EM constraints. In addition, PDNs compete for scarce on-chip wiring resources with signal and clock nets, and resource-sensitive PDN optimization can aid design closure through the optimal allocation of interconnect resources between signal and power wires. Optimization must invoke circuit analysis, which involves the solution of a large system of equations. While fast ML-based methods can be used under the hood of an optimizer, the PDN optimization task can still be made faster by the direct use of ML in optimization.

5.4.1   Template-Driven PDN Optimization

OpeNPDN [11, 13] is an open-source neural network-based framework for PDN synthesis. This scheme completely bypasses the expensive PDN analysis step in the inner loop of PDN optimization methods by leveraging ML to construct a correct-by-construction optimal PDN for a given design. By encapsulating the PDN analysis into a one-time neural network training step, the PDN optimization problem is reduced to a neural network inference with a very low computational cost.

OpeNPDN uses a composition of PDN *templates* to enable correct-by-construction, optimized, ML-based PDN synthesis. The PDN templates are predefined, DRC-correct building blocks of the PDN that vary in their metal layer utilizations. The templates are DRC-correct building blocks that vary in their PDN utilizations and place restrictions on the optimization search space. In each metal layer, wires are unidirectional; this is consistent with gridded design rules in FinFET nodes. The power grid in the lower layers (M1/M2) lines up with the standard cells and is already regular. Each template provides a different choice of wire pitches, which are constant in each metal layer but may vary across layers. A critical requirement in the construction of the PDN templates is their *stitchability*, i.e., if two templates are placed side by side, they should align at the edges.

The work in [11] uses eight templates for both 65LP and 12LP technologies with template IDs 0–7. The templates are labeled in decreasing order of PDN utilization, where template ID 0 has the highest PDN utilization and template ID 7 has the least. As an example, Fig. 5.10 shows a chip that consists of four regions, and each region is assigned a different template. Therefore, each region varies in PDN metal utilization.

The chip is tessellated into regions such that each region can use one of the predefined templates; different regions may use different templates, and the templates are designed to guarantee connectivity when abutted, i.e., when any two different templates are assigned to adjacent regions. The concept resembles the locally regular, globally irregular grids in [51] in the top two metal layers, but

Fig. 5.10 A template-based PDN with piecewise uniform pitches

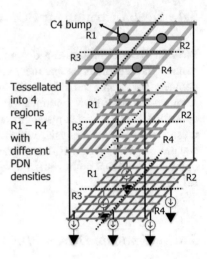

the templates span the entire interconnect stack. Moreover, a trained ML model is employed that decides which template must be assigned to each region based on distributions of (i) currents, (ii) congestion, (iii) macros, and (iv) C4 bumps. The PDN design problem then reduces to mapping templates to regions of the chip as a classification task.

OpeNPDN overcomes the shortfalls of several other methods. In [20] a multilayered perceptron is used to predict the width of the power stripes based on the current and its location as features but builds a power grid that is congestion-unaware. Since this work uses testcases that are small perturbations of the training set, it is unclear if the multilayered perceptron generalizes across a wide range of real test designs. The work in [9] proposes an iterative method for PDN synthesis that calls a fast under-the-hood ML-based post-route wirelength and IR drop predictor. However, the iterative nature of this makes it slow. Both of these works construct uniform grids to meet worst-case IR drop, but such PDNs are likely to be overdesigned and may use more wiring resources than a region-wise-uniform PDN.

5.4.2 PDN Synthesis as an Image Classification Task

Based on the template/region abstraction, the optimization problem of finding a PDN that is most parsimonious in using routing resources, while meeting IR drop and EM constraints and avoiding high densities in regions of high signal congestion, reduces to assigning a template to every region on the chip. OpeNPDN maps this problem to a CNN-based classification task that assigns a template (class) to each region.

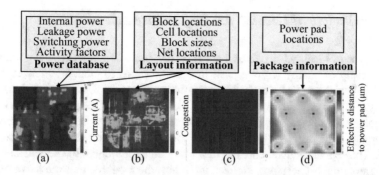

Fig. 5.11 Input feature representation of OpeNPDN: spatial maps showing (**a**) current, (**b**) congestion, (**c**) macro locations, and (**d**) effective distance from C4 bumps

As shown in Fig. 5.11, OpeNPDN considers four features that dictate IR drop values and thereby the selection of the correct PDN: (i) current distribution patterns (Fig. 5.11a), (ii) C4 bump locations (Fig. 5.11d), (iii) congestion distribution patterns, and (iv) macro locations. All the input features of the circuit are extracted from a standard design-flow environment and are represented as 2D spatial distributions, as shown in Fig. 5.11. The first two features are extracted from the standard design-flow environment in the same way as [15] and as explained in Sect. 5.2.2. The last two features are explained below.

Congestion Distributions The congestion information is obtained after performing an early global route of the signal nets using [7]. The congestion estimates are obtained from the layout database on a per global cell (gcell) basis for horizontal and vertical directions. A single congestion value is obtained by summing the vertical and horizontal congestions for each gcell. Based on the gcell locations, a 2D spatial congestion map is constructed as shown in Fig. 5.11b.

Macro Maps The location of the macros are extracted from the layout database to create binary macro map distributions. All areas on the chip which are covered by macros are filled with ones, and the rest of the map is filled with zeros (Fig. 5.11c). Macro blocks are typically treated as blockages[2] to standard cell power stripes, since they have presynthesized power grids within the blocks, effectively splitting the PDN stripes, which impacts the equivalent resistance between regions and C4 bumps.

---

[2] In principle, the macro grid can be connected to the PDN through the edges of the macro. However, in practice, the regions around the macro are halos of placement and routing blockages, to avoid abutment and DRC issues.

5.4.3 Principle of Locality for Region Size Selection

The principle of locality [19] states that the current paths to a node depend primarily on the density of nearby regions. This idea is leveraged to predict templates on a per-region basis, which helps build CNNs independent of the chip size, i.e., the input layer of the neural network is now of fixed dimension irrespective of the chip size. Thus, it is adequate to train a model based on the features in nine regions, enabling the tiling of power grid templates over a chip with an arbitrary number of regions. This has the added benefit of faster training as it reduces the dimensionality of CNN input data. Therefore, for each inference, the input features are the current, congestion, macro, and the effective distance to power bump maps in nine regions. The CNN predicts the template ID of the central region in consideration. Thus, the inference is performed for each region on the chip to predict the entire IR- and EM-safe PDN.

5.4.4 ML-Based PDN Synthesis and Refinement Through the Design Flow

OpeNPDN targets PDN synthesis at various stages of the design flow. Early planning of PDNs occurs at the floorplan stage [5, 27] of physical implementation, and the PDN is refined at the placement stage. At the floorplan stage, coarse estimates of the spatial current distributions are available from block-level power estimates, and signal congestion estimates are approximate. Detailed, fine-grained spatial current distributions are only available after placement [9] and may deviate from floorplan-level assumptions.

Figure 5.12 shows the inference flow of the two-stage OpeNPDN PDN synthesizer. It consists of two CNNs, one applicable to the early floorplanning stage of the design and another for the later placement stage. Both CNNs are trained

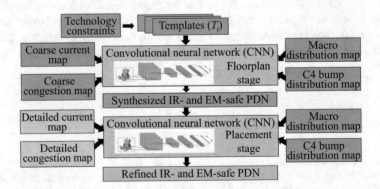

Fig. 5.12 Two-stage inference flow of OpeNPDN

to synthesize a *safe-by-construction* PDN. Design predictability is maintained by ensuring the synthesized PDN at the floorplan stage is only incrementally refined by the placement-stage CNN. This is achieved by feeding the PDN synthesized at the floorplan stage as an additional input feature to the placement-stage CNN. Thus, the two CNNs are devised to operate self-consistently so that placement-stage PDN design corresponds to an incremental refinement of floorplan-stage design, i.e., a small perturbation. This provides predictability in PDN congestion, which aids design closure.

5.4.5   Neural Network Architectures for PDN Synthesis

For both the floorplan- and placement-stage CNN, a standard LeNet [35] CNN topology is used, with four convolutional layers and four max pool layers. The CNN input is a 3×3 region window of the features listed earlier in this section in the form of channels. The first convolutional and max pool layers use 5×5 and 3×3 filters, respectively. The rest of the convolutional and max pool layers use 3×3 and 2×2 filters. The final fully connected layers feed the output classes, corresponding to the templates.

5.4.6   Transfer Learning-Based CNN Training

Training the CNNs consists of two steps: (i) training data or "golden" data generation and (ii) CNN training. OpeNPDN uses a transfer learning (TL)-based method for training and simulated annealing (SA) for optimized training data generation. Finding adequate training data is a significant problem. This small volume of available data alone is inadequate for training the CNNs. OpeNPDN leverages TL-based training, described in Fig. 5.13, to overcome this challenge. The

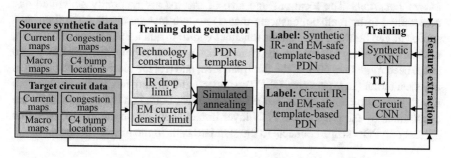

Fig. 5.13 PDN optimization scheme: the training flow produces the "golden" data for the synthetic and real circuits. The golden real circuit data is used to train the circuit CNN using TL. The inference flow uses the trained circuit CNN to synthesize the PDN on the testcase

flow, which is applied to both the floorplan- and placement-stage CNNs, proceeds as follows:

- The first part (blue boxes) uses a large synthetic training set that uses a simulated annealing (SA) engine to generate labels. The labeled synthetic data trains the synthetic CNN, represented by the blue box at the top right.
- The second part (green boxes) shows the transfer of knowledge from the synthetic CNN to train the circuit CNN using a small population of real design examples.

5.4.6.1   Synthetic Input Feature Set Generation

The synthetic benchmarks include current and congestion maps, macro maps, and C4 bump distributions that adhere to standard circuit design guidelines. These benchmarks are generated using randomization techniques for each feature, where the randomization is constrained between appropriate lower and upper bounds. The technology-specific upper and lower bounds are listed in Table 5.4 for each feature. The synthetic structures are generated as follows.

Current and Congestion Map Generation Synthetic current maps are generated using GSTools [42], which generates 2D random spatial fields corresponding to a Gaussian covariance model parameterized by variance and correlation length scale. Several current maps are generated with random values of variance and length scale constrained between an upper bound and lower bound listed in Table 5.4.

C4 Bump Model and Power Bump Locations Synthetic power bump locations are generated by adopting modern flip-chip/C4 technology package conventions. Power bump assignments can be arbitrary or on predefined bump sites [26]. Both predefined bump patterns (e.g., the checkerboard pattern [2, 59] in Fig. 5.11d) and arbitrary power bump assignments are generated. For the arbitrary assignment, to prevent obviously unrealistic assignments, it is ensured that exactly one bump in a 3×3 bump subarray is connected to VDD or GND.

Macro Locations The locations and size of macros are randomly generated and constrained to be realistic, ensuring that (i) no two macros overlap; (ii) there is a sufficient gap between two macros (channel width) to add a PDN stripe in between, ensuring that every instance in the channel has a power supply; (iii) the macros do not exceed more than 60% of the floorplan area, and (iv) the macros can be placed in any orientation (an aspect ratio 0.3). The upper and lower bounds for the randomized macro generation are listed in Table 5.4. The lower bound on the macro widths are constrained by the pitch of the PDN layer immediately above the macro, such that at least one PDN stripe connects to the standard cells in the macro channels.

Table 5.4 Parameters used for synthetic feature set generation

		65LP		12LP	
Feature	Parameter	Lower bound	Upper bound	Lower bound	Upper bound
Current	Mean scaling	$0.5\,\mu A$		$2\,\mu m$	
	Variance	1	3	0.5	2
	Length scale	50	100	20	80
	Macro currents	$200\,\mu A$	$500\,\mu A$	$10\,\mu A$	$250\,\mu A$
Macros	Number	0	6	0	10
	Min channel width	$14\,\mu m$		$4.2\,\mu m$	
	Width	$5\,\mu m$	$300\,\mu m$	$5\,\mu m$	$100\,\mu m$

5.4.6.2 Transfer Learning Model

A supervised inductive network-based approach [46] is used for TL to train both the floorplan- and placement-stage CNN, where a portion of the CNN in the source domain is transferred to the target domain.

A network-based TL strategy is leveraged from [45] which reuses the front layers, i.e., the convolutional and max pool layer pairs, trained by the CNN on the ImageNet dataset to compute the intermediate representation for images in other datasets. The trained CNN in the source domain learns a low-dimensional representation (extracted features) from the images that can efficiently be transferred to another image recognition task in the target domain with a limited target dataset. OpeNPDN transfers the weights of the first four convolutional, and max pool layers are transferred from the source domain, and the two fully connected layers are trained from scratch to select a specified label in the target domain. Intuitively, this strategy finds success for the following reasons: (i) inherently, the synthetic dataset and the target dataset share similar low-dimensional representations, (ii) the tasks in both domains are classification-based, and (iii) both the source and target domains are identical with the same possible class/template set.

For network training, the data from the golden SA optimizer is divided into training (80% of the data), validation (10%), and test (10%). The training dataset is normalized to ensure that both inputs are on the same scale and neither dominates the other. An Adam optimizer [33] and cross-entropy loss function are used, with an L2 regularizer with a dropout factor of 0.5 after each fully connected layer.

5.4.6.3 Training Data Generation

An SA-based optimizer that generates labels for the "golden" data used to train the CNNs determines an optimal power grid for a given chip configuration. The use of SA is justified due to its ability to deliver near-optimal solutions over large discrete solution spaces. For example, for 16 regions, with 1 of 8 possible templates per region, the solution space has 8^{16} possible configurations. The SA computation is a one-time cost, and it is vital to be accurate; computational efficiency is not a significant consideration.

The SA solver finds a solution that optimizes, across the full-chip, the utilization of the PDN, the maximum IR drop for better power integrity, and maximum current density for greater EM safety, given awareness of bump and macro locations. For placement-stage training, in addition to these constraints, the SA solver must ensure the proximity of the solution to the floorplan-stage solution to ensure consistency.

5.4.7 Evaluation of OpeNPDN for PDN Synthesis

The improvement in wiring resource utilization is measured based on the widely used ACE metric [55], which estimates the improvement in congestion only if the region is critical, i.e., if it has an average signal congestion value greater than a certain threshold, set here to 50%. The congestion improvement in a region r is defined as $\Delta c_r = \frac{u_{b,r} - u_{t,r}}{s_r + u_{b,r}} \times 100$, where $u_{b,r}$ is the utilization of the baseline uniform PDN in the region, $u_{t,r}$ is the utilization of the predicted template in the region, and the rest of the terms are as defined before. Based on the ACE metric, the total percentage improvement in congestion of a design is given by $\Delta c_t = \sum_r \Delta c_r \ \forall \ r$ where $s_r > 0.5$.

5.4.7.1 Justification for Transfer Learning

The real circuit testset (116 datapoints in 65LP and 241 datapoints in 12LP) is grossly insufficient for both training and test. The use of transfer learning is justified by showing that direct use of the synthetic CNN on real testcases is unsatisfactory. Figure 5.14 shows the confusion matrices on the real testcases using the synthetic

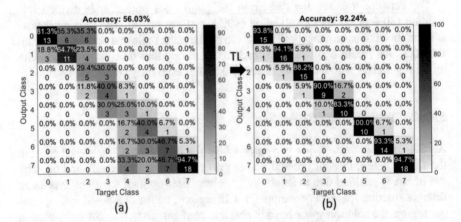

Fig. 5.14 Placement-stage CNN confusion matrix for 116 regions real circuit testcases in 65LP: (**a**) before TL and (**b**) after TL

CNN at the placement stage before and after transfer learning. The figure shows that placement-stage CNN's overall accuracy is low before transfer learning (below 60%). TL helps recover the gap in the information between the synthetic dataset and real circuit dataset to improve the classification accuracy (above 90%).

5.4.7.2   Validation on Real Design Testcases

OpeNPDN methodology is validated on the nine open-source designs implemented in commercial 65LP and 12LP technologies as listed in Table 5.5. The designs have approximately between 40,000 and 500,000 instances and up to 50 macros. Each design is evaluated using leave-one-out cross-validation, where the rest of the designs are used for training the CNN, and the design under consideration is used for testing.

The results for all testcases are summarized in Table 5.5, which lists the Δc_t, worst IR drop (d_r) across all regions, and the worst-case normalized current density ($J_{r,norm}$), at both placement and floorplan (obtained from [8, 12]. The table also lists the runtimes for synthesizing a PDN, including feature extraction, data preparation, and ML inference. The runtimes show that an optimized correct-by-construction PDN can be synthesized rapidly without slow analysis checks.

The CNN-synthesized PDN is compared against the SA-synthesized PDN for each testcase listed in Table 5.5. Due to the high accuracy of the ML model, there are very few mispredictions in comparison with the SA-generated ground-truth, and 100% of the circuits are found to be IR- and EM-safe. The comparison highlights two key insights: (i) the mispredictions are often pessimistic, and (ii) an optimistic misprediction does not imply the circuit fails, i.e., the worst-case IR drop is less than the specified limits, since the pessimistic mispredictions and surrounding correctly predicted templates compensate for it.

5.5   DL for PDN Benchmark Generation

5.5.1   Introduction

The availability of large datasets has been a driving force for the widespread applications of DL in applications, such as image classification. However, in IC design and EDA, the availability of datasets is extremely limited due to IP-related issues, making evaluating novel EDA algorithms and tools extremely challenging. Moreover, with the recent adoption of ML-based techniques to solve complex PDN-related challenges, as described in the rest of this section, there is a need for not just a few sets of benchmarks but a large and diverse set of benchmarks to successfully train, test, and fairly evaluate each model.

Table 5.5 Evaluation of both floorplan- and placement-stage CNN on a set of testcases across 65LP and 12LP technologies

Testcase	Tech. node	#cells	#regions	Feature extraction time	Floorplan							Placement							
					Uniform grid		CNN-synthesized					Uniform grid		CNN-synthesized					
					Worst d_r	Worst $J_{r,norm}$	Worst d_r	Worst $J_{r,norm}$	Δc_t	#Tracks saved	GPU time	Worst d_r	Worst $J_{r,norm}$	Worst d_r	Worst $J_{r,norm}$	Δc_t	#Tracks saved	GPU time	Total time
BP_BE	65LP	57,900	12	110 s	10.1 mV	96.8%	11.5 mV	98.7%	0.89%	1322	3 s	9.9 mV	96.4%	11.6 mV	98.6%	1.4%	1920	3 s	116 s
BP_FE		39,315	12	79 s	10.4 mV	89.8%	11.2 mV	92.0%	2.32%	916	3 s	10.5 mV	89.3%	11.3 mV	91.9%	2.11%	1190	3 s	85 s
BP		159,389	36	214 s	9.2 mV	89.7%	9.8 mV	93.1%	1.15%	884	6 s	10.1 mV	92.2%	10.6 mV	93.8%	2.35%	1120	6 s	226 s
BP_single	12LP	518,808	225	267 s	5.7 mV	91.1%	6.7 mV	93.9%	1.03%	754	6 s	6.1 mV	92.7%	6.9 mV	94.6%	1.42%	1062	6 s	279 s
PI_G		108,836	12	104 s	5.9 mV	90.4%	6.7 mV	95.8%	1.69%	1274	3 s	6.6 mV	93.8%	7.1 mV	96.3%	2.59%	1954	3 s	110 s
GW_ERV		149,958	4	126 s	5.5 mV	87.2%	6.2 mV	95.2%	1.37%	1108	3 s	6.3 mV	91.9%	6.9 mV	96.5%	2.31%	1868	3 s	138 s

Benchmarks provide standards for evaluation and serve as linchpins in advancing EDA research. For example, the release of PDN benchmarks [38, 43] has driven research in the areas of PDN analysis (IR drop and electromigration) and PDN synthesis. However, circuit benchmarks become obsolete and must be constantly updated to reflect scaling, new design classes, and constraints. Moreover, releasing public domain benchmarks is labor-intensive, requires intense volunteer effort, and must overcome IP-related challenges before release. BeGAN [17], an automated technology- and design-portable PDN benchmark generator, utilizes generative adversarial networks (GANs) [23] to overcome these challenges.

5.5.2 GANs for PDN Benchmark Generation

The two components of PDN benchmarks are the current maps (CMs) and power grids. Power grids can be generated using a PDN synthesizer, such as OpeN-PDN [11]. However, the challenge lies in generating CMs, which contain design placement and power information. The public release of these benchmarks poses challenges on two fronts: (1) the current maps encapsulate design- and library-specific IP, and (2) even for the industry, it is not easy to curate a sufficiently large and diverse dataset (thousands of datapoints) of CMs that can be used to train ML models. BeGAN, inspired by the immense success GANs have in generating images of faces, leverages GANs to generate a diverse synthetic yet-realistic set of CMs.

GANs have been used in the ML world to synthesize images based on training sets of many thousands of images that are as realistic to the human eye as real images. In the context of benchmark generation, there are further constraints that the GAN-based approach must be trained on limited data: unlike image GANs that are trained on numerous widely available images, it is nearly impossible to curate a database of thousands of real circuits even in industry. Therefore, BeGAN leverages recent advances [44, 63] on image generation from small datasets that are motivated by the high expense of data collection or privacy issues. The approach is summarized in Fig. 5.15a, where the GAN is first pretrained using a large set of synthetic process-technology-independent images in a *source dataset* that have similar characteristics as on-chip CMs. Next, the pretrained GAN is *transferred* from a source dataset to a *target dataset* of CMs from a small set of real circuit designs at a specific technology node generated by the OpenROAD flow [54]. The pretraining of the GAN with the source dataset is process-technology-independent, i.e., the same model can be reused across technologies. A new process node only demands fine-tuning the pretrained weights for the new target dataset, ensuring our methodology is process-portable with a low computational cost. During inference, the GAN can rapidly generate thousands of CMs that maintain the key features of the real maps.

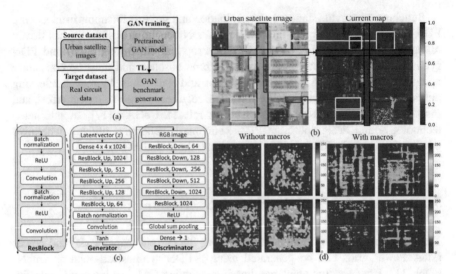

Fig. 5.15 BeGAN: (**a**) TL framework, (**b**) mapping between urban satellite images and current maps, (**c**) SNGAN model architecture, and (**d**) generated current maps

5.5.2.1 Synthetic Image Generation for GAN Pretraining

To guide the choice of synthetic images, some key characteristics of chip-level CMs are examined:

- Chips are placed and routed using Manhattan geometries and often contain rectangular macros or blockages.
- Due to well-known limitations of on-chip power and dark-silicon trends, the number of current hotspots is small.
- Modern chips are packaged using C4 technology with area IO, due to which chip hotspots have a relatively small footprint and may occur anywhere in the chip.

Many of the features of chips with macros map well to satellite images of urban areas in the United States, where many features have straight edges, as illustrated by the example in Fig. 5.15b. For example, trees can be mapped to current hotspots, building tops to macros, and roads to macro channels. The urban satellite images are preprocessed with appropriate color transformation, as shown in the figure to represent a chip-like CM.

5.5.2.2 GAN Architecture and Training

A SNGAN [41] model architecture, as shown in Fig. 5.15c, is used for current map generation as it is easy to adopt compared to other GAN architectures such as BigGAN [6], StyleGAN [32], or ProGAN [31] and also has modest power requirements. The model uses urban satellite image sizes of 128×128 as input and contains 90 million trainable parameters. The SNGAN is trained using approximately 4000 urban satellite images collected from Google Maps and then fine-tuned with a small dataset of real circuit CMs from [10].

The curated dataset of real circuit CMs is divided into two sets based on the presence of macro blocks in the design. Each set of CMs is separately used to fine-tune the weights of the pretrained GAN. Since each circuit CM in the set is of a different size representing the die area of the chip, however, for training the GAN, the CMs must be of a fixed size to maintain consistency with the urban satellite images for the successful application of TL. Therefore, the circuit CM representing the die area is converted into a CM matrix with a fixed size of 128×128, where each matrix value is calculated by dividing the total current value within each region by the area of the region. This matrix is converted to a three-channel jet colormap image with pixel values between 0 and 255.

5.5.2.3 GAN Inference for Current Map Generation

During inference, the GAN generates a three-channel RGB image of size 128×128 with intensity values from 0 to 255, representing a CM image. Eight example BeGAN-generated CM images are shown in Fig. 5.15d for a 45 nm technology highlighting CMs with and without macros. A scaling and coarsening technique for each GAN-generated CM image was used to move to a "chip-like" current map, where each pixel represents the current in a fixed size region of the generated benchmark. The scaling method translates the intensity value distributions of the GAN-generated CMs to chip current magnitudes, while the coarsening technique deals with mapping the 128×128-sized CM to a die area. The scaling technique first converts the three-channel input to a single-channel distribution by converting the RGB CM to a gray scale. Next, the CM is multiplied by a chosen current value (constrained by a minimum and maximum value specified in the technology library file) to generate current distributions of the correct order for that technology.

5.5.3 Evaluation of GAN-Generated PDN Benchmarks

For each GAN-generated CM, OpeNPDN is used to build power grids. The PDN benchmarks (current maps and power grids) are evaluated for both diversity and realism using IR drop as a metric. Figures 5.16 and 5.17 shows the IR drop of the real circuit benchmarks and BeGAN benchmarks, respectively. The BeGAN benchmarks have similar full-chip IR drop hotspot characteristics such as the percentage of hotspot regions across the chip, worst-case IR drop, and average IR drop. While full-chip characteristics are similar, the detailed distributions of instance-level IR drops, such as the locations of hotspots and the shape of the hotspots, are different, showing the diversity in the BeGAN benchmarks.

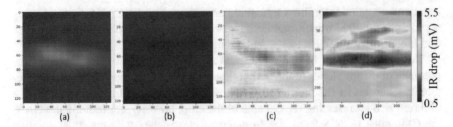

Fig. 5.16 Instance-level IR drop heatmaps for (**a**) AES, (**b**) Dynamic_node, (**c**) Ibex, and (**d**) JPEG in GF 12 nm technology

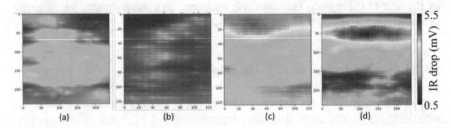

Fig. 5.17 Instance-level IR drop heatmaps for (**a**) BeGAN1, (**b**) BeGAN2, (**c**) BeGAN3, and (**d**) BeGAN4 in GF 12 nm technology

Thousands of BeGAN benchmarks (current maps and power grids) have been released in the public domain [16] across three open-source technologies (ASAP 7nm, FreePDK45 with Nangate Open Cell Library, and SkyWater 130 nm(HD)). The benchmarks include both current maps and PDNs in SPICE format.

5.6   Conclusion

This chapter has provided an overview of techniques for analyzing and optimizing PDNs and analyzing thermal networks using ML techniques. These methods provide fast solutions that show high fidelity to exact analysis. These fast ML-based analyses and optimizations have opened doors to fast thermal-aware floorplan iterations, IR-aware placement, and PDN optimizations. Such high-quality optimizations are critical in designing today's and next-generation high-performance ICs. Moreover, with future technologies such as gate-all-around FETs, chiplets, heterogeneous packaging, and backside PDNs creating new challenges in the areas of PDN and thermal analysis, the role ML will play in future EDA tools for these areas is crucial.

References

1. Allec, N., Hassan, Z., Shang, L., Dick, R.P., Yang, R.: ThermalScope: multi-scale thermal analysis for nanometer-scale integrated circuits. In: Proceedings of the IEEE/ACM International Conference on Computer-Aided Design, pp. 603–610 (2008)
2. Amick, B.W., Gauthier, C.R., Liu, D.: Macro-modeling concepts for the chip electrical interface. In: Proceedings of the ACM/IEEE Design Automation Conference, pp. 391–394 (2002)
3. Ansys: Icepak (2018). https://www.ansys.com/products/electronics/ansys-icepak
4. Badrinarayanan, V., Kendall, A., Cipolla, R.: SegNet: A deep convolutional encoder-decoder architecture for image segmentation. IEEE Trans. Pattern Anal. Mach. Intell. **39**(12), 2481–2495 (2017)
5. Bhooshan, R.: Novel and efficient IR-drop models for designing power distribution network for sub-100nm integrated circuits. In: Proceedings of the IEEE International Symposium on Quality Electronic Design, pp. 287–292 (2007)
6. Brock, A., Donahue, J., Simonyan, K.: Large scale GAN training for high fidelity natural image synthesis. In: Proceedings of the International Conference on Learning Research (2019)
7. Cadence: Innovus Implementation System (2018). https://www.cadence.com/en_US/home/tools/digital-design-and-signoff/soc-implementation-and-floorplanning/innovus-implementation-system.html
8. Cadence: Voltus IC Power Integrity Solution (2018). https://www.cadence.com/en_US/home/tools/digital-design-and-signoff/silicon-signoff/voltus-ic-power-integrity-solution.html
9. Chang, W., Lin, C., Mu, S., Chen, L., Tsai, C., Chiu, Y., Chao, M.C.: Generating routing-driven power distribution networks with machine-learning technique. IEEE Trans. Comput. Aided Des. Integr. Circuits Syst. **36**(8), 1237–1250 (2017)
10. Chhabria, V.A., Sapatnekar, S.S.: OpeNPDN (2018). https://github.com/The-OpenROAD-Project/OpeNPDN
11. Chhabria, V.A., Sapatnekar, S.S.: OpeNPDN: a neural-network-based framework for power delivery network synthesis. IEEE Trans. Comput. Aided Des. Integr. Circuits Syst. **41**(10), 3515–3528 (2022)
12. Chhabria, V.A., Sapatnekar, S.S.: PDNSim (2021). https://github.com/The-OpenROAD-Project/OpenROAD/tree/master/src/psm
13. Chhabria, V.A., Kahng, A.B., Kim, M., Mallappa, U., Sapatnekar, S.S., Xu, B.: Template-based PDN synthesis in floorplan and placement using classifier and CNN techniques. In: Proceedings of the Asia-South Pacific Design Automation Conference, pp. 44–49 (2020)
14. Chhabria, V.A., Ahuja, V., Prabhu, A., Patil, N., Jain, P., Sapatnekar, S.S.: Encoder-decoder networks for analyzing thermal and power delivery networks. ACM Trans. Des. Autom. Electron. Syst. (in press)
15. Chhabria, V.A., Ahuja, V., Prabhu, A., Patil, N., Jain, P., Sapatnekar, S.S.: Thermal and IR drop analysis using convolutional encoder-decoder networks. In: Proceedings of the Asia-South Pacific Design Automation Conference, pp. 690–696 (2021)
16. Chhabria, V.A., Kunal, K., Zabihi, M., Sapatnekar, S.S.: BeGAN-benchmarks (2021). https://github.com/UMN-EDA/BeGAN-benchmarks
17. Chhabria, V.A., Kunal, K., Zabihi, M., Sapatnekar, S.S.: BeGAN: Power grid benchmark generation using a process-portable GAN-based methodology. In: Proceedings of the IEEE/ACM International Conference on Computer-Aided Design (2021)
18. Chhabria, V.A., Zhang, Y., Ren, H., Keller, B., Khailany, B., Sapatnekar, S.S.: MAVIREC: ML-aided vectored IR-drop estimation and classification. In: Proceedings of the Design, Automation & Test in Europe, pp. 1825–1828 (2021)
19. Chiprout, E.: Fast flip-chip power grid analysis via locality and grid shells. In: Proceedings of the IEEE/ACM International Conference on Computer-Aided Design, pp. 485–488 (2004)
20. Dey, S., Nandi, S., Trivedi, G.: PowerPlanningDL: reliability-aware framework for on-chip power grid design using deep learning. In: Proceedings of the Design, Automation & Test in Europe, pp. 1520–1525 (2020)

21. Dumoulin, V., Visin, F.: A guide to convolution arithmetic for deep learning. arXiv:1603.07285 (2016)
22. Fang, Y.C., Lin, H.Y., Su, M.Y., Li, C.M., Fang, E.J.W.: Machine-learning-based dynamic IR drop prediction for ECO. In: Proceedings of the IEEE/ACM International Conference on Computer-Aided Design (2018)
23. Goodfellow, I.J., Pouget-Abadie, J., Mirza, M., Xu, B., Warde-Farley, D., Ozair, S., Courville, A., Bengio, Y.: Generative adversarial networks. In: Proceedings of the Conference on Neural Information Processing Systems, pp. 2672–2680 (2014)
24. Ho, C.T., Kahng, A.B.: IncPIRD: fast learning-based prediction of incremental IR drop. In: Proceedings of the IEEE/ACM International Conference on Computer-Aided Design (2019)
25. Hochreiter, S., Schmidhuber, J.: Long short-term memory. Neural Comput. **9**(8), 1735–1780 (1997)
26. Hsu, H., Chen, M., Chen, H., Li, H., Chen, S.: On effective flip-chip routing via pseudo single redistribution layer. In: Proceedings of the Design, Automation & Test in Europe, pp. 1597–1602 (2012)
27. Jakushokas, R., Friedman, E.G.: Methodology for multi-layer interdigitated power and ground network design. In: Proceedings of the IEEE International Symposium on Circuits and Systems, pp. 3208–3211 (2010)
28. Jin, W., Sadiqbatcha, S., Sun, Z., Zhou, H., Tan, S.X.D.: EM-GAN: data-driven fast stress analysis for multi-segment interconnects. In: Proceedings of the IEEE International Conference on Computer Design, pp. 296–303 (2020)
29. Jin, W., Sadiqbatcha, S., Zhang, J., Tan, S.X..D.: Full-chip thermal map estimation for commercial multi-core CPUs with generative adversarial learning. In: Proceedings of the IEEE/ACM International Conference on Computer-Aided Design (2020)
30. Juan, D., Zhou, H., Marculescu, D., Li, X.: A learning-based autoregressive model for fast transient thermal analysis of chip-multiprocessors. In: Proceedings of the Asia-South Pacific Design Automation Conference, pp. 597–602 (2012)
31. Karras, T., Aila, T., Laine, S., Lehtinen, J.: Progressive growing of GANs for improved quality, stability, and variation. In: Proceedings of the International Conference on Learning Research (2018)
32. Karras, T., Laine, S., Aila, T.: A style-based generator architecture for generative adversarial networks. In: Proceedings of the IEEE Conference on Computer Vision and Pattern Recognition, pp. 4396–4405 (2019)
33. Kingma, D., Ba, J.: Adam: a method for stochastic optimization. In: Proceedings of the International Conference on Learning Research (2015)
34. Kozhaya, J.N., Nassif, S.R., Najm, F.N.: Multigrid-like technique for power grid analysis. In: Proceedings of the IEEE/ACM International Conference on Computer-Aided Design, pp. 480–487 (2001)
35. LeCun, Y., Bottou, L., Bengio, Y., Haffner, P.: Gradient-based learning applied to document recognition. Proc. IEEE **86**(11), 2278–2324 (1998)
36. Li, P.: Power grid simulation via efficient sampling-based sensitivity analysis and hierarchical symbolic relaxation. In: Proceedings of the ACM/IEEE Design Automation Conference, pp. 664–669 (2005)
37. Li, P., Pileggi, L., Asheghi, M., Chandra, R.: IC thermal simulation and modeling via efficient multigrid-based approaches. IEEE Trans. Comput. Aided Desig. Integr. Circuits Syst. **25**(9), 1763–1776 (2006)
38. Li, Z., Balasubramanian, R., Liu, F., Nassif, S.: 2012 TAU power grid simulation contest: benchmark suite and results. In: Proceedings of the IEEE/ACM International Conference on Computer-Aided Design, pp. 478–481 (2012)
39. Lin, S.Y., Fang, Y.C., Li, Y.C., Liu, Y., Yang, T., Lin, S.C., Li, C.M.J., Fang, E.J.W.: IR drop prediction of ECO-revised circuits using machine learning. In: Proceedings of the VLSI Test Symposium (2018)
40. Mao, X.J., Shen, C., Yang, Y.B.: Image restoration using very deep convolutional encoder-decoder networks with symmetric skip connections. In: Proceedings of the Conference on Neural Information Processing Systems (2016)

41. Miyato, T., Kataoka, T., Koyama, M., Yoshida, Y.: Spectral normalization for generative adversarial networks. In: Proceedings of the International Conference on Learning Research (2018)
42. Müller, S., Schüler, L., Zech, A., Heße, F.: GSTools v1.3: a toolbox for geostatistical modelling in Python. Geosci. Model Dev. **15**, 3161–3182 (2022)
43. Nassif, S.: Power grid analysis benchmarks. In: Proceedings of the Asia-South Pacific Design Automation Conference, pp. 376–381 (2008)
44. Noguchi, A., Harada, T.: Image generation from small datasets via batch statistics adaptation. In: Proceedings of the IEEE International Conference on Computer Vision, pp. 2750–2758 (2019)
45. Oquab, M., Bottou, L., Laptev, I., Sivic, J.: Learning and transferring mid-level image representations using convolutional neural networks. In: Proceedings of the IEEE Conference on Computer Vision and Pattern Recognition, pp. 1717–1724 (2014)
46. Pan, S.J., Yang, Q.: A survey on transfer learning. IEEE Trans. Knowl. Data Eng. **22**(10), 1345–1359 (2010)
47. Ronneberger, O., Fischer, P., Brox, T.: U-Net: Convolutional networks for biomedical image segmentation. In: Proceedings of the International Conference on Medical Image Computing and Computer Assisted Intervention, pp. 234–241 (2015)
48. Sadiqbatcha, S., Zhang, J., Amrouch, H., Tan, S.X.D.: Real-time full-chip thermal tracking: a post-silicon, machine learning perspective. IEEE Trans. Comput. **71**(6), 1411–1424 (2022)
49. Shelhamer, E., Long, J., Darrell, T.: Fully convolutional networks for semantic segmentation. IEEE Trans. Pattern Anal. Mach. Intell. **39**(4), 640–651 (2017)
50. Shi, X., Chen, Z., Wang, H., Yeung, D.Y., Wong, W.K., Woo, W.C.: Convolutional LSTM network: a machine learning approach for precipitation nowcasting. In: Proceedings of the Conference on Neural Information Processing Systems (2015)
51. Singh, J., Sapatnekar, S.S.: Partition-based algorithm for power grid design using locality. IEEE Trans. Comput. Aided Des. Integr. Circuits Syst. **25**(4), 664–677 (2006)
52. Su, H., Hu, J., Sapatnekar, S.S., Nassif, S.R.: Congestion-driven codesign of power and signal networks. In: Proceedings of the ACM/IEEE Design Automation Conference, pp. 64–69 (2002)
53. Tan, X.D., Shi, C.R., Lungeanu, D., Lee, J.C., Yuan, L.P.: Reliability-constrained area optimization of VLSI power/ground networks via sequence of linear programmings. In: Proceedings of the ACM/IEEE Design Automation Conference, pp. 78–83 (1999)
54. The OpenROAD project (2022). https://github.com/The-OpenROAD-Project
55. Wei, Y., Sze, C., Viswanathan, N., Li, Z., Alpert, C.J., Reddy, L., Huber, A.D., Tellez, G.E., Keller, D., Sapatnekar, S.S.: GLARE: global and local wiring aware routability evaluation. In: Proceedings of the ACM/IEEE Design Automation Conference, pp. 768–773 (2012)
56. Wen, J., Pan, S., Chang, N., Chuang, W., Xia, W., Zhu, D., Kumar, A., Yang, E., Srinivasan, K., Li, Y.: DNN-based fast static on-chip thermal solver. In: Proceedings of the IEEE Symposium on Semiconductor Thermal Measurment, Modeling, and Management, pp. 65–75 (2020)
57. Wu, X., Hong, X., Cai, Y., Luo, Z., Cheng, C.K., Gu, J., Dai, W.: Area minimization of power distribution network using efficient nonlinear programming techniques. IEEE Trans. Comput. Aided Desig. Integr. Circuits Syst. **23**(7), 1086–1094 (2004)
58. Xie, Z., Ren, H., Khailany, B., Sheng, Y., Santosh, S., Hu, J., Chen, Y.: PowerNet: transferable dynamic IR drop estimation via maximum convolutional neural network. In: Proceedings of the Asia-South Pacific Design Automation Conference, pp. 13–18 (2020)
59. Yazdani, F.: Foundations of Heterogeneous Integration: An Industry-Based, 2.5D/3D Pathfinding and Co-Design Approach. Springer, Boston (2018)
60. Zhan, Y., Kumar, S.V., Sapatnekar, S.S.: Thermally-aware design. Found. Trends Electron. Desig. Automat. **2**(3), 255–370 (2008)
61. Zhang, K., Guliani, A., Ogrenci-Memik, S., Memik, G., Yoshii, K., Sankaran, R., Beckman, P.: Machine learning-based temperature prediction for runtime thermal management across system components. IEEE Trans. Parallel Distrib. Syst. **29**(2), 405–419 (2018)
62. Zhao, M., Panda, R., Sapatnekar, S., Blaauw, D.: Hierarchical analysis of power distribution networks. IEEE Trans. Comput. Aided Des. Integr. Circuits Syst. **21**(2), 159–168 (2002)

63. Zhao, M., Cong, Y., Carin, L.: On leveraging pretrained GANs for limited-data generation. In: International Conference on Machine Learning, pp. 11340–11351 (2020)
64. Zhong, Y., Wong, M.D.F.: Fast algorithms for IR drop analysis in large power grid. In: Proceedings of the IEEE/ACM International Conference on Computer-Aided Design, pp. 351–357 (2005)
65. Zhuang, H., Yu, W., Weng, S., Kang, I., Lin, J., Zhang, X., Coutts, R., Cheng, C.: Simulation algorithms with exponential integration for time-domain analysis of large-scale power delivery networks. IEEE Trans. Comput. Aided Des. Integr. Circuits Syst. 35(10), 1681–1694 (2016)

Chapter 6
Machine Learning for Testability Prediction

Yuzhe Ma

6.1 Introduction

As the technology node keeps scaling down, there are billions of transistors on a single die, which is much more prone to defects than ever. Therefore, the testing process for manufacturing defects is of great significance in the production cycle of integrated circuits (ICs), which accounts for the reliability and development cost. In those early ages of semiconductor industry, design steps and test steps were totally decoupled, which were handled separately by different groups. As the integration went up, it was realized that the decoupled design flow of function and testing suffered a lot from the test cost as well as the quality, i.e., poor testability. More testing concerns should be considered in design steps, which calls for design-for-testing (DFT) methodologies.

Testability is a measure of the effort of testing a circuit. Basically, circuit testing involves two aspects which are controlling internal signals from primary inputs and observing internal signals from primary outputs. In order to consider the effort of circuit testing in early design steps, firstly, engineers need to quantitatively measure the testability of the given design, which motivates various approaches for testability measurements, providing different ways to calculate a set of values to indicate the relative effort of testing. There are some classical testability measurements that have been widely used for a few decades, most of which are developed based on the expertise of designers to formulate a set of rules [1–3]. Similar to most engineering problems, these measurements also suffer from a trade-off between accuracy and efficiency.

Y. Ma (✉)
The Hong Kong University of Science and Technology (Guangzhou), Guangzhou, China
e-mail: yuzhema@ust.hk

© The Author(s), under exclusive license to Springer Nature Switzerland AG 2022
H. Ren, J. Hu (eds.), *Machine Learning Applications in Electronic Design
Automation*, https://doi.org/10.1007/978-3-031-13074-8_6

In recent years, machine learning (ML) and deep learning (DL) techniques have emerged as powerful solutions in many fields. In the field of electronic design automation (EDA), learning approaches also have been leveraged to derive knowledge from a large amount of historical data and provide predictions, estimations, or even content generation in various problems, including testability predication. This chapter will introduce typical approaches for testability measurements. Then a set of learning-based methods are introduced for testability-related prediction problems, including node-level testability prediction, fault coverage prediction, test cost prediction, and X-sensitivity prediction. In addition, several considerations on applying machine learning models for practical testability improvement will also be discussed.

6.2   Classical Testability Measurements

Testability of a design can be measured at different abstraction levels [4], while the most commonly seen measurement is performed on the gate level. The circuit testability contains two metrics, controllability and observability. Controllability indicates the difficulty of setting internal circuit lines to 0 or 1 by setting primary inputs (PIs) of a circuit, while observability indicates the difficulty of observing internal circuit lines through primary outputs (POs).

There are several types of methodologies of calculating quantitative values for the testability. Intuitively, the testability of a circuit is highly associated with its topology of component (gates) connection in a circuit as well as their types. Therefore, some algorithms have been proposed to derive the measurement by analyzing the topological information of the circuit, in which the test patterns (vectors) are not needed; thus, it is also called static measurements or approximate measurements. On one hand, these algorithms are in linear complexity; thus, it is computationally efficient and commonly used in various DFT applications. On the other hand, the obtained measurements by static analysis may not be accurate due to a lack of behavior information of the circuits under practical operating scenario. In contrast, more accurate measurements can be generated by simulating the circuit behavior with a large amount of test patterns, which may lead to a severe runtime overhead.

In this section, we will introduce more details regarding these two types of testability measurements.

6.2.1   Approximate Measurements

6.2.1.1   SCOAP

Sandia Controllability/Observability Analysis Program (SCOAP) is a classical approximate testability measurement [2]. It defines a set of rules for calculating the controllability and observability metrics, which applies to both sequential circuits and combinational circuits. For combinational measures, CC_0 and CC_1 denote the 0-controllability and 1-controllability, respectively, and CO denotes the observability. Similarly, SC_0, SC_1, SO denote three sequential measures accordingly. Basically, all the measures are represented as integers in SCOAP measurements, and higher numbers indicate poorer testability, i.e., more difficult to control or observe.

Controllability calculation proceeds forward from the primary inputs to the outputs. For combinational measures, the controllability of the output line of each gate is calculated as a function of the controllability of its input lines, which depends on the function type of the gate. Given a gate with n input ports, denote the i-th input port X_i and the output port Y. Essentially, the idea is to list the conditions for a successful observation or control, and then calculate the total effort. If there are multiple conditions to make a successful observation or control, select the one with minimum effort. Note that the value is increased by 1 for every pass through a gate in SCOAP combinational measurement.

To illustrate the principle, we first introduce the *controlling value* and *noncontrolling value* of a gate. The *controlling value* of a specific output of a gate is the value that can be assigned to only one input of a gate to determine the output of this gate. For instance, the controlling value of the OR gate to produce output 1 is 1, while the controlling value of AND gate to produce output 0 is 0. The output controllability of the produced output d can be derived as:

$$CC_k(Y) = \min CC_l(X_i) + 1, 1 \le i \le n, \quad (6.1)$$

where l is the controlling value and k is the output value.

The *noncontrolling value* of a specific output of a gate is the value that should be assigned to all the inputs to determine the output of the gate. For instance, the noncontrolling value of OR gate to produce output 0 is 0, while the noncontrolling value of AND gate to produce output 1 is 1. Then the output controllability of the produced output d can be derived using noncontrolling value as:

$$CC_k(Y) = \sum_{i=1}^{n} CC_m(X_i) + 1, \quad (6.2)$$

where m is the noncontrolling value and k is the output value. Table 6.1 summarizes the rules of the combinational controllability calculation for typical gates.

Sequential controllability measures are calculated following a similar fashion as combinational measures, except for a few differences. Instead of increasing the measures by 1 for every pass through a gate in combinational controllability, the sequential controllability is increased by 1 only when passing through a flip-flop. Take the D flip-flop as an example. There are four ports in the flip-flop, including the data input port D, clock port CLK, reset signal port $RESET$, and data output port Q. According to the principle of the flip-flop, the conditions of making output port as 0 or 1 can be enumerated. Thus, the rules of sequential controllability measures can be derived, as listed in Table 6.2.

Table 6.1 SCOAP combinational controllability calculation rules

Type	$CC_0(Y)$	$CC_1(Y)$
INV	$CC_1(X_1) + 1$	$CC_0(X_1) + 1$
AND	$\min_{1 \leq i \leq n} (CC_0(X_i)) + 1$	$\sum_{i=1}^{n} CC_1(X_i)$
OR	$\sum_{i=1}^{n} CC_0(X_i)$	$\min_{1 \leq i \leq n} (CC_1(X_i)) + 1$
NAND	$\sum_{i=1}^{n} CC_1(X_i)$	$\min_{1 \leq i \leq n} (CC_0(X_i)) + 1$
NOR	$\min_{1 \leq i \leq n} (CC_1(X_i)) + 1$	$\sum_{i=1}^{n} CC_0(X_i)$
XOR	$\min\{\sum_{i=1}^{n} CC_0(X_i), \sum_{i=1}^{n} CC_1(X_i)\} + 1$	$\min\{CC_0(X_1) + CC_1(X_2),$ $CC_1(X_1) + CC_0(X_2)\} + 1$
XNOR	$\min\{CC_0(X_1) + CC_1(X_2),$ $CC_1(X_1) + CC_0(X_2)\} + 1$	$\min\{\sum_{i=1}^{n} CC_0(X_i), \sum_{i=1}^{n} CC_1(X_i)\} + 1$

Table 6.2 SCOAP sequential controllability calculation rules

Port	SC_0	SC_1
Q	$\min\{SC_1(RESET) + SC_0(CLK),$ $SC_0(D) + SC_1(CLK) +$ $SC_0(CLK) + SC_0(RESET)\} + 1$	$SC_1(D) + SC_1(CLK) +$ $SC_0(CLK) + SC_0(RESET) + 1$

Different from the controllability measures which proceeds from primary input to primary output, the observability calculation proceeds backward from the primary outputs to the inputs. Essentially, it requires to make other input values noncontrolling and also observe the output. Thus, the observability calculation involves both observability values and controllability values. Table 6.3 summarizes the rules of the combinational observability calculation for typical gates.

The observability of sequential circuits can be derived in a similar manner. Table 6.4 summarizes the rules of the sequential observability calculation rules for a D flip-flop.

6.2.1.2 Random Testability

Another approach for testability measurement is random testability. Similar to SCOAP measurement, random testability measurement also contains three measures to represent controllability and observability. The difference is that random testabil-

Table 6.3 SCOAP combinational observability calculation rules

Type	$CO(X_j)$
INV	$CO(Y) + 1$
AND	$CO(Y) + \sum\limits_{1 \leq i \leq n, i \neq j} CC_1(X_i) + 1$
OR	$CO(Y) + \sum\limits_{1 \leq i \leq n, i \neq j} CC_0(X_i) + 1$
NAND	$CO(Y) + \sum\limits_{1 \leq i \leq n, i \neq j} CC_1(X_i) + 1$
NOR	$CO(Y) + \sum\limits_{1 \leq i \leq n, i \neq j} CC_0(X_i) + 1$
XOR	$CO(Y) + \min\limits_{1 \leq i \leq n, i \neq j} \{CC_0(X_i), CC_1(X_i)\} + 1$
XNOR	$CO(Y) + \min\limits_{1 \leq i \leq n, i \neq j} \{CC_0(X_i), CC_1(X_i)\} + 1$

Table 6.4 SCOAP sequential observability calculation rules

Port	SO
D	$SO(Q) + SC_1(CLK) + SC_0(CLK) + SC_0(RESET) + 1$
$RESET$	$SO(Q) + SC_1(Q) + SC_0(CLK) + 1$
CLK	$SO(Q) + SC_0(CLK) + SC_1(CLK) + SC_0(RESET) + \min\{SC_0(D) + SC_1(Q), SC_1(D) + SC_0(Q)\} + 1$

ity measurement uses signal probability to analyze the random testability of the circuit, which has several properties to apply. Firstly, the values are within the range of $[0, 1]$. The smaller a probability-based testability measure of a signal, the more difficult it is to control or observe the signal. Secondly, the sum of 0-controllability and 1-controllability of any port should be 1. Considering that the 0-controllability and 1-controllability of a primary input are assumed to be equal, both of them are set to 0.5. The observability of a primary output is set as 1. A simple method to calculate the probability-based testability measures is similar to the one used for calculating combinational testability measures in SCOAP, which can be derived by identifying the controlling value and noncontrolling value of a specific type of gate. However, the accumulation is replaced by multiplication.

6.2.2 Simulation-Based Measurements

It can be seen that the aforementioned measurements can be calculated using the topological information only, without any consideration of the practical scenarios. These kinds of static testability analysis are efficient, usually in linear time complexity, while the evaluation accuracy would be sacrificed. To capture the real behavior of the signal lines within a circuit, the stimulus vectors can be used such that the status of the internal signal lines can be better reflected. Simulation-based

Fig. 6.1 Simulation
signature

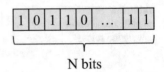

$$N \text{ bits}$$

measurement is such a way that uses real stimulus vectors, also known as test
patterns, for testability analysis.

Firstly, a set of test patterns is selected. The functional simulation drives the
circuit with the test patterns [5]; then the response of the internal nodes is collected,
which is also known as signature [6]. The number of occurrences of 0s, 1s, 0-to-
1 transitions, and 1-to-0 transitions can all be extracted from the signature. The
signature at each node $SIG(j)$ is a bit vector as shown in Fig. 6.1. With the collected
response or signature, the 1-controllability and 0-controllability can be calculated
as:

$$C_1 = \frac{Ones(SIG(j))}{N}, C_0 = \frac{Zeros(SIG(j))}{N}, \tag{6.3}$$

where $Ones(SIG(j))$ and $Zeros(SIG(j))$ are the number of 1s and 0s in the
signature, respectively. It can also be seen that $C_1 + C_0 = 1$. Very few transitions or
even no transitions might indicate that the signal is with poor controllability.

Unlike previous approaches, which have a single value representing observability, simulation-based methods estimate the observability with two values to
differentiate the difficulty of observing two signals, i.e., 0 and 1. A concept of
observability don't-care (ODC) is introduced to facilitate the calculation [6–8]. In
addition to the vector of simulation signature, another vector called *ODC mask* is
calculated, which is also a bit vector that has the same length as the simulation
signature. A bit in the ODC mask is set when the value of a node does not affect the
output for a specific input pattern, which indicates that the corresponding simulation
signature cannot be observed. For instance, an AND(a, b) circuit, port b is not
observable if $a = 0$. Therefore, if port a has a signature $SIG(a) = \{0, 1, 1, 0\}$,
the port b has an ODC mask $ODC(b) = \{0, 1, 1, 0\}$. Note that this is just a local
ODC mask for a node. A global ODC mask calculation requires considering all the
nets a node involves and the ODC mask of its fan-out. Therefore, a global ODC
mask can be calculated by reversely traversing the circuit. For each node input, a
local ODC mask is computed and then bitwise ANDed with the global ODC mask
at the node's output. Algorithm 6.1 presents the computation steps.

A detailed example is shown in Fig. 6.2. With ODC mask, the observability can
be estimated as

$$O_1 = \frac{Ones(ODC(j)\&SIG(j))}{Ones(SIG(j))}, O_0 = \frac{Ones(ODC(j)\&\neg SIG(j))}{Zeros(SIG(j))}. \tag{6.4}$$

Algorithm 6.1 ODC calculation

1: **for** net n in the circuit **do**
2:     **for** sink n_k of n **do**
3:         **if** n_k is PO or register input **then**
4:             $ODC(n_k) = \{1, ..., 1\}$;
5:         **else**
6:             $ODC(n_k) = ODC_L(n_k)$;
7:         **end if**
8:     **end for**
9: **end for**
10: **for** net n in the circuit in reversed topological order **do**
11:     **for** sink n_k of n **do**
12:         Find the output net m of n_k
13:         $ODC(n) = ODC(n)|(ODC_L(n_k)\&ODC(m))$
14:     **end for**
15: **end for**

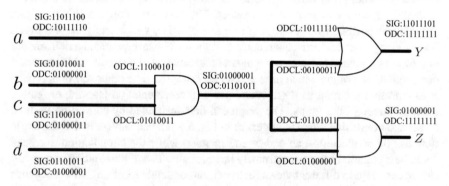

Fig. 6.2 An ODC example

6.3 Learning-Based Testability Prediction

Deep learning has been considered a powerful and promising technique for the next-generation circuit design tools [9, 10]. A deep learning model can provide accurate and fast prediction and even content generation to enable an agile circuit design flow by training on a large amount of historical data. To obtain a high-performance deep learning model, the task should be appropriately defined, including training data preparation, data representation, algorithm selection, and criteria for evaluation, among which data representation and algorithm selection are of great significance. There are various recent advances being presented in the deep learning domain within the context of circuit testing [11]. This section will cover typical learning-based methods for various testability prediction problems, including node-level testability prediction, fault coverage prediction, test cost prediction, and X-sensitivity prediction, where the first one is on node level and the remaining ones are on the circuit level.

6.3.1   Node-Level Testability Prediction

Predicting the testability of internal locations in a circuit is of great importance in circuit testing, which facilitates an efficient design flow for testability improvement. Since gate attributes and circuit structure are both taken into consideration in calculating the approximate testability measurements as introduced before, e.g., COP [3] and SCOAP [2], they are also investigated as an effective way of circuit feature extraction for testability prediction on internal locations.

When dealing with circuits in a deep learning problem, the representation has always been a major challenge in achieving extraordinary performance. It can be noticed from the traditional testability measurements that the testability of a circuit is closely related to its structure and various characteristics, e.g., the number of primary inputs and primary outputs, logic level, the types of used cells, and the connection among the cells. To form an informative representation of a circuit, typically there are two sorts of approaches. The first sort of approach is to extract the relevant attributes by leveraging human expertise, which can facilitate future feature analysis and improvement correspondingly. However, there is a rich amount of attributes contained in each component (gate or wire) of a circuit, which makes it nontrivial to determine which ones should be extracted. The other sort of approach is to represent a circuit as a graph and perform representation learning or feature learning directly on a graph. The graph can be constructed by converting ports to nodes and converting wires to edges, as in Fig. 6.3. The advantage is that the entire flow can be conducted in an end-to-end manner, while the computation overhead could be a potential issue for extremely large graphs. The representation of a circuit can be extracted in different views (entire circuit or single location), which depends on the particular downstream task.

6.3.1.1   Conventional Machine Learning Methods

In [13, 14], the COP measurements and the gate types are combined as the extracted features. Firstly, each gate is associated with several attributes containing the

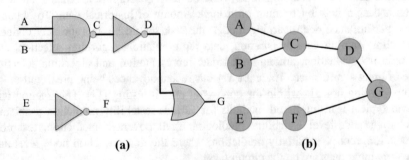

Fig. 6.3 Convert a circuit to a graph [12]. (**a**) A circuit; (**b**) The graph representation

corresponding COP measurements and a one-hot encoding denoting its gate type, such that each type of gate has a unique encoding. Thus, the length of the attribute vector for each gate depends on the number of gate types in the circuit. Next, the circuit structure is encoded by combining the attributes of a sub-circuit centered around a gate, which is essentially its fan-in and fan-out cones. By traversing each sub-circuit in a breadth-first manner, the closely located neighborhood nodes are visited, whose attributes are concatenated together to form the feature representation of the node of interest. Note that the maximum number of visited nodes in the fan-in cone and fan-out cone can be adjusted, represented by N_{fi} and N_{fo}, respectively. The workflow is illustrated in Algorithm 6.2.

A concrete example is given in Fig. 6.4. Each node in the graph is associated with SCOAP values as its attribute. In addition, a logic level value is also included. Denote the attribute of node i as x_i. It can be represented as

$$x_i = [CC_0(i), CC_1(i), O(i), LL(i)], \tag{6.5}$$

Algorithm 6.2 Neighborhood encoding

Require: Graph $G(V, E)$; Node attribute $x_0, x_1, \cdots, x_{|V|-1}$; Maximum number of visited nodes N_{fi}, N_{fo};
Ensure: Feature representation of node in G;
    for node i in graph G **do**
        $X_i \leftarrow x_i$;
        $G_{fi} \leftarrow$ fan-in cone of i;
        $G_{fo} \leftarrow$ fan-out cone of i;
        $Q_{fi} \leftarrow$ Breadth-First-Search(G_{fi}, i, N_{fi});
        $Q_{fo} \leftarrow$ Breadth-First-Search(G_{fo}, i, N_{fo});
        for node j in $Q_{fi} \cup Q_{fo}$ **do**
            $X_i \leftarrow$ Concatenate(X_i, x_j)
        end for
    end for

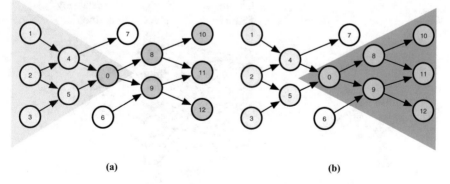

(a)   (b)

Fig. 6.4 (a) Fan-in cone of the node-0; (b) fan-out cone of the node-0

where $CC_0(i)$, $CC_1(i)$, $O(i)$, $LL(i)$ represent 0-controllability, 1-controllability, observability, and logic level, respectively.

Suppose node-0 is the location of interest, whose local neighborhood will be visited for feature extraction. Starting with a node-0, a breadth-first-search is performed in the fan-in cone first until a maximum of N_{fi}-gates have been visited. If the gates in the fan-in cone are fewer than N_{fi}, some padding strategies can be utilized. In Fig. 6.4, five gates are included here and ordered as "4, 5, 1, 2, 3". Similarly, a breadth-first-search is performed in the fan-out cone until N_{fo}-gates have been visited. Also, five gates are visited with the order "8, 9, 10, 11, 12".

Following the visiting order of the gates, their attributes are concatenated to form a long vector as the final feature representation X:

$$X_0 = [x_0, x_4, x_5, x_1, x_2, x_3, x_8, x_9, x_{10}, x_{11}, x_{12}], \tag{6.6}$$

which has a length of $L \times (N_{fi} + N_{fo} + 1)$, where L is the number of the attributes of each node. In the example Fig. 6.4, $N_{fi} = N_{fo} = 5$ and $L = 4$.

Next, the extracted feature representation will be fed into a learning model. Artificial neural networks (ANNs), more commonly known as neural networks (NNs), are computing systems inspired by the biological neural networks that constitute animal brains. Essentially, it is composed of input layers, hidden layers, and output layers, as shown in Fig. 6.5. Denote the input feature of layer l as $x^{(l)} \in \mathbb{R}^{D_l}$, the weight of layer l as $\mathbf{W} \in \mathbb{R}^{D_l \times D_{l+1}}$. In each layer, a nonlinear transformation is performed, including a linear projection and an activation operation, which can be formulated as follows:

$$x^{(l+1)} = \sigma(\mathbf{W}x^{(l)}), \tag{6.7}$$

where $\sigma(\cdot)$ is the activation function, e.g., ReLU [15]:

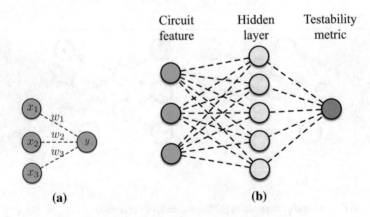

Fig. 6.5 (a) Example of a single neuron; (b) typical neural network

Table 6.5 Accuracy of ANN model

Design	arbiter	b15	b17	b18	b19	b21	log2	memctrl	sin	s38584
#Nodes	12,223	14,712	27,814	82,528	12,489	13,073	25,970	49,269	19,026	13,894
Valid. loss	0.085	0.052	0.025	0.048	0.046	0.029	0.03	0.049	0.091	0.052
Test loss	0.271	0.061	0.168	0.142	0.048	0.125	0.149	0.157	0.095	0.059

$$\sigma(x_i) = \text{ReLU}(x_i) = \begin{cases} x_i, & \text{if } x_i \geq 0; \\ 0, & \text{otherwise.} \end{cases} \tag{6.8}$$

The performance of ANN model on predicting the fault coverage change of each position in a netlist is listed in Table 6.5 according to [16]. The ANN model contains an input layer, a hidden layer with a dimension of 128, and an output layer. It is trained using the mean square error (MSE) loss that is to minimize the difference between the real fault coverage change and the predicted one of each node, which can be written as

$$MSE = \frac{1}{N} \sum_{i=1}^{N} \left(y_{real}^{(i)} - y_{predict}^{(i)} \right)^2, \tag{6.9}$$

where N is the total number of training data samples. $y_{real}^{(i)}$ and $y_{predict}^{(i)}$ are the real fault coverage change and predicted fault coverage change resulted by the i-th sample, respectively. Eight netlists are used for training, one netlist is used for validation, and one netlist is used for testing. It can be observed in Table 6.5 that the performance of ANN model can achieve fairly accurate predictions with the MSE being as small as 5%.

6.3.1.2 Graph-Based Deep Learning Methods

Instead of performing feature extraction and encoding manually, deep learning has been demonstrated to ease the manual efforts for feature extraction since it can take high-dimensional input and perform automatic feature extraction during training. Typically, testability analysis or prediction is conducted on the netlist that is more naturally modeled as a graph. There are a few issues in performing conventional learning tasks on a graph. Particularly, it is not straightforward to transfer conventional grid-based operations (e.g., convolution, pooling, etc.) in convolutional neural networks to tackle irregularly structured data like graphs. Recently graph neural networks (GNNs) have been considered as a promising approach for building machine learning models on a graph [17]. Before performing a particular task, a representation of a node or graph should be obtained first, which is known as embedding and can be fed to downstream models. There are a collection of representative approaches that generate node embeddings by leveraging node feature information from the neighborhood [18–20]. Inspired by these strategies, GNN for testability analysis has also been explored [16, 21].

Model Structure In [21], a methodology using graph neural networks for netlist representation and testability prediction is proposed. A netlist is represented as a directed graph in which each node and each edge correspond to a gate and a wire, respectively. In addition, the source nodes and sink nodes correspond to primary inputs and primary outputs, respectively. The predicted metric is a binary value that indicates that a location is easy to test or difficult to test. The workflow is shown in Fig. 6.6. To classify a node in the graph, a graph neural network first generates its node embedding, which is not only based on its attributes but also the structural information of the local neighborhood. Then, a classification model takes the node embedding as input and predicts a label. To achieve this, three kinds of modules are designed in the graph neural network, which are aggregator, encoder, and classifier. Aggregators and encoders are used to generate the node embeddings by exploiting node attributes and the neighborhood information. The classifier predicts the label for each node in the graph based on its embedding, which is a set of fully connected layers in Fig. 6.6.

Denote a graph $\mathcal{G}(\mathcal{V}, \mathcal{E})$, node attributes $\{x^{(v)} : \forall v \in \mathcal{V}\}$, the node embeddings $\{e^{(v)} : \forall v \in \mathcal{V}\}$ are generated as in Algorithm 6.3. The local neighborhood information is aggregated by a weighted sum function,

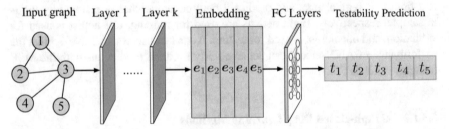

Fig. 6.6 Graph neural network for testability prediction. $t_1, t_2, ..., t_5$ are the predicted testability metrics for each node

---

Algorithm 6.3 Node embedding computation

Require: Graph $\mathcal{G}(\mathcal{V}, \mathcal{E})$; node attributes $\{x^{(v)} : \forall v \in \mathcal{V}\}$; Search depth D; non-linear activation function $\sigma(\cdot)$; Weight matrices W_d of encoders $E_d, d = 1, ..., D$;

Ensure: Embedding of for each node $e_D^{(v)}, \forall v \in \mathcal{V}$.

1: $e_0^{(v)} \leftarrow x^{(v)}, \forall v \in \mathcal{V}$;
2: **for** $d = 1, ..., D$ **do**
3:   **for all** $v \in \mathcal{V}$ **do**
4:     $g_d^{(v)} = e_{d-1}^{(v)} + w_{pr} \times \sum_{u \in \text{PR}(v)} e_{d-1}^{(u)} + w_{su} \times \sum_{u \in \text{SU}(v)} e_{d-1}^{(u)}.$

5:     $e_d^{(v)} \leftarrow \sigma(W_d \cdot g_d^{(v)});$
6:   **end for**
7: **end for**

Table 6.6 Accuracy comparison on balanced dataset

Design	LR	RF	SVM	MLP	GCN
B1	0.778	0.790	0.813	0.860	0.928
B2	0.767	0.785	0.809	0.845	0.929
B3	0.779	0.793	0.814	0.856	0.930
B4	0.782	0.801	0.821	0.862	0.935
Average	0.777	0.792	0.814	0.856	0.931

$$g_d^{(v)} = e_{d-1}^{(v)} + w_{pr} \times \sum_{u \in PR(v)} e_{d-1}^{(u)} + w_{su} \times \sum_{u \in SU(v)} e_{d-1}^{(u)}, \qquad (6.10)$$

where w_{pr} and w_{su} are the weights for predecessors and successors, respectively. Next, a nonlinear transformation is performed to encode the aggregated representation using a weight matrix $W_d \in \mathbb{R}^{K_{d-1} \times K_d}$ and an activation function $\sigma(\cdot)$, which is a feature projection step similar to conventional neural networks. Essentially, after d iterations, the embedding of a node combines the information of its d-hop neighborhood. Usually, it is pre-defined how large a local neighborhood should be considered. When maximum depth D is reached, the final embeddings are obtained and fed to the fully connected layers for classification. Parameters that need to be trained include w_{pr}, w_{su}, $W_1, \ldots, W_D$, and parameters in fully connected layers.

Performance Demonstration The graph neural network shows good performance for testability prediction [21]. The experiments are performed on four industrial designs implemented in 12 nm technology. Table 6.6 lists performance comparison among GNN and various classical machine learning models, including logistic regression (LR), random forest (RF), support vector machine (SVM) and multi-layer perceptron (MLP), in which the feature representation for each node is extracted based on Algorithm 6.2 introduced in Sect. 6.3.1. The maximum number of visited nodes is 500 for both fan-in cone and fan-out cone, i.e., $N_{fi} = N_{fo} = 500$. The accuracy is obtained on a balanced dataset, which aims to show the representation capability of different models. It can be observed that GCN achieves 93.1% accuracy on average, outperforming other conventional machine learning models on all designs.

The region of the "local neighborhood" is vital to the performance of a graph neural network. On the one hand, a larger region can encode more information locally. On the other hand, it may lead to over-fitting and over-smoothing if the region is too large. In typical GNN models [18, 21], the number of layers actually corresponds to the local search depth. More specifically, a k-layer GNN aggregates the information from k-hop neighborhood to generate the node embedding. For the testability prediction task [21], the record of training accuracy and validation accuracy during learning for 300 epochs with different local regions is recorded, as shown in Fig. 6.7. It can be seen that the performance of the GNN model improves to some extent as the search depth increases.

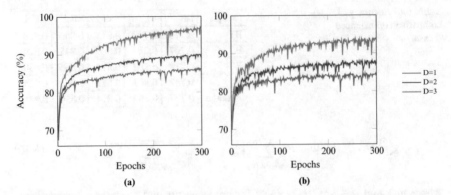

Fig. 6.7 Performance with different search depths [21]. (**a**) Training accuracy. (**b**) Testing accuracy

6.3.2 Circuit-Level Testability Prediction

Apart from predicting testability on the node level, there are prior arts measuring the testability on an entire circuit [22–24] using machine learning approaches, which aim at testability metrics defined on the entire circuit, e.g., fault coverage, test cost, X-sensitivity, etc.

6.3.2.1 Fault Coverage Prediction

Fault coverage is the ability of a collection of test patterns to detect a given fault that may occur on the circuit, which can be formulated as follows:

$$\text{Fault coverage} = \frac{\text{\#detected faults}}{\text{\#possible faults}} \qquad (6.11)$$

A learning-based approach was presented to predict the fault coverage [23]. Generally, two groups of features or attributes should be taken into consideration for testability prediction, which should cover both component characteristics and circuit structures. In [23], there are in total 17 parameters that are used as the representation, i.e., a vector with a length of 17. Most of these parameters are highly correlated with the testability, while some may need additional explanation. Naturally, the number of loops in a circuit must be taken into account. The number of flip-flops per loop (FF/L) gives an idea about the cyclic structure of the design, while the number of loops per flip-flop (L/FF) indicates how different loops are tied together in a flip-flop. A SAD gate in a sequential circuit is that the gate is neither self-hidden nor delay reconvergence. A D flip-flop is non-SAD if the D flip-flop is on paths that pass through a different number of delay elements and have different inversion parities and reconverge. These attributes are listed in Table 6.7 and are used to

Table 6.7 Features proposed in [23]

Notation	Meaning	Notation	Meaning	Notation	Meaning
FF	# of FFs	G	# of gates	PI	# of PIs
PO	# of POs	FF/L	# of FFs per loop	L/FF	# of loops per FF
L	# of loops	FS	# of fan-out stems	TF	# of possible faults
CON	# of interconnection	NOG	# of non observable gates	NOFF	# of non-observable flip-flops
SD	sequential depth	PFB	# of prime fan-out branches	SAPI	# of PIs controlling circuits state
NSG	# of non-SAD gates	NSFF	# of non-SAD FFs		

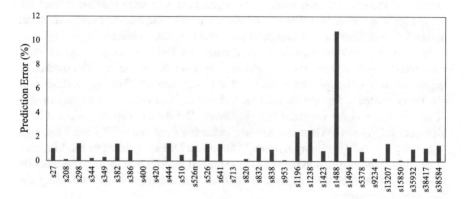

Fig. 6.8 Fault coverage prediction error on ISCAS'89 benchmarks [23]

predict the fault coverage and the number of test patterns on the entire circuit with neural network models.

The input is a flatten vector containing all the characteristics extracted from the entire circuit (Table 6.7). ANN model has been applied to construct the prediction model [23], whose basic structure is introduced in Sect. 6.3.1.1. The performance of ANN-based fault coverage prediction on ISCAS'89 benchmark circuits is shown in Fig. 6.8. It can be seen that NN can obtain a pretty accurate prediction for fault coverage on most of the circuits, whose error can be as low as less than 0.1%. However, a large variance can also be observed. The error can be greater than 2% or even 10% on some designs.

6.3.2.2 Test Cost Prediction

As the design complexity keeps growing, the test cost is increasing tremendously as well in terms of the test-data volume and test-application time, which makes it very difficult to comply with the budgets of test cost with automatic test pattern generation (ATPG). One of the commonly applied techniques of reducing the test cost is scan compression [25–27]. Typically, the architecture of a scan compression can be parameterized using a set of primary attributes, e.g., the length of pseudorandom pattern generator (PRPG), the length of scan-chain, etc.

It has been shown that different parameters can lead to significantly different results on test cost. For instance, the difference between the maximum and the minimum test-data volume and test-application time can reach as much as 25% with different lengths of PRPG [22], indicating that a good selection of the parameters can reduce the test cost substantially. One can always select the best parameters with running ATPG exhaustively, while it is not practical with large designs. Therefore, a machine learning model that can efficiently predict the test cost under different parameters will facilitate the design of scan compression architectures.

In [22], the task is defined as predicting the test cost under a given scan compression architecture that is represented by a set of parameters. The prediction target contains test-application time and test-data volume. Test-application time can be measured using the number of test cycles, and test-data volume consists of test stimuli and the expected test responses. The dataset can be collected from historical ATPG runs. Then features are extracted from the ATPG log files, e.g., "#primary inputs," "#primary outputs," "#nonscan DFFs," "#nonscan DLATs," etc. To maximize the correlation between the extracted features and the target prediction value, a feature selection scheme is applied. The criterion is defined using the following formulation:

$$\text{Merit}_s = \frac{k\overline{r_{tf}}}{\sqrt{k + k(k - 1)\overline{r_{ff}}}}, \tag{6.12}$$

where Merit_s is the defined merit of a feature subset S containing k features. r_{tf} is the average value of the Pearson's correlation coefficient [28] between each individual feature and the target value, and r_{ff} is the average value of the Pearson's correlation coefficient between feature pairs.

A support vector machine (SVM) is leveraged for the test cost prediction [22], which is to find an optimal hyperplane to fit data points within a certain margin. The samples whose prediction is more than ϵ away from the ground truth will be penalized. The penalties are ζ and ζ^* for the points lying above and below the tube, respectively.

$$\min_{w,b,\zeta,\zeta^*} C \sum_{i=1}^{n} \left(\zeta_i + \zeta_i^*\right) + \frac{1}{2} w^T w$$

$$\text{s.t. } y_i - w^T x_i - b \leq \varepsilon + \zeta_i \qquad (6.13)$$

$$w^T x_i + b - y_i \leq \varepsilon + \zeta_i^*$$

$$\zeta_i, \zeta_i^* \geq 0, i = 1, \dots, n$$

These variables are determined during training. w and b are coefficients and bias of the hyperplane, respectively. $\zeta_i(\zeta_i^*)$ denotes the error for the i-th data sample. It is commonly seen that ζ_i and ζ_i^* are set to be equal. The objective function consists of error minimization (the first term) and L_2 regularization (the second term). C is a weighting coefficient between two terms.

The prediction results of test cost using SVM are presented in Fig. 6.9. Two metrics are applied, including average relative error (ARE) and correlation coefficient

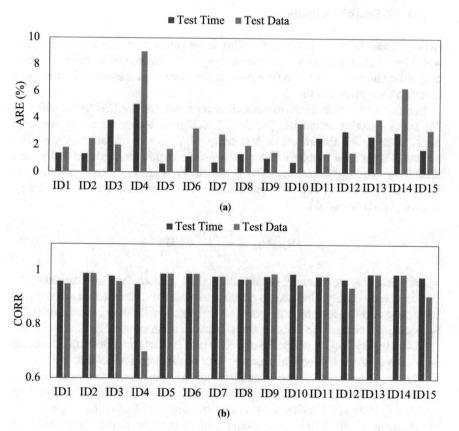

Fig. 6.9 Prediction results of test cost [22]: (**a**) average relative error; (**b**) correlation coefficient

(CORR), which are defined as

$$ARE = \frac{1}{N} \sum_{i=1}^{N} \left(|e_i - p_i| / e_i \right),$$ (6.14)

$$CORR = \frac{\sum_{i=1}^{N} (p_i - \bar{p})(e_i - \bar{e})}{\sqrt{\sum_{i=1}^{N} (p_i - \bar{p})^2 \sum_{i=1}^{N} (e_i - \bar{e})^2}},$$ (6.15)

where p_i and e_i represent the predicted value and the ground truth, respectively. Variables $\bar{p}$ and $\bar{e}$ are the mean value for p_i and e_i, respectively. N is the number of designs used for evaluation, which is 15 in [22]. As shown in Fig. 6.9, ARE is less than 5%, and CORR is larger than 0.9 for most of the designs, which demonstrates the high accuracy of the SVM-based predictor.

6.3.2.3 X-Sensitivity Prediction

Another scenario that affects the testability is the presence of unknown (X) values, which can be caused by tri-state elements, unspecified inputs, etc. Since these values cannot be computed during ATPG runs, it may cause an inevitable loss in fault coverage for a given test set.

In order to deal with the X-sources, there are handcrafted ATPG techniques [29, 30] and X-masking techniques [31–34]. To achieve effective testability enhancement, investigating the effects or sensitivity of X-sources on fault coverage has become a significant problem. The most sensitive should be identified and prioritized for X-masking/elimination. The X-sensitivity of an X-source s_i is defined as the loss of detectability due to the presence of an X-value at s_i, which is denoted by $D_{Loss}(s_i)$ and computed by

$$D_{Loss}(s_i) = \frac{|UF_i|}{|F|} \times 100\%,$$ (6.16)

where F and UF_i are the entire set of faults and the undetectable faults caused by X-source s_i, respectively. One way to estimate the loss for different X-sources would be running ATPG. However, this would be computationally time-consuming. It has been studied that machine learning can provide an accurate and efficient prediction for X-sensitivity [24], in which the objective is to estimate the impact of X-sources on coverage based on structural characteristics of the circuit-under-test.

The circuit-under-test is partitioned into three disjoint sets before extracting the features from the circuit. Partition P_1 consists of output cone of an X-source s_i. Partition P_2 is the sub-circuit that can propagate a signal to P_1. Partition P_3 is the rest of the circuit. Figure 6.10 gives an example of the partition result. The features are extracted for each vertex in each cone, including the level, depth, X-depth, etc. The

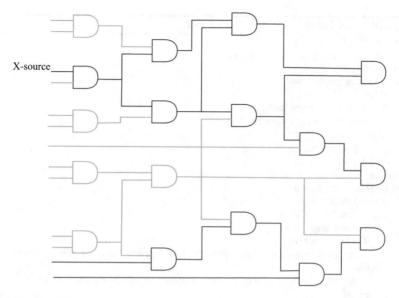

Fig. 6.10 A partition example [24]. P_1 is colored red. P_2 is colored green. P_3 is colored blue

level of a vertex is the length of the longest path through which it is reachable from any input port, while the depth of a vertex is the length of the shortest path to it from an input port. Considering that the INV, BUF, XOR, and XNOR are X-insensitive, which let an X-signal to pass whenever it appears at input, such elements may be ignored from the structure of the circuit when analyzing the D_{Loss}. In this context, *node* refers to the vertices that do not belong to these types. The level and depth computation procedure is presented in Algorithm 6.4, which is based on the traversal techniques. Then X-depth is computed, which is the length of the shortest path from the X-source. It reflects the topological relation with the X-source. Only the X-cone of the X-source needs to be traversed. The time complexity for computing the features is $\mathcal{O}(|E| + |V|)$. The procedure is introduced in Algorithm 6.5.

Similar to test cost prediction introduced in Sect. 6.3.2.2, SVM model is utilized for X-sensitivity prediction [24]. The label is obtained from the ATPG tool, TetraMAX [35]. The circuits from the ISCAS'89 and ITC'99 benchmark are used for evaluation. To measure the prediction performance, the coefficient of determination (CD) is defined as

$$CD = 1 - \frac{\sum_{i=1}^{N} \left(y_i - \hat{y}_i \right)^2}{\sum_{i=1}^{N} \left(y_i - \bar{y} \right)^2}, \tag{6.17}$$

where y_i and $\hat{y}_i$ are real D_{Loss} and predicted D_{Loss}, respectively. $\bar{y}$ is the mean of the y_i. The CD values for the different circuits are shown in Fig. 6.11, which shows that the SVM has better CD than linear regression for all the circuits. It can also

---

Algorithm 6.4 Level and depth computation

Require: Graph $G(V, E)$;
Ensure: Level and depth of the nodes in G;
  Initialize a queue Q;
  for node v in graph G **do**
      $v.level \leftarrow 0$;
      if v is a primary input **then**
          $v.depth \leftarrow 0$, $Q.enqueue(v)$
      else
          $v.depth \leftarrow \infty$
      end if
  end for
  while Q is not empty **do**
      $v \leftarrow Q.dequeue()$
      for each child w of v **do**
          Remove edge (v, w);
          if w is a node **then**
              $w.level \leftarrow \max(w.level, v.level + 1)$, $w.depth \leftarrow \min(w.depth, v.depth + 1)$
          else
              $w.level \leftarrow \max(w.level, v.level)$, $w.depth \leftarrow \min(w.depth, v.depth)$
          end if
          if indegree of w is 0 **then**
              $Q.enqueue(w)$
          end if
      end for
  end while

Algorithm 6.5 X-depth computation

Require: Graph $G(V, E)$, X-source s;
Ensure: X-depth of the vertices in the X-Cone of an X-source s;
  Initialize a queue Q;
  $s.x\_depth \leftarrow 0$, $Q.enqueue(s)$
  while Q is not empty **do**
      $q \leftarrow Q.dequeue()$;
      SUB_BFS(q, Q)
  end while
  procedure SUB_BFS(q, Q)
      for each child r of q **do**
          if $r.visited$ is False **then**
              $r.visited \leftarrow$ True
              if r is a node **then**
                  $r.x\_depth \leftarrow q.x\_depth + 1$;
                  $Q.enqueue(r)$;
              else
                  $r.x\_depth \leftarrow q.x\_depth$;
                  SUB_BFS(r, Q)
              end if
          end if
      end for
  end procedure

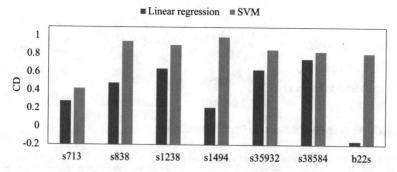

Fig. 6.11 X-sensitivity prediction with linear regression and SVM [24]

be observed that the variance of the prediction performance of linear regression is substantial. The CD can even become negative on some circuits, which indicates that the underlying correlation between the target and features is highly nonlinear. In contrast, SVM can achieve nearly a CD of 0.8 on most designs.

6.4 Additional Considerations

6.4.1 Imbalanced Dataset

For a typical design, it is common to have many more negative nodes than positive nodes, which is not desirable for training machine learning models. Training a single classification model can lead to poor overall performance since significant bias would be introduced toward the majority class. Similar issues also occur in many other machine learning applications in EDA, including lithography hotspot detection [36, 37], design rule violation prediction [38, 39], routability prediction [40], etc. Typical ways for dealing with the imbalance issue include enlarging datasets, generating synthetic samples, data augmentation, and changing performance metrics.

In [36], a deep bias learning algorithm is proposed to handle the imbalanced issue. A bias value ϵ is applied to the label of the training data, i.e., the one-hot encoding of the positive and negative label becomes $[1 - \epsilon, \epsilon]$ and $[\epsilon, 1 - \epsilon]$, respectively. In [37], surrogate loss functions for direct area-under-the-ROC-curve (AUC) maximization are proposed as a substitute for the conventional cross-entropy loss, which provides a more holistic measure for imbalanced datasets. In [38, 39], the neural network model is modified by assigning weights to the instances of the minority class in the loss function to ease the bias toward the majority class.

For the testability prediction problem, a multistage classification approach is proposed to handle the data imbalance issues [21]. In each stage, a graph neural network is trained and only filters out negative cases with high confidence and

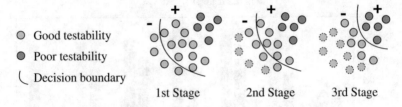

Fig. 6.12 An example of three-stage GNN classification

Table 6.8 F1-score comparison on imbalanced dataset

Design	#POS	#NEG	GNN			GNN-MS		
			Precision	Recall	F1-score	Precision	Recall	F1-score
B1-imb.	8894	1,375,370	0.096	0.909	0.174	0.568	0.742	0.643
B2-imb.	9755	1,446,698	0.112	0.916	0.199	0.541	0.746	0.627
B3-imb.	9043	1,407,338	0.102	0.923	0.184	0.529	0.730	0.613
B4-imb.	8978	1,388,608	0.099	0.923	0.179	0.555	0.759	0.641

passes the remaining nodes to the next stage, which is illustrated in Fig. 6.12. This is achieved by imposing a large weight on the positive nodes such that the penalty of misclassifying them would be large. In this way, most positive points remain on the right side of the decision boundary until negative points are substantially reduced. After a few stages, the remaining nodes should become relatively balanced, and a network can make the final predictions.

In highly imbalanced classification, *F1-score* is commonly used since accuracy would be misleading. Statistics of designs are summarized in Table 6.8. #POS and #NEG indicate the number of difficult-to-observe nodes and easy-to-observe nodes, respectively. Table 6.8 shows a three-stage classification results, in which GNN-MS refers to multistage GNN model. The prediction results of each stage are combined to obtain the F1-score on the entire dataset. It can be observed that multistage GNN achieves much higher F1-scores than single-stage GNN on all of these imbalanced datasets.

6.4.2 Scalability of Graph Neural Networks

Unlike conventional graph learning tasks, graph learning for EDA problems is prone to runtime overhead considering that the scale of circuits keeps soaring. Similar to conventional CNNs, the most time-consuming process in the computation of a GNN is the embedding generation.

To tackle the issue of scalability, several attempts have been made for efficient graph representation learning. In [21], a simple yet efficient inference computation is applied to enable GNN to process millions of nodes. The key idea is to leverage the adjacency matrix of the graph, denoted by $A \in \mathbb{R}^{N \times N}$. N is the total number of nodes in the graph. A matrix $E_d \in \mathbb{R}^{N \times K_d}$ can also be obtained, in which the v-th row represents the embedding of node v after the d-th iteration, i.e., $E_d[v, :] = e_d^{(v)}$. Take the input graph in Fig. 6.6 as an example. The weighted sum aggregation in iteration d is equivalent to Eq. 6.18.

$$
G_d = A \cdot E_{d-1} = \begin{array}{c} \\ 1 \\ 2 \\ 3 \\ 4 \\ 5 \end{array} \begin{array}{ccccc} 1 & 2 & 3 & 4 & 5 \\ \left[\begin{array}{ccccc} 1 & w_1 & w_1 & 0 & 0 \\ w_2 & 1 & w_1 & 0 & 0 \\ w_2 & w_2 & 1 & w_1 & w_1 \\ 0 & 0 & w_2 & 1 & 0 \\ 0 & w_2 & w_2 & 0 & 1 \end{array} \right] \end{array} \times \begin{bmatrix} e_{d-1}^{(1)} \\ e_{d-1}^{(2)} \\ e_{d-1}^{(3)} \\ e_{d-1}^{(4)} \\ e_{d-1}^{(5)} \end{bmatrix} \tag{6.18}
$$

Here $A \in \mathbb{R}^{5 \times 5}$, $G_d \in \mathbb{R}^{5 \times K_{d-1}}$, and the v-th row is the representation for node v after aggregation in the d-th iteration. The inner loop in Algorithm 6.3 (lines 3–6) can be simply formulated as

$$
E_d = \sigma(G_d \cdot W_d) = \sigma((A \cdot E_{d-1}) \cdot W_d). \tag{6.19}
$$

Then, all the computations can be formulated as a series of matrix multiplications that can be efficiently computed, and duplicated computation can be avoided. One potential issue is the dimension of the adjacency matrix A is $N \times N$, which is extremely large and cannot be stored in memory directly. Generally, it can be observed from the fact that the A is a sparse matrix if the graph is abstracted from a circuit netlist, which can be represented in a compressed sparse format, e.g., coordinate (COO) format which stores a list of (value, row_index, column_index) tuple. The matrix can be stored in the memory to enable matrix multiplication.

By taking the advantage of the sparse representation of the adjacency matrix and matrix multiplication-based computation, the GNN can scale to process designs with millions of cells. The inference runtime comparison between recursion and sparse computation is shown in Fig. 6.13. For a design with one million cells, the recursive computation costs more than one hour to complete the inference. While sparse computation only costs 1.5 s, which is three orders of magnitude speedup compared to recursion.

Besides, there are several systematic approaches proposed to enhance the scalability of graph neural networks. The most straightforward way is to reduce the overhead in aggregation steps. In [41], a forward computation method based on personalized PageRank is investigated to incorporate neighborhood features without conventional aggregation procedures. In GraphSAGE [18], duplicated computation is considered to be the bottleneck of the computing efficiency [42]. To address this, PinSAGE [42] is proposed to select important neighbors by random walk instead of aggregating all the neighbors. Then a MapReduce pipeline is leveraged for maxi-

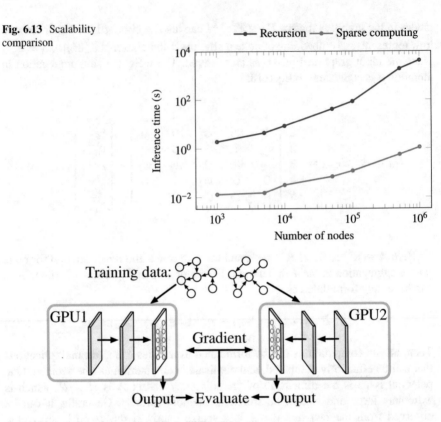

Fig. 6.13 Scalability comparison

Fig. 6.14 Parallel training with multiple GPUs [21]

mizing the inference throughput of the trained GNN model. Recently, GraphZoom [43] is proposed for improving both accuracy and scalability of unsupervised graph embedding algorithms, which is a multi-level spectral framework.

Moreover, the scalability can also be improved from the perspective of implementation. Considering there are many graphs (netlists) in the training set, a feasible solution to improve the scalability is parallel training with multiple GPUs. Conventionally, a batch of input data is split into equal chunks, and each GPU processes a chunk. However, the input of GNN model includes an adjacency matrix and node representation matrix which cannot be split. Different from conventional data-parallel scheme for regular data, each GPU processes one graph in GNN training, and all of the output is gathered to calculate the loss and then do back-propagation to update the model [21], as illustrated in Fig. 6.14. At present, there are also a few third-party libraries like DGL [44] and PyG [45] to support parallel training and inference and enable scalable computation of GNN models.

6.4.3   Integration with Design Flow

An accurate testability prediction model can facilitate to improve the testability through different design methodologies. A typical approach is test point insertion (TPI) [6, 46–49] which involves adding extra control points (CPs) or observation points (OPs) to the circuit. Integrating a testability prediction model to a DFT flow requires more consideration in addition to model accuracy. Inserting more test points to the circuit will increase the area and hence increase the overall cost and degrade the quality of results. Therefore, the testability prediction model should be leveraged to identify the most "valuable" locations as test point insertion candidates.

In [13, 14], the ANN models are trained to predict the fault coverage, which can further be used to evaluate the change of the fault coverage. More specifically, the prediction model can estimate the fault coverage before and after a test point is inserted; thus, the fault coverage difference can be obtained and used as an indicator of the impact of the inserted test point, which is written as

$$\Delta FC = FC_{p,t} - FC_p, \tag{6.20}$$

where $FC_{p,t}$ and FC_p are the predicted fault coverage of the circuit after and before a test point t is inserted, respectively. The test point keeps being inserted until one of the following four conditions is satisfied: (1) No test points are predicted to increase the fault coverage. (2) The number of inserted test points reached budget. (3) The predicted fault coverage reaches the target. (4) The overall runtime is reached.

In [21], a graph neural network model is trained to predict the difficult-to-test locations of a circuit. The impact of each location is defined as the positive prediction reduction in a local neighborhood after inserting a test point, as shown in Fig. 6.15. An iterative flow for test points insertion is developed, which is shown in Fig. 6.16. In each iteration, every positive prediction is evaluated to get its impact. Finally, a list containing observation points' locations is obtained. Then they are sorted based on their impact, and the top-ranked locations are selected. Next, the

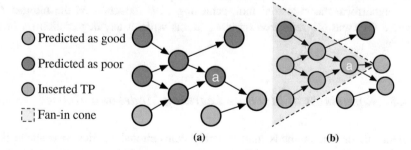

Fig. 6.15 An illustration to compute the impact for a test point [21]. (**a**) Prediction on original graph with 5 poor nodes in fan-in cone of a. (**b**) Prediction on graph after inserting a test point with 1 poor node. The impact of node a is $5 - 1 = 4$

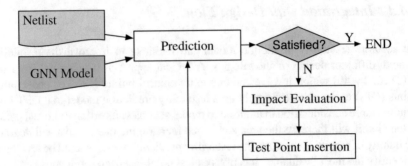

Fig. 6.16 Iterative flow for test points insertion with GNN model [21]

Table 6.9 Testability results comparison

Design	Industrial Tool			GCN-Flow		
	#OPs	#PAs	Coverage	#OPs	#PAs	Coverage
B1	6063	1991	99.31%	5801	1687	99.31%
B2	6513	2009	99.39%	5736	2215	99.38%
B3	6063	2026	99.29%	4585	1845	99.29%
B4	6063	2083	99.30%	5896	1854	99.31%
Average	6176	2027	99.32%	5505	1900	99.32%
Ratio	1.00	1.00	1.00	0.89	0.94	1.00

graph is modified, and model inference is performed for testability prediction on the updated netlist. The positive predictions will become the candidates in the next iteration. The exit condition is that there are no poor predictions left or the number of inserted test points reached budget. Three metrics are used for evaluation, including the total number of OPs inserted, the fault coverage, and the number of test patterns required. The results is presented in Table 6.9, with a commercial testability analysis tool as the baseline. Column "#OPs" represents the total number of OPs inserted. "#PAs" represents the number of test patterns. "Coverage" represents fault coverage. It can be seen that the OPs recommended by the proposed GNN model and iterative flow outperform the industrial tool, achieving 11% reduction on the number of inserted OPs and 6% reduction on test patterns without any degradation on fault coverage.

6.4.4 Robustness of Machine Learning Model and Metrics

Current DL or ML methods tend to apply conventional metrics to evaluate the usefulness of the model to the VLSI test problems, including accuracy, MSE, F1-score, etc. It has been argued that these metrics may not be meaningful with respect to the objective of VLSI testability prediction [16]. Using these metrics tends to

Table 6.10 Fault coverage results of TPI based on classification task

Design	b20	b22	log2	memctrl	voter	div	arbiter	b21	b17	b18	s38584
No TPI	96.02	94.87	85.26	95.71	87.31	99.31	99.80	99.28	98.14	99.24	98.19
GNN-TPI	98.34	97.25	93.41	98.41	95.33	99.33	99.91	99.89	99.12	99.65	98.24
GNN-TPI-Alt.	98.97	98.64	93.88	98.65	95.35	99.52	99.96	99.89	99.25	99.75	98.50

Table 6.11 Fault coverage results of TPI based on regression task

Design	arbiter	b15	b17	b18	b19	b21	log2	memctrl	sin	s38584
No TPI	20.39	79.34	76.80	80.76	77.49	87.76	85.43	54.42	94.26	96.41
ANN-TPI	33.67	83.27	81.35	82.51	89.19	91.35	86.67	70.08	95.24	97.53
ANN-TPI-Alt.	42.43	85.16	85.98	86.78	90.48	93.12	88.72	75.52	96.21	97.45

guide the model to learn the heuristics of a particular tool that generates the label of the data. Essentially this is a proxy objective instead of an ultimate objective in improving the testability of a circuit. Chowdhury et al. [16] explored an alternative metric to replace the metric used in [21] for node classification. The "1-hop F1-score" is utilized instead of the conventional F1-score, *which considers that a positive prediction is true as long as it is within 1-hop of a real positive node.* The intuition is that a positive prediction, e.g., hard-to-observe node, would also be useful for testability improvement if it is close to real hard-to-observe nodes. To validate the alternative metric, it is used to guide the GNN training as in [21], and then TPI is conducted. The fault coverage results are presented in Table 6.10 in which "GNN-TPI-Alt." represents using the alternative metric. It can be seen that using alternative metric for classification leads to higher fault coverage than using the conventional F1-score.

In addition to the classification task, similar issue is also observed in the regression task. MSE is used as the evaluation metric to assess the quality of the model in previously mentioned fault coverage prediction task. However, it has been investigated that even a large MSE does not necessarily imply a poor model [50]. An alternative metric *Spearman ranking correlation coefficient* [51] is investigated to better evaluate the capability of the regression model for fault coverage prediction, which captures the consistency between the ranking of the prediction values and the ground truth, instead of the raw values. From Table 6.11, it can be seen that ranking correlation is much more effective than MSE when training a testability prediction model for TPI.

Another concern regarding the robustness is the trained model. Ideally, a model should be robust to the redundant patterns which do not change the functionality of a netlist. A typical case is the redundant re-convergent fan-out (RRF) patterns. Unfortunately, it is observed in [16] that a typical model cannot capture the redundancy of the functionality. If RRF patterns are manually inserted into the circuit and then apply the DL-based TPI, the resulting fault coverage degrades significantly. It can be expected in the sense that current DL-based models leverage approximated metrics like SCOAP as the input features, which fundamentally is not robust to the perturbation. Therefore, exploring more structural and fundamental features is worthwhile.

6.5 Summary

In this chapter, we introduce conventional testability measurements and several machine learning methodologies for testability prediction. Specifically, we present how to conduct machine learning tasks on circuit netlist, where feature extraction is the key process. Two different approaches are introduced, including direct feature extraction with traditional learning models and representation learning on a graph with graph neural network models. Several testability-related problems are investigated, including node-level testability prediction, fault coverage prediction, test cost prediction, and X-sensitivity prediction. We further discuss how to deal with the computation efficiency issues on graph neural networks, in which the sparse adjacency matrix is leveraged. Moreover, we also investigate a few practical concerns for testability prediction models. A multistage classification model is introduced to alleviate the data-imbalance issue, as well as an iterative test point insertion flow is introduced to apply the testability prediction model to DFT flow considering the real quality of result of the designs.

In addition, machine learning for testability prediction is still at its early stage. Manual selection of important features from a design is still challenging and nontrivial. It is also not clear how precise the automatic feature extraction is for a VLSI design. Furthermore, machine learning is a data-hungry technique after all. However, it is usually very difficult and expensive to obtain sufficient data in VLSI design to train an accurate model. Therefore, it is essential to develop more advanced learning paradigms where data requirements can be eased and modeling accuracy can be maintained.

References

1. Grason, J.: TMEAS, a testability measurement program. In: ACM/IEEE Design Automation Conference (DAC), pp. 156–161 (1979)
2. Goldstein, L.H., Thigpen, E.L.: SCOAP: Sandia controllability/observability analysis program. In: ACM/IEEE Design Automation Conference (DAC), pp. 190–196 (1980)
3. Brglez, F.: On testability of combinational networks. In: IEEE International Symposium on Circuits and Systems (ISCAS), vol. 1 (1984)
4. Karimi, N., Riyahi, P., Navabi, Z.: A survey of testability measurements at various abstraction levels. In: IEEE North Atlantic Test Workshop (NATW), pp. 26–33 (2003)
5. Jain, S.K., Agrawal, V.D.: Statistical fault analysis. IEEE Design Test Comput. 2(1), 38–44 (1985)
6. Ren, H., Kusko, M., Kravets, V., Yaari, R.: Low cost test point insertion without using extra registers for high performance design. In: IEEE International Test Conference (ITC), pp. 1–8 (2009)
7. Krishnaswamy, S., Plaza, S.M., Markov, I.L., Hayes, J.P.: Enhancing design robustness with reliability-aware resynthesis and logic simulation. In: IEEE/ACM International Conference on Computer-Aided Design (ICCAD), pp. 149–154 (2007)
8. Plaza, S.M., Chang, K.h., Markov, I.L., Bertacco, V.: Node mergers in the presence of don't cares. In: IEEE/ACM Asia and South Pacific Design Automation Conference (ASPDAC), pp. 414–419 (2007)

9. Huang, G., Hu, J., He, Y., Liu, J., Ma, M., Shen, Z., Wu, J., Xu, Y., Zhang, H., Zhong, K., et al.: Machine learning for electronic design automation: a survey. ACM Trans Design Autom Electron Syst **26**(5), 1–46 (2021)

10. Elfadel, I.M., Boning, D.S., Li, X.: Machine Learning in VLSI Computer-Aided Design. Springer, Berlin (2019)

11. Pradhan, M., Bhattacharya, B.B.: A survey of digital circuit testing in the light of machine learning. Wiley Interdiscip. Rev. Data Mining Knowl. Discovery **11**(1), e1360 (2021)

12. Ma, Y., He, Z., Li, W., Zhang, L., Yu, B.: Understanding graphs in EDA: from shallow to deep learning. In: ACM International Symposium on Physical Design (ISPD), pp. 119–126 (2020)

13. Millican, S., Sun, Y., Roy, S., Agrawal, V.: Applying neural networks to delay fault testing: test point insertion and random circuit training. In: 2019 IEEE 28th Asian Test Symposium (ATS), pp. 13–135 (2019)

14. Sun, Y., Millican, S.: Test point insertion using artificial neural networks. In: 2019 IEEE Computer Society Annual Symposium on VLSI (ISVLSI), pp. 253–258 (2019)

15. Nair, V., Hinton, G.E.: Rectified linear units improve restricted Boltzmann machines. In: International Conference on Machine Learning (ICML), pp. 807–814 (2010)

16. Chowdhury, A.B., Tan, B., Garg, S., Karri, R.: Robust deep learning for ic test problems. IEEE Trans Comput.-Aided Design Integr. Circuits Syst. **41**(1), 183–195 (2021)

17. Cai, H., Zheng, V.W., Chang, K.: A comprehensive survey of graph embedding: problems, techniques and applications. IEEE Trans. Knowl. Data Eng. **30**(9), 1616–1637 (2018)

18. Hamilton, W., Ying, Z., Leskovec, J.: Inductive representation learning on large graphs. In: Conference on Neural Information Processing Systems (NIPS), pp. 1024–1034 (2017)

19. Kipf, T.N., Welling, M.: Semi-supervised classification with graph convolutional networks. In: International Conference on Learning Representations (ICLR) (2016)

20. Veličković, P., Cucurull, G., Casanova, A., Romero, A., Liò, P., Bengio, Y.: Graph attention networks. In: International Conference on Learning Representations (2018)

21. Ma, Y., Ren, H., Khailany, B., Sikka, H., Luo, L., Natarajan, K., Yu, B.: High performance graph convolutional networks with applications in testability analysis. In: ACM/IEEE Design Automation Conference (DAC), pp. 18:1–18:6 (2019)

22. Li, Z., Colburn, J.E., Pagalone, V., Narayanun, K., Chakrabarty, K.: Test-cost optimization in a scan-compression architecture using support-vector regression. In: 2017 IEEE 35th VLSI Test Symposium (VTS), pp. 1–6. IEEE, Piscataway (2017)

23. Xu, S., Dias, G.P., Waignjo, P., Shi, B.: Testability prediction for sequential circuits using neural networks. In: IEEE Asian Test Symposium (ATS), pp. 48–53 (1997)

24. Pradhan, M., Bhattacharya, B.B., Chakrabarty, K., Bhattacharya, B.B.: Predicting X-sensitivity of circuit-inputs on test-coverage: a machine-learning approach. IEEE Trans. Comput.-Aided Design Integr. Circuits Syst. **38**(12), 2343–2356 (2018)

25. Arai, M., Fukumoto, S., Iwasaki, K., Matsuo, T., Hiraide, T., Konishi, H., Emori, M., Aikyo, T.: Test data compression of 100x for scan-based bist. In: IEEE International Test Conference (ITC), pp. 1–10 (2006)

26. Barnhart, C., Brunkhorst, V., Distler, F., Farnsworth, O., Keller, B., Koenemann, B.: Opmisr: The foundation for compressed ATPG vectors. In: IEEE International Test Conference (ITC), pp. 748–757 (2001)

27. Kumar, A., Kassab, M., Moghaddam, E., Mukherjee, N., Rajski, J., Reddy, S.M., Tyszer, J., Wang, C.: Isometric test data compression. IEEE Trans. Comput.-Aided Design Integr. Circuits Syst. **34**(11), 1847–1859 (2015)

28. Benesty, J., Chen, J., Huang, Y., Cohen, I.: Pearson correlation coefficient. In: Noise Reduction in Speech Processing, pp. 1–4. Springer, Berlin (2009)

29. Erb, D., Kochte, M.A., Reimer, S., Sauer, M., Wunderlich, H.J., Becker, B.: Accurate QBF-based test pattern generation in presence of unknown values. IEEE Trans. Comput.-Aided Design Integr. Circuits Syst. **34**(12), 2025–2038 (2015)

30. Scheibler, K., Erb, D., Becker, B.: Accurate CEGAR-based ATPG in presence of unknown values for large industrial designs. In: IEEE/ACM Proceedings Design, Automation and Test in Europe (DATE), pp. 972–977 (2016)

31. Touba, N.A.: X-canceling MISR—an X-tolerant methodology for compacting output responses with unknowns using a MISR. In: IEEE International Test Conference (ITC), pp. 1–10 (2007)

32. Tang, Y., Wunderlich, H.J., Vranken, H., Hapke, F., Wittke, M., Engelke, P., Polian, I., Becker, B.: X-masking during logic BIST and its impact on defect coverage. In: IEEE International Test Conference (ITC), pp. 442–451 (2004)

33. Mitra, S., Kim, K.S.: X-compact: An efficient response compaction technique. IEEE Trans. Comput.-Aided Design Integr. Circuits Syst. 23(3), 421–432 (2004)

34. Chickermane, V., Foutz, B., Keller, B.: Channel masking synthesis for efficient on-chip test compression. In: IEEE International Test Conference (ITC), pp. 452–461 (2004)

35. Synopsys: TetraMAX ATPG: Automatic Test Pattern Generation (2013)

36. Yang, H., Su, J., Zou, Y., Ma, Y., Yu, B., Young, E.F.Y.: Layout hotspot detection with feature tensor generation and deep biased learning. IEEE Trans. Comput.-Aided Design Integr. Circuits Syst. 38(6), 1175–1187 (2019)

37. Ye, W., Lin, Y., Li, M., Liu, Q., Pan, D.Z.: LithoROC: lithography hotspot detection with explicit ROC optimization. In: IEEE/ACM Asia and South Pacific Design Automation Conference (ASPDAC), pp. 292–298 (2019)

38. Tabrizi, A.F., Darav, N.K., Rakai, L., Bustany, I., Kennings, A., Behjat, L.: Eh? predictor: A deep learning framework to identify detailed routing short violations from a placed netlist. IEEE Trans. Comput.-Aided Design Integr. Circuits Syst. 39(6), 1177–1190 (2019)

39. Tabrizi, A.F., Rakai, L., Darav, N.K., Bustany, I., Behjat, L., Xu, S., Kennings, A.: A machine learning framework to identify detailed routing short violations from a placed netlist. In: ACM/IEEE Design Automation Conference (DAC), pp. 1–6 (2018)

40. Cheng, C.K., Kahng, A.B., Kim, H., Kim, M., Lee, D., Park, D., Woo, M.: Probe2. 0: a systematic framework for routability assessment from technology to design in advanced nodes. IEEE Trans. Comput.-Aided Design Integr. Circuits Syst. 41(5), 1495–1508 (2021)

41. Bojchevski, A., Klicpera, J., Perozzi, B., Blais, M., Kapoor, A., Lukasik, M., Günnemann, S.: Is pagerank all you need for scalable graph neural networks? In: ACM KDD, MLG Workshop (2019)

42. Ying, R., He, R., Chen, K., Eksombatchai, P., Hamilton, W.L., Leskovec, J.: Graph convolutional neural networks for web-scale recommender systems. In: ACM International Conference on Knowledge Discovery and Data Mining (KDD), pp. 974–983 (2018)

43. Deng, C., Zhao, Z., Wang, Y., Zhang, Z., Feng, Z.: Graphzoom: a multi-level spectral approach for accurate and scalable graph embedding. In: International Conference on Learning Representations (ICLR) (2020)

44. Wang, M., Yu, L., Da Zheng, Q.G., Gai, Y., Ye, Z., Li, M., Zhou, J., Huang, Q., Ma, C., et al.: Deep graph library: Towards efficient and scalable deep learning on graphs. In: ICLR Workshop on Representation Learning on Graphs and Manifolds (2019)

45. Fey, M., Lenssen, J.E.: Fast graph representation learning with pytorch geometric (2019). arXiv preprint arXiv:1903.02428

46. Hayes, J.P., Friedman, A.D.: Test point placement to simplify fault detection. IEEE Trans. Comput. 100(7), 727–735 (1974)

47. Touba, N.A., McCluskey, E.J.: Test point insertion based on path tracing. In: IEEE VLSI Test Symposium (VTS), pp. 2–8 (1996)

48. Cheng, K.T., Lin, C.J.: Timing-driven test point insertion for full-scan and partial-scan BIST. In: IEEE International Test Conference (ITC), pp. 506–514 (1995)

49. Nakao, M., Kobayashi, S., Hatayama, K., Iijima, K., Terada, S.: Low overhead test point insertion for scan-based BIST. In: IEEE International Test Conference (ITC), pp. 348–357 (1999)

50. Rousseeuw, P., Leroy, A.: Robust regression and outlier detection. In: Wiley Series in Probability and Statistics (1987)

51. Spearman, C.: The proof and measurement of association between two things. Int. J. Epidemiol. 39(5), 1137–50 (2010)

Part II
Machine Learning-Based Design Optimization Techniques

Chapter 7
Machine Learning for Logic Synthesis

Rajarshi Roy and Saad Godil

7.1 Introduction

Logic synthesis is the EDA process that converts the behavioral description of a circuit to a transistor level implementation. The typical logic synthesis process for integrated circuits (IC) consumes register-transfer level (RTL) description and produces a gate-level netlist describing the connectivity of standard cells from a technology library. For lookup table (LUT)-based field-programmable gate array (FPGA) devices, logic synthesis produces a network of LUTs. Generally, logic synthesis consists of two steps: logic optimization and technology mapping.

Logic optimization transforms the logic functions of the behavioral description to equivalent logic functions that are optimized for specified objectives. Modern logic synthesis frameworks such as ABC [1] and Mockturtle [2] first convert the behavioral description into and-inverter-graph (AIG) [3] or majority-inverter-graph (MIG) [4] directed-acyclic-graph (DAG) logic networks, respectively. Then they apply sequence of network-level logic transformations (Table 7.2, Table 7.5) that transform the logic network to more optimal logic networks for objectives such as minimizing logic network node count and level count. The order of these transformations and the network-level (vs node-level) nature of these transformations are heuristics. Node counts and levels are proxy heuristic for the final area and delay of the logic network after technology mapping and physical design.

Technology mapping maps the optimized logic functions into a standard-cell netlist for ICs or LUTs for FPGAs. The goal of logic synthesis is to produce implementations that are optimized for quality-of-results (QoR) metrics such as area, power consumption, and delay. Modern technology mapping algorithms for

R. Roy (✉) · S. Godil
NVIDIA, Santa Clara, CA, USA
e-mail: rajarshir@nvidia.com; sgodil@nvidia.com

© The Author(s), under exclusive license to Springer Nature Switzerland AG 2022
H. Ren, J. Hu (eds.), *Machine Learning Applications in Electronic Design Automation*, https://doi.org/10.1007/978-3-031-13074-8_7

FPGAs [5] and ICs [6] operate on DAG logic networks such as AIGs. These algorithms compute cuts or subgraphs at the fan-in of every node. FPGA mapping picks and maps optimal cuts to LUTs in a topological order. Mapping for IC picks and maps optimal cuts to optimal matching standard cells in the process library. Cut choices and standard cell choices are picked using heuristics such as the number of cut inputs and local arrival times for matching standard cells. The FPGA technology mapping algorithm in [5] uses a heuristic of multiple mapping passes with various prioritizations of depth and area objectives.

Certain operators in datapath RTL descriptions such as shift, addition, multi-plication, and others have structured logic that are synthesized with specialized generators. These datapath logic generators typically rely on well-engineered heuristic algorithms. For example, [7] minimizes the node count of parallel prefix graphs from specified graph levels for parallel prefix circuits. Node counts and levels are proxy heuristic for the final area and delay of the parallel prefix circuit after technology mapping and physical design.

Logic synthesis algorithms typically operate on designs that have large and often intractable implementation choices. Since exact algorithms may be too com-putationally intensive, these algorithms usually have heuristic-driven components that are engineered by humans. Since the EDA flow is composed of multiple steps, within logic synthesis (logic optimization, technology mapping) and after (physical synthesis optimizations, floor planning, place and route), it is often intractable for logic synthesis algorithms to optimize for objective metrics that can be obtained after later steps are run. Modern machine learning models have the ability to learn from large amounts of complex data and make accurate predictions on unseen data. Predictions from supervised learning models can guide logic synthesis algorithms by predicting the result of downstream EDA algorithms early (Sect. 7.3). Reinforcement learning models can directly learn optimization policies from exploring the design space automatically (Sect. 7.4). These learned policies can adapt to new designs and outperform fixed heuristics that are engineered for specific benchmark designs.

Section 7.2 presents an overview of supervised and reinforcement learning. Section 7.3 presents various applications of supervised learning in logic synthesis. Section 7.4 presents various applications of reinforcement learning in logic syn-thesis. Section 7.5 discusses scalability considerations for reinforcement learning applications in logic synthesis.

7.2 Supervised and Reinforcement Learning

The machine learning applications for logic synthesis in this chapter utilize machine learning models that are trained using supervised learning or reinforcement learning approaches.

7.2.1 Supervised Learning

Supervised learning is generally applied to prediction tasks, where machine learning models are trained to output predictions for given inputs. Trained models can make fast predictions about computationally intensive logic synthesis algorithmic steps and allow for exploring design spaces that are intractable otherwise. This is especially true when models are trained to predict quality-of-results (QoR) metrics that would after multiple EDA tool flow steps. Predictions from these model can guide early optimizations during logic synthesis for better final QoR metrics.

In order to train a model that takes an input x and makes a prediction y, supervised learning requires a dataset of i datapoints with example inputs x_i and corresponding labels y_i. For example, training a model that can predict the circuit area of Boolean logic functions after technology mapping and physical design will require a dataset of various Boolean logic functions and their corresponding circuit areas after technology mapping and physical design.

The supervised learning models in this chapter are either regression models or classification models. Regression models predict a number, such as the area of a circuit. Classification models predict a class, such as the class of a logic function among the classes (control logic, datapath logic, memory logic).

The supervised learning algorithm for training a machine learning model given a labeled dataset is dependent on the type of machine learning model. The supervised learning applications in this chapter utilize neural network models. These models are typically trained using gradient descent family of algorithms. Details of these algorithms are largely out of the scope of this chapter. The various neural network model architectures in this chapter are described from input to output layers using the layer naming convention from Table 7.1. For example, the convolutional architecture of the model in Fig. 7.1 is specified as follows:

- Q-network: PRERES→32×[RESBLOCK]→POSTRES
- PRERES: CONV2D(3X3,I,256,1)→BNORM→LRELU
- POSTRES: CONV2D(1X1,256,256,1)→BNORM→LRELU→
  CONV2D(1X1, 256,4,1)
- RESBLOCK: x→CONV2D(5X5,256,256,1)→BNORM→LRELU→
  CONV2D(5X5,256,256,1)→BNORM→ADD(x)→LRELU
  where x is the input to RESBLOCK.

Neural network models support a fine-tuning style of supervised training. A neural network model that is already trained on a large dataset can be further trained, or fine-tuned, on a new dataset. During fine-tuning, some layers of the neural network may be fixed. Fine-tuning has the advantage of requiring much fewer datapoints in the new dataset if the inputs are similar to the old dataset. For example, a model trained to predict the circuit area of logic in a 45 nm process may be fine-tuned with a smaller dataset of circuit areas in a 10 nm process and achieve better accuracy than training a model using just the small 10 nm dataset. Yu and Zhou [9] utilizes this technique in Sect. 7.3.2.

Table 7.1 Neural network layers and operators

Layer	Description
FC(in,out)	Fully connected layer with specified input and output lengths
CONV2D(height×width,in,out,stride)	2D convolution layer with specified kernel height, width, input channels, output channels and stride
MPOOL(height×width)	Maximum pooling layer with specified pooling height, width
LSTM(in,out)	Long-short term memory layer with specified input and output channels
RELU	Rectified linear unit activation layer
LRELU	Leaky rectified linear unit activation layer
SELU	Scaled exponential linear unit activation layer
SIG	Sigmoid activation layer
SOFT	Softmax activation layer
DROP	Dropout layer
BNORM	Batch normalization layer
FLAT	Flatten operator
ADD	Add operator
CONCAT	Concatenate operator

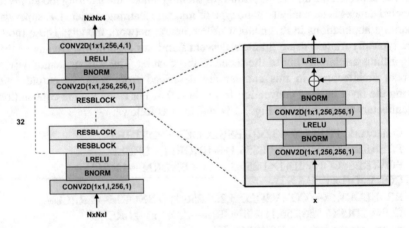

Fig. 7.1 Convolutional Q-network architecture of [8] from Sect. 7.4.3

7.2.2 Reinforcement Learning

Reinforcement learning is generally applied to optimization tasks, where machine learning models are trained to learn policies that optimize specified objectives. Logic synthesis algorithms are primarily optimization tasks that transform logic toward objectives such as minimizing circuit area, power consumption, and delay.

Thus, various logic synthesis algorithms may be formulated as reinforcement learning tasks.

The typical reinforcement learning formulation consists of an *agent* and an *environment*. The environment maintains the current *state* of the task, while the agent can observe the environment state and take an *action* (make a move) in the environment. Taking an action may possibly transition (update) the state to a new state. For a transition, the agent may receive a *reward* that is related to the objective. After interacting with the environment, the agent learns strategies to take actions that maximize the rewards it receives from the environment. Typically, the interactions between the agent and the environment is divided into episodes. The agent can take actions during an episode until specified end conditions reset the environment and end the episode. Often, a reinforcement learning formulation is designed such that the agent aims to maximize the sum of rewards during the span of an episode. These agents learn strategies that may take sequences of multiple actions with possibly smaller immediate rewards in order to get to a state that produces a large reward at the end.

Formulating an algorithm as a reinforcement learning task requires the specification of a Markov decision process (MDP). The MDP defines the following properties of the reinforcement learning environment:

- S: the state space, or the set of all states. For example, in a reinforcement learning environment for technology mapping, S could contain the various possible combinations of cells and connections used to represent a logic function.
- A: the action space, or the set of all possible actions that the agent can take in the environment. In the technology mapping example, a possible action could be to replace a cluster of simple cells like two-input AND gates with a complex cell like a four-input AND gate. One consideration for a well-designed action space is that the agent should be able to reach any state in the state space by taking actions from the action space. Otherwise the reinforcement learning environment may restrict the agent from reaching optimal solutions.
- R: the reward specification for all transitions. Typically this reward is related to the objective. If the objective of the technology mapping example is to reduce the area, then for every transition from state s to next state s', the reward could be reduction in area: $area(s) - area(s')$, which would be negative if the area increased and positive if the area decreased.
- P: the state transitions and associated probabilities. In a stochastic environment, a particular action a applied to a particular state s could yield various possible next states $s'_1 \ldots s'_n$ with various probabilities $P(s'_1|s, a) \ldots P(s'_n|s, a)$. In a deterministic environment, action a taken at state s transitions to a specific next state s' with probability 1. The logic synthesis applications discussed in this chapter map to deterministic reinforcement learning environments.

There are other considerations for designing a reinforcement learning solution. Some of them are the following:

- Task horizons: a finite horizon task has ending or done states which do not have any next states to transition to. Infinite horizon tasks do not have done states and can transition forever. For example, a reinforcement learning environment that allows logic transformations on a netlist in an input-to-output order is a finite horizon since the environment is done after the output logic is transformed. Meanwhile, a reinforcement learning environment that allows logic transformations anywhere in a netlist and allows transformations to logic that have already been transformed is an infinite horizon since the logic transformation can always be applied to any state forever.
- Density of rewards: environments with dense rewards produce rewards at every transition. For example, in a previous example where the reward for reducing area is $area(s) - area(s')$. Dense rewards provide a granular feedback to the agent but may be computationally intensive if the process of reward calculation, such as measuring area, is expensive. Environments with finite horizons may choose to produce sparse rewards, where every transition produces 0 rewards except for the transition to the done state. In the area objective example, this reward could be $area(s_0) - area(s_d)$ which is the area reduction from the initial state s_0 to the done state s_d.
- Generalization: reinforcement learning agents may be trained for a fixed task or for generalizing to unseen tasks. For example, a reinforcement learning agent that is trained to optimize the logic of a specific multiplier circuit, operates in an environment containing states that are always logically equivalent to that particular multiplier circuit. However, it may be desirable to train a reinforcement learning agent for optimizing any circuit logic. In that case, the agent would be trained with a variety of different circuits and ideally learn general optimization strategies that it can apply to any circuit. One mechanism for training such a general reinforcement learning agent would be to reset the environment to the logic of a different circuit at the start of every episode.

Reinforcement learning agents are trained to learn a strategy of taking actions in the environment that maximizes reward. At every step of an episode, the agent should observe the current state of the environment as input and output an action based on the current state. Value-based agents learn the value of every action at a current state and select the action with the highest value. Policy-based agents directly sample actions from learned action probabilities based on the current state. There are several algorithms for training value-based [10] or policy-based [11, 12] reinforcement learning agents. Details of these algorithms are largely out of the scope of this chapter.

The reinforcement learning agents in this chapter utilize neural network models. Their architecture is described using the same layer naming convention (Table 7.1) as the supervised learning models (Sect. 7.2.1).

7.3 Supervised Learning for Guiding Logic Synthesis Algorithms

Various logic synthesis algorithms benefit from guidance by machine learning models that are trained using supervised learning. In the following subsections, we will discuss supervised learning formulations for four logic synthesis algorithms.

7.3.1 Guiding Logic Network Type for Logic Network Optimization

An efficient representation of large Boolean logic functions is the directed acyclic graph (DAG) logic networks such as and-inverter graphs (AIGs) [3] and majority-inverter-graphs (MIGs) [4]. During logic optimization, certain types of logic, such as arithmetic logic, may be better suited to MIG optimizations versus AIG optimizations.

Neto et al. [13] demonstrates that a heterogenous logic synthesis framework of AIGs and MIGs guided by a learned model can offer gains of up to 10.60% and 6.27% in the final technology mapped area-delay product of designs as compared to pure AIG- or pure MI-based optimization, respectively. In this supervised learning approach, a model is trained to choose whether a partition of a logic network should be optimized as an AIG or a MIG.

The inputs to the model are various 16-input Boolean logic functions representing logic network partitions. Each function is represented using a truth-table with 2^{16} entries. More specifically, the representation uses a 2^{16}-element vector of 0s and 1s with the element ordering based on the flattened 256X256 Karnaugh map of the truth-table. This representation is independent of AIGs or MIGs.

A classification model is trained in a supervised manner on a dataset of Boolean logic functions that are labelled with the (AIG, MIG) classes based on whether MIG or AIG optimization on the functions achieve smaller network size. The logic function classification model has a feedforward MLP architecture:

- FC(65536,250)→DROP→FC(250,2)

When a design is synthesized, the synthesis framework partitions the logic of the design into 16-input Boolean logic functions. For each partition, the trained model predicts whether AIG or MIG optimizations would lead to smaller network size. With the objective of area reduction, the synthesis framework chooses AIG or MIG optimizations to apply to each partition based on the model predictions. After the partitions are optimized, the synthesis framework merges back all the optimized partitions.

7.3.2 Guiding Logic Synthesis Flow Optimization

The ABC synthesis framework [1] offers various algorithms for logic transformations on and-inverter-graph (AIG) logic networks [3]. Each transformation is applied to an entire logic network and may modify multiple nodes in the AIG. For example, the *balance* transform minimizes AIG by balancing AND trees in the AIG such as $(a \cdot (b \cdot (c \cdot (d))))$ to $(a \cdot b) \cdot (c \cdot d)$. Table 7.2 lists commonly used AIG transformations in ABC.

A *synthesis flow* is a sequence of k transformations applied to a logic network. For example, a popular heuristic synthesis flow provided be ABC is *resyn2*: $b; rw; rf; b; rw; rwz; b; rfz; rwz; b$. Since the synthesis flow transformations scale to large logic networks, several supervised learning approaches [9, 14] and reinforcement learning approaches [15–17] aim to optimize synthesis flows for logic network optimization. We will discuss the supervised learning approaches [9, 14] in this subsection and the reinforcement learning approaches [15–17] in Sect. 7.4.2.

If there are n available transformations, there are n^k possible synthesis flows. Yu et al. [14] demonstrates that the best synthesis flow varies significantly from design to design due to varying logic network structures. Yu et al. [14] uses learned models to select a small number of optimal flows from large number of randomly generated flows. While the quality of flows that this method can produce is limited to the quality of random flows, Yu et al. [14] demonstrates that the models can successfully select the best quality flows from a large sample of random flows.

The inputs to the models are various *synthesis flows* for a specific design. Each synthesis flow is represented as a matrix of size $k \times n$ where $k = 24$ is the transformation sequence length of the synthesis flow and $n = 6$ is the number of available transformations. Each transformation in the sequence is represented as a one-hot vector of length k indicating the transformation out of all the available transformations.

For a specific design, 10,000 random synthesis flows (random sequences of synthesis transformations) are run. Each flow is labeled with the area and delay of the design after the flow is run and technology mapping is performed. The area and delay labels are both binned into seven classes with the percentile thresholds {5% 15%, 40%, 65%, 90%, 95%}. A multi-class classification model is trained in

Table 7.2 AIG network-level transformations in ABC [1]

Transformation	Description
b	Balance
rw	Rewrite
rwz	Rewrite with zero-cost replacements
rf	Refactor
rfz	Refactor with zero-cost replacements
rs	Resubstitute
rsz	Resubstitute with zero-cost replacements

a supervised manner with the area labels to predict the area class for a synthesis flow. Another model is trained with the delay labels to predict the delay class for a synthesis flow. The flow classification models have a convolutional architecture:

- CONV2D(6X12,1,200,1)→SELU→MPOOL(2X2)→
  CONV2D(6X12,200,200,1)→SELU→MPOOL(2X2)→FLAT→
  FC(,64)→SELU→DROP→FC(64,7)→SOFT

The area and delay models are used to select 200 optimal flows for area and delay, each from 100,000 new randomly generated flows. For each area or delay objective, the flows that have the highest classification score of the <5% class (lowest 5% area or delay) from the respective models are chosen.

Yu and Zhou [9] changes the supervised learning model to regression model that directly predicts the area and delay of new synthesis flows for a design. When trained with 20,000 datapoints and tested on 80,000 datapoints for a design in a 14 nm process, the model demonstrates high accuracy with an average mean absolute percentage error (MAPE) of ≤0.37%.

Moreover, if the model is trained on a particular process technology for a particular design with a large dataset, it can make accurate predictions for new technologies or new designs by fine-tuning on few additional datapoints collected with the new technology or design. For example, the model trained for a design at 14 nm with 20,000 datapoints can be fine-tuned with just 100 additional datapoints for the same design at 7 nm and obtain ≤0.5% MAPE when tested on 100,000 7 nm datapoints. The fine-tuning updates the weights of all layers. When fine-tuning for the same technology to new designs, the model achieves ≤2.0% MAPE, and when fine-tuning to both a new technology and new designs, the model achieves ≤3.7% MAPE.

The synthesis flow representation for this model is a $n \times k$ matrix where $k = 24$ is the transformation sequence length of the synthesis flow and $n = 6$ is the number of available transformations. The flow quality prediction model has a recurrent architecture:

- LSTM(6,128)→BNORM→LSTM(128,128)→BNORM→FC(128,30)→
  RELU→BNORM→FC(30,30)→RELU→BNORM→DROP→FC(30,1)

7.3.3 Guiding Cut Choices for Technology Mapping

The technology mapping stage of logic synthesis for ICs maps a logic network in the and-inverter graph (AIG) or majority-inverter graph (MIG) format into a netlist of standard cells. The state-of-the-art open-source ABC mapper [1] uses a cut-based mapping algorithm to map AIGs to standard cells. At a high level, the mapper first computes K-feasible cuts for each node in the AIG. The K-feasible cuts for a node in the AIG are all the possible subgraphs from the inputs of the overall AIG to the node with up to K unique subgraph inputs. After the K-feasible cuts for each node are computed, the mapper matches each cut for each node to a standard cell with the equivalent Boolean function to the cut. Finally, the mapper assigns the best standard cells to cuts based on an objective. Since there can be a large number of cuts ($O(n^k)$ in the worst case for a graph with n nodes), the ABC mapper applies a heuristic of sorting cuts by number of cut inputs and then keeping up to 250 cuts per node available for matching. Even if this filter is removed, and all cuts are exhaustively considered, the ABC mapper matches a cut to the standard cell that minimizes arrival time to the node. Since mapping a certain cut reduces the choices of cuts for downstream nodes, this heuristic may not lead to the minimum delay solution for the final mapped netlist.

Neto et al. [18] presents a supervised learning approach where a learned model guides the filtering of cut choices that are available to the standard cell matching based on cut structure and properties. This method can improve delay by an average of 10% (up to 18%) over the ABC filtering heuristic and by an average of 6% over exhaustively considering all cuts. This method is able to improve overall delay over exhaustively considering all cuts by making certain cuts unavailable that to that matching heuristic that would otherwise be picked for minimizing local node arrival time but not lead to global delay reduction.

The inputs to the model are various five-feasible cuts (same as ABC) for nodes in the AIG. Each five-feasible cut is represented as a 10×15 embedding matrix (Table 7.3) where 6 of the 15 rows are embeddings for the root node and the 5 input nodes to the cut. If the cut has fewer than five nodes, the embeddings are padded with zeros. Each node, n, is a two-input AND with two parents p_1 and p_2 with connecting edges e_1 and e_2 that may have an inverter or not. Each node embedding is a vector of size 10 with the following features: $inv(e), lvl(n), FO(n), rlvl(n), inv(e_1), lvl(p_1), FO(p_1), inv(e_2), lvl(p_2), and$ $FO(p_2)$. inv indicates whether there is an inverter on the edge, and $inv(e)$ indicates

Table 7.3 Cut embedding

Index	Feature
0	Root node embedding
1	Input 0 node embedding
2	Input 1 node embedding
3	Input 2 node embedding
4	Input 3 node embedding
5	Input 4 node embedding
6	Any inverters on root node outgoing edges
7	Number of cut inputs
8	Number of nodes in cut
9	Min level of cut inputs
10	Max level of cut inputs
11	Sum of cut input levels
12	Min fanout of cut inputs
13	Max fanout of cut inputs
14	Sum of cut input fanouts

whether there is an inverter on any of the outgoing edges of the node. FO is the fan out of node. lvl and $rlvl$ are the level and reverse level of the node from the overall AIG inputs and outputs. The remaining nine rows of the cut embedding are cut features that are repeated to a vector of size 10 each.

Logic networks of a set of circuits are technology mapped 50,000 times each using random five-feasible cut choices for nodes, leading to various standard cell netlists with various overall delays. The delays are normalized and binned into 10 classes with 0 indicating the lowest delay and 9 indicating the highest delay. For each netlist, all the chosen cuts are labeled with the same binned delay class of the overall netlist delay. A multi-class classification model is then trained on this dataset in a supervised manner to predict the delay classes of cuts. The cut classification model has a convolutional architecture:

- CONV2D(15X1,1,128,1)→FLAT→FC(1280,10)→SOFT

During the technology mapping of a netlist, all computed five-feasible cuts for each node are classified by the trained model. The class predictions from the model are used to decide if a cut will be kept or discarded. The list of kept cuts for a node is made available for matching with standard cells, while discarded cuts are not available for matching. If there are any cuts predicted to be in the classes ≤ 3, then only those cuts are kept. If there are no cuts predicted to be in classes ≤ 3, then only cuts predicted to be in classes ≤ 6 are kept. If there are no cuts predicted to be in classes ≤ 6, then all cuts are discarded, and the matching is done with a trivial cut containing just the root node.

7.3.4 Guiding Delay Constraints for Technology Mapping

After the logic optimization and technology mapping steps of logic synthesis, the overall EDA flow for integrated circuits performs physical synthesis optimizations and physical design steps such as floor planning and place and route (PnR). Ultimately, the quality-of-results (QoR) metrics after PnR is of actual importance.

Neto et al. [19] presents a supervised learning approach for guiding logic synthesis for post-PnR objectives using a learned model. In this approach, a learned model identifies post-PnR timing critical paths early in the logic synthesis flow (after logic optimization). The model predictions guide later stages of logic synthesis (technology mapping) to focus on reducing the delay of these paths. The benefits of this strategy are up to 15.53% area-delay-product improvements and 18.56% power-delay-product improvements post-PnR as compared to standard commercial tool flows.

The inputs to the model are various paths through the netlist of generic gates after logic optimization. Each path is represented as a grid where each row corresponds to a generic gate and the three columns correspond to the features: target clock period, gate type, normalized output load/fan out. Note that the target clock period feature is the same for all gates on the path. The gate-type feature uses a fixed embedding scheme of {not:1.0, nand2/nor2:2.0, and2/or2:3.0, complex2:4.0, flop:5.0}. The number of rows is fixed to the length L of the longest path in the dataset. The representation for shorter paths are padded with 0s. Thus, all paths have a $L \times 3$ grid representation.

As logic undergoes logic optimization, technology mapping, and then physical synthesis optimizations with floor planning and PnR, paths can be tracked using their endpoints. Thus, paths after the logic optimization step can be labeled specifying whether they are critical post-PnR or not. The threshold of labeling a path critical is if its post-PnR delay is >90% of the target clock period. A classification model is trained on this dataset in a supervised manner to predict whether paths after logic optimizations will be critical or not post-PnR. The path classification model has a convolutional architecture:

- CONV2D(3X3,1,32,1)→SIG→DROP→FLAT→FC(32,1)→SIG

During logic synthesis, all paths after the logic optimization step are classified by the trained model. If the model predicts that a path will be critical post-PnR, then an aggressive timing constraint is applied to the path during technology mapping and later steps. The aggressive timing constraints set a maximum delay value of 0 to the path. If the model predicts that the path will not be critical post-PnR, then the default maximum delay of the clock period will be applied to the path.

7.4 Reinforcement Learning Formulations for Logic Synthesis Algorithms

Machine learning models can learn optimization policies for various logic synthesis algorithms using reinforcement learning. In the following subsections, we will discuss reinforcement learning formulations for three logic synthesis algorithms.

7.4.1 Logic Network Optimization

The technology-independent logic optimization step in logic synthesis can be formulated as a reinforcement learning problem. Haaswijk et al. [20] demonstrates that reinforcement learning can offer significant improvements over existing state-of-the-art heuristic-based algorithms such as up to 86% size improvements and up to 47.4% depth improvements for logic networks. We will discuss the reinforcement learning formulation from [20] in this subsection.

This formulation uses MIG [4] logic networks where every node in the DAG is the three-input majority operator $M(x, y, z) = xy + xz + yz$ of the inputs x, y, z and every edge is marked by a regular or complemented attribute. The *size* of a MIG is the number of nodes in the MIG, and the *depth* of a MIG is the maximum number

of levels in the MIG. The objective of the logic optimization on MIGs is to either
reduce the size or depth. Minimizing these properties may correlate to minimizing
the area or delay of the technology-mapped implementations from the MIGs. The
reinforcement learning formulation trains general MIG optimization agents that can
optimize logic that the agent is not trained on. Haaswijk et al. [20] demonstrates that
agents trained on all four-input logic (2^{16} functions) can learn to optimize a much
larger set of six-input logic (2^{64}) that are unseen during training.

The state space S for a particular episode are all the possible MIGs that can
represent the logic function that is selected for the episode. Since the agent is trained
to be general, different episodes pick different logic functions. The starting states
of initial episodes are MIGs that correspond to the Shannon decomposition of the
functions. Later episodes are initialized with MIG representations of states that are
randomly selected among states that are encountered during previous episodes.

The MIG for state s is directly represented as a graph for the agent. Each node
in the graph is a majority node from the MIG, and the connectivity between nodes
is represented with a graph adjacency matrix of size $p \times p$ where p is the number
of nodes in the graph. Each node also has a feature vector of size p that one-hot
identifies the node. Whether an edge has an inverter or not may be encoded as an
edge feature (Fig. 7.2).

The action space A available to the agent is the set of all logic transformations
on state s such that the next state s' is logically equivalent to s. MIGs have a well-
defined set of such transformations [4] that can be applied to each node in the MIG.
If the agent takes an action (applying transformation on a node of the MIG), the
transition to the next state s' is deterministic. All possible MIGs for a logic function
are reachable using the MIG transformation action space. Table 7.4 lists the MIG
transformations in the action space. Note that the identity transformation is a pseudo

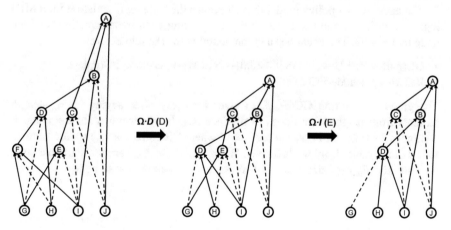

Fig. 7.2 Logic transformation actions on majority-inverter-graphs (MIGs). Each node is a three-
input majority operator. Dashed edges are inverted

Table 7.4 MIG node-level
transformations [4]

Transformation	Description
$\Omega \cdot C$	Commutativity
$\Omega \cdot M$	Majority
$\Omega \cdot A$	Associativity
$\Omega \cdot D$	Distributivity
$\Omega \cdot I$	Inverter propagation
$\Psi \cdot R$	Relevance
$\Psi \cdot C$	Complementary associativity
$\Psi \cdot S$	Substitution
ϵ	Identity

transformation that does not change the MIG : $s' = s$ but provides the agent a legal action when all the other transformations are illegal.

The rewards for both size and depth minimization agents are dense rewards that represent the improvement in a score function $score(s') - score(s)$ from state s to next state s'. For the size minimization agent, $score(s) = 1/size(s)$, and for the depth minimization agent, $score(s) = 1/depth(s)$. Note that other possible score functions for minimization of properties could be functions with negative correlations to properties such as $score(s) = -size(s)$ or $score(s) = -depth(s)$. The task horizon is finite over a fixed number of actions per episode. The agent is trained to maximize rewards that can be gathered in those fixed number of actions. The agent is trained using the policy gradient algorithm [11]. The agent produces a probability distribution of MIG transformation actions for all of the nodes in the MIG. The agent samples this probability distribution to pick an action. The agent uses a $p \times 9$ tensor to represent the probability of picking each of the nine transformation actions for each of the p nodes in the MIG.

The agent uses a policy model that consumes the state representation for a MIG logic network and outputs a probability for picking each transformation for each node in the MIG. The agent has a graph neural network architecture:

- GC(p,d)$\rightarrow$RELU$\rightarrow$(L-1)$\times$[GC(d,d)$\rightarrow$RELU]$\rightarrow$per-node-FC(d,d)$\rightarrow$
  RELU$\rightarrow$per-node-FC(d,9)

The graph convolution GC(d_{in},d_{out}) layer for every node aggregates d_{in}-length features from neighboring nodes to produce d_{out}-length features for the node. It computes the function $F_{out} = A \times F_{in} \times W$ where F (of size $p \times d_{in}$) are the features for all of the nodes from the last layer, W (of size $d_{in} \times d_{out}$) are the learnable layer weights, and F_{out} (of size $p \times d_{out}$) are the output features for all nodes.

7.4.2 Logic Synthesis Flow Optimization

While direct logic network optimization using reinforcement learning such as the method [20] from Sect. 7.4.1 is demonstrated to be effective for small logic networks at the scale of six-input logic networks, larger networks need more scalable techniques. Network-level logic transformations (Table 7.2, Table 7.5) scale to larger networks despite lacking the granularity of the node-level transformations (Table 7.4).

In Sect. 7.3.2, we discussed several supervised learning approaches [9, 14] for selecting optimal synthesis flows (sequence of network-level transformations) for a design. However, these approaches are limited to selecting from a set of randomly generated flows. We will discuss several reinforcement learning approaches [15–17] in this subsection that train agents to directly generate optimal synthesis flows. Note that [21] is also a promising approach of tuning synthesis flows and other logic algorithms such as conjunctive-normal-form (CNF) simplification. This approach uses a multi-armed bandit (MAB) [22] formulation that does not utilize a learned policy that considers the logic network state and has to be run separately for different designs.

The state space S in all of the approaches are all the possible and-inverter-graph (AIG) logic networks [3] of the target design for which the agent optimizes the synthesis flow. The action space A consists of the list of available AIG transformations (Table 7.2) from the ABC synthesis framework [1]. Transitions to next states are deterministic since a transformation applied to the current state s's AIG produces a specific AIG for the next state s'. Note that [17] also demonstrate their method on majority-inverter-graph (MIG) logic networks [4] with similar transformations (Table 7.5) from the Mockturtle synthesis framework [2].

7.4.2.1 Synthesis Flow Optimization for Circuit Area and Delay

Hosny et al. [15] optimizes synthesis flows for circuit area and delay after technology mapping of the logic network. More specifically, this approach tries to minimize area while meeting a certain delay constraint. This method can obtain up to 13% lower area than the initial design while lowering delay to meet delay

Table 7.5 MIG network-level transformations in Mockturtle [2]

Transformation	Description
b	balance
b-c	Balance only the critical path of the MIG
rw	Rewrite
rw -udc	Rewrite with don't cares
rw -azg	Rewrite with zero-cost replacements
rw -udc -azg	Rewrite with don't cares and zero-cost replacements

Table 7.6 Formulation of reward function for reducing area under delay constraints

			Area		
			Decrease	None	Increase
Delay	Met		+3	0	−1
	Not met	Decrease	+3	+2	+1
		None	+2	0	−2
		Increase	−1	−2	−3

constraints. $resyn2$ in comparison increased area by 26% in order to meet the same delay constraints.

A dense reward system is used (Table 7.6) that compares whether area or delay from state s to next state s' increased or decreased, and whether s' meets the delay constraint.

The agent is trained with the actor-critic algorithm [12]. Like [20] from Sect. 7.4.1, the agents sample actions from a probability distribution across actions given the current state.

The AIG for state s is represented using a vector of seven extracted and normalized features: (#input/output pins, #nodes, #edges, #levels, #latches, %AND, %NOT). Since this formulation uses actor-critic style reinforcement learning [12], the agent has an actor model and a critic model. The state representation is consumed by both models. The actor model outputs a vector of seven probabilities for sampling the seven actions. The critic model outputs the state value. Both models have feedforward multilayer perceptron (MLPs) architectures:

- actor: FC(7,20)→RELU→ FC(20,20)→RELU→FC(20,7)
- critic: FC(7,10)→RELU→FC(10,1)

7.4.2.2 Synthesis Flow Optimization for Logic Network Node and Level Counts

Zhu et al. [16] and Peruvemba et al. [17] aim to minimize the number of nodes and the number of levels of the logic network instead of the area or delay after technology mapping. Peruvemba et al. [17] also aims to minimize runtime of a synthesis flow since a synthesis flow using longer running transformations will have a longer runtime than a synthesis flow with the same number of shorter running transformations. While [16] can achieve fewer nodes and levels than $resyn2$, it has a longer runtime and is trained separately for different designs. Peruvemba et al. [17] is a more recent method that achieves even fewer nodes and levels (9.5%) while having a shorter runtime. Once trained on a set of designs together, the agent in [17] can generate synthesis flows on-the-fly (inference only) for the various designs. We will focus our discussion on [17].

The dense reward system which considers improvement in the number of nodes and levels, runtime, and performance of the baseline script (such as $resyn2$). Taking a transformation action a_k with runtime t_k transitions a logic network from state s_k

to the next state s_{k+1}. The value of a state $v(s_k)$ (Eq. 7.1) is a weighted sum of its node count $s_k.n$ and level count $s_k.l$.

$$v(s_k) = \frac{c_n * s_k.n + c_l * s_k.l}{c_1 + c_2} \qquad (7.1)$$

The reward is then calculated as the *betterment* $v(s_k) - v(s_{k+1})$ in a value normalized by the transformation runtime (Eq. 7.2). The reward is baselined for better convergence by subtracting the *betterment* in the baseline synthesis flow such as $resyn2$

$$r(k) = \frac{v(s_k) - v(s_{k+1})}{t_k} - baseline \qquad (7.2)$$

Each environment is defined as a combination of a baseline synthesis flow such as $resyn2$, the sequence length of that baseline, and the transformations used in that baseline. The agent is with episode lengths and available actions matching the baseline. For each environment, various agents with various weight pairs $(c_n, c_l) = \{(1, 0), (1, 1), (2, 1), (2, 3), (2, 7), (2, 9)\}$ to balance node and level counts are trained separately. For each environment and weight pair, a single agent is trained on various designs in a random order. After training the results are evaluated for the various designs by running the common agent with different designs as inputs.

The AIG or MIG for state s is represented using both extracted features and the graph itself. The representation also captures a history of few previous states and actions. An input vector of length i is the concatenation of the following scalars and vectors of features:

- Node and level counts of current state
- Node and level count of previous state
- One hot vectors for last three actions
- Runtime of expected sequence

The AIG or MIG graphs are represented using node features and node connectivity. A graph with p nodes has its connectivity represented with a graph adjacency matrix of size $p \times p$. Node features are represented using one-hot vectors of length f representing the node type. For AIGs, the two-input AND nodes are typed {constant, PI, PO, 0_inv, 1_inv, 2_inv}. For MIGs, the three-input majority nodes are typed {constant, PI, PO, 0_inv, 1_inv, 2_inv, 3_inv}. PI and PO are primary inputs and outputs of the logic network, while x_inv are graph nodes with x inverters on their inputs.

The agent is trained using the policy gradient algorithm (REINFORCE with baseline) [11]. The agent produces a probability distribution of transformation actions. The agent samples this probability distribution to pick an action. The agent uses a vector of size n to represent the probability of picking each of the n transformation actions available in the environment. Most of the environments are

AIG based. One of the environments is MIG based with six transformation actions (Table 7.5).

The agent uses a policy model that consumes the state representation for a logic network and outputs a probability for each transformation action. The agent also uses a state-value function model that outputs a scalar with the same input state representation. This output is used as a baseline in the training algorithm. The agent uses a mix of graph neural network and feedforward multilayer perceptron (MLPs) architectures:

- Policy model: BODY→FC(32,n)→SOFT
- State-value model: BODY→FC(32,1)
- BODY: FC(i,28)→RELU→CONCAT(g)→FC(32,32)→RELU
- GNN: GC(f,12)→RELU→GC(12,12)→RELU→GC(12,4)→MEAN→g

Refer to Sect. 7.4.1 for details of the GC operation. The input to the GNN block is the logic network graph representation with node features of length f (6 for AIGs and 7 for MIGs). The GNN block outputs a vector g of length 4 that is the mean of all the final node embeddings after the GC layers. The input to the BODY block is the state feature vector of length i.

7.4.3 Datapath Logic Optimization

Datapath logic optimization for datapath logic synthesis can be formulated as a reinforcement learning problem. Roy et al. [8] formulates a reinforcement learning approach to designing an important class of datapath circuits called (parallel) prefix circuits that are used for adders, priority encoders, inc(dec)rementers, and gray-to-binary code converters. Roy et al. [8] demonstrates that reinforcement learning agents can synthesize 64b prefix adder circuits that have 30.2% lower area than circuits generated by state-of-the-art algorithms. We will discuss the reinforcement learning formulation from [8] in this subsection.

A prefix graph is a directed acyclic graph that defines the architecture of a prefix circuit. Each node of a prefix graph computes an associative operator ∘ with two inputs (upper u and lower l) and one output o. The edges of a prefix graph have no attributes and simply define the flow of computation from parent to child nodes. A legal N-input prefix graph with inputs $x_{n-1} \ldots x_0$ must compute outputs $y_{n-1} \ldots y_0$ such that every output $y_i = x_i \circ x_{i-1} \circ \ldots \circ x_0$. A circuit can be implemented from the prefix graph by implementing the operator ∘ with logic gates. For example, adder circuits implement the carry propagate/generate operator $(P_o, G_o) = (P_u \cdot P_l, G_u + P_u \cdot G_l)$ for every prefix graph node using logic gates. Since the operator ∘ is associative, there are on the order of $\mathcal{O}(2^{N^2})$ legal N-input prefix graphs. The reinforcement learning state space S consists of all the legal N-input prefix graphs for the input width N that the agent is trained to optimize.

The action space A available to the agent is to either add a node to the prefix graph at state s or delete a node that exists in the prefix graph of state s. Each node is referenced using the most and the least significant bit (MSB,LSB) of the range of inputs that are in the node's predecessor, such that the node's output computes $o_{MSB,LSB} = x_{MSB} \circ \dots \circ x_{LSB}$. Every input node (i,i) and output node (i,0) must always exist in a legal prefix graph; the agent is allowed to add or delete other nodes (MSB,LSB) whose MSB>LSB. The deterministic next state s' is a prefix graph with a node added to or deleted from state s based on the action. The environment may add more nodes to s' as a legalization procedure in order to always maintain a legal prefix graph state. Since the agent is always able to modify any state s by either adding or deleting nodes, the task horizon is infinite.

The prefix graph of state s is represented using a two-dimensional grid. Since every node in the prefix graph can be referenced using the (MSB,LSB) notation, the grid is populated with 1s at (row:MSB, column:LSB) if the node (MSB,LSB) exists in the prefix graph. Otherwise the grid is populated with 0s. Additional channels on the input grid may also be used to encode normalized node features such as level and fan out instead of 1. Thus, the input to the agent is a $N \times N \times I$ tensor where N is the width of the prefix graph and I is the number of input channels. The agent represents actions to either add or delete a node (MSB,LSB) on the same grid system (Fig. 7.3).

There exists exact algorithms that can minimize the size and depth of prefix graphs [7]. However, these properties of prefix graphs do not directly correlate to the area, power consumption, or delay properties of the final prefix circuit after technology mapping and physical synthesis optimizations such as gate sizing and buffer insertion. In [8] the objective of the reinforcement learning agent is to minimize the final area and delay of prefix circuits after physical synthesis. However, since there are two objectives, area and delay, these objectives are combined with a scalarization weight: $\omega \cdot delay + (1 - \omega) \cdot area; 0 \leq \omega \leq 1$. A specialized agent is trained for a specific weight ω. For example, an agent trained

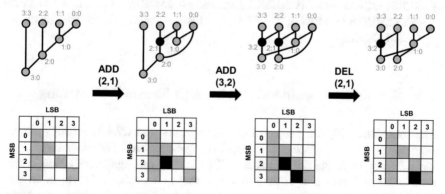

Fig. 7.3 Node addition and deletion actions on four-input prefix graph states. Corresponding two-dimensional grid representation is shown below each prefix graph

with $\omega = 0$ is focused on minimizing area, an agent trained with $\omega = 1$ is focused on minimizing delay, and an agent trained with $\omega = 0.5$ tries to balance the minimization of area and delay equally. The Pareto set of solutions from the various agents yield a minimal area circuit at various delay targets.

Since physical synthesis optimizations are not performed by the agent, the reinforcement learning environment measures the area and delay of every state by generating a technology mapped circuit from the prefix graph of the state and performing physical synthesis optimizations using a physical synthesis tool [23]. The physical synthesis tool itself can generate various area optimized circuits under various delay constraints. Thus, the same prefix graph state can lead to a Pareto set of circuits with numerous area-delay trade-offs. So, when training an agent for the scalarization weight ω, the environment selects the area and delay of the ω optimal circuit. The ω optimal circuit is the circuit that minimizes $\omega \cdot delay + (1 - \omega) \cdot area$ among the circuits produced by physical synthesis. The reinforcement learning environment produces dense rewards for area and delay separately: $r_{area} = area(s) - area(s')$ and $r_{delay} = delay(s) - delay(s')$. The scalarized version of the double-DQN algorithm [10] trains the agent to predict Q values for area and delay separately. The agent then picks the action with the largest weighted Q value: $argmax_a[\omega \cdot Q_{delay}(s, a) + (1 - \omega) \cdot Q_{area}(s, a)]$.

The agent uses a $\hat{Q}$-network model that consumes the state representation of prefix graph and outputs separate Q-values for area and delay, for (add, delete) actions for every node. Thus, the output of the model is a $N \times N \times 4$ tensor with the 4 output channels:

$(Q_{AREA}(ADD), Q_{AREA}(DELETE), Q_{DELAY}(ADD), Q_{DELAY}(DELETE))$

The model has a convolutional architecture:

- Q-network: PRERES→32×[RESBLOCK]→POSTRES
- PRERES: CONV2D(3X3,I,256,1)→BNORM→LRELU
- POSTRES: CONV2D(1X1,256,256,1)→BNORM→LRELU→CONV2D(1X1, 256,4,1)
- RESBLOCK: x→CONV2D(5X5,256,256,1)→BNORM→LRELU→ CONV2D (5X5,256,256,1)→BNORM→ADD(x)→LRELU where x is the input to RESBLOCK.

7.5 Scalability Considerations for Reinforcement Learning

The formulation [8] for datapath logic optimization (Sect. 7.4.3) requires circuit synthesis of the prefix graph of each state to compute rewards. The synthesis process has a latency of 35.56s for 64b adders. This latency can significantly slow down reinforcement learning algorithms if every step of an episode has to await the circuit synthesis latency. Reward computation latency may be a bottleneck for other reinforcement learning formulations of logic synthesis since the ideal quality-of-results (QoR) metrics are typically obtainable after running other EDA algorithms that may be of high latency. We will discuss some strategies from [8] that may relieve such reward computation bottlenecks.

In typical reinforcement learning algorithms, the agent only requires to observe the state in order to select an action. Rewards from previous actions are typically not consumed by the agent. Thus, if reward computation is decoupled from the agent's interaction with the environment, the agent can progress rapidly through the environment without awaiting the latency for reward computation. The decoupled reward computation can then be parallelized with high throughput across multiple processes to account for its latency.

If episodes always reset to the same state for training an agent on a fixed problem, it is likely that the early states in a training episode may be encountered repeatedly across episodes. As training progresses, more states may be encountered repeatedly across episodes as the agent converges on a strategy. Thus, to prevent recomputing rewards for the same states, the reward computation results may be cached. If a previous episode had already computed the QoR metrics for a state, the cached results can be reused for reward computation.

Another strategy that may be possible but not implemented in [8] is training a separate machine learning model for predicting the QoR metrics for a state. Generally the inference latency of a machine learning model is much less than the latency of running EDA flow steps. The accuracy of the machine learning model would be the key consideration as inaccurate reward predictions may misguide the reinforcement learning agent. A related consideration to accuracy would be the training dataset for the model. The model may be pretrained before reinforcement learning or trained on data gathered from initial iterations of reinforcement learning that are run without the model. Yu and Zhou [9] presents compelling results for a QoR predictor for synthesis flows (Sect. 7.3.2) which is both accurate and can be fine-tuned to new designs or technology processes with few additional data.

Finally, off-policy algorithms that utilize an experience buffer such as deep-Q-networks (DQN) can further decouple training the agent's machine learning model from the agent's interactions with the environment. Even if the agent's interaction with the environment is slow and produces experiences at a slow rate into the experience buffer, the machine learning model training can progress without bottlenecks on the data in the experience buffer.

References

1. Abc: A system for sequential synthesis and verification. Berkeley Logic Synthesis and Verification Group. http://www.eecs.berkeley.edu/~alanmi/abc/
2. Soeken, M., Riener, H., Haaswijk, W., De Micheli, G.: The EPFL logic synthesis libraries. In: International Workshop on Logic Synthesis (2018). Pre-print available at arXiv:1805.05121
3. Mishchenko, A., Chatterjee, S., Jiang, R., Brayton, R.K.: FRAIGs: a unifying representation for logic synthesis and verification. In: ERL Technical Report (2005)
4. Amarù, L.G., Gaillardon, P.E., De Micheli, G.: Majority-inverter graph: a novel data-structure and algorithms for efficient logic optimization. In: Design Automation Conference, pp. 194:1–194:6 (2014)

5. Mishchenko, A., Cho, S., Chatterjee, S., Brayton, R.: Combinational and sequential mapping with priority cuts. In: 2007 IEEE/ACM International Conference on Computer-Aided Design, pp. 354–361 (2007). https://doi.org/10.1109/ICCAD.2007.4397290

6. Chatterjee, S., Mishchenko, A., Brayton, R.K., Wang, X., Kam, T.: Reducing structural bias in technology mapping. In: IEEE Trans. Comput.-Aided Design Integr. Circuits Syst. **25**(12), 2894–2903 (2006). https://doi.org/10.1109/TCAD.2006.882484

7. Zimmermann, R.: Non-heuristic optimization and synthesis of parallel-prefix adders. In: Proc. of IFIP Workshop. Citeseer (1996)

8. Roy, R., et al.: PrefixRL: optimization of parallel prefix circuits using deep reinforcement learning. In: 2021 58th ACM/IEEE Design Automation Conference (DAC), pp. 853–858 (2021). https://doi.org/10.1109/DAC18074.2021.9586094

9. Yu, C., Zhou, W.: Decision making in synthesis cross technologies using LSTMs and transfer learning. In: 2020 ACM/IEEE 2nd Workshop on Machine Learning for CAD (MLCAD), pp. 55–60 (2020). https://doi.org/10.1145/3380446.3430638

10. van Hasselt, H., Guez, A., Silver, D.: Deep reinforcement learning with double q-learning. In: AAAI'16 Proceedings of the Thirtieth AAAI Conference on Artificial Intelligence, pp. 2094–2100 (2016)

11. Sutton, R.S., McAllester, D.A., Singh, S.P., Mansour, Y., et al.: Policy gradient methods for reinforcement learning with function approximation. In: NIPS, vol. 99, pp. 1057–1063 (1999)

12. Konda, V.R., Tsitsiklis, J.N.: Actor-critic algorithms. In: Advances in Neural Information Processing Systems, pp. 1008–1014 (2000)

13. Neto, W.L., Austin, M., Temple, S., Amaru, L., Tang, X., Gaillardon, P.-E.: LSOracle: a logic synthesis framework driven by artificial intelligence: invited paper. In: 2019 IEEE/ACM International Conference on Computer-Aided Design (ICCAD), pp. 1–6 (2019). https://doi.org/10.1109/ICCAD45719.2019.8942145

14. Yu, C., Xia, H., De Micheli, G.: Developing synthesis flows without human knowledge. In: 2018 55th ACM/ESDA/IEEE Design Automation Conference (DAC), pp. 1–6 (2018). https://doi.org/10.1109/DAC.2018.8465913

15. Hosny, A., Hashemi, S., Shalan, M., Reda, S.: DRiLLS: deep reinforcement learning for logic synthesis. In: 2020 25th Asia and South Pacific Design Automation Conference (ASP-DAC), pp. 581–586 (2020). https://doi.org/10.1109/ASP-DAC47756.2020.9045559

16. Zhu, K., Liu, M., Chen, H., Zhao, Z., Pan, D.Z.: Exploring logic optimizations with reinforcement learning and graph convolutional network. In: 2020 ACM/IEEE 2nd Workshop on Machine Learning for CAD (MLCAD), pp. 145–150 (2020). https://doi.org/10.1145/3380446.3430622

17. Peruvemba, Y.V., Rai, S., Ahuja, K., Kumar, A.: RL-guided runtime-constrained heuristic exploration for logic synthesis. In: 2021 IEEE/ACM International Conference On Computer Aided Design (ICCAD), pp. 1–9 (2021). https://doi.org/10.1109/ICCAD51958.2021.9643530

18. Neto, W.L., Moreira, M.T., Li, Y., Amarù, L., Yu, C., Gaillardon, P.-E.: SLAP: a supervised learning approach for priority cuts technology mapping. In: 2021 58th ACM/IEEE Design Automation Conference (DAC), pp. 859–864 (2021). https://doi.org/10.1109/DAC18074.2021.9586230

19. Neto, W.L., Trevisan Moreira, M., Amaru, L., Yu, C., Gaillardon, P.-E.: Read your circuit: leveraging word embedding to guide logic optimization. In: 2021 26th Asia and South Pacific Design Automation Conference (ASP-DAC), pp. 530–535 (2021)

20. Haaswijk, W., et al.: Deep learning for logic optimization algorithms. In: 2018 IEEE International Symposium on Circuits and Systems (ISCAS), pp. 1–4 (2018). https://doi.org/10.1109/ISCAS.2018.8351885

21. Yu, C.: FlowTune: practical multi-armed bandits in boolean optimization. In: 2020 IEEE/ACM International Conference On Computer Aided Design (ICCAD), pp. 1–9 (2020)

22. Auer, P., Cesa-Bianchi, N., Fischer, P.: Finite-time analysis of the multiarmed bandit problem. Mach. Learn. **47**(2), 235–256 (2002)

23. Agiza, A., Reda, S.: Openphysyn: an open-source physical synthesis optimization toolkit. In: 2020 Workshop on Open-Source EDA Technology (WOSET) (2020)

Chapter 8
RL for Placement and Partitioning

Anna Goldie and Azalia Mirhoseini

8.1 Introduction

Designing a chip's physical layout is an important problem in the overall chip design process. Modern computer chips may contain upwards of tens of thousands of macros (memory nodes) and billions of standard cells (logic gates such as NAND, NOR, XOR). Larger chips are partitioned into dozens of blocks, where blocks typically represent a specific submodule, such as a compute unit, control logic, interconnect, or a memory subsystem. At the placement stage, each block is described by a netlist, which is a hypergraph of macros and standard cells connected by wires. Chip placement is the task of placing this netlist onto a grid (chip canvas) with the objective of optimizing key metrics, such as performance, power, and area, while meeting requirements for density, timing, and routing congestion.

Six decades have been dedicated to the chip placement problem, and approaches can be categorized into four broad categories: partitioning-based methods [7, 14, 37], hill-climbing and stochastic approaches [26, 38, 42], analytic solvers [11, 16, 23, 29, 30, 48], and learning-based methods [33].

The main focus of this chapter is a deep reinforcement learning approach to the chip placement problem proposed in [33]. Unlike prior methods, this RL-based approach has the ability to improve as it solves more instances of the placement problem and can become both better and faster at this task over time.

This chapter first provides a history of research conducted on chip placement research. It then introduces reinforcement learning as a tool for decision-making optimization. Next, it discusses how to formulate chip placement as a reinforcement learning problem and how to achieve generalization (i.e., the transfer of knowledge

A. Goldie (✉) · A. Mirhoseini
Google Brain, Mountain View, CA, USA
e-mail: agoldie@stanford.edu; azalia@stanford.edu

© The Author(s), under exclusive license to Springer Nature Switzerland AG 2022
H. Ren, J. Hu (eds.), *Machine Learning Applications in Electronic Design Automation*, https://doi.org/10.1007/978-3-031-13074-8_8

across chips). Lastly, this chapter describes a number of future directions and the broader implications of RL for the chip design process.

8.2 Background

Since the 1960s, several approaches to chip placement have been proposed, which fall into three broad categories: (1) partitioning-based methods, (2) stochastic/hill-climbing methods, and (3) analytic solvers. Most recently, a fourth category based on deep learning-based methods has been proposed. In the following, these four categories are briefly discussed.

Early approaches to chip placement were largely focused on partitioning-based [7, 13, 14] and resistive-network-based methods [12, 46]. These methods take a hierarchical divide-and-conquer strategy, iteratively partitioning the netlist and canvas. Each of the subproblems then becomes tractable and can be more easily solved in isolation. While these approaches can scale to large chip blocks, they cannot achieve globally optimal performance, due to the partitioning decisions made early on. They especially struggle in modern chips, where congestion is an important, difficult to optimize, and holistic constraint.

Later on, hill-climbing algorithms, such as simulated annealing [26, 38, 42], were proposed and shown to be more effective. Simulated annealing was inspired by metallurgy, where metals are heated and then slowly cooled to create energy-stable crystalline structures. Simulated annealing works as follows: random perturbations are applied to an initial placement. The perturbations are operations, such as shifting the location of a node, swapping the location of two nodes, or flipping the orientation of a node. After either one or a number of such perturbations, the change in the objective function is calculated, and a decision is made about whether to keep the perturbations or revert them. The decision is controlled by a parameter referred to as temperature. If the perturbation improves the objective function, it will be immediately accepted. If the perturbation degrades the placement quality, then it is only accepted with probability related to temperature. As this iterative optimization process continues, the temperature is slowly reduced until local regression is no longer permitted. Simulated annealing is a well-known and effective algorithm for chip placement, but suffers from sample inefficiency, meaning that it requires many iterations and many measurements of the objective function to achieve convergence.

Multilevel partitioning methods were popular in the 1990s and 2000s [2, 37], as well as analytic techniques [3], such as force-directed methods [16, 30, 35, 45, 47, 48] and nonlinear optimizers [10, 17, 19, 20]. These methods achieved renewed success in large part due to the growing complexity and size of modern chips, which now contained 10s or 100s of millions of nodes. This increased complexity justified approximating the area of standard cells as zero, unlocking analytical approaches which were heavily focused on wirelength reduction. While these approaches are quite computationally efficient, they tend to produce placements that are inferior to those of nonlinear approaches.

Nonlinear methods are a class of optimization methods that employ nonlinear function approximations to model placement objectives. For example, the log-sum-exp [49] has been used to model wirelength, and Gaussian [9] and Helmholtz models to approximate density. A weighted sum of these approximate costs can then be used as a single objective function. However, the increased complexity of these methods requires a hierarchical approach, where clusters of nodes are placed, rather than individual ones.

Finally, in the past decade, there has been a resurgence in analytic techniques, including advanced quadratic methods [6, 23–25, 27] and electrostatics-based approaches, such as ePlace [29] and RePlAce [11]. ePlace draws an analogy to electrostatic systems, modeling each node as a charged particle whose charge is proportional to its area. The repulsive force between the nodes is thus a function of their area, and the system's potential energy is related to the density function. RePlAce [11] is an extension of ePlace that enables local density function optimization. DREAMPlace [28] further builds on this direction by taking a deep learning approach to model and solve the optimization problem, while using accelerators to considerably speed up the process as compared to prior methods.

Most recently, a deep reinforcement learning approach [33] was proposed for this classic problem. This approach can directly optimize for any target metric, including those which are non-differentiable. Furthermore, the learning-based approach leverages knowledge gained from previous examples to get better and faster at solving placement problems over time. The deep RL policy also exhibits interesting zero-shot behavior, meaning that a pre-trained policy is capable of generating reasonable placements for a previously unseen block with little or no fine-tuning.

8.3 RL for Combinatorial Optimization

There is a recent trend toward using deep reinforcement learning for combinatorial optimization. For example, a deep reinforcement learning method has been used to design neural network architectures (NAS) [53], solve the traveling salesman problem [5, 22], and perform hardware mapping and model parallelism [31, 32]. Following this trend is recent work on domain-adaptive compiler optimization [1, 36, 51, 52] and a new approach to the maximum cut problem [4]. A 2021 *Nature* publication [33] proposes Circuit Training, a deep reinforcement learning method for chip placement that has been open-sourced [15] and used to design Google's latest AI accelerator (TPU). This method generated TPU placements that were frozen all the way to tapeout and are currently running in Google datacenters.

Unlike traditional supervised learning approaches, reinforcement learning methods do not require access to labeled data, making them well-suited to this challenging, low-resource domain. Historically, most successful applications of machine learning have been examples of supervised learning, meaning that the model is trained to predict an output label given an input example (e.g., label the species for a photo of an animal). Today, most top-performing models are neural networks, where a deep learning model learns to map the input examples to the output labels through optimization of a differentiable loss function via gradient descent.

In contrast, reinforcement learning (RL) involves a model (which may be parameterized as a deep neural network) learning to maximize a reward function through actions it takes within an environment (which may be simulated or real). If the RL agent is parameterized as a deep neural network, then one would refer to this method as deep reinforcement learning. One famous example is AlphaGo [44], which is a deep reinforcement learning agent that learned to play the game of Go. In this case, the actions consist of all feasible moves at a given game state, and the reward is the final victory or loss after the game ends.

8.3.1 How to Perform Decision-Making with RL

The following section describes how to pose a decision-making problem as a reinforcement learning task.

One way to formulate an RL task is as a Markov decision process (MDP), which makes the assumption that a given state s_t at time t depends only on its immediately preceding state s_{t-1} at time $t - 1$ and the action a_t taken at time t and is otherwise conditionally independent from all previous states and actions.

More concretely, the following equation must hold:

$$P_a(s_t|s_0 \ldots s_{t-1}) = P_a(s_t|s_{t-1})$$

Generally, an RL task thus consists of the following elements:

- States: the set of valid states for the environment (e.g., the set of all possible board states in Go)
- Actions: the set of moves or decisions that can be taken from the current state (e.g., all feasible moves that can be taken from a given board state in Go)
- State transition probabilities: the transition probability between one state to another.
- Reward: the metric or objective function to maximize (e.g., 1 for victory, 0 for loss)
- Reward discount factor: the extent to which future reward is considered less valuable than immediate reward.

The RL agent takes an iterative approach to optimizing its choice of actions. This optimization is performed over multiple episodes, where each episode consists of a finite number of state transitions terminating when either the target reward is reached or a maximum number of transitions have occurred. In each optimization episode, the agent starts from an initial state s_0. At each time step t, the RL agent takes action (a_t), transitions from state s_t to a new state (s_{t+1}), and collects a reward (r_t) from interacting with the environment as shown in Fig. 8.1. By performing many such

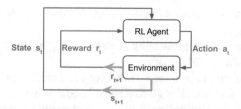

Fig. 8.1 In reinforcement learning, the goal of an agent is to take actions that maximize cumulative reward. At each time step, the RL agent takes an action given the current state of the environment, transitions to a new state given the chosen action, and receives a reward from the environment

episodes and using the resulting reward to update the parameters of its policy, the agent gradually learns to maximize cumulative reward by taking better actions.

At a high level, the goal of the RL agent is to maximize reward, but how can this best be achieved? One common strategy is to carefully model the value of a given state (meaning the expected reward from that state) and then to have the agent move toward states that have the highest predicted value. This strategy is referred to as value iteration.

Another strategy is to train a network which directly predicts the action which will produce the highest reward. These methods are referred to as policy gradient methods and include popular algorithms such as REINFORCE [50], A3C (Asynchronous Advantage Actor-Critic) [34], TRPO (Trusted Region Policy Optimization) [39], and Proximal Policy Optimization (PPO) [40].

RL is especially suitable in settings where there is little or no labeled data available or where the objective function is non-differentiable. Furthermore, unlike brute force or greedy search methods, RL has been shown to scale to massive design search spaces, such as those in chip floorplanning [33] and Go [44].

8.4 RL for Placement Optimization

The following section describes how RL policy optimization can be used to solve the challenging problem of chip placement, as originally proposed in the 2021 *Nature* paper [33].

The chip placement problem can be formulated as a sequential Markov decision process (MDP):

- **States** represent the state of the placement so far, including information about the netlist (the connectivity between nodes), features of each node (e.g., width and height) and each edge (connection count), index of the next node to be placed, and other properties of the netlist (total count of macros, number of standard cell clusters, and wire count, as well as routing allocations, which are a function of technology node size).

- **Actions** are all valid locations (on the target chip canvas) where the current node can be placed next.
- **State transitions** are the probability of transitioning between two states, given a current state and a candidate action.
- **Rewards** are defined to be 0 for all intermediate actions within an episode aside from the final action, which is given a reward equal to the negative weighted sum of approximate (proxy) wirelength, density, and routing congestion.

The RL agent is modeled by a neural network parameterized by θ. The parameters of this neural network are then optimized using PPO [41]. The Circuit Training algorithm optimizes the placement decisions by minimizing the following cost function $J(\theta, G)$:

$$J(\theta, G) = \frac{1}{K} \sum_{g \sim G} E_{g, p \sim \pi_\theta}[R_{p,g}] \tag{8.1}$$

The policy optimizes placements for netlists within the dataset G. Each individual netlist is denoted by g. The number of netlists is shown by k. For a given placement p for netlist g, the reward is denoted by $R_{p,g}$:

$$R_{p,g} = -Wirelength(p, g) \tag{8.2}$$
$$-\lambda \, Congestion(p, g) - \gamma \, Density(p, g)$$

In each episode, the RL agent places the macros of the netlist one at a time onto the chip canvas. After placing all macros, a force-directed method [35, 45, 47, 48] is used to perform coarse placement of the standard cell clusters. At the end of each episode, the reward is calculated by taking a weighted sum of proxy wirelength, routing congestion, and cell density.

8.4.1 The Action Space for Chip Placement

In the Circuit Training algorithm [15, 33], the canvas is defined to be an $m \times n$ grid, where m and n are integers representing the number of rows and columns, respectively. Limiting the number of actions to a discrete number is helpful for policy optimization. At each time step, the set of actions are possible grid cells onto which no macro has yet been placed. In practice, in order to ensure that a newly placed macro does not overlap with previously placed macros, a mask is used to limit the action space to grid cells that satisfy this constraint. Thus, unlike methods such as ePlace or RePlace, this method maintains legalization at each point in the placement process. The RL policy generates a probability distribution over all of the $m \times n$ grid cells, and the next action is selected by sampling from the masked probability distribution. Over the process of training, this distribution converges toward optimized placements that maximize the reward function.

Circuit Training took a hybrid approach to the placement problem inspired by the production workflow employed by physical engineers. In these settings, human experts typically place the larger components (macros) and then allow a commercial autoplacer tool to place standard cells. The hybrid approach works as follows: (1) In each policy iteration, the macros are placed one at a time onto the canvas, (2) a force-directed method is then used to place clusters of standard cells, and (3) a reward signal is calculated and used to update the parameters of the RL agent, as described in the following section.

8.4.2   Engineering the Reward Function

Reward is meant to capture the quality of a given placement, meaning that an agent which is able to maximize this quantity should be able to generate high-quality chip layouts. Reinforcement learning is typically sample inefficient, requiring many iterations to achieve policy convergence. Therefore, not only is it critical that the reward functions correlate with the true objective function (PPA), but it must also be possible for them to be calculated quickly. The output of a commercial electronic design automation (EDA) tool is a highly accurate measurement of key metrics, such as wirelength, routing congestion, cell density, timing, and power, and is closest to ground truth at this stage of the chip design process. However, these measurements can take tens of hours to generate and require expensive licenses, so it is not possible for this to be used as a reward signal in the loop of RL. Therefore, Circuit Training defines approximate cost functions that can be quickly calculated and which correlate with the true reward.

8.4.2.1   Wirelength

The wirelength reward is calculated by taking the half perimeter of the bounding boxes containing all components of the netlist. This is a standard way to approximate wirelength [8, 18, 21, 43]. See the equation below for a definition of HPWL for a given net (hyper-edge) i of the netlist:

$$HPWL(i) = (MAX_{b \in i}\{x_b\} - MIN_{b \in i}\{x_b\} + 1)$$
$$+ (MAX_{b \in i}\{y_b\} - MIN_{b \in i}\{y_b\} + 1) \qquad (8.3)$$

The x and y coordinates of net i are defined as x_b and y_b. To calculate the wirelength for the entire netlist, a sum is taken over the half-perimeters of each bounding box:

$$HPWL(netlist) = \sum_{i=1}^{N_{netlist}} HPWL(i) \qquad (8.4)$$

Not only can HPWL be calculated quickly, but it also correlates with important metrics, such as power and timing, which are far more expensive to approximate. As a result, Circuit Training is able to generate placements that meet power and timing requirements, in spite of the fact that it does not directly optimize for these metrics.

8.4.2.2 Routing Congestion

The congestion reward function is another key metric for placement. Congestion is important because after placement is performed, the nodes must be routed (connected together with wires). There are limited routing resources present at each position in the grid, which is a function of the technology node size and the number of wafers. For a given placement, if no routing can be found that satisfies these constraints at each position, that placement is invalid and cannot be manufactured.

The Circuit Training method uses an approximate approach to quickly calculate congestion. It calculates the congestion for each horizontal and vertical cell. It then employs a 5×1 convolutional filter to smooth this estimate, as a routing violation may be acceptable if there are nearby routing resources available to spill into. The congestion cost (negative of reward function) is then the average of the top 10% most congested cells.

8.4.2.3 Density and Macro Overlap

Another key metric is density. A manufacturable placement cannot contain macros which overlap. For standard cells, there is a threshold on the number of standard cells that can be located in a given area, meaning that there is a threshold on the maximum density allowed.

In Circuit Training, standard cell density is treated as a soft constraint by modeling it as a weighted component of the reward function shown in Eq. 8.2. On the other hand, no macro overlap is permitted and this hard constraint is enforced via a mask function. At each time step, the mask is updated to show only available grid locations onto which the current node can be placed without any macro overlap. This approach reduces the number of infeasible placements proposed by the policy, especially early in the training process. The mask function also enables accounting for physical constraints, such as blockages and clockstraps.

8.4.2.4   State Representation

At each time step, the state must contain the information necessary to make reward-maximizing decisions about where to place the next node. In Circuit Training, state consists of an adjacency matrix representing the connectivity of all nodes within the netlist (described in Sect. 8.4.3), features of each node (e.g., their width, height, and type), features of each hyper-edge (e.g., number of connections), the index of the node to place next, and chip metadata (e.g., total number of macros and clusters of standard cells, routing allocations, and the underlying technology node size). This information is aggregated and then processed by a deep neural network architecture (described in Sect. 8.4.4), which learns useful representations of the netlist and its partial placement from which to make optimized placement decisions.

8.4.3   Generating Adjacency Matrix for a Chip Netlist

This section describes the process by which an input chip netlist (hypergraph) is transformed into an adjacency matrix that can be consumed by the neural encoder described in the next section. For each pair of nodes (macros or standard cell clusters) with register distance 4 or less, an edge is generated connecting them. The weight for each edge decays exponentially with register distance, starting with a weight of 1 for register distance 0 and halved for each incremental increase in register distance.

8.4.4   Learning RL Policies that Generalize

Learning-based methods are capable of transferring knowledge gained from prior experience and leveraging that knowledge to better optimize previously unseen examples. Generalization is relatively well understood and more easily achieved in the context of supervised learning, where there are labels for each training example. The same does not hold for chip placement, where the number of data examples (chip netlists) is very limited and labels must be generated by physical design experts or expensive electronic design automation tools. Furthermore, for a single netlist, there are a combinatorially large number of possible placements, each with their own reward.

For chip placement, even learning to solve a single instance of the problem (how to place a given netlist) is both very challenging and of high value. Placing a given netlist is analogous to a game with its own particular board, pieces, rules, and win conditions. Netlists can vary greatly in the number and dimension of their macros, the number and type of standard cells, the connectivity between nodes (macros and standard cells), the size and aspect ratio of their canvas, and the relative importance of different metrics (e.g., power consumption may matter far more for a chip in a mobile phone).

Although it can be useful and challenging to train a policy to place just one netlist (e.g., if there is only one netlist in G in Eq. 8.1, there is even greater value and challenge in training a policy capable of performing the more general task of placing any given netlist. In practice, training a policy from scratch for a given netlist takes about 24 h. From the perspective of a physical design engineer, such a method would be of far more practical value if it could return a high-quality result in just a few hours or at least be run overnight. Furthermore, if the runtime could be reduced to subseconds, powerful new design space explorations become viable, as discussed in the Future Work section.

Circuit Training [33] approached the problem of developing generalizable placement policies by first finding neural architectures that could predict the reward function for partial placements of a wide variety of chip blocks. This reward prediction architecture was then used as the encoder of the deep RL policy, taking in information about the current state of the problem and converting that information into embeddings that aid in the policy's decision-making. The intuition behind this approach was that, in order to be able to generate high-quality placements for any given netlist (i.e., generalize on the placement task), the RL policy must be capable of recognizing the quality of any given placement (i.e., generalize on the task of predicting reward). In this sense, predicting reward is a strictly easier subtask of the overall placement problem and can be conveniently formulated as a supervised learning task.

To generate the dataset, Circuit Training ran a vanilla policy on 5 different chip netlists and collected a total of 10,000 different placement and their corresponding labels, i.e., the reward function as described in Eq. 8.2. It then proposed a new graph neural network architecture, called Edge-GNN, to solve the supervised task of predicting the labels, given the partially placed netlists. The graph neural network works as follows:

$$e_{ij} = fc_e(concat(v_i | v_j | w_{ij}^e))$$
$$v_i = mean_{j \in \mathcal{N}(v_i)}(e_{ij})$$
(8.5)

Here v_i for $1 <= i <= N$ is the node embedding, and N is the total number of netlist nodes (e.g., macros and standard cell clusters). e_{ij} denotes the vector representation (embedding) of the edge connection nodes v_i and v_j.

The node embeddings are initially generated by concatenating several features, such as node type (e.g., macro or standard cell cluster), width, height, and grid position (x-y coordinate). The Edge-GNN then iteratively updates edge and node embeddings by applying Eq. 8.5. In practice, running this algorithm for seven iterations produces effective node representations.

The supervised model has three components, (1) the Edge-GNN described above which generates edge and node embeddings, (2) a feedforward network to embed netlist metadata, and (3) a feedforward network that takes as input a vector representation of the partially placed netlist (i.e., the average of edge embeddings

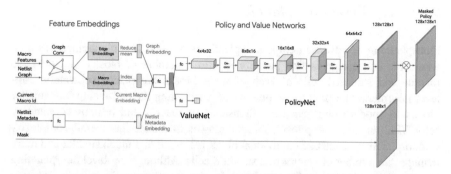

Fig. 8.2 The RL policy is modeled with a neural network. This figure is from [33], which describes the policy architecture in more detail. At each time step, the policy inputs are the netlist and the current node to be placed. The policy encoder uses Edge-GNN, an edge-based graph neural network to embed information about the netlist and the current node. The policy decoder uses a cascade of deconvolutional layers. The policy output is a probability distribution over grid locations to place the current node. A mask is used to prevent the policy from generating infeasible placements. The mask is updated at each step and captures physical constraints, such as blockages and macro overlaps

generated in step 1) and the embedding of netlist metadata (i.e., the output of step 2), and predicts the reward as output. This supervised model is trained using an ℓ_2 regression loss.

The RL agent architecture is shown in Fig. 8.2. The supervised model described above is used as the RL agent's encoder. The RL agent's decoder consists of two components: (1) a policy network which is composed of a number of deconvolutional layers that map to a logit layer whose size is equal to that of the grid and (2) a value network that maps the output of the encoder to a single number, which is a prediction of the partial reward at any given state.

The policy is pre-trained on a training set of chip netlists. To find optimized placements for a new netlist, the pre-trained weights of the policy are loaded and fine-tuned for that netlist. A placement can also be achieved with no fine-tuning. In the deep learning literature, this is referred to as a zero-shot approach. Circuit Training showed that the quality of fine-tuned placements is superior to those achieved by zero-shot; however it also showed that as the training set size becomes larger, the gap between the quality of zero-shot placements and fine-tuned placements diminishes, which encourages further research toward enabling zero-shot to match fine-tuned results.

8.5 Future Directions

This final section discusses a number of future directions on RL for physical design and its broader implications for the entire chip design process.

8.5.1 Top-Level Floorplanning

Floorplanning of advanced chips is typically performed in a hierarchical fashion, where the entire chip is first partitioned into a number of blocks, and then the macros and standard cells of each of these blocks are placed separately. Top-level floorplanning refers to the process of partitioning the entire chip into smaller blocks. A modern chip can have thousands of macros and billions of standard cells. The purpose of top-level floorplanning is to reduce the overall complexity of floorplanning, since each individual block is limited to a much smaller and more manageable number of macros and standard cells. Although top-level floorplanning has an outsized impact on final quality, it is often done manually, and heuristics are used to shape and place the blocks. To evaluate a given top-level floorplan, one must place all the nodes (macros and standard cells) within all the individual blocks.

Circuit Training could thus be used to greatly speed up the top-level floor-planning problem by automating the lower-level placements. This hierarchical optimization could have an outer loop which optimizes the partitioning of the chip and an inner loop which uses Circuit Training for placement of the nodes within the blocks. To reduce iteration time, the time that Circuit Training uses for fine-tuning the placement could also be limited.

Circuit Training could also be used to further optimize the top-level floorplanning by performing pin placement optimization and block shaping. The pin placement problem requires finding the optimized placement of pins on the boundaries of each block. The block shaping problem requires determining the width and height of rectangular blocks or potentially inclusion of non-rectilinear blocks. The action space of Circuit Training can be extended to find optimized decisions about placement of pins, as well as the shapes of blocks.

8.5.2 Netlist Design Space Exploration

Different design choices in creating the netlist can drastically affect the quality of the chip. For example, a given memory node in the netlist can be implemented using a number of smaller macros or one larger macro, from a library of macros. Circuit Training's zero-shot and few-shot modes enable quick ways to place a netlist and evaluate a design choice. The zero-shot model generates a placement at inference (in sub-seconds), and the few-shot mode provides a way to improve the quality of placement while limiting the time needed for fine-tuning.

8.5.3 Broader Discussions: ML for Co-optimization Across the Overall Chip Design Process

Automating placement, previously a time-consuming and labor-intensive step, has implications for the overall chip design process. For example, the zero-shot or few-shot capabilities of learning-based methods like Circuit Training could enable co-optimization with earlier stages of chip design process such as architecture design and synthesis.

Deep RL methods can also be developed to solve other combinatorial optimization and decision-making problems across the chip design stack, but would require formulating new action spaces, reward functions, and state representations that are effective in these domains. There is enormous potential for learning-based method to improve the speed and quality of future hardware. These learning-based methods only grow stronger as they are exposed to more data, and as their neural architectures, learning algorithms, and reward functions are further optimized.

References

1. Addanki, R., Venkatakrishnan, S.B., Gupta, S., Mao, H., Alizadeh, M.: Learning generalizable device placement algorithms for distributed machine learning. In: Advances in Neural Information Processing Systems, vol. 32, pp. 3981–3991 (2019)
2. Agnihotri, A., Ono, S., Madden, P.: Recursive bisection placement: Feng Shui 5.0 implementation details. In: Proceedings of the International Symposium on Physical Design, pp. 230–232 (2005). https://doi.org/10.1145/1055137.1055186
3. Alpert, C., Caldwell, A., Chan, T., Huang, D.H., Kahng, A., Markov, I., Moroz, M.: Analytical engines are unnecessary in top-down partitioning-based placement. VLSI Design 10 (2002). https://doi.org/10.1155/1999/93607
4. Barrett, T.D., Clements, W.R., Foerster, J.N., Lvovsky, A.I.: Exploratory combinatorial optimization with reinforcement learning (2020). arXiv preprint arXiv:1909.04063v2
5. Bello, I., Pham, H., Le, Q.V., Norouzi, M., Bengio, S.: Neural Combinatorial Optimization with Reinforcement Learning (2016)
6. Brenner, U., Struzyna, M., Vygen, J.: BonnPlace: placement of leading-edge chips by advanced combinatorial algorithms. Trans. Comp.-Aided Des. Integr. Circuits Syst. 27(9), 1607–1620 (2008). https://doi.org/10.1109/TCAD.2008.927674
7. Breuer, M.A.: A class of min-cut placement algorithms. In: Proceedings of the 14th Design Automation Conference, DAC 1977, pp. 284–290. IEEE Press (1977)
8. Caldwell, A.E., Kahng, A.B., Mantik, S., Markov, I.L., Zelikovsky, A.: On wirelength estimations for row-based placement. IEEE Trans. Comput.-Aided Design Integr. Circuits Syst. 18(9), 1265–1278 (1999). https://doi.org/10.1109/43.784119
9. Chen, T., Jiang, Z., Hsu, T., Chen, H., Chang, Y.: NTUplace3: an analytical placer for large-scale mixed-size designs with preplaced blocks and density constraints. IEEE Trans. Comput.-Aided Design Integr. Circuits Syst. 27(7), 1228–1240 (2008). https://doi.org/10.1109/TCAD.2008.923063
10. Chen, T.C., Jiang, Z.W., Hsu, T.C., Chen, H.C., Chang, Y.W.: A high-quality mixed-size analytical placer considering preplaced blocks and density constraints. In: Proceedings of the 2006 IEEE/ACM International Conference on Computer-Aided Design, pp. 187–192. Association for Computing Machinery (2006)

11. Cheng, C., Kahng, A.B., Kang, I., Wang, L.: Replace: Advancing Solution Quality and Routability Validation in Global Placement, pp. 1717–1730 (2019)
12. Chung-Kuan Cheng, Kuh, E.S.: Module placement based on resistive network optimization. IEEE Trans. Comput.-Aided Design Integr. Circuits Syst. 3(3), 218–225 (1984). https://doi.org/10.1109/TCAD.1984.1270078
13. Dunlop, A.E., Kernighan, B.W.: A procedure for placement of standard-cell VLSI circuits. IEEE Trans. Comput.-Aided Design Integr. Circuits Syst. 4(1), 92–98 (1985). https://doi.org/10.1109/TCAD.1985.1270101
14. Fiduccia, C.M., Mattheyses, R.M.: A linear-time heuristic for improving network partitions. In: 19th Design Automation Conference, pp. 175–181 (1982). https://doi.org/10.1109/DAC.1982.1585498
15. Guadarrama, S., Yue, S., Boyd, T., Jiang, J.W., Songhori, E., Tam, T., Mirhoseini, A.: Circuit training: an open-source framework for generating chip floor plans with distributed deep reinforcement learning (2021). https://github.com/google_research/circuit_training. Accessed 21 Dec 2021
16. Hu, B., Marek-Sadowska, M.: Multilevel fixed-point-addition-based VLSI placement. IEEE Trans. Comput.-Aided Design Integr. Circuits Syst. 24(8), 1188–1203 (2005). https://doi.org/10.1109/TCAD.2005.850802
17. Kahng, A.B., Qinke Wang: Implementation and extensibility of an analytic placer. IEEE Trans. Comput.-Aided Design Integr. Circuits Syst. 24(5), 734–747 (2005). https://doi.org/10.1109/TCAD.2005.846366
18. Kahng, A.B., Reda, S.: A tale of two nets: Studies of wirelength progression in physical design. In: Proceedings of the 2006 International Workshop on System-Level Interconnect Prediction, SLIP '06, pp. 17–24. Association for Computing Machinery, New York (2006). https://doi.org/10.1145/1117278.1117282
19. Kahng, A.B., Reda, S., Qinke Wang: Architecture and details of a high quality, large-scale analytical placer. In: ICCAD-2005. IEEE/ACM International Conference on Computer-Aided Design, pp. 891–898 (2005). https://doi.org/10.1109/ICCAD.2005.1560188
20. Kahng, A.B., Wang, Q.: An analytic placer for mixed-size placement and timing-driven placement. In: IEEE/ACM International Conference on Computer Aided Design, ICCAD-2004, pp. 565–572 (2004). https://doi.org/10.1109/ICCAD.2004.1382641
21. Kahng, A.B., Xu, X.: Accurate pseudo-constructive wirelength and congestion estimation. In: Proceedings of the 2003 International Workshop on System-Level Interconnect Prediction, SLIP '03, pp. 61–68. Association for Computing Machinery, New York (2003). https://doi.org/10.1145/639929.639942
22. Khalil, E., Dai, H., Zhang, Y., Dilkina, B., Song, L.: Learning combinatorial optimization algorithms over graphs. In: Advances in Neural Information Processing Systems, vol. 30, pp. 6348–6358 (2017)
23. Kim, M., Lee, D., Markov, I.L.: SimPL: an effective placement algorithm. IEEE Trans. Comput.-Aided Design Integr. Circuits Syst. 31(1), 50–60 (2012). https://doi.org/10.1109/TCAD.2011.2170567
24. Kim, M.C., Markov, I.L.: ComPLx: a competitive primal-dual lagrange optimization for global placement. In: Design Automation Conference 2012, pp. 747–755 (2012)
25. Kim, M.C., Viswanathan, N., Alpert, C.J., Markov, I.L., Ramji, S.: MAPLE: multilevel adaptive placement for mixed-size designs. In: Proceedings of the 2012 ACM International Symposium on International Symposium on Physical Design, ISPD 2012, pp. 193–200. Association for Computing Machinery, New York (2012). https://doi.org/10.1145/2160916.2160958
26. Kirkpatrick, S., Gelatt, C.D., Vecchi, M.P.: Optimization by simulated annealing. Science 220(4598), 671–680 (1983)
27. Lin, T., Chu, C., Shinnerl, J.R., Bustany, I., Nedelchev, I.: POLAR: placement based on novel rough legalization and refinement. In: Proceedings of the International Conference on Computer-Aided Design, pp. 357–362. IEEE Press (2013)

28. Lin, Y., Dhar, S., Li, W., Ren, H., Khailany, B., Pan, D.Z.: Dreamplace: deep learning toolkit-enabled GPU acceleration for modern vlsi placement. In: Proceedings of the 56th Annual Design Automation Conference 2019, DAC '19. Association for Computing Machinery, New York (2019)
29. Lu, J., Chen, P., Chang, C.C., Sha, L., Huang, D.J.H., Teng, C.C., Cheng, C.K.: ePlace: Electrostatics-Based Placement Using Fast Fourier Transform and Nesterov's Method (2015)
30. Luo, T., Pan, D.Z.: DPlace2.0: A stable and efficient analytical placement based on diffusion. In: 2008 Asia and South Pacific Design Automation Conference, pp. 346–351 (2008). https://doi.org/10.1109/ASPDAC.2008.4483972
31. Mirhoseini, A., Goldie, A., Pham, H., Steiner B., Le, Q. V. & Dean, J.: A hierarchical model for device placement. In: Proceedings of the International Conference on Learning Representations (2018)
32. Mirhoseini, A., Pham, H., Le, Q. V., Steiner, B., Larsen, R., Zhou, Y., Kumar, N., Norouzi, M., Bengio, S., Dean, J.: Device placement optimization with reinforcement learning. In: Proceedings of the International Conference on Machine Learning (2017)
33. Mirhoseini and Goldie, Yazgan, M., Jiang, J.W., Songhori, E., Wang, S., Lee, Y.J., Johnson, E., Pathak, O., Nazi, A., Pak, J., Tong, A., Srinivasa, K., Hang, W., Tuncer, E., V. Le, Q., Laudon, J., Ho, R., Carpenter, R., Dean, J.: A graph placement methodology for fast chip design. Nature **594**(7862), 207–212 (2021)
34. Mnih, V., Badia, A.P., Mirza, M., Graves, A., Lillicrap, T.P., Harley, T., Silver, D., Kavukcuoglu, K.: Asynchronous methods for deep reinforcement learning (2016)
35. Obermeier, B., Ranke, H., Johannes, F.M.: Kraftwerk: a versatile placement approach. In: Proceedings of the 2005 International Symposium on Physical Design, pp. 242–244. Association for Computing Machinery, New York (2005). https://doi.org/10.1145/1055137.1055190
36. Paliwal, A., Gimeno, F., Nair, V., Li, Y., Lubin, M., Kohli, P., Vinyals, O.: Reinforced genetic algorithm learning for optimizing computation graphs. In: Proceedings of International Conference on Learning Representations (2020)
37. Roy, J.A., Papa, D.A., Markov, I.L.: Capo: Congestion-Driven Placement for Standard-cell and RTL Netlists with Incremental Capability, pp. 97–133. Springer, Boston (2007)
38. Sarrafzadeh, M., Wang, M., Yang, X.: Dragon: A Placement Framework, pp. 57–89. Springer, Boston (2003)
39. Schulman, J., Levine, S., Moritz, P., Jordan, M.I., Abbeel, P.: Trust region policy optimization (2015)
40. Schulman, J., Wolski, F., Dhariwal, P., Radford, A., Klimov, O.: Proximal policy optimization algorithms (2017)
41. Schulman, J., Wolski, F., Dhariwal, P., Radford, A., Klimov, O.: Proximal Policy Optimization Algorithms (2017)
42. Sechen, C.M., Sangiovanni-Vincentelli, A.L.: TimberWolf3.2: a new standard cell placement and global routing package. In: DAC, pp. 432–439. IEEE Computer Society Press (1986). https://doi.org/10.1145/318013.318083
43. Shahookar, K., Mazumder, P.: VLSI cell placement techniques. ACM Comput. Surv. **23**(2), 143–220 (1991). https://doi.org/10.1145/103724.103725
44. Silver D., H.A.M.C.: Mastering the game of go with deep neural networks and tree search. Nature (2016)
45. Spindler, P., Schlichtmann, U., Johannes, F.M.: Kraftwerk2-A fast force-directed quadratic placement approach using an accurate net model. IEEE Trans. Comput.-Aided Design Integr. Circuits Syst. **27**(8), 1398–1411 (2008). https://doi.org/10.1109/TCAD.2008.925783
46. Tsay, R.S., Kuh, E., Hsu, C.P.: Proud: A Fast Sea-of-Gates Placement Algorithm, pp. 318–323 (1988)
47. Viswanathan, N., Nam, G.J., Alpert, C., Villarrubia, P., Ren, H., Chu, C.: RQL: global placement via relaxed quadratic spreading and linearization. In: Proceedings of Design Automation Conference, pp. 453–458 (2007). https://doi.org/10.1145/1278480.1278599
48. Viswanathan, N., Pan, M., Chu, C.: FastPlace: An Efficient Multilevel Force-Directed Placement Algorithm, pp. 193–228. Springer, Berlin (2007). https://doi.org/10.1007/978-0-387-68739-1_8

49. William, N., Ross, D., Lu, S.: Non-linear optimization system and method for wire length and delay optimization for an automatic electric circuit placer. In: Patent US6301693B1 (2001)
50. Williams, R.: Simple Statistical Gradient-Following Algorithms for Connectionist Reinforcement Learning. Mach Learn (1992)
51. Zhou, Y., Roy, S., Abdolrashidi, A., Wong, D., Ma, P.C., Xu, Q., Zhong, M., Liu, H., Goldie, A., Mirhoseini, A., Laudon, J.: GDP: Generalized Device Placement for Dataflow Graphs (2019)
52. Zhou, Y., Roy, S., Abdolrashidi, A., Wong, D., Ma, P., Xu, Q., Liu, H.. Phothilimtha, M. P., Wang, S., Goldie, A., Mirhoseini, A., & Laudon, J.: Transferable graph optimizers for ML compilers. In: Advances in Neural Information Processing Systems (2020)
53. Zoph, B., Le, Q.V.: Neural architecture search with reinforcement learning. In: Proceedings of International Conference on Learning Representations (2017)

Chapter 9
Deep Learning Framework for Placement

Yibo Lin, Zizheng Guo, and Jing Mai

9.1 Introduction

Placement is an important stage in modern VLSI design automation [1]. It determines the locations of millions of instances, optimizing for multiple objectives such as wirelength, routability, timing, etc. Its performance and efficiency can significantly impact the design closure and turnaround time of the backend design flow. Thus, both industry and academia have been exploring efficient and high-performance placement engines.

The increasing design scale and complexity challenge placement algorithms in both efficiency and performance. We need to deal with tens of millions of instances and optimize for multiple objectives, considering complicated design constraints from both high-level architecture and low-level manufacturing technology such as region constraints and design rules. As modern placement algorithms follow iterative procedures in optimization, complicated objectives and constraints with large problem sizes can result in low computing efficiency and slow convergence.

State-of-the-art placement engines usually include wirelength-driven analytical placement kernels, which can be categorized into quadratic placement and nonlinear placement. The kernel placement can be extended to consider other objectives and constraints. Quadratic placement formulates the wirelength optimization problem into a quadratic programming [2–6]. As simply minimizing the wirelength can cause overlaps between instances, quadratic placement adopts iterative quadratic programming and rough legalization to spread instances in the layout. Nonlinear placement formulates a nonlinear wirelength objective, subjecting to a density constraint for overlap minimization [7–14]. By relaxing the density constraint into

Y. Lin (✉) · Z. Guo · J. Mai
Peking University, Beijing, China
e-mail: yibolin@pku.edu.cn; gzz@pku.edu.cn; magic3007@pku.edu.cn

© The Author(s), under exclusive license to Springer Nature Switzerland AG 2022
H. Ren, J. Hu (eds.), *Machine Learning Applications in Electronic Design Automation*, https://doi.org/10.1007/978-3-031-13074-8_9

the objective and gradually increasing the overlap penalty, we can spread the instances with minimum wirelength. Many commercial tools like Cadence Innovus and Synopsys IC Compiler adopt nonlinear placement algorithms for ASIC [15, 16], so we focus on nonlinear placement in this book chapter.

Nonlinear placement algorithms may need thousands of gradient descent iterations for convergence, which is computationally expensive. The literature has investigated multi-threading on CPUs using partitioning [17–19]. The speedup ratio typically saturates at $5\times$ with 2–6% quality degradation. GPU acceleration has been explored to accelerate a placement algorithm consisting of clustering, declustering, and nonlinear optimization [10]. Around $15\times$ speedup has been reported with less than 1% quality degradation. The speedup ratio is mostly limited by the sequential clustering and declustering steps in the algorithm. With the recent advances in nonlinear placement [13, 14] and deep learning [20], Lin et al. establish an analogy between the nonlinear placement problem and the neural network training problem [21]. With such an analogy, deep learning frameworks/toolkits like `PyTorch` [22] can be adopted to develop placement algorithms with native support to GPU acceleration and highly optimized kernel operators. Eventually, 30–$40\times$ speedup can be achieved for the nonlinear placement kernel. In this book chapter, we target this line of studies [21, 23–27] and survey how deep learning frameworks help placement development with high efficiency and performance. We also introduce how to customize kernel operators for further speedup [28] as well as how to handle cutting-edge objectives and constraints with both conventional and machine learning techniques in such a framework [21, 25, 26].

9.2 DL Analogy for the Kernel Placement Problem

Modern placement consists of three steps: global placement (GP), legalization (LG), and detailed placement (DP). Global placement determines the rough locations of instances by relaxing the placement problem into continuous optimization. It can achieve roughly legal solutions with small amount of overlaps. Legalization removes the remaining overlaps and satisfies all design rules. Detailed placement refines the results by local perturbation to further improve the solution quality. The performance of global placement is critical to the eventual design closure.

The kernel problem of global placement is to solve a wirelength minimization problem subjecting to density constraints [11]:

$$\min_{\mathbf{x},\mathbf{y}} \quad \sum_{e \in E} \mathrm{WL}(e; \mathbf{x}, \mathbf{y}), \tag{9.1a}$$

$$\text{s.t.} \quad d(\mathbf{x}, \mathbf{y}) \le d_t, \tag{9.1b}$$

where $\mathrm{WL}(\cdot; \cdot)$ denotes the wirelength cost function that takes any net instance e and returns the wirelength, $d(\cdot)$ denotes the density of a location in the layout, and

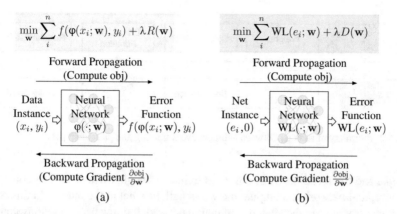

$$\min_{\mathbf{w}} \sum_{i}^{n} f(\varphi(x_i; \mathbf{w}), y_i) + \lambda R(\mathbf{w}) \qquad \min_{\mathbf{w}} \sum_{i}^{n} \text{WL}(e_i; \mathbf{w}) + \lambda D(\mathbf{w})$$

Forward Propagation (Compute obj) ⟶

Forward Propagation (Compute obj) ⟶

Data Instance (x_i, y_i) ⇨ Neural Network $\varphi(\cdot; \mathbf{w})$ ⇨ Error Function $f(\varphi(x_i; \mathbf{w}), y_i)$

Net Instance $(e_i, 0)$ ⇨ Neural Network $\text{WL}(\cdot; \mathbf{w})$ ⇨ Error Function $\text{WL}(e_i; \mathbf{w})$

⟵ Backward Propagation (Compute Gradient $\frac{\partial \text{obj}}{\partial \mathbf{w}}$)

⟵ Backward Propagation (Compute Gradient $\frac{\partial \text{obj}}{\partial \mathbf{w}}$)

(a) (b)

Fig. 9.1 Analogy between neural network training and analytical placement [21]. (**a**) Train a neural network for weights $\mathbf{w}$. (**b**) Solve a placement, for instance, locations $\mathbf{w} = (\mathbf{x}, \mathbf{y})$

d_t is a given target density. The target of solving this problem is to spread instances in the layout with minimum wirelength.

As the density constraints are non-convex and nonlinear, it is difficult to solve the constrained optimization directly. The widely adopted nonlinear placement algorithm relaxes the constraints into the objective and formulates an unconstrained nonlinear optimization:

$$\min_{\mathbf{x}, \mathbf{y}} \left(\sum_{e \in E} \text{WL}(e; \mathbf{x}, \mathbf{y}) \right) + \lambda D(\mathbf{x}, \mathbf{y}), \qquad (9.2)$$

where $D(\cdot)$ is the density penalty to spread cells out in the layout. We satisfy the density constraints by gradually increasing the density weight λ.

Since both solving the global placement and training a neural network are essentially solving a nonlinear optimization problem, we examine the underlying similarity between these two problems [21]. We first review the neural network training process, as shown in Fig. 9.1a. Given a neural network with trainable weights, we feed data samples to the network and compute error functions (or loss functions) at the output. The target of neural network training is to optimize a non-linear objective consisting of two terms, the total error term and the regularization term. The total error term is related to both the data samples and the weights of the network, while the regularization term, adopted to avoid overfitting, is usually only related to the weights of the network. By iterative forward propagation to compute the objective and backward propagation to compute the gradient, we can find the best weights that minimize the objective.

If we write the placement objective in the same format as neural network training, as shown in Fig. 9.1b, we can observe that the wirelength term can correspond to the total error term in neural network training and the density term can correspond to the regularization term. More specifically, we can view each net instance as a data

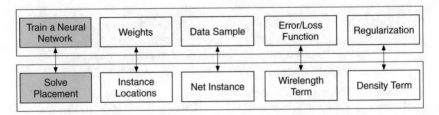

Fig. 9.2 Summary of analogy between neural network training and placement [21]

sample and the locations of instances as trainable weights. Then, we can construct a conceptual network to compute the wirelength of a net at the output. In this way, the total error term in the objective of neural network training becomes equivalent to the total wirelength term in the placement objective. The density term is only related to the locations of instances, which is similar to the regularization term. Figure 9.2 summarizes the analogy between the two problems.

By casting the placement problem into neural network training, we can solve the placement problem following the procedure of neural network training, with iterative forward and backward propagation. We can then leverage highly optimized deep learning frameworks to develop placement engines. Note that with this analogy, each time when we solve a placement problem, we essentially perform neural network training once. In other words, there is no testing phase like that in machine learning applications.

Figure 9.3 illustrates the software architecture of the placement developed with deep learning frameworks [21, 24]. Deep learning frameworks like TensorFlow and PyTorch consist of three stacks, i.e., low-level operators (OPs), automatic gradient derivation, and optimization engines. The low-level OPs are usually implemented and optimized on both CPU and GPU. The frameworks also provide APIs to extend the existing set of OPs so that we can customize the OPs for wirelength and density computation. The other parts of the frameworks are usually implemented in Python, which enables convenient development and customization of optimization engines. Thus, we can easily ensemble placement flows with low-level OPs, automatic gradient derivation, and optimization engines in deep learning frameworks.

9.3 Speedup Kernel Operators

DREAMPlace adopts the ePlace algorithm where each instance is analogous to an electric particle in an electrostatics system, which is summarized in Fig. 9.4. In this analogy, instance density corresponds to charge density, density penalty corresponds to electric potential energy, and density gradient corresponds to electric field. We can rewrite the objective in Eq. (9.2) as the following:

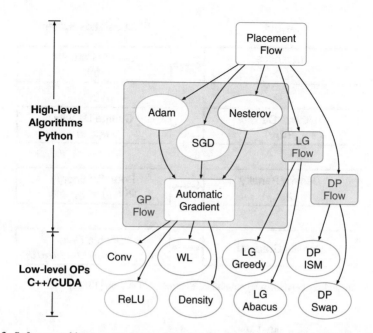

Fig. 9.3 Software architecture consisting of low-level operators written in C++/CUDA and high-level algorithms written in Python [24]. In the low-level operators, "Conv" and "ReLU" denote the existing convolution and activation OPs in the deep learning toolkit, respectively; "WL" denotes the custom OP to compute wirelength term in the placement objective; "Density" denotes the custom OP to compute the density term; "LG Greedy" and "LG Abacus" are two custom OPs for legalization algorithms leveraging a greedy approach [11] and a row-based dynamic programming method [29], respectively; "DP ISM" and "DP Swap" are two custom OPs for detailed placement algorithms leveraging independent set matching [11] and global swap [30], respectively. In the high-level algorithms, placement flows, i.e., GP flow for global placement, LG flow for legalization, and DP flow for detailed placement, can be ensembled using low-level operators with the automatic gradient derivation component and optimizers such as Adam [31], stochastic gradient descent (SGD), and Nesterov's accelerated gradient descent method

$$\min_{\mathbf{x,y}} \left(\sum_{e \in E} \mathrm{WL}(e; \mathbf{x}, \mathbf{y}) \right) + \lambda \Phi(\mathbf{x}, \mathbf{y}), \qquad (9.3)$$

where $\Phi(\cdot)$ is the electric potential energy. The electric potential energy can be computed by solving a Poisson's equation from the instance density $\rho(x, y)$ by spectral methods. Given an $M \times M$ grid of bins and $w_u = \frac{2\pi u}{M}$ and $w_v = \frac{2\pi v}{M}$ with $u = 0, 1, \ldots, M - 1$, $v = 0, 1, \ldots, M - 1$, the solution can be computed as the following [13]:

$$a_{u,v} = \frac{1}{M^2} \sum_{x=0}^{M-1} \sum_{y=0}^{M-1} \rho(x, y) \cos(w_u x) \cos(w_v y), \qquad (9.4a)$$

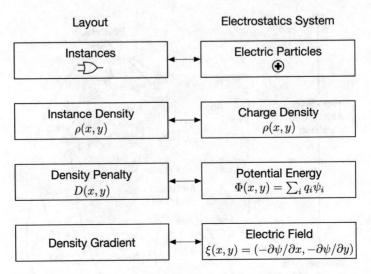

Fig. 9.4 An analogy between a layout and an electrostatics system [13, 32]

$$\psi_{\text{DCT}}(x, y) = \sum_{u=0}^{M-1} \sum_{v=0}^{M-1} \frac{a_{u,v}}{w_u^2 + w_v^2} \cos(w_u x) \cos(w_v y), \qquad (9.4b)$$

$$\xi_{\text{DSCT}}^{X}(x, y) = \sum_{u=0}^{M-1} \sum_{v=0}^{M-1} \frac{a_{u,v} w_u}{w_u^2 + w_v^2} \sin(w_u x) \cos(w_v y), \qquad (9.4c)$$

$$\xi_{\text{DCST}}^{Y}(x, y) = \sum_{u=0}^{M-1} \sum_{v=0}^{M-1} \frac{a_{u,v} w_v}{w_u^2 + w_v^2} \cos(w_u x) \sin(w_v y), \qquad (9.4d)$$

where ψ_{DCT} denotes the numerical solution of the potential function and ξ_{DSCT}^{X} and ξ_{DCST}^{Y} denote the solution of the electric field in horizontal and vertical directions, respectively. Eq. (9.4) requires discrete cosine transformation (DCT) and its variations to solve efficiently.

`DREAMPlace` also adopts the weighted-average wirelength (WAWL) as a smooth approximation to the half-perimeter wirelength (HPWL) [33, 34]:

$$\text{WA}_e = \frac{\sum_{i \in e} x_i e^{\frac{x_i}{\gamma}}}{\sum_{i \in e} e^{\frac{x_i}{\gamma}}} - \frac{\sum_{i \in e} x_i e^{-\frac{x_i}{\gamma}}}{\sum_{i \in e} e^{-\frac{x_i}{\gamma}}}, \qquad (9.5)$$

where γ is a parameter to control the smoothness and accuracy. The smaller γ is, the more accurate it is to approximate HPWL, but the less smooth.

The efficiency of the wirelength and density operators is critical to the overall efficiency of placement, as the placement algorithm needs to iteratively perform

Table 9.1 Notations

Notation	Description	Notation	Description
V	Set of cells	E	Set of nets
P	Set of pins	B	Set of bins
x_e^+	$\max_{i \in e} x_i, \forall e \in E$	x_e^-	$\min_{i \in e} x_i, \forall e \in E$
a_i^+	$e^{\frac{x_i - x_e^+}{\gamma}}, \forall i \in e, e \in E$	a_i^-	$e^{-\frac{x_i - x_e^-}{\gamma}}, \forall i \in e, e \in E$
b_e^+	$\sum_{i \in e} a_i^+, \forall e \in E$	b_e^-	$\sum_{i \in e} a_i^-, \forall e \in E$
c_e^+	$\sum_{i \in e} x_i a_i^+, \forall e \in E$	c_e^-	$\sum_{i \in e} x_i a_i^-, \forall e \in E$
$\mathbf{x}^+$	$\{x_e^+\}, \forall e \in E$	$\mathbf{x}^-$	$\{x_e^-\}, \forall e \in E$
$\mathbf{a}^+$	$\{a_i^+\}, \forall i \in P$	$\mathbf{a}^-$	$\{a_i^-\}, \forall i \in P$
$\mathbf{b}^+$	$\{b_e^+\}, \forall e \in E$	$\mathbf{b}^-$	$\{b_e^-\}, \forall e \in E$
$\mathbf{c}^+$	$\{c_e^+\}, \forall e \in E$	$\mathbf{c}^-$	$\{c_e^-\}, \forall e \in E$

gradient descent on Eq. (9.3). In the following subsections, we introduce the optimized kernels for these operators.

9.3.1 Wirelength

When implementing the wirelength operators, DREAMPlace has developed several strategies to remedy the latent numerical problems, to compare the efficiency of different parallelism schemes, to profile the performance bottlenecks, and to further squeeze the GPU performance via operator fusions.

The direct implementation of Eq. (9.5) can potentially cause numerical overflow problems. Thus, they convert $e^{\frac{x_i}{\gamma}}$ to $e^{\frac{x_i - \max_{j \in e} x_j}{\gamma}}$ and $e^{-\frac{x_i}{\gamma}}$ to $e^{-\frac{x_i - \min_{j \in e} x_j}{\gamma}}$ in an equivalent manner. With the annotations in Table 9.1, the wirelength gradient w.r.t. a pin location can be written as

$$\frac{\partial \mathrm{WL}_e}{\partial x_i} = \frac{(1 + \frac{x_i}{\gamma})b_e^+ - \frac{1}{\gamma}c_e^+}{(b_e^+)^2} \cdot a_i^+ - \frac{(1 - \frac{x_i}{\gamma})b_e^- + \frac{1}{\gamma}c_e^-}{(b_e^-)^2} \cdot a_i^-. \tag{9.6}$$

DREAMPlace proposes three parallelism schemes and discusses their pros and cons. The first one is the naive *net-by-net* parallelism scheme that allocates one thread for each net. The speedup of this scheme, however, remains limited because the maximum number of allocatable threads is $|E|$, and the heterogeneity of net degrees also leads to imbalanced workload for each thread.

The second one is the *pin-by-pin* scheme. Figure 9.5 illustrates the dependency graph for WA wirelength forward and backward based on Eq. (9.6). Because of the local nature of $\mathbf{a}^\pm$, $\mathbf{b}^\pm$, and $\mathbf{c}^\pm$ at pin level, a straightforward implementation of this pin-level parallelism is to compute $\mathbf{a}^\pm$, $\mathbf{b}^\pm$, and $\mathbf{c}^\pm$ in separate CUDA kernels by using multiple CUDA streams. This method has the probability to outperform the

Fig. 9.5 Forward and
backward dependency graph
for weighted-average
wirelength

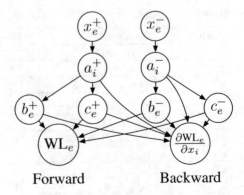

Forward   Backward

net-by-net scheme, because the number of pins $|P|$ is much larger than the number
of nets $|E|$. However, the pin-level parallelism still has drawbacks such as expensive
CUDA streams, frequent kernel synchronization, and heavy global memory access
hindering performance improvements.

Based on the elaborated profiling, Lin et al. further point out that the runtime
performance is essentially memory bounded. In other words, the major runtime
bottleneck lies in frequent writing to intermediate variables $x^{\pm}$, $a^{\pm}$, $b^{\pm}$, and $c^{\pm}$,
so they propose the third scheme that can remove all the intermediate variables by
merging the forward and backward functions, as shown in Algorithm 9.1. Instead of
storing $x^{\pm}$, $a^{\pm}$, $b^{\pm}$, and $c^{\pm}$ in global memory, they only create local variables in the
kernel function and directly compute the wirelength for each net and the gradient
for each pin. Experimental results show that this *operator fusion* scheme can
significantly alleviate the memory pressure and eventually achieve 3.7× speedup
over the first scheme and 1.8× speedup over the second scheme with `float32`.

9.3.2 Density Accumulation

The density operator requires to compute density distribution ρ in the layout
(Eq. (9.4)), which is a special case of the density accumulation operation. Density
accumulation is a general operation widely used in placement and routing. It can be
used for characterizing the density distributions of rectangular shapes on an $M \times N$
grid system. There are two typical variations of density accumulation as shown in
Fig. 9.6: the *forward* accumulation for computing the density map inside a set of
rectangles and the *backward* one to sum up the weights to the rectangles from a
predefined density map. Their definitions are presented as follows.

Problem 1 (Forward Density Accumulation) *Given a set of instances* V, *an* $M \times$
N *grid system* G, *and the weight* w_v^{inst} *of each instance* v, *compute the density
map* ρ^{grid} *on the grid system. Density* $\rho_{x,y}^{grid}$ *for each grid* $g_{x,y} \in G$ *is defined as
follows* [27]:

Algorithm 9.1 Wirelength forward and backward merged [21]

Require: A set of nets E, a set of pins P, and pin locations x;
Ensure: Wirelength cost and gradient;
1: **function** FORWARD_BACKWARD(E, P, x)
2:     **for** each thread $0 \leq t < |E|$ **do**   ▷ $\mathrm{WL}_e, \frac{\partial \mathrm{WL}_e}{\partial x_p}$ kernel
3:         Define e as the net corresponds to thread t;
4:         $x_e^+ \leftarrow \max_{p \in e} x_p$;   ▷ $x_e^\pm$ are local in the kernel
5:         $x_e^- \leftarrow \min_{p \in e} x_p$;
6:         $b_e^\pm \leftarrow 0, c_e^\pm \leftarrow 0$;   ▷ $b_e^\pm, c_e^\pm$ are local in the kernel
7:         $\mathrm{WL}_e \leftarrow 0$;   ▷ WL_e is in the global memory
8:         **for** each pin $p \in e$ **do**
9:             $a_p^\pm \leftarrow e^{\pm \frac{x_p - x_e^\pm}{\gamma}}$;   ▷ $a_p^\pm$ is local in the loop
10:            $b_e^\pm \leftarrow b_e^\pm + a_p^\pm$;
11:            $c_e^\pm \leftarrow c_e^\pm + x_p \cdot a_p^\pm$;
12:        **end for**
13:        $\mathrm{WL}_e \leftarrow \frac{\mathbf{c^+}}{\mathbf{b^+}} - \frac{\mathbf{c^-}}{\mathbf{b^-}}$;
14:        **for** each pin $p \in e$ **do**
15:            $a_p^\pm \leftarrow e^{\pm \frac{x_p - x_e^\pm}{\gamma}}$;   ▷ Compute $a_p^\pm$ again
16:            Compute $\frac{\partial \mathrm{WL}_e}{\partial x_p}$;   ▷ $\frac{\partial \mathrm{WL}_e}{\partial x_p}$ is in the global memory
17:        **end for**
18:    **end for**
19:    **return** reduce($\{\mathrm{WL}_e\}$), $\{\frac{\partial \mathrm{WL}_e}{\partial x_p}\}, \forall p \in P, e \in E$;
20: **end function**

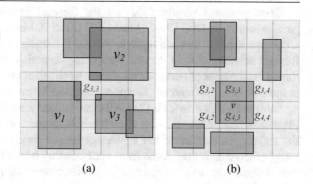

Fig. 9.6 An illustration on (a) forward and (b) backward density accumulation [27]

$$\rho_{x,y}^{grid} = \sum_{v \in V} w_v^{inst} \times \frac{OA(Box_v, g_{x,y})}{area_{x,y}^{grid}}, \quad \forall g_{x,y} \in G, \qquad (9.7)$$

where w_v^{inst} is the weight of instance v, $area_{x,y}^{grid}$ is the area of the grid $g_{x,y}$, and $OA(Box_v, g_{x,y})$ is the overlapping area between the bounding box of instance v and grid $g_{x,y}$.

Problem 2 (Backward Density Accumulation) *Given a set of instances V, an $M \times N$ grid system G, and the weight $w_{x,y}^{grid}$ of each grid $g_{x,y}$, compute the* *density array ρ^{inst} for all instances. Density ρ_v^{inst} for each instance v is defined as follows [27]:*

$$\rho_v^{inst} = \sum_{g_{x,y} \in G} w_{x,y}^{grid} \times \frac{OA(Box_v, g_{x,y})}{area_v^{inst}}, \forall v \in V, \tag{9.8}$$

where $w_{x,y}^{grid}$ is the weight of grid $g_{x,y}$, $area_v^{inst}$ is the area of instance v, and $OA(Box_v, g_{x,y})$ is the overlapping area between the bounding box of instance v and grid $g_{x,y}$.

Density accumulation is widely used in many placers such as POLAR [6] and ePlace series [32, 35]. A generalized density accumulation with bell-shaped functions is also used in NTUplace series [11, 12]. In DREAMPlace, density accumulation is used to evaluate and optimize the electric potential term $\Phi(\mathbf{x}, \mathbf{y})$ in Eq. (9.3). Specifically, the cell density is first distributed through a forward accumulation process to yield the density distribution of the grid system $\rho(x, y)$ in Eq. (9.4a). Then after DCT, the computed electric field $\xi_{DSCT}^X(x, y)$ and $\xi_{DCST}^Y(x, y)$ is accumulated by cells through their bounding boxes to yield their moving direction through a backward accumulation process. It has been shown in DREAMPlace that the density computation for every iteration of gradient descent takes up to 60% runtime on designs with millions of cells, especially when macro placement is considered where cells have large sizes [21, 27].

Existing works mostly focus on the forward density accumulation through parallelization on CPU [36, 37] and GPU [21, 36]. Lin et al. [36] explores efficient CPU atomic primitives and reproducible GPU kernels to solve the race condition of forward accumulation in parallel computing scenarios. Gessler et al. [37] allocate thread local copies of partial density maps to alleviate synchronization burden while at the cost of larger memory footprint. DREAMPlace assigns multiple threads for updating each shape on GPU [21]. All these experiments are based on a standard cell placement flow, where the size of each cell is comparable to the grid size. However, the techniques mentioned before cannot handle much larger rectangles covering many grids, which may come from net bounding boxes in routability modeling or macro placement, as the number of primitive operations is correlated to how many grids a shape covers.

To overcome the challenges in accelerating density accumulation, Guo et al. [27] propose a generic CPU/GPU algorithm for both forward and backward accumulation. They decompose the problem into two phases: a constant-time density collection phase for each instance and a linear-time prefix sum phase. They have a linear runtime complexity that does not depend on the size of the instance bounding boxes.

As both the forward and the backward algorithms in [27] require to compute two-dimensional (2D) prefix sum on a matrix A, we first introduce it as a building block to the algorithms. The result matrix P of the 2D prefix sum has the same dimension as the input matrix A, with element $P_{i,j}$ equals to the sum of all values in A which are above it or on left of it. Each element in matrix P can be written as

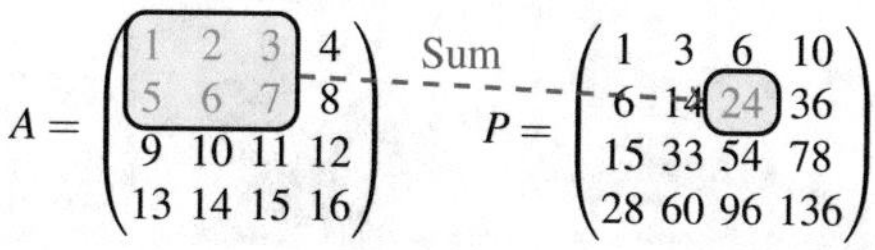

Fig. 9.7 An example of 2D prefix sum on a 4 × 4 matrix [27]

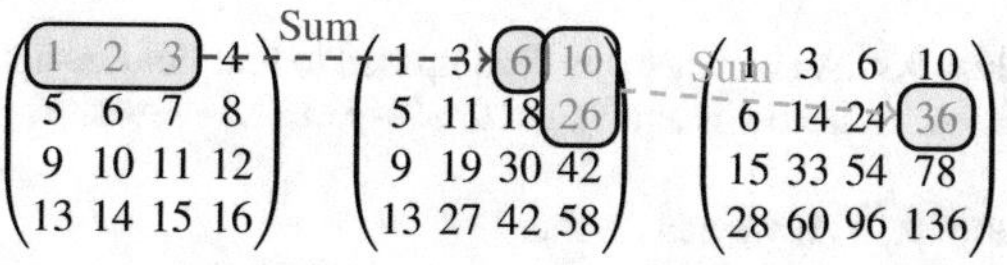

$$A = \begin{pmatrix} 1 & 2 & 3 & 4 \\ 5 & 6 & 7 & 8 \\ 9 & 10 & 11 & 12 \\ 13 & 14 & 15 & 16 \end{pmatrix} \xrightarrow{\text{Sum}} P = \begin{pmatrix} 1 & 3 & 6 & 10 \\ 6 & 14 & 24 & 36 \\ 15 & 33 & 54 & 78 \\ 28 & 60 & 96 & 136 \end{pmatrix}$$

Fig. 9.8 A row sum followed by a column sum is equivalent to a 2D prefix sum [27]

$$\begin{pmatrix} 1 & 2 & 3 & 4 \\ 5 & 6 & 7 & 8 \\ 9 & 10 & 11 & 12 \\ 13 & 14 & 15 & 16 \end{pmatrix} \xrightarrow{\text{Sum}} \begin{pmatrix} 1 & 3 & 6 & 10 \\ 5 & 11 & 18 & 26 \\ 9 & 19 & 30 & 42 \\ 13 & 27 & 42 & 58 \end{pmatrix} \xrightarrow{\text{Sum}} \begin{pmatrix} 1 & 3 & 6 & 10 \\ 6 & 14 & 24 & 36 \\ 15 & 33 & 54 & 78 \\ 28 & 60 & 96 & 136 \end{pmatrix}$$

Algorithm 9.2 `compute2DPrefixSum(A)` [27]

1: $m, n \leftarrow A.size$;
2: $P \leftarrow M \times N$ zero matrix
3: **for** $i = 1$ to M **do**
4:     **for** $j = 1$ to N **do**
5:         $P_{i,j} \leftarrow P_{i,j-1} + A_{i,j}$;   ▷ Define $P_{i,0} = 0$
6:     **end for**
7: **end for**
8: **for** $j = 1$ to N **do**
9:     **for** $i = 1$ to M **do**
10:        $P_{i,j} \leftarrow P_{i-1,j} + P_{i,j}$;   ▷ Define $P_{0,j} = 0$
11:     **end for**
12: **end for**
13: **return** P;

$$P_{i,j} = \sum_{x=1}^{i} \sum_{y=1}^{j} A_{x,y}. \tag{9.9}$$

Figure 9.7 shows an example of 2D prefix sum on a 4 × 4 matrix, where $i = 1, 2, \ldots, M$, $j = 1, 2, \ldots, N$, $A \in \mathcal{R}^{M \times N}$, and $P \in \mathcal{R}^{M \times N}$. Figure 9.8 shows that 2D prefix sum can be computed by performing one-dimensional (1D) prefix sum along rows and then along columns. A detailed algorithm for computing the 2D prefix sum is presented in Algorithm 9.2.

Based on the 2D prefix sum algorithm, the forward and backward accumulation can be made much more efficient by transforming the operations into equivalent forms. Specifically, in a forward accumulation, each instance is transformed into four bottom-right instances (i.e., instances at the bottom-right corner of the grid system), as shown in Fig. 9.9. It is straightforward to deduce the equivalence between the increment/decrements on the 4 bottom-right instances and the increment on the original rectangle. It turns out that the bottom-right instances are easier to handle in forward accumulation. Guo et al. propose an equivalent way to increase the values on a bottom-right subregion of a matrix, as illustrated in Fig. 9.10, by firstly incrementing a single value in a temporary matrix and then computing the 2D

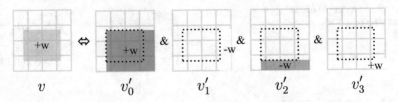

Fig. 9.9 An instance v is decomposed to four bottom-right instances v'_0, v'_1, v'_2, v'_3. The increment w on instance v is equivalent to $v'_0+ = w, v'_1- = w, v'_2- = w, v'_3+ = w$ [27]

Fig. 9.10 If we add a value 1 to $D_{2,2}$ and let P denote the 2D prefix sum of D, we will get that value propagated to the bottom-right region in P

$$D = \begin{pmatrix} 0 & 0 & 0 & 0 \\ 0 & 1 & 0 & 0 \\ 0 & 0 & 0 & 0 \\ 0 & 0 & 0 & 0 \end{pmatrix} \qquad P = \begin{pmatrix} 0 & 0 & 0 & 0 \\ 0 & 1 & 1 & 1 \\ 0 & 1 & 1 & 1 \\ 0 & 1 & 1 & 1 \end{pmatrix}$$

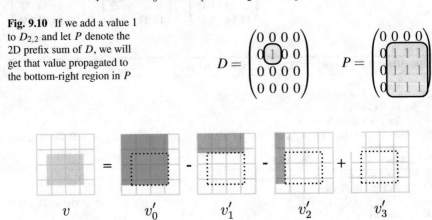

Fig. 9.11 A instance v can be decomposed to four top-left instances: v'_0, v'_1, v'_2, and v'_3 [27]

prefix sum. This method has the advantage that multiple bottom-right increments can be batched to the same temporary matrix leveraging the linearity of the 2D prefix sum. This directly yields a two-phase algorithm consisting of constant-time density collections for instances and a linear-time 2D prefix sum on the resulting matrix. These algorithms do not depend on the sizes of instances as it only needs to increment a constant amount on matrix elements.

The backward accumulation adopts a similar analysis. Different from the forward algorithm, the first step of the backward algorithm is to compute the prefix sum, and then density collection is performed on the prefix sum matrix. Each instance in backward accumulation is decomposed into four independent top-left instances, as shown in Fig. 9.11. Similar ideas can be seen in image processing literature like Crow et al. [38]. The sum of a top-left instance is just the corresponding matrix element in the precomputed prefix sum matrix and can be retrieved in $O(1)$ time.

In placement, bounding boxes in density accumulation may adopt non-integer coordinates. The weights need to be assigned proportional to the proportion of the grids within the bounding box. To deal with such scenarios, they further decompose each top-left or bottom-right instance to four equivalent instances with integer coordinates. For bottom-right instances used in the forward accumulation, this is illustrated in Fig. 9.12, and the top-left instances are processed similarly.

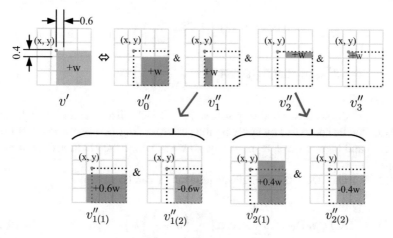

Fig. 9.12 An increment w on a bottom-right instance v' is split into four instances v_0'', v_1'', v_2'', v_3''. Of them, v_0'' is itself a bottom-right submatrix increment; v_1'' and v_2'' can be each split into two bottom-right submatrix increments; v_3'' consists of a single grid and can be handled individually [27]

The accelerated algorithms for the forward and backward accumulation can be parallelized using GPU. The 2D prefix sum is parallelized by first completing a batched 1D prefix sum at all rows, and then another batched 1D prefix sum at columns. The density collection phase of both forward and backward density accumulation consists of computation tasks for different instances, and these tasks are mostly independent so they can be parallelized. However, as one grid may be simultaneously updated by multiple threads in the forward accumulation, this algorithm needs the `atomicAdd` primitive to avoid data race. Guo et al. [27] have implemented these CPU/GPU operators in the `DREAMPlace` framework. According to their experiments, the accelerated algorithms can achieve up to 22× speedup on CPU and up to 64× speedup on GPU compared with naive versions due to balanced workload.

9.3.3 Discrete Cosine Transformation

Given the density distribution computed in Sect. 9.3.2, we still need to compute electric potential and fields according to Eq. (9.4), which can be broken down into discrete cosine transformation and its variations like the following [21]:

$$a_{u,v} = \text{DCT}(\text{DCT}(\rho)^T)^T, \tag{9.10a}$$

$$\psi_{\text{DCT}} = \text{IDCT}\left(\text{IDCT}\left(\left\{\frac{a_{u,v}}{w_u^2 + w_v^2}\right\}\right)^T\right)^T, \tag{9.10b}$$

$$\xi_{DSCT}^X = \text{IDXST(IDCT}\left(\left\{\frac{a_{u,v}w_u}{w_u^2 + w_v^2}\right\}\right)^T)^T,$$ (9.10c)

$$\xi_{DCST}^Y = \text{IDCT(IDXST}\left(\left\{\frac{a_{u,v}w_v}{w_u^2 + w_v^2}\right\}\right)^T)^T,$$ (9.10d)

where $(\cdot)^T$ denotes matrix transposition. As 2D DCT, IDCT, and IDXST transformations can be computed by applying the corresponding 1D transformations first to columns and then to rows, respectively, so we illustrate the 1D transformations for simplicity.

1D DCT and IDCT for a length-N sequence x can be written as

$$\text{DCT}(\{x_n\})_k = \sum_{n=0}^{N-1} x_n \cos\left(\frac{\pi}{N}\left(n + \frac{1}{2}\right)k\right),$$ (9.11a)

$$\text{IDCT}(\{x_n\})_k = \frac{1}{2}x_0 + \sum_{n=1}^{N-1} x_n \cos\left(\frac{\pi}{N}n\left(k + \frac{1}{2}\right)\right),$$ (9.11b)

where $k = 0, 1, \ldots, N - 1$. IDXST can be further derived as

$$\text{IDXST}(\{x_n\})_k = \sum_{n=0}^{N-1} x_n \sin\left(\frac{\pi}{N}n\left(k + \frac{1}{2}\right)\right),$$ (9.12a)

$$= (-1)^k \sum_{n=0}^{N-1} x_n(-1)^k \sin\left(\frac{\pi n(k + \frac{1}{2})}{N}\right),$$ (9.12b)

$$= (-1)^k \sum_{n=0}^{N-1} x_n \cos\left(\frac{\pi(N-n)(k + \frac{1}{2})}{N}\right),$$ (9.12c)

$$= (-1)^k \sum_{n=0}^{N-1} x_{N-n} \cos\left(\frac{\pi}{N}n(k + \frac{1}{2})\right),$$ (9.12d)

$$= (-1)^k \text{IDCT}(\{x_{N-n}\})_k,$$ (9.12e)

where $x_N = 0$. The equality between Eqs. (9.12d) and (9.12e) can be derived by incorporating x_{N-n} into Eq. (9.11b). Since all the computations are broken down into the 1D DCT/IDCT kernels with proper transformations, the performance of DCT/IDCT kernels is critical to the overall efficiency of the entire placement.

DCT and its variations are correlated to fast Fourier transformations (FFT), so we can leverage the highly optimized FFT kernels provided by many deep learning frameworks to compute DCT with linear-time additional processing. Empirically, the most efficient way in the placement problem is to adopt the N-point real

Algorithm 9.3 DCT/IDCT with N-point FFT [21]

Require: An even-length real sequence x;
Ensure: An even-length transformed real sequence y;

1: **function** DCT(x)
2:     $N \leftarrow |x|$;
3:     **for** each thread $0 \le t < N$ **do** $\triangleright$ Reorder kernel
4:         **if** $t < \frac{N}{2}$ **then**
5:             $x'_t \leftarrow x_{2t}$;
6:         **else**
7:             $x'_t \leftarrow x_{2(N-t)-1}$;
8:         **end if**
9:     **end for**
10:    $x'' \leftarrow \text{RFFT}(x')$; $\triangleright$ One-sided real FFT kernel
11:    **for** each thread $0 \le t < N$ **do** $\triangleright$ $e^{-\frac{j\pi t}{2N}}$ kernel
12:         **if** $t \le \frac{N}{2}$ **then**
13:             $y_t \leftarrow \frac{2}{N}\Re(x''_t e^{-\frac{j\pi t}{2N}})$; $\triangleright$ get real part
14:         **else**
15:             $y_t \leftarrow \frac{2}{N}\Re(\overline{x''_{(N-t)}}e^{-\frac{j\pi t}{2N}})$; $\triangleright$ get real part
16:         **end if**
17:    **end for**
18:    **return** y;
19: **end function**
20: **function** IDCT(x)
21:    $N \leftarrow |x|$;
22:    **for** each thread $0 \le t < \frac{N}{2} + 1$ **do** $\triangleright$ Complex kernel
23:         $x'_t \leftarrow (x_t - jx_{(N-t)})e^{\frac{j\pi t}{2N}}$; $\triangleright$ let $x_N \leftarrow 0$
24:    **end for**
25:    $x'' \leftarrow \text{IRFFT}(x')$; $\triangleright$ One-sided real IFFT kernel;
26:    **for** each thread $0 \le t < N$ **do** $\triangleright$ Reverse kernel
27:         **if** $t \mod 2 == 0$ **then**
28:             $y_t \leftarrow \frac{N}{4}x''_{\frac{t}{2}}$;
29:         **else**
30:             $y_t \leftarrow \frac{N}{4}x''_{(N-\frac{t+1}{2})}$;
31:         **end if**
32:    **end for**
33:    **return** y;
34: **end function**

1D FFT/IFFT proposed by Makhoul [39] leveraging the symmetric property of FFT for real input sequences, as shown in Algorithm 9.3. It is reported that this implementation can achieve $2.1\times$ faster than the $2N$-point implementations adopted by Tensorflow on DCT with map sizes from 512×512 to 4096×4096 and float32. To further speed up the kernels in practice, DREAMPlace also proposes to directly leverage 2D FFT [21] with more complicated pre-processing and post-processing, which can boost the speedup ratio to $5.0\times$ on DCT. This version of implementations can eliminate the redundant computations with a one-time call to 2D FFT kernels. Meanwhile, the pre-processing and the post-processing routines are fully parallelizable.

9.4 Handle Region Constraints

Region constraints are specified by designers to improve performance, isolate voltage regions, leave space for later optimization, reduce power consumption, and so on [40]. Designers can set fence regions that are *member-hard* and *nonmember-hard*. In other words, instances assigned to a fence region have to be placed within the region, while instances not assigned to the region must be placed outside. Figure 9.13 plots an example of fence regions. Note that a fence region can consist of multiple disjoint rectangular subregions in the layout and instances assigned to the region can be placed in any of these subregions. Such a property necessitates global optimization techniques to ensure instances properly distributed to different subregions with minimized wirelength.

The literature has explored various methods to incorporate region constraints into the state-of-the-art placement algorithms. NTUplace4dr [12] proposes to handle region constraints with a region-aware clustering during nonlinear placement iterations. Eh?Placer [41] and RippleDR [42] follow the quadratic placement algorithm, i.e., an upper-bound-lower-bound optimization method with look-ahead rough legalization. They honor the region constraints during the rough legalization. These approaches essentially regard region assignment as a separate step and thus cannot smoothly integrate the constraints into the continuous optimization.

DREAMPlace 3.0 [25] proposes to handle region constraints by extending the electrostatics-based placement model in ePlace series [13, 14] to a multi-electrostatic system. The ePlace family casts the placement problem into an energy minimization problem of an electrostatic system, where instances are analogous to electric charges. Minimizing the electric potential energy of the system contributes to the spreading of instances in the layout. Based on such a formulation, the key idea to consider region constraints is as follows. If we construct a separate electrostatic field for each region constraint and minimize the total

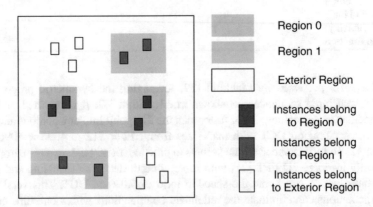

Fig. 9.13 An example of two fence regions in a layout

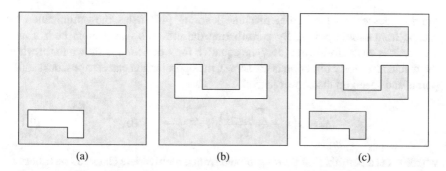

Fig. 9.14 Electric fields for (**a**) region 0, (**b**) region 1, and (**c**) the exterior region in Fig. 9.13. Gray areas denote blockages

potential energy of all fields, then the region constraints can be naturally handled during the placement optimization.

Figure 9.14 provides one example of the multi-field construction for two fence regions. In this case, we need three fields, two for each fence region and one for the exterior region. For each field, we insert blockage to the area outside the fence region such that the electric potential stays high and the related instances only move within the region. Similar settings apply to the exterior region. Based on this setup, we adapt the placement objective in Eq. (9.3) to the following:

$$\min_{\mathbf{x},\mathbf{y}} \quad \sum_{e \in E} \mathrm{WL}(e; \mathbf{x}, \mathbf{y})) + \langle \lambda, \Phi \rangle, \tag{9.13}$$

where the density weight $\lambda \in \mathbb{R}^{K+1}$ is a vector $(\lambda_0, \cdots, \lambda_K)$, K is the number of fence regions, and $\Phi = (\Phi_0, \cdots, \Phi_K)$ is a potential energy vector that considers fence regions into density calculation.

Besides the placement model, DREAMPlace 3.0 also proposes two techniques to facilitate convergence: a quadratic density penalty to speed up convergence at plateau and an entropy injection method to avoid saddle points, as optimizing a multi-field system challenges the optimizer and often requires more iterations. The placement objective with the quadratic density penalty can be written as the following:

$$\min_{\mathbf{x},\mathbf{y}} \quad \sum_{e \in E} \mathrm{WL}(e; \mathbf{x}, \mathbf{y})) + \langle \lambda, \Phi + \frac{1}{2}\mu\Phi^2 \rangle, \tag{9.14}$$

where μ is the coefficient balancing the first-order and second-order density penalty terms. The basic idea of the quadratic penalty comes from the *augmented Lagrangian relaxation* technique that can avoid ill-conditioning and numerical instabilities [43]. In practice, the quadratic term Φ^2 needs to be normalized by the Φ at iteration 0 [25]. The entropy injection technique inspired by the perturbed gradient descent method from machine learning [44] helps the optimization to escape from saddle points. By perturbating the movable instances in both x and y directions as the following (only the equation for x direction is shown for brevity; the mechanism for y direction is the same), the optimization can escape from saddle points and speed up the convergence:

$$\hat{\mathbf{x}} = s\left(\mathbf{x} - \frac{\sum_{i \in v} x_i}{|v|}\right) + \frac{\sum_{i \in v} x_i}{|v|} + \Delta\mathbf{x}, \tag{9.15}$$

where $\Delta\mathbf{x}$ is a perturbation vector sampled from a multivariate Gaussian distribution $\Delta\mathbf{x} \sim \mathcal{N}(0, \sigma^2)$. The shrinking factor $s \in (0, 1)$ rolls back the spreading process and increases the density overflow, forcing the optimizer to re-optimize the perturbed wirelength. The above techniques can be easily and efficiently integrated into the DREAMPlace framework by only writing Python codes, indicating the flexibility of the methodology.

The experimental results on ISPD 2015 contest benchmarks [40] demonstrate that DREAMPlace 3.0 can achieve more than 13% HPWL reduction and 11% global routing overflow reduction compared with NTUplace4dr and Eh?Placer on designs with region constraints.

9.5 Optimize Routability

Routability optimization is required for modern placement engines to achieve high-quality results after routing. In this section, we introduce the typical routability optimization strategy based on instance inflation as well as a recent strategy integrating a machine learning-based routability penalty into the placement objective.

9.5.1 Instance Inflation

A widely adopted technique for routability optimization is instance inflation [14], because routability issues usually come from congested regions. Increasing the area of instances in these regions can leave more space and thus can help resolve routing congestion. DREAMPlace [21] adopts this idea and leverages an external global router, NCTUgr [45], to evaluate the congestion during placement iterations. Figure 9.15 plots the placement flow for routability optimization based on instance inflation. When the density overflow drops to 20%, it invokes the global router to evaluate the routing congestion map and adjust the areas of instances. For each metal layer, it computes the ratios between routing demand and capacity at each routing tile and chooses the maximum ratio across all layers to compute the ratio:

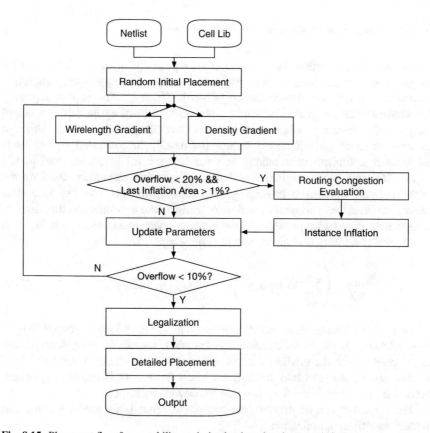

Fig. 9.15 Placement flow for routability optimization based on instance inflation

$$ratio = \min \left(\left(\max_{\forall l \in L} \frac{demand_l}{capacity_l} \right)^{2.5}, 2.5 \right), \hspace{1cm} (9.16)$$

where L is the set of metal layers. The exponent and maximum limits can be adjusted according to the benchmarks. The inflation ratios need to be carefully controlled, as there may not be enough whitespace in the layout to digest the inflated areas. The inflation stops when the total increased area is less than 1% of the total instance area.

Experiments on DAC 2012 routability-driven placement contest benchmarks [46] demonstrate that DREAMPlace can achieve 10% better scaled HPWL and 0.5% smaller routing congestion compared with RePlAce [14]. Meanwhile, with the power of GPU acceleration, DREAMPlace is 9× faster than RePlAce in global placement and 5× faster in the entire placement flow. It also reports that more than 75% of the runtime in global placement is taken by the external router to obtain congestion maps, which should be optimized in the future.

9.5.2   Deep Learning-Based Optimization

To overcome the runtime issue raised from repeatedly invoking global routing in placement, many studies have explored machine learning-based congestion estimation to improve the efficiency [47–50]. These studies aim at replacing the conventional congestion estimator with machine learning models, while still adopting the instance inflation strategy for routability optimization. Although instance inflation can effectively reduce the density in congested regions, it is an indirect optimization technique and requires careful parameter tuning. Liu et al. [26] propose an explicit routability optimization technique based on the DREAMPlace framework leveraging deep learning models. The key idea is to obtain a differentiable congestion estimator using neural networks and then directly integrate the congestion penalty into the nonlinear placement objective in Eq. (9.3) for explicit routability optimization as the following:

$$\min_{\mathbf{x}, \mathbf{y}} \left(\sum_{e \in E} \mathrm{WL}(e; \mathbf{x}, \mathbf{y}) \right) + \lambda \Phi(\mathbf{x}, \mathbf{y}) + \eta \mathrm{L}(\mathbf{x}, \mathbf{y}), \tag{9.17}$$

where $\mathrm{L}(\cdot)$ is the congestion penalty calculated by the deep learning-based congestion estimator. Due to the differentiability of neural networks, the congestion penalty is compatible with the gradient descent optimizers in placement. Meanwhile, integrating neural networks into the DREAMPlace framework is natively supported, since it is developed in the deep learning framework PyTorch.

The algorithm can be divided into two phases, routability model training and routability-driven optimization.

Routability model training is to obtain the congestion estimator. They adopt RouteNet [47] as the network architecture for congestion map prediction. RouteNet is a fully convolutional neural network that takes image-like feature maps such as rectangular uniform wire density (RUDY) [51], pin RUDY, and snapshots of macros as input. It outputs a congestion map given the features. They train the network using a dataset with at least 1400 placement solutions from ISPD 2015 contest benchmarks [40] generated by DREAMPlace. For brevity, we denote the congestion map prediction task as f_R:

$$f_R : \mathbf{M} \subset \mathbb{R}^{M \times N \times 3} \longrightarrow \mathcal{Y} \subset \mathbb{R}^{M \times N}, \tag{9.18}$$

where $\mathbf{M}$ denotes the features and $\mathcal{Y}$ denotes the congestion map.

Routability-driven optimization explicitly minimizes the congestion penalty defined as the mean squared Frobenius norm of congestion map:

$$L(\mathbf{x}, \mathbf{y}) = \frac{1}{MN} \| f_R(\mathbf{M}) \|_2^2, \tag{9.19}$$

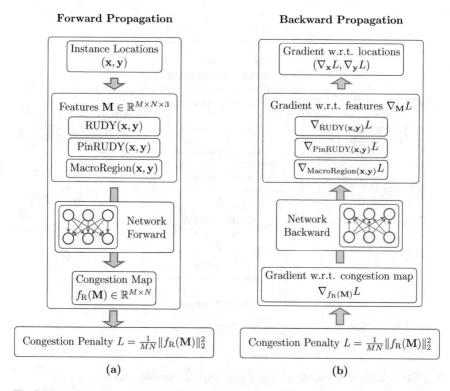

Fig. 9.16 (a) Forward and (b) Backward propagation for congestion penalty [26]

where the feature map $\mathbf{M}$ is derived from instance locations $\mathbf{x}, \mathbf{y}$ using RUDY, pin RUDY, and locations of macros. It is also differentiable to $\mathbf{x}, \mathbf{y}$, so we can also follow the procedure of forward and backward propagation to optimize the penalty in placement. Figure 9.16 sketches the forward and backward propagation procedure for the congestion penalty. Different from neural network training where backward propagation computes the gradient to weights, here it computes the gradients to the inputs, i.e., essentially the instance locations. The overall placement flow is shown in Fig. 9.17. The major difference from DREAMPlace lies in the congestion gradient computation. Eventually, up to 9.05% reduction in congestion and 5.30% reduction in routed wirelength are reported on ISPD 2015 contest benchmarks compared to the state-of-the-art placer NTUplace4dr [12] with more than 20× speedup.

As the routability experiments of [21] and [26] are conducted on different benchmarks suites and different machines, it is difficult to have direct comparison. However, we observe that on million-size designs, both algorithms report similar overall runtime, which is out of expectation. This is because [21] only invokes the external global router for about 6 times, while [26] needs to compute the congestion gradient in every iteration (around 1000 times in total). Although deep learning-based congestion estimator is much faster than the global router, frequent invocation

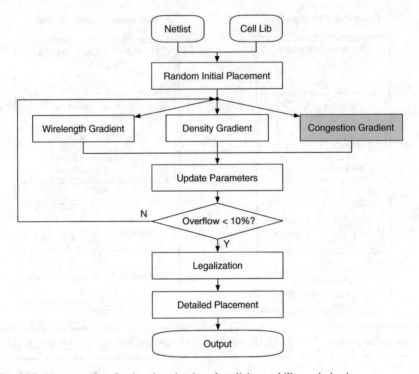

Fig. 9.17 Placement flow for deep learning-based explicit routability optimization

can significantly slow down the placement. There are several directions to reduce the overhead, e.g., simplifying the network architecture or only updating the congestion gradient for a certain amount of iterations, which is worth exploring in the future. Overall speaking, the studies on routability-driven optimization demonstrate the efficiency and flexibility of deep learning frameworks-enabled placement engines and the compatibility with deep learning-assisted optimization.

9.6 Conclusion

In this chapter, we introduce a series of studies leveraging deep learning frameworks to develop placement engines. We summarize the major advantages as follows:

- Efficiency. Developing placement with deep learning framework enables native support to GPU acceleration, which brings significant speedup compared with conventional placement engines on CPU.
- Extensibility. The framework is highly extensible to incorporate additional optimization objectives like routability and natively compatible with machine learning-assisted placement techniques.

- Friendly to beginners. Due to the wide adoption of deep learning techniques, frameworks like `TensorFlow` and `PyTorch` are well-known packages with abundant tutorials, lowering the bar for beginners to explore new ideas.

Despite the current progress, there are still directions worth exploring in the future.

- Timing- and power-driven optimization.
- Integration with other design stages like routing, gate sizing, and clock tree synthesis.
- Application to other scenarios like FPGA placement and routing [52].

As `DREAMPlace` has been open-source, we believe it can stimulate more efforts for placement research and open new directions in both acceleration and optimization.

References

1. Markov, I.L., Hu, J., Kim, M.C.: Progress and challenges in VLSI placement research. Proc. IEEE **103**(11), 1985–2003 (2015)
2. Viswanathan, N., Chu, C.C.: Fastplace: Efficient analytical placement using cell shifting, iterative local refinement, and a hybrid net model. IEEE TCAD **24**(5), 722–733 (2005)
3. Viswanathan, N., Pan, M., Chu, C.: FastPlace 3.0: a fast multilevel quadratic placement algorithm with placement congestion control. In: IEEE/ACM Asia and South Pacific Design Automation Conference (ASPDAC), pp. 135–140. IEEE, Piscataway (2007)
4. Kim, M.C., Lee, D.J., Markov, I.L.: Simpl: An effective placement algorithm. IEEE TCAD **31**(1), 50–60 (2012)
5. He, X., Huang, T., Xiao, L., Tian, H., Young, E.F.Y.: Ripple: a robust and effective routability-driven placer. IEEE TCAD **32**(10), 1546–1556 (2013)
6. Lin, T., Chu, C., Shinnerl, J.R., Bustany, I., Nedelchev, I.: POLAR: a high performance mixed-size wirelengh-driven placer with density constraints. IEEE TCAD **34**(3), 447–459 (2015)
7. Kahng, A.B., Reda, S., Wang, Q.: Architecture and details of a high quality, large-scale analytical placer. In: ICCAD, pp. 891–898. IEEE, Piscataway (2005)
8. Kahng, A.B., Wang, Q.: A faster implementation of APlace. In: ISPD, pp. 218–220. ACM, New York (2006)
9. Chan, T., Cong, J., Sze, K.: Multilevel generalized force-directed method for circuit placement. In: ISPD, pp. 185–192. ACM (2005)
10. Chan, T.F., Sze, K., Shinnerl, J.R., Xie, M.: mPL6: Enhanced multilevel mixed-size placement with congestion control. In: Modern Circuit Placement. Springer, Berlin (2007)
11. Chen, T.C., Jiang, Z.W., Hsu, T.C., Chen, H.C., Chang, Y.W.: Ntuplace3: an analytical placer for large-scale mixed-size designs with preplaced blocks and density constraints. IEEE TCAD **27**(7), 1228–1240 (2008)
12. Huang, C., Lee, H., Lin, B., Yang, S., Chang, C., Chen, S., Chang, Y., Chen, T., Bustany, I.: NTUplace4dr: a detailed-routing-driven placer for mixed-size circuit designs with technology and region constraints. IEEE TCAD **37**(3), 669–681 (2018)
13. Lu, J., Zhuang, H., Chen, P., Chang, H., Chang, C.C., Wong, Y.C., Sha, L., Huang, D., Luo, Y., Teng, C.C., et al.: ePlace-MS: electrostatics-based placement for mixed-size circuits. IEEE TCAD **34**(5), 685–698 (2015)
14. Cheng, C.K., Kahng, A.B., Kang, I., Wang, L.: RePlAce: Advancing solution quality and routability validation in global placement. IEEE TCAD (2018)
15. Cadence Innovus. http://www.cadence.com

16. Synopsys IC Compiler. http://www.synopsys.com
17. Ludwin, A., Betz, V., Padalia, K.: High-quality, deterministic parallel placement for FPGAs on commodity hardware. In: FPGA, pp. 14–23. ACM, New York (2008)
18. Lin, T., Chu, C., Wu, G.: Polar 3.0: An ultrafast global placement engine. In: ICCAD, pp. 520–527 (2015)
19. Li, W., Li, M., Wang, J., Pan, D.Z.: Utplacef 3.0: a parallelization framework for modern FPGA global placement. In: ICCAD, pp. 908–914 (2017)
20. Goodfellow, I., Bengio, Y., Courville, A.: Deep Learning. MIT Press, Cambridge (2016)
21. Lin, Y., Jiang, Z., Gu, J., Li, W., Dhar, S., Ren, H., Khailany, B., Pan, D.Z.: Dreamplace: deep learning toolkit-enabled GPU acceleration for modern VLSI placement. IEEE TCAD (2020)
22. Paszke, A., Gross, S., Massa, F., Lerer, A., Bradbury, J., Chanan, G., Killeen, T., Lin, Z., Gimelshein, N., Antiga, L., et al.: PyTorch: an imperative style, high-performance deep learning library. In: Conference on Neural Information Processing Systems (NIPS), pp. 8024–8035. Curran Associates (2019)
23. Lin, Y., Li, W., Gu, J., Ren, H., Khailany, B., Pan, D.Z.: Abcdplace: accelerated batch-based concurrent detailed placement on multithreaded cpus and GPUs. IEEE Trans. Comput.-Aided Design Integr. Circuits Syst. **39**(12), 5083–5096 (2020)
24. Lin, Y., Pan, D.Z., Ren, H., Khailany, B.: Dreamplace 2.0: Open-source GPU-accelerated global and detailed placement for large-scale VLSI designs. In: 2020 China Semiconductor Technology International Conference (CSTIC), pp. 1–4 (2020)
25. Gu, J., Jiang, Z., Lin, Y., Pan, D.Z.: Dreamplace 3.0: multi-electrostatics based robust VLSI placement with region constraints. In: 2020 IEEE/ACM International Conference On Computer Aided Design (ICCAD), pp. 1–9 (2020)
26. Liu, S., Sun, Q., Liao, P., Lin, Y., Yu, B.: Global placement with deep learning-enabled explicit routability optimization. In: DATE. Virtual Conference (2021)
27. Guo, Z., Mai, J., Lin, Y.: Ultrafast CPU/GPU kernels for density accumulation in placement. In: DAC. San Francisco (2021)
28. Lin, Y.: GPU acceleration in VLSI back-end design: overview and case studies. In: Proceedings of the 39th International Conference on Computer-Aided Design, ICCAD '20. Association for Computing Machinery, New York (2020)
29. Spindler, P., Schlichtmann, U., Johannes, F.M.: Abacus: Fast legalization of standard cell circuits with minimal movement. In: ISPD, ISPD '08, pp. 47–53. Association for Computing Machinery, New York (2008)
30. Pan, M., Viswanathan, N., Chu, C.: An efficient and effective detailed placement algorithm. In: ICCAD, pp. 48–55 (2005)
31. Kingma, D.P., Ba, J.: Adam: A method for stochastic optimization. In: ICLR (Poster) (2015)
32. Cheng, C.K., Kahng, A.B., Kang, I., Wang, L.: Replace: advancing solution quality and routability validation in global placement. IEEE TCAD (2018)
33. Hsu, M.K., Chang, Y.W., Balabanov, V.: TSV-aware analytical placement for 3D IC designs. In: DAC, pp. 664–669. ACM, New York (2011)
34. Hsu, M.K., Balabanov, V., Chang, Y.W.: TSV-aware analytical placement for 3-D IC designs based on a novel weighted-average wirelength model. DAC **32**(4), 497–509 (2013)
35. Lu, J., Chen, P., Chang, C.C., Sha, L., Huang, D.J.H., Teng, C.C., Cheng, C.K.: ePlace: Electrostatics-based placement using fast fourier transform and Nesterov's method. ACM TODAES **20**(2), 17 (2015)
36. Lin, C.X., Wong, M.D.: Accelerate analytical placement with GPU: a generic approach. In: DATE, pp. 1345–1350. IEEE, Piscataway (2018)
37. Gessler, F., Brisk, P., Stojilović, M.: A shared-memory parallel implementation of the replace global cell placer. In: International Conference on VLSI Design, pp. 78–83. IEEE, Piscataway (2020)
38. Crow, F.C.: Summed-area tables for texture mapping. In: SIGGRAPH '84, pp. 207–212. ACM, New York (1984)
39. Makhoul, J.: A fast cosine transform in one and two dimensions. IEEE Trans. Signal Process. **28**(1), 27–34 (1980)

40. Bustany, I.S., Chinnery, D., Shinnerl, J.R., Yutsis, V.: ISPD 2015 benchmarks with fence regions and routing blockages for detailed-routing-driven placement. In: ISPD, pp. 157–164 (2015)
41. Darav, N.K., Kennings, A., Tabrizi, A.F., Westwick, D., Behjat, L.: Eh?Placer: a high-performance modern technology-driven placer. ACM TODAES **21**(3), 1–27 (2016)
42. Chow, W., Kuang, J., Tu, P., Young, E.F.Y.: Fence-aware detailed-routability driven placement. In: ACM Workshop on System Level Interconnect Prediction (SLIP), pp. 1–7 (2017)
43. Birgin, E.G., Martínez, J.M.: Practical augmented Lagrangian methods for constrained optimization. SIAM (2014)
44. Jin, C., Ge, R., Netrapalli, P., Kakade, S.M., Jordan, M.I.: How to escape saddle points efficiently. In: International Conference on Machine Learning (ICML), pp. 1724–1732. PMLR (2017)
45. Liu, W.H., Li, Y.L., Koh, C.K.: A fast maze-free routing congestion estimator with hybrid unilateral monotonic routing. In: ICCAD, pp. 713–719 (2012)
46. Viswanathan, N., Alpert, C., Sze, C., Li, Z., Wei, Y.: The DAC 2012 routability-driven placement contest and benchmark suite. In: DAC, pp. 774–782. ACM, New York (2012)
47. Xie, Z., Huang, Y.H., Fang, G.Q., Ren, H., Fang, S.Y., Chen, Y., Hu, J.: Routenet: routability prediction for mixed-size designs using convolutional neural network. In: 2018 IEEE/ACM International Conference on Computer-Aided Design (ICCAD), pp. 1–8. IEEE, Piscataway (2018)
48. Kirby, R., Godil, S., Roy, R., Catanzaro, B.: Congestionnet: routing congestion prediction using deep graph neural networks. In: 2019 IFIP/IEEE 27th International Conference on Very Large Scale Integration (VLSI-SoC), pp. 217–222. IEEE, Piscataway (2019)
49. Alawieh, M.B., Li, W., Lin, Y., Singhal, L., Iyer, M.A., Pan, D.Z.: High-definition routing congestion prediction for large-scale FPGAs. In: 2020 25th Asia and South Pacific Design Automation Conference (ASP-DAC), pp. 26–31. IEEE, Piscataway (2020)
50. Liang, R., Xiang, H., Pandey, D., Reddy, L., Ramji, S., Nam, G.J., Hu, J.: DRC hotspot prediction at sub-10 nm process nodes using customized convolutional network. In: Proceedings of the 2020 International Symposium on Physical Design, pp. 135–142 (2020)
51. Spindler, P., Johannes, F.M.: Fast and accurate routing demand estimation for efficient routability-driven placement. In: DATE, pp. 1226–1231 (2007)
52. Meng, Y., Li, W., Lin, Y., Pan, D.Z.: elfPlace: electrostatics-based placement for large-scale heterogeneous FPGAs. IEEE TCAD (2021)

Chapter 10
Circuit Optimization for 2D and 3D ICs with Machine Learning

Anthony Agnesina, Yi-Chen Lu, and Sung Kyu Lim

10.1 Introduction

The complex characteristics of advanced technology nodes and the ever-increasing scale of industrial designs challenge achievable power, performance, and area (PPA) results under traditional two-dimensional design flows. Unfortunately, this slowing of Moore's law also prolongs design time and, therefore, the costs of the chip design cycle. Replacing traditional two-dimensional (2D) design flows with three-dimensional (3D) design flows for both monolithic and heterogeneous integrated circuits has been proposed, but power and thermal issues remain using currently available electronic design automation (EDA) tools and practices.

This chapter introduces seven recent machine learning (ML)-driven techniques to optimize 2D and 3D integrated circuits (ICs) implementations.

The first three research articles describe the application of one such philosophy, graph neural networks (GNNs), to physical design. Through graph embedding and neighbor aggregation, the GNNs exploit the intrinsic graph nature of very-large-scale integration (VLSI) netlists to learn feature embeddings, which are then used as input to placement optimization, 3D IC tier partitioning, and multi-classifier for cell-level threshold voltage assignment.

The following three works adopt reinforcement learning (RL) to solve computationally intractable multistage decision problems. They can place decoupling capacitors on 2.5D/3D power delivery networks, conduct gate sizing, and perform timing closure through sequential exploration led by an agent represented as a deep policy network or a multi-armed bandit.

A. Agnesina (✉) · Y.-C. Lu · S. K. Lim
School of Electrical and Computer Engineering, Georgia Institute of Technology, Atlanta, GA, USA
e-mail: agnesina@gatech.edu; yclu@gatech.edu; limsk@ece.gatech.edu

© The Author(s), under exclusive license to Springer Nature Switzerland AG 2022
H. Ren, J. Hu (eds.), *Machine Learning Applications in Electronic Design Automation*, https://doi.org/10.1007/978-3-031-13074-8_10

The last work touches on the critical problem of parameter tuning for EDA. It shows how Bayesian optimization, a sequential decision-making strategy, helps search an ample state space to efficiently optimize the thermal and electrical characteristics of a complex black-box-like 3D system.

10.2 Graph Neural Network-Based Methods

10.2.1 2D Placement Optimization

The traditional objective of 2D placement is to minimize the half-perimeter wire length (HPWL) by placing cells efficiently on a fixed canvas. However, timing and power are equally important placement objectives as wire length in modern EDA flows. Therefore, it is crucial to have an algorithmic way to determine cell locations leading to desirable timing and power results.

The authors of [1] leverage a GNN to guide the placement of a commercial placer. First, the initial features of the standard cells in the netlist, i.e., the nodes in the GNN, are mapped to a higher-dimensional space by aggregating information around local neighborhoods. Next, the weighted k-means clustering algorithm [2] partitions cells into clusters based on the learned representations. These clusters then constitute soft placement constraints to the commercial placer, *Synopsys IC Compiler II (ICC2)*, which is used to perform global and detailed placements while honoring the clustering recommendations. This work aims to provide designers with a placement optimization framework that achieves high-quality placements for diverse designs by distilling underlying design knowledge.

10.2.1.1 Solution

Traditional gate-level placers usually reduce the runtime complexity of global placement with multilevel clustering schemes [3]. However, this clustering often performed with suboptimal k-way partitioning heuristics can, among other things, result in routing congestion, which heavily degrades achievable post-route PPA. Appreciating the already excellent placement quality achieved by existing commercial engines, this work enhances them by proposing intelligent *placement guidance* to the commercial tool. For example, in *ICC2*, the command *create_placement_attraction* allows telling the placer where groups of cells should be in respect to each other. This user-provided clustering, if well-executed, can prevent the fragmentation of cells found *similar* and improve PPA compared with traditional clustering methods. The overview of the proposed framework is shown in Fig. 10.1.

The proposed methodology is a three-step process. First, the original hypergraph netlist is transformed into a clique-based undirected and unweighted graph $G = (V, E)$, a representation well-suited to apply standard GNN algorithms. Each node

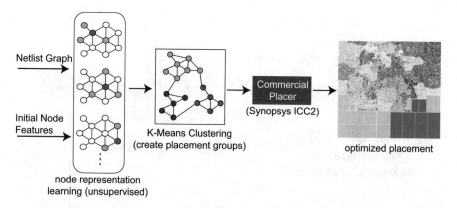

Fig. 10.1 Overview of the ML framework for placement guidance. The convolutional GNN transforms the initial node features of the netlist graph into more useful representations, helping characterize the underlying design knowledge accurately. Then, a weighted k-means clustering groups the cells according to the learned node embeddings. The resulting clusters are used as placement guidance for the commercial placer during global and detailed placements

$v \in V$ is associated with a vector of features influential to the placement results. For example, the features chosen here to focus on commercial CPU designs are shown in Fig. 10.2. They are based on hierarchy information—as instances in the same hierarchy are usually more interconnected—and the logical affinity with the memory macros in the design, as the logic to memory paths is often timing-critical. The instances' hierarchies are encoded with a trie whose keys correspond to the different components of an instance name, and the logical levels to the M memory macros define M values per node v. The t-SNE plot [4] in Fig. 10.2 shows that a naive clustering using the initial features would be unsuccessful, as no distinct order emerges from the plot.

Thus, the second step is graph learning starting from the previously defined features and the clique-based graph. Information of each cell is aggregated within their three-hop neighborhood using unsupervised GraphSAGE [5]. A fixed-sampling technique at each level of the aggregation helps prevent overfitting and computational instability due to the exponential growth of the number of neighbors with the hop count. The dimension of the final representations of each design instance is subject to the number of neurons of the last layer of the GNN, here, 128. Compared with GNNs used for supervised learning, the proposed unsupervised GNN lacks a multilayer perceptron (MLP) following the graph convolutions. The representation at level $l = L$ of a node $v \in V$ is obtained recursively:

$$h^l_{N_l(v)} = \text{REDUCEMEAN} \left(\left\{ W^{\text{Agg.}}_l h^{l-1}_u, \forall u \in N_l(v) \right\} \right),$$

$$h^l_v = \sigma \left(W^{\text{Proj.}}_l \cdot \text{CONCAT} \left[h^{l-1}_v, h^l_{N_l(v)} \right] \right),$$

$$h^l_v = \text{NORM}(h^l_v), \tag{10.1}$$

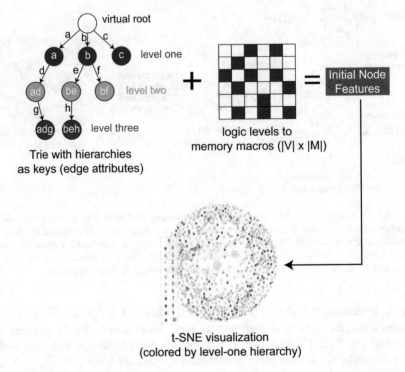

Fig. 10.2 Construction and visualization [4] of the initial node features, obtained from design hierarchy and logical affinity to memory macros

where σ is the sigmoid activation function, $N_l(v)$ is the set of neighbors of v sampled at hop l, and $\boldsymbol{W}_l^{\text{Agg.}}$ and $\boldsymbol{W}_l^{\text{Proj.}}$ denote the aggregation and projection matrices, respectively, which together form the neural layer at level l.

Next, one must define a loss function to train this neural network. However, there are no evident target labels for the nodes. Therefore, we posit that a good placement solution should place *logically connected instances closely*, reasonably minimizing wire length. This view naturally defines a score where connected cells should have *similar* representations, in terms of some pairwise distance reused in k-means. Thus, the unsupervised loss per node v is

$$\mathcal{L}(v) = - \underbrace{\sum_{u \in N(v)} \log\left(\sigma(h_v^{L\top} h_u^L)\right)}_{\text{neighboring cells}} - \underbrace{\sum_{w \in \text{Neg}(v)} \log\left(\sigma(-h_v^{L\top} h_w^L)\right)}_{\text{distant cells}}, \quad (10.2)$$

where $\text{Neg}(v)$ is the set of repeatedly resampled nodes away from v. The neural network is updated through gradient descent of the global loss $\sum_{v \in V} \mathcal{L}(v)$ until it stabilizes. A different GNN is trained per design, from scratch, in about 30 min.

Table 10.1 Placement quality on commercial CPU designs of the *ICC2* default placement flow, the *Louvain* [7] modularity-based clustering, and the proposed ML flow

Design	Method	# of clusters	WL (m)	WNS (ns)	TNS (ns)	Power (mW)	# of buffers
CPU-A	ICC2	–	4.37	−0.07	−0.22	142	5942
	Louvain [7]	82	4.34	−0.10	−0.62	141	5826
	Proposed	22	4.20	−0.01	−0.03	138	5371
CPU-B	ICC2	–	11.66	−0.24	−240	582	2728
	Louvain [7]	58	11.65	−0.38	−297	578	2689
	Proposed	32	11.55	−0.18	−62	574	2274

Finally, the final representations $\{h_v^{L=2}\}_{v\in V}$ are used by the weighted k-means clustering algorithm to group cells into clusters. Each node receives a weight $w_v = \text{area}(v)$ in the k-means formulation to achieve area-balanced clusters. The algorithm solves iteratively

$$\arg\min_{\{C_1,\dots,C_k\}} \sum_{i=1}^{k} \sum_{v\in C_i} w_v \|h_v^L - c_i\|_2^2 \,, \text{ where } c_i = \frac{\sum_{v\in C_i} w_v h_v^L}{\sum_{v\in C_i} w_v} \tag{10.3}$$

is the centroid of cluster C_i. The number k of clusters is selected to maximize the Silhouette score [6], a popular measurement of how good clustering is. The obtained clusters are then used, as explained above, to perform placement with the commercial placer.

10.2.1.2   Results

The proposed method is compared in Table 10.1 with *ICC2* used without any placement guidance and a heuristic technique called Louvain [7], which groups the nodes by minimizing the connections between clusters. Louvain is purely connection-based and does not incorporate design features. In contrast, the proposed ML methodology uses well-defined features and the powerfulness of graph representation learning, achieving superior PPA as a result.

10.2.2   3D IC Tier Partitioning

Tier partitioning is a critical stage in the physical design of 3D ICs. It is an algorithmic process of assigning each design instance, whether an individual or set of standard cells, macros, or more generally modules, to a specific vertical tier (=partition).

Modern 3D IC flows are *pseudo-3D* flows where a preliminary 2D implementation is first obtained with a place-and-route (P&R) step performed with commercial 2D physical design tools. Then, the design is transformed into a 3D layout through a tier-partitioning method named *bin-based min-cut* [8]. This method slices the 2D floorplan into bins, and the Fiduccia-Mattheyses (FM) algorithm [9] directly partitions all the standard cells into two tiers while respecting the area balance constraints of each bin. However, this approach has a severe drawback: long wires crossing different bins are not accounted for and randomly partitioned, degrading PPA substantially.

The authors of [10] propose a completely different tier-partitioning scheme based on a GNN, referred to as TP-GNN. Very similar to the previously presented work, the primary assumption is that the graph nature of the netlist can help transform the manually defined initial node features into meaningful high-dimensional representations, which can guide the tier assignment performed with k-means.

10.2.2.1 Problem Statement

Figure 10.3 shows the overall modern pseudo-3D placement flow along with the standard and proposed tier-partitioning methods. For a n-tier 3D design (here, we assume $n = 2$), all cells, wires, and chip footprint are first shrunk down by n in size compared to the original 2D technology and design. Next, the design is implemented in 2D with a commercial tool. Finally, after inflating the cells and wires back, one needs to find a 3D partitioning solution that leads to a suitable PPA of the resulting 3D IC. This is an arduous task, given the (x, y) cell locations are optimized for 2D, and extending them with a new vertical coordinate z, assigned by the partitioning engine, should preferentially not undo the high-quality optimizations of the 2D design.

10.2.2.2 Solution

The overview of the proposed framework is summarized in Fig. 10.4. A placed and routed flat netlist is first transformed into a clique-based graph, where the Manhattan distance is used as edge weight. Then, a hierarchy-aware edge-contraction algorithm (10.1) merges nodes and creates supernodes with new locations and edge weights, similar to Karger's algorithm [11]. This algorithm is run twice before graph learning to reduce the hierarchy depth drastically. Despite being heuristic and greedy, this algorithm helps partition cells onto different tiers without transforming short nets into long nets and causing severe timing degradation. Furthermore, this transformation substantially refines the 2D space partitioning of the bin-based scheme into one that simultaneously considers hierarchy, locations, and connectivity.

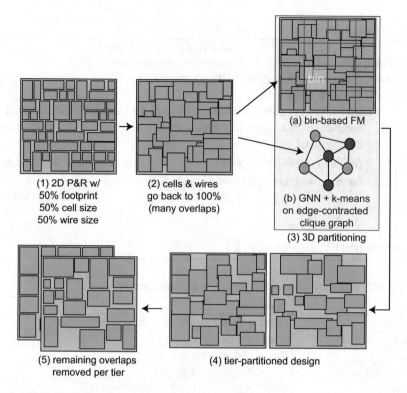

Fig. 10.3 Illustration of 3D placement in *pseudo-3D* flows with (3a) standard bin-based FM [9] partitioning vs. (3b) proposed GNN-based partitioning

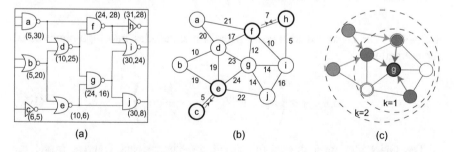

Fig. 10.4 Proposed TP-GNN partitioning framework on a small example. (**a**) Input netlist with two design hierarchies: $\{a, b, d, f, h\}$ and $\{c, e, g, i, j\}$. Pairs (v_x, v_y) represent the 2D cell locations. (**b**) Hierarchy-aware edge contractions on the transformed clique-based graph. Edge weights represent the Manhattan distance. (**c**) For the target node g, sampling and aggregating features from its k-hop neighbors

An initial feature vector (Table 10.2) is defined for every supernode in the transformed graph. These features include design hierarchy (i.e., an integer index of the module list) and timing information such as slack, transition, and delay. Finally,

---

Algorithm 10.1: Hierarchy-aware edge-contraction algorithm

Input: $G = (V, E)$: original clique-based graph of flattened netlist
Output: $G^C = (V^C, E^C)$: edge-contracted clique-based graph
1 $E \leftarrow$ AscendingSortOfEdgesByWeight(E) ;
2 **for** $e = (u, v) \in E$ **do**
3   **if** edge (u, v) not contracted **and** cells u, v in the same RTL module **then**
4   contract (u, v) to form new vertex v^C ;   `// in-place contraction`
5          $(v_x^C, v_y^C) = (\frac{u_x+v_x}{2}, \frac{u_y+v_y}{2})$;   `// update location`
6   **for** $n \in \{$neighbors$(u) \cup$ neighbors$(v)\}$ **do**
7              $w(v^C, n) = \|v_x^C - n_x\|_1 + \|v_y^C - n_y\|_1$;   `// assign edge weight`

8 **return** $G^C(V^C, E^C) \leftarrow G(V, E)$

Table 10.2 Initial features of each *supernode* in the edge-contracted graph G^C

Features	Descriptions
hierarchy	"Module" (encoded) defined in the synthesized netlist
sum_slack	Sum of worst slacks of all cells
sum_slew	Sum of maximum pin slew of all cells
sum_delay	Sum of worst delay of all cells
dist2source	Length of the shortest path to clock source in G^C
1-hop degree	# of 1-hop neighbors in G^C
2-hop degree	# of 2-hop neighbors in G^C

a GNN [12] is leveraged to transform the initial features into high-dimensional representations:

$$h_v^l = \sigma \left(h_v^{l-1} + W_l \sum_{u \in N_l(v)} \frac{h_u^{l-1}}{|N_l(v)|} \right). \tag{10.4}$$

The GNN algorithm dilutes the initial low-dimensional features inside the network graph and extends them with connectivity and neighbors' information. The increased dimensionality of the GNN representations helps separability but, in return, renders generalization harder, a phenomenon known as the *curse of dimensionality*. The weighted k-means algorithm further utilizes these representations and the total area of each supernode as a weight to determine the tier assignment. The weighting scheme is motivated by the fact that it is preferable in a 3D design to balance the silicon area to minimize the full-chip power consumption and decrease the footprint.

Table 10.3 PPA on RISC-V designs of 2D, Shrunk-2D (S2D) [8], and the proposed ML flow

Design	Method	Footprint (mm^2)	WL (m)	Freq. (MHz)	Energy/ cycle (pJ)	# of MIVs	Crit. path WL (um)	Partitioning time
OpenPiton	2D	1.22	6.33	289	344	–	543	–
	S2D [8]	0.61	4.91	270	340	76K	579	9 min
	Proposed	0.61	4.56	344	271	99K	292	26 min
RocketCore	2D	0.28	1.78	832	126	–	314	–
	S2D [8]	0.14	1.62	921	107	39K	289	5 min
	Proposed	0.14	1.51	964	101	23K	129	22 min

Similar to the previous article, the proposed GNN is trained unsupervised by minimizing the loss function of the formula (10.2). Here, we posit that a good partitioning solution should partition *logically connected supernodes on the same tier*, reasonably minimizing the disruption from the starting 2D implementation and improving PPA with the added benefits of 3D integration. Furthermore, the unsupervised framework does not require generating a database to train the model. Therefore, it is conceptually able to generalize to any design because it does not assume a predefined netlist structure.

10.2.2.3 Results

The proposed framework only targets the tier-partitioning stage. Therefore, it is incorporated with a complete design flow named Shrunk-2D [8] to perform full-chip analysis. The previous bin-based FM min-cut tier partitioning is replaced with the proposed ML model. The results are shown in Table 10.3. It is observed that the proposed framework demonstrates consistent improvement over various benchmarks in terms of full-chip PPA. In addition, the reported partitioning time per netlist, which includes the training of the GNN from scratch, is reasonable.

10.2.3 Threshold Voltage Assignment

The authors of [13] target the threshold voltage (Vth) assignment of standard cells for power optimization during the engineering change orders (ECO) step of the physical design flow. It casts the problem as a supervised multi-class classification problem, where the ground truth is given by a commercial signoff tool.

High-quality features of the candidate cells for discrete Vth assignment are obtained using a GNN, enabling the proposed approach's generalization and scalability. The assignment decision results are analyzed, and critical local factors are extracted with a GNN interpretation scheme based on subgraph mutual information.

10.2.3.1 Problem Statement

During signoff, the Vth assignment of cells is modified to optimize power (leakage and dynamic) and timing [14]. Unfortunately, to solve this NP problem [15], state-of-the-art commercial tools rely on the iterative use of expensive heuristics and algorithms, such as linear programming. This work replaces this long step by first learning the Vth assignment of the commercial tool with an ML-based model, offering during testing or deployment a rapid inference, with up to 14× faster runtime, for comparable optimization results.

10.2.3.2 Solution

This work relies again on a GNN to learn high-quality features for each standard cell in the clique-based graph representation of the netlist $G = (V, E)$, where nodes represent cells and edges represent net connections. However, in contrast with the two previous works, these features are input to an MLP softmax classifier to assign one of four discrete Vth values. The overall flow is depicted in Fig. 10.5. The end-to-end model of GNN plus softmax is trained in a supervised fashion to minimize a cross-entropy loss between the model predictions (four probabilities $P_{v,c}$ for each Vth class c and cell v) and the ground truths Y obtained from Synopsys *PrimeTime* ($Y_{v,c} = 1$ when cell v is assigned Vth $= c$, and else 0), i.e.,

$$\mathcal{L} = -\sum_{v \in V} \sum_{c=1}^{4} Y_{v,c} \log(P_{v,c}). \tag{10.5}$$

The 20 initial handcrafted features for the GNN learning are listed in Table 10.4, focusing on timing information obtained from the cell library and post-route static timing analysis (STA) results. These initial features are transformed using the same recursion as in the formula (10.1). A critical difference is that the GNN embeddings $\{h_v^{L=3}\}_{v \in V} \in \mathbb{R}^{128}$ are inputs to a one-layer MLP with softmax activation

$$P_{v,c} = \frac{e^{z_c}}{\sum_{c=1}^{4} e^{z_c}}, \text{ where } z_c = W_L h_v^L. \tag{10.6}$$

10.2.3.3 Results

The ML model is trained on nine close-to-signoff designs and validated/tested on five unseen designs obtained from the ISPD contest and targeting a TMSC 28 nm technology. The four types of possible Vth assignments are ULVT (ultralow Vth, the starting type), LVT, HVT (high), and UHVT. Training takes 12 h with TensorFlow [16] through gradient descent with Adam optimizer [17].

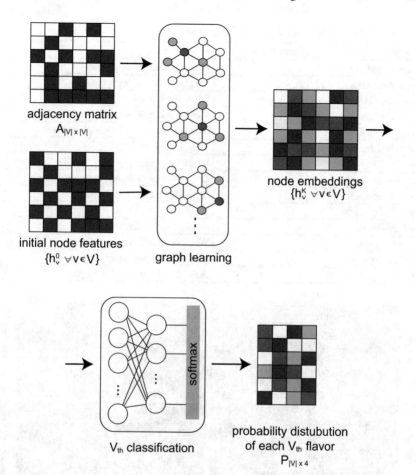

Fig. 10.5 Illustration of the learning process for Vth assignment. First, the GNN generates cell embeddings that represent the netlist better than the initial features. Next, a softmax-based classifier determines the final Vth assignment from the cell embeddings to optimize the signoff power

The results show that the ML model can generalize its Vth assignments to unseen designs with high fidelity to *PrimeTime* (F1-score >0.85), in a fraction of the runtime (14× faster) and with similar power optimization quality. Figure 10.6 shows the unseen designs' instance-based leakage power consumption maps before and after applying the proposed ML framework. Across all designs, the overall leakage power is reduced effectively without introducing extra hotspots.

This paper proposes to use a novel GNN explanation method to explain the Vth assignment and thereby confirm the ML model's reliability. The model's prediction is a probability distribution over the four threshold classes $P_\theta(v|G)$, where θ denotes the parameters of the trained ML model. However, as highlighted in the conditional probability, the entire graph G is needed to predict the Vth assignment of a single cell v, resulting in a very complex and difficultly interpretable prediction. The idea

Table 10.4 Starting cell features for Vth assignment, obtained with an STA

Features	Dimensions	Descriptions
max output slew	1	Max transition of output pin
max input slew	1	Max transition of input pins
wst output slack	1	Worst slack of output pin
wst input slack	1	Worst slack of input pins
Output cap limit	4	Max driving cap of output pin per Vth
max leakage	4	Max leakage per Vth
tot input cap	1	Sum of input pin cap
tot fanout cap	1	Output net cap + input pin cap of fanouts
tot fanout slack	1	Sum of worst slack of fanouts
wst fanout slack	1	Worst slack of fanouts
avg fanin cap	1	Average cap of fanins
wst fanin slack	1	Worst slack of fanins
tot sibling cap	1	Sum of input pin cap of siblings
tot sibling slack	1	Sum of worst slack of siblings

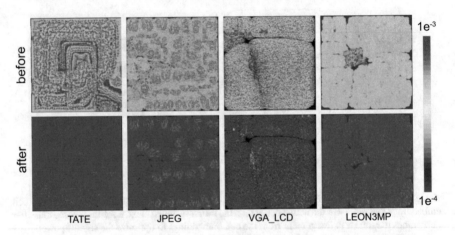

Fig. 10.6 Leakage power consumption (in mW) of each cell before and after using the proposed ML Vth optimization. The designs are unseen during training

of [18] is to find a much smaller subgraph G_S of G around v that can approximate the model prediction well, i.e., it searches to

$$\text{Find } G_S \subset G \text{ s.t. } P_\theta(v|G_S) \approx P_\theta(v|G) \text{ and } |G_S| \ll |G|, \qquad (10.7)$$

where the closeness of distributions ($\approx$) is obtained by minimizing the entropy $-\sum_c \mathbb{1}_{Y=c} \log P_\theta(Y = y|G = G_S)$, i.e., the gap from the ground truth and predicted class probabilities. The subgraphs G_S are randomly sampled in the one-hop vicinity of v (fanins plus fanouts). The critical feedback from the analysis shown in Fig. 10.7 is that the Vth assignment performed by the model is reliable,

Fig. 10.7 GNN
explanation [18] of the red
cell's Vth assignment by
extracting a small subgraph

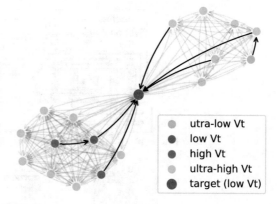

utra-low Vt
low Vt
high Vt
ultra-high Vt
target (low Vt)

aligning with common design knowledge. Indeed, as observed through the subgraph
extraction, the minority neighbors in lower Vth influence the final Vth type of a
given node more; the larger capacitance in lower Vth cells requires imposing tighter
constraints on their drivers than cells in high Vth.

10.3 Reinforcement Learning-Based Methods

10.3.1 Decoupling Capacitor Optimization

The authors of [19] apply deep RL to insert and place decoupling capacitors
(decaps) on a multitier hierarchical power delivery network (PDN) for 2.5D and
3D ICs. Compared with the traditional solution of full search relying on analytical
formulas, the proposed method is much simpler and relies uniquely on proper
definitions of the PDN structures, decaps, and reward. Furthermore, the deep RL
agent achieves target impedance of the self-PDN along with minimum area occupied
by decaps in just a few minutes, a 1000× runtime improvement.

10.3.1.1 Problem Statement

High-density and high-performance 3D ICs experience heavy simultaneous switching noises (SSN) [20], high-voltage fluctuations created by input/output drivers
switching simultaneously, which can cause logic failure and full-system degradation
through the PDN. The most common strategy to reduce SSN is to use decoupling
capacitors: placed near the load capacitances, they replace the supply charge coming
from the PDN [21]. On average, more than 10% of the total chip area is dedicated
to decoupling capacitors, which makes the area and position of these an important
optimization problem. The issue is exacerbated in 3D ICs as space available for

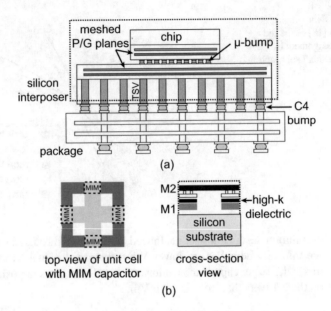

Fig. 10.8 (a) Conceptual view of the silicon interposer-based 2.5D/3D IC for decap design. The multiple-level PDN includes the target chip PDN and interposer PDN. (b) Top and cross-section views of the unit cell (UC) of the PDN with unit decap

decap reduces. Current solutions use complicated and human-engineered power integrity formulas to determine the optimal locations of the decaps.

Here we consider a two-tier PDN made of two PDNs, one for the interposer and one for the chip, as shown in Fig. 10.8. Both PDNs are decomposed into a regular grid of unit cells (UCs)—defined from the meshed P/G planes, where each UC can accommodate at most one unit decap, made of high-k metal-insulator-metal (MIM).

10.3.1.2 RL Setting

This work considers the decap assignment as an RL problem. The key components in the proposed RL formulation are as follows:

- *Environment State*: The two PDN structures are represented as two 3D matrices: the unit cell matrix (=distribution of UCs configuring the PDNs) and the decap matrix (=distribution of the decaps on the PDNs). Each matrix possesses two channels corresponding to the interposer and on-chip PDNs (target), respectively. Numerical values in these matrices represent the presence/absence of a UC/decap.
- *Agent/Observable State*: The 3D decap matrix only, given to the RL agent.
- *Action*: Assigning a unit decap to an available unit cell of one of the two PDNs.

- *Reward*: Improvement of the PDN self-impedance in the number of frequency points (N) meeting the target impedance. This one-port impedance is measured between a fixed point on the power planes (the probing port) and the ground.
- *State Transition*: Update of the available decap locations.

The reward is expressed formally as

$$r_t = \left(\frac{N_{t+1} - N_t}{\text{total \# of frequency points}}\right) + \mathbf{1}_{\text{achieve target over whole range}}, \qquad (10.8)$$

where the last term boosts the reward when the impedance is under target for the whole frequency range. The reward value is obtained using models of the P/G planes, TSV, and micro-bumps, through segmentation of cells with fast impedance matrix calculation [22].

10.3.1.3   Deep Q-Learning

The RL agent is represented by a deep Q-network (DQN). The advantage of a deep network resides in its ability to handle computationally large state and action spaces. The proposed agent architecture depicted in Fig. 10.9 comprises convolution layers followed by fully connected layers. The DQN outputs an estimation of $Q(s, a)$ for all possible states and discrete actions. Formally, the Q-value of a policy π is defined as

$$Q^\pi(s, a) = \mathbb{E}\left[\sum_{t=0}^{\infty} \gamma^t r(s_t, a_t)|s_0 = s, a_0 = a\right], \qquad (10.9)$$

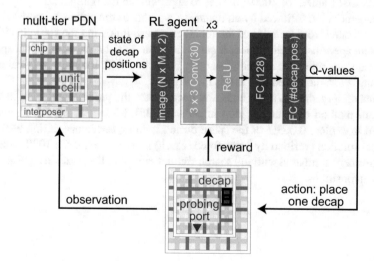

Fig. 10.9 The proposed deep Q-learning-based decap inserter. The PDN self-impedance is measured at the fixed probing port

where γ is the discount rate and t the discrete timestep. The small value of $\gamma = 0.5$ in the experiments forces the agent to find a solution quickly, effectively accomplishing a budget for the number of inserted decaps. The Q-value represents the expected cumulative reward when performing action a in the state s and then following policy π. The goal of Q-learning is to find a policy π that maximizes the Q-values of each state-action pair, i.e., $Q^\star(s, a) = \max_\pi Q^\pi(s, a)$.

The DQN is trained with value iteration using the mean squared temporal difference $TD(0)$ error [23] as a loss function. Moreover, the two important modern techniques of *experience replay* with replay memory and *periodically updated target* with a cloned network [24] are used to train the DQN efficiently. The loss function for the training of the DQN through gradient descent is then

$$\mathcal{L}(\theta) = \mathbb{E}_{(s,a,r,s') \sim U(D)} \left[(r + \gamma \max_{a'} Q(s', a'; \theta^-) - Q(s, a, \theta))^2) \right], \quad (10.10)$$

where $U(D)$ is a uniform distribution over the pool of stored samples D and θ, θ^- are the parameters of the Q-network and frozen target Q-network, respectively. During training, an epsilon-greedy action selection is used, where we select a random action with probability ϵ, and else the action with the highest estimated Q-value of the online network. This scheme helps in the tradeoff between exploration and exploitation.

10.3.1.4 Results

Different RL agents are built and trained from scratch per PDN configuration, as the state and action spaces' dimensions differ for different PDN structures. The RL agent is first trained for 1000 episodes to approximate the optimal policy by a loop of data generation followed by an update of weights of the online network with the loss calculated from samples of the replay memory, following the formula (10.10). Once the agent has been trained, it starts from the initial state without decaps and inserts decaps one after another, selecting the action with the highest Q-value and stopping when the impedance meets the target.

Interestingly, the RL agent adds unit decaps near the probing port, similar to the standard method of reducing loop impedance [25]. Moreover, the RL agent only needed to explore 0.003% of the space during training to achieve during testing an optimal solution (verified by brute-force search) in a few minutes, a 1000× runtime improvement compared with full search simulations of all the possible combinations of decap positions.

10.3.2 Gate Sizing

This work [26] presents a methodology that leverages RL to solve the traditional gate sizing problem. Specifically, the focus is on the problem of gate sizing for timing optimization, an NP-hard problem [15] where the goal is to optimize the TNS at the post-route stage through gate sizing.

The classical gate sizing problem is formulated as a control process by properly defining state, action, and reward. Then, a policy gradient-based algorithm named deep deterministic policy gradient (DDPG) [27] is leveraged to solve the problem by maximizing the total reward. In addition, to accurately characterize netlist characteristics, a GNN is leveraged to encode the neighborhood information of each design instance and make the sizing decision in a better context.

10.3.2.1 Problem Statement

The study's goal is to assign a feasible size to each gate to minimize the TNS of a placed and routed design. However, time-honored gate sizing algorithms of EDA tools use pseudo-linear heuristics or analytical methods driven by (statistical) STA that result in globally suboptimal sizing solutions. This work builds a high-dimensional RL framework focusing on timing only, called RL-Sizer, which formulates the classic gate sizing problem as an RL process and solves it by applying advanced RL algorithms equipped with GNNs.

10.3.2.2 Solution

The RL framework repeats its gate sizing procedure until all timing violations are fixed. A *single* gate sizing iteration ends with an STA update and is mapped into an RL trajectory, whose individual components are:

- *State*: A state s represents an individual *cell* in the netlist. It is the cell currently considered for gate sizing in the gate sizing iteration. The state is constructed by concatenating the GNN embedding of the cell's local graph (=three-hop neighborhood) with the cell-specific technology features extracted from libraries. The embedding is obtained with a GNN applied to the initial features of Table 10.5.
- *Action*: An action a gives the *new gate size* assigned to the cell in state s. It is continuous and realized as the floating-point change Δd in the current driving strength. Formally, if the cell s has a current size with strength d, taking the action $a = \Delta d$ will change the gate size to the one in the technology with the strength closest to $d + \Delta d$ among all possible choices.
- *Reward*: The reward $r = -\Delta_{local}$ TNS of performing action a on instance s is the TNS change in the instance's local graph computed after all gates in the trajectory have been resized. This reward focuses on fixing timing issues instead of a traditional timing-power tradeoff.

Table 10.5 Initial node features for GNN encoding of the state in the RL-Sizer framework

Features	Descriptions
slack	Worst slack of paths through instance
in_slew	Worst input pin slew
out_slew	Output pin slew
arc_delay	Worst cell arc (input to output pin) delay
nom_delay	Nominal delay (fanout of 4)
cell_cap	Cell capacitance
drv_length	Driving (output) net length
drv_load	Sum of driving capacitance (net + cell)
drv_res	Sum of driving resistance
fanin_cap	Average capacitance of fanins
sibling_cap	Sum of capacitance of siblings

An RL trajectory is created as depicted in the toy example of Fig. 10.10. First, the paths with a negative slack are extracted with an STA. Then, only the worst path (red) and all negative paths that intersect with it (blue) are looked at, drastically reducing the number of considered cells. Finally, these T cells sorted in topological order define an RL trajectory from timestep $t = 0$ to $t = T - 1$.

When traversing the trajectory sequentially, the local graph of each cell is extracted to define the state s_t given to the RL agent, which then predicts an action a_t (the cell is not resized yet). Next, the sizing decisions/actions are applied to the design at the end of the trajectory. An updated STA is performed there, which provides the reward r_t for each action. Finally, the network of the RL agent is updated by the DDPG algorithm through the backpropagation of tuples $\{(s_t, a_t, r_t)\}$ sampled in the replay buffer of past trajectories.

The GNN scheme uses the equations in the formula (10.1). The resulting representations $\{h_v^{L=2}\}_{v \in V(v)} \in \mathbb{R}^{64}$ of the local graph of v, namely, $V(v)$, are reduced with mean pooling and concatenated with the technology features (e.g., driving strength, capacitance, and slew constraints of the current gate size) to yield the state vector

$$s = \text{CONCAT}\left[\text{MEANPOOLING}(\{h_u^{L=2}\}_{u \in V(v)}), \text{tech}(v)\right]. \tag{10.11}$$

This local graph approximation to define the state is motivated by the timing impact on the overall netlist of a gate resize diminishes as the hop count increases.

10.3.2.3 Results

The RL agent network in Fig. 10.10c is trained end to end from scratch on each benchmark since the technology node varies between circuits. Training stops when the total design TNS no longer improves across ten consecutive iterations. Table 10.6 shows that the RL-based gate sizing framework results are promising.

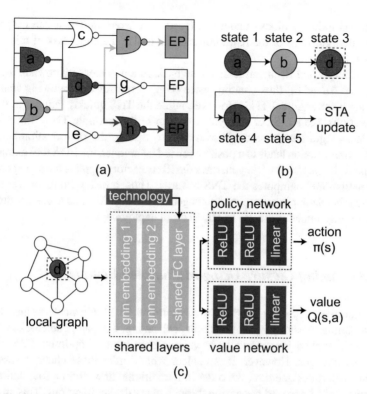

Fig. 10.10 Overview of the RL framework for timing-driven gate sizing. (**a**) Input netlist with three endpoints (EPs). The worst path (red) and its overlapping negative slack paths (e.g., blue) are extracted. The instances on these paths are selected for one sizing iteration, (**b**) whose topological order defines an RL trajectory. (**c**) The new gate sizes are sequentially determined by considering each cell and its local graph as an RL state passed to the RL agent action prediction. (**a**) Instance selection. (**b**) RL trajectory (**c**) Action prediction by RL agent

Table 10.6 TNS optimization results comparison between RL-Sizer and Synopsys *ICC2* on six commercial blocks. WNS is the worst negative slack, TNS the total negative slack, and #vio. EPs the number of violating endpoints. The commercial tool and RL-Sizer runtimes are measured on the same machine without GPU support

Design (tech)	SYNOPSYS ICC2					RL-SIZER				
	WNS (ns)	TNS (ns)	#vio. EPs	Power (mW)	Run-time	WNS (ns)	TNS (ns)	#vio. EPs	Power (mW)	Run-time
B1 (5 nm)	−0.08	−46.6	1728	68.7	30 m	−0.07	−44.7	1631	68.8	6 h
B2 (5 nm)	−0.05	−1.2	182	205.3	1 h	−0.04	−0.8	116	205.7	14 h
B3 (12 nm)	−0.02	−0.07	4	44.9	1 h	−0.07	−0.4	39	45.1	15 h
B4 (12 nm)	−0.21	−8.7	348	123.9	1 h	−0.20	−8.1	201	124.1	22 h
B5 (16 nm)	−0.79	−90.1	383	743.0	10 m	−0.78	−80.5	379	718.4	5 h
B6 (16 nm)	−0.02	−0.03	5	25.5	24 m	−0.04	−0.7	74	26.0	6 h

For various industry blocks under advanced technology nodes, it can outperform the default algorithms of the *ICC2* commercial tool at the target task of minimizing design TNS.

Interestingly, the RL agent learns to setback on a trajectory by downsizing some cells ($a = \Delta d < 0$), thus creating more sizing room for the following states and achieving a better global TNS. However, while the TNS quickly reduces in the first few iterations, the RL-Sizer struggles to optimize the last-mile TNS. Since the RL agent lacks rigid heuristics to close near-zero timing, the optimization stagnates when no sizing action leads to a positive reward. Moreover, the EDA tool extracts the worst paths in the framework, generates the RL trajectory, applies the sizing changes to the netlist, and computes the TNS rewards. Thus, a heavy runtime overhead in the switch between programming languages (Python for ML and C++ for the EDA tool) currently impedes the use of RL-Sizer in deployment.

10.3.3 Timing Closure Using Multiarmed Bandits

Many iterative steps are needed in the current VLSI physical design flow to achieve timing closure [28]. For example, cell sizing, datapath, and clock buffering are usually applied incrementally in various flow parts to optimize PPA without functional changes. However, it is unclear which order these changes should be applied. Expert designers set the order in a traditional flow, from a foundation flow customized and improved per design through many design iterations. This approach can be expensive, limiting practical exploration from humans who often prefer to trust their own experience. However, the search space is most likely highly complex, nonlinear, and non-convex. Therefore, human heuristics are not always tenable.

The authors of [29] propose to use RL to learn from experience to decide online, at a given point of the flow, which optimization heuristic to apply on the design so that the final PPA is optimized. The focus is on gate sizing techniques, i.e., determining transistor widths, buffer insertion techniques, and the order in which they are applied to the design, which is essential to the final result.

10.3.3.1 Problem Statement

First, a fixed set of simple timing optimization algorithms and heuristics is defined. Then, the RL agent selects one of the techniques, applies it to the current design, keeps the changes if PPA improves, and observes a reward, a mixture of PPA and runtime. This process repeats until satisfying a stopping criterion (e.g., reward <1% for ten consecutive iterations). The multiarmed bandit (MAB) dictates the RL agent behavior, maximizing the accumulated reward through a balance of exploration vs. exploitation. The overall proposed flow is shown in Fig. 10.11. The key RL components of the proposed approach are as follows:

Fig. 10.11 Autonomous timing-driven design optimization. The algorithm selected by the multiarmed bandit (MAB) repeatedly optimizes the design. The reward of each algorithm and the MAB policy determine which algorithm will be applied next to the restructured and partially optimized design

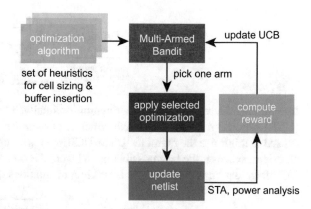

Algorithm 10.2: UCB algorithm for regret minimization in the MAB setting

Input: Set of arms $\mathcal{A}$ and stopping time T
1 **for** $t \in 1, \ldots, T$ **do**
2 $\quad$ Choose action $a_t = \arg\max_{a \in \mathcal{A}} \text{UCB}(a, t - 1)$;
3 $\quad$ Observe reward r_t and update UCB ;

- *State*: None, a particularity of the MAB's setting.
- *Action*: Pick one arm a among the seven optimization algorithms: three Lagrange relaxation-based gate sizers plus four buffer inserters. These only apply *locally* on a small set of worst paths and in reduced iterations.
- *Reward*: Encapsulation of the TNS, power, and runtime. The reward increases when one of these three metrics reduces, and its details will be provided later in formula (10.14).
- *State Transition*: The design is updated if the reward is positive, and the score for the chosen arm is updated.

10.3.3.2 Solution

The MAB targets the problem of sequential resource allocation. Let $\mathcal{A} = \{1, 2, \ldots, K\}$ denote the fixed set of actions/arms, i.e., the seven timing optimizations. These are associated with probability distributions of rewards streams with means $\{\mu_a\}_{a \in \mathcal{A}}$. While the optimal policy is to play the fixed action $a^\star = \arg\max_{a \in \mathcal{A}} \mu_a$, the means are unknown; thus, one must estimate them all reasonably (*exploration*) but also play good actions simultaneously (*exploitation*). The latter objective can be written in terms of minimization of the regret over T rounds, defined as

$$R(\mathcal{A}, T) = \underbrace{T \max_{a \in \mathcal{A}} \mu_a}_{\substack{\text{sum of rewards} \\ \text{of optimal policy}}} - \underbrace{\mathbb{E}\left[\sum_{t=1}^{T} r_t\right]}_{\substack{\text{sum of rewards} \\ \text{of agent's strategy}}}. \tag{10.12}$$

It represents the rewards lost by acting non-optimally. The upper confidence bound (UCB) Algorithm 10.2 based on the principle of *optimism in the face of uncertainty* is used to minimize the regret [30]. The UCB$(a, \cdot)$ is an upper bound of μ_a, and the algorithm acts as if the best possible model were the true model. The most common UCB formulation is derived from Hoeffding inequalities:

$$\text{UCB}(a, t-1) = \underbrace{\hat{\mu}_a(t-1)}_{\text{exploitation term}} + \underbrace{\sqrt{\frac{2\log(t)}{N_a(t-1)}}}_{\text{exploration bonus}}, \tag{10.13}$$

where $\hat{\mu}_a(t-1)$ is the empirical mean of rewards of arm a and $N_a(t-1)$ the number of times arm a has been chosen after $t-1$ iterations. The regret of this algorithm is $O(\log(T))$, assuming rewards are bounded.

10.3.3.3 Innovative Ideas

Three main innovative ideas help optimize the performance of the MAB to the given problem:

1. *Composite Reward*: The designer's metrics of interest, i.e., TNS, leakage power, and runtime, are encapsulated into a singular scalar reward

$$r_t(a) = \text{runtime}(a)\left(\delta\Delta\,\text{TNS} + (1-\delta)\Delta\,\text{Power}\right), \tag{10.14}$$

   where δ is dynamic, changed from 0.8 to 0.2 once timing closure is achieved. The runtime factor prioritizes fast algorithms, while the normalized TNS/power relative improvements favor TNS and power reduction.
2. *Address drift of UCB*: Using the empirical mean in the formula (10.13) is inappropriate when the rewards get smaller and N_a increases with time. Indeed, for the former, because arms apply sequentially and the design improves with time as it is updated through the optimizations, the magnitude of potential rewards reduces. For example, it is simpler to improve a bad starting TNS than improve it when violations are minor. Therefore, the first part of (10.13) is replaced by

$$\hat{\mu}_a(t-1) = \alpha\hat{\mu}_a(t-2) + (1-\alpha)r_{t-1}(a), \tag{10.15}$$

   to weigh more the most recent rewards. The experiments use a value of 0.5 for α.

3. *Replace expensive algorithms with faster suboptimal heuristics*: A small set of seven heuristics that change the netlist minimally is devised to allow for limited restructuring between iterations. First, the gate sizing algorithm minimizes leakage power given late/early constraints by a linear program rewritten with Lagrange relaxation. Standard EDA tools usually resize all gates one at a time following the topological order and until TNS or power stagnate. Here, the resizing applied on all cells or only the top 1000 timing-critical cells stops after one or two iterations. Moreover, only a small set of critical data and clock nets are considered for buffer insertion.

10.3.3.4 Results

The MAB is integrated inside the open-source Rsyn [31] physical design framework and applied on ISPD and TAU design contests benchmarks. In addition, the STA is carried out with OpenTimer [32]. The baseline for comparison is the Lagrange relaxation optimization running for 120 iterations. The results are promising, reaching comparable PPA results in many fewer iterations ($>2\times$ less). Interestingly, the all-cells one iteration is the most promising approach for gate sizing, and the critical path isolation that adds buffers at noncritical sinks is the best option for buffer insertion.

10.4 Other Methods

10.4.1 3D IC Optimization Using Bayesian Optimization

3D ICs are challenged by reliability and performance issues accompanying increased temperature densities [33]. For example, clock skew is heavily impacted by large temperature gradients over the die surface. In addition, many parameters of the 3D structure and package (e.g., geometries, fan speed, materials) directly impact the temperature profiles.

The authors of [34] use Bayesian optimization [35] to co-optimize the electrical and thermal performances of a 3D package. The complex relationship between the high-dimensional configuration of the 3D structure and corresponding performances is modeled as a black box. Bayesian optimization is a robust and widely used optimization algorithm for such problems. The advantage of this method is sampling efficiency, which is deeply needed when querying expensive simulation frameworks for 3D IC. An electrical-thermal solver is first built and validated on a prototype in this work. The solver is then used to simulate the quality of the 3D package parameters, which is used as an optimization target for the Bayesian model.

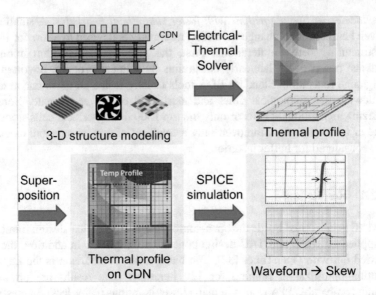

Fig. 10.12 Electrical-thermal simulation flow for 3D system design. The 3D structure is made of stacked dies, interposer, and PCB. First, an iterative solver [36] solves the thermal-electrical differential equations considering the heat sink, airflow, and superimposed power map. Then, a circuit solver computes skew, noise, and impedance of the CDN, considering the extracted temperature gradients and PDN response [37]

10.4.1.1 Problem Statement

The considered 3D package comprises three stacked dies, an interposer, and a printed circuit board (PCB), as shown in Fig. 10.12. An H-tree clock distribution network (CDN) is inserted in the center die. Various power maps for each tier are considered. Five tunable input parameters of the design referred to as variable x, namely, airflow, thermal interface material (TIM), TIM thickness, underfill material, and PCB material, are selected in predefined well-chosen ranges to minimize the composite target function

$$f(x) = 0.34 \times T_{\max} + 4.5 \times T_{\mathrm{grad}}, \qquad (10.16)$$

where $T_{\max}$ is the maximum temperature and T_{grad} is the temperature gradient in the clock die. The temperature gradient receives a higher weight as it is most influential to reducing clock skew, the study's overall goal.

10.4.1.2 Solution

The black-box function f is expensive to query as it requires thermal and electrical simulation of the entire 3D package, following the numerical finite volume method [36]. The analytical form of f is also unknown, and its evaluations may be noisy, defined on a subset $\mathcal{X}$ of $\mathbb{R}^d$ (here, $d = 5$ input parameters). We wish to minimize the value of f based only on function evaluations, a problem often referred to as derivative-free optimization. The idea is to choose points $x_1, \ldots, x_N \in \mathcal{X}$ to query the function and output a new candidate $x_{N+1} \in \mathcal{X}$ such that $f(x_{N+1}) - \inf_{x \in \mathcal{X}} f(x)$ is small.

Bayesian optimization builds a surrogate model $\mathcal{M}$ of f that is cheap to evaluate (a prior). The posterior obtained from the combination of prior and likelihood is then used to select the subsequent evaluation according to an acquisition function. Every time a new candidate $\hat{x}$ is selected, the original function f is queried at $\hat{x}$, returning a noisy evaluation y_i, and the model $\mathcal{M}$ is updated using the history of the N evaluations $\{(x_i, y_i)\}_{i \in \{1, \ldots, N\}}$. The multivariate Gaussian process (GP) with zero mean is used as prior:

$$f(x) \sim \text{GP}(0, k(x, \cdot)), \tag{10.17}$$

with a squared-exponential kernel $k(x, \cdot) = \exp\left(-\alpha \|x - \cdot\|_2^2\right)$ as covariance, defining a probability measure over random functions. Then, the posterior given the previous observations $\mathbf{y} = [y_1, \ldots, y_N]^T$ is also a GP:

$$f(x)|\mathbf{y} \sim \text{GP}(\mu(x), \sigma^2(x)), \tag{10.18}$$

where the value of posterior mean $\mu(x)$ and variance $\sigma(x)$ can be computed explicitly from the dataset:

$$\mu(x) = k_N(x)^T K_N^{-1} \mathbf{y} \tag{10.19}$$

$$\sigma(x)^2 = k(x, x) - k_N(x)^T K_N^{-1} k_N(x) \tag{10.20}$$

where $K_N = \{k(x_i, x_j)\}_{(i,j) \in \{1, \ldots, N\}^2}$ is the kernel matrix and $k_N(x) = [k(x_1, x), \ldots, k(x_N, x)]^T$.

An acquisition function then evaluates all the candidates. It ranks them and picks the best one. In this paper, the lower confidence bound (LCB) is used to explore-exploit:

$$x_{N+1} = \arg \min_{x \in \mathcal{X}} \{\mu(x) - \kappa \sigma(x)\} \quad (\kappa \geq 0). \tag{10.21}$$

The complete optimization framework is shown in Fig. 10.13.

10.4.1.3 Results

A full-fledged electrical-thermal simulation framework for 3D IC is built. The temperature profiles are first computed using an iterative solver based on the finite

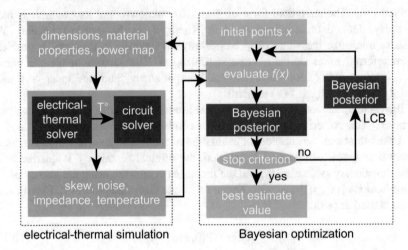

Fig. 10.13 Proposed Bayesian-driven flow for electrical-thermal optimization of the 3D package. The loop exits if the value of the objective in (10.16) is close to the target

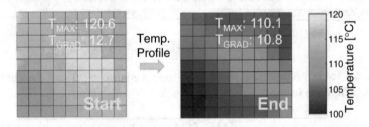

Fig. 10.14 Improvement in the temperature distribution of the clock die after applying the proposed Bayesian optimization procedure

volume method. Then, power maps are superimposed on each tier to simulate the 3D structure [38], and the CDN is modeled using temperature-dependent nonlinear clock buffers and interconnects [37]. Finally, the electrical-thermal solver is calibrated and validated on a 3D chip benchmark fabricated in 180 nm CMOS technology.

After adjusting the system parameters with Bayesian optimization, the maximum temperature, temperature gradient, and resulting clock skews are reduced substantially, by 9%, 12%, and 15%, respectively, and consistently after 100 function queries for all 3 considered power maps. For example, Fig. 10.14 shows how the maximum temperature and temperature gradient are improved for a given power map. Furthermore, the Bayesian method outperforms two traditional optimizers: pattern search and nonlinear solver, with up to 10% reduction of the clock skew and 1.5× faster runtime.

10.5 Conclusions

This chapter provided an overview of seven research works that utilize machine learning (ML) algorithms for circuit optimization. These works are broadly categorized into graph neural networks, reinforcement learning, and Bayesian optimization-based methods. The history of ML-based electronic design automation (EDA) is short, but this new breed of approaches for circuit optimization is gaining momentum. Nevertheless, more works are needed in this exciting area of research and development to further the state-of-the-art and practical impact on everyday design activities.

References

1. Lu, Y.C., Pentapati, S., Lim, S.K.: The law of attraction: affinity-aware placement optimization using graph neural networks. In: Proceedings of the 2021 International Symposium on Physical Design, pp. 7–14 (2021)
2. De Amorim, R.C., Mirkin, B.: Minkowski metric, feature weighting and anomalous cluster initializing in K-Means clustering. Pattern Recognit. **45**(3), 1061–1075 (2012)
3. Alpert, C.J., Kahng, A.B.: Recent directions in netlist partitioning: a survey. Integration **19**(1–2), 1–81 (1995)
4. Van der Maaten, L., Hinton, G.: Visualizing data using t-SNE. J. Mach. Learn. Res. **9**(11) (2008)
5. Hamilton, W.L., Ying, R., Leskovec, J.: December. Inductive representation learning on large graphs. In: Proceedings of the 31st International Conference on Neural Information Processing Systems, pp. 1025–1035 (2017)
6. Rousseeuw, P.J.: Silhouettes: a graphical aid to the interpretation and validation of cluster analysis. J. Comput. Appl. Math. **20**, 53–65 (1987)
7. Blondel, V.D., Guillaume, J.L., Lambiotte, R., Lefebvre, E.: Fast unfolding of communities in large networks. J. Stat. Mech. Theory Exp. **2008**(10), P10008 (2008)
8. Panth, S., Samadi, K., Du, Y., Lim, S.K.: Shrunk-2-D: a physical design methodology to build commercial-quality monolithic 3-D ICs. IEEE Trans. Comput.-Aided Design Integr. Circuits Syst. **36**(10), 1716–1724 (2017)
9. Fiduccia, C.M., Mattheyses, R.M.: A linear-time heuristic for improving network partitions. In: 19th Design Automation Conference, pp. 175–181. IEEE, Piscataway (1982)
10. Lu, Y.C., Pentapati, S.S.K., Zhu, L., Samadi, K., Lim, S.K.: TP-GNN: a graph neural network framework for tier partitioning in monolithic 3D ICs. In: 2020 57th ACM/IEEE Design Automation Conference (DAC), pp. 1–6. IEEE, Piscataway (2020)
11. Karger, D.R.: Global min-cuts in RNC, and other ramifications of a simple min-cut algorithm. In: SODA, vol. 93, pp. 21–30 (1993)
12. Kipf, T.N., Welling, M.: Semi-supervised classification with graph convolutional networks (2016). arXiv preprint arXiv:1609.02907
13. Lu, Y.C., Nath, S., Pentapati, S.S.K., Lim, S.K.: A fast learning-driven signoff power optimization framework. In: 2020 IEEE/ACM International Conference On Computer Aided Design (ICCAD), pp. 1–9. IEEE, Piscataway (2020)
14. Mok, S., Lee, J., Gupta, P.: Discrete sizing for leakage power optimization in physical design: a comparative study. ACM Trans. Design Autom. Electron. Syst. **18**(1), 1–11 (2013)
15. Ning, W.: Strongly NP-hard discrete gate-sizing problems. IEEE Trans. Comput.-Aided Design Integr. Circuits Syst. **13**(8), 1045–1051 (1994)

16. Abadi, M., Agarwal, A., Barham, P., Brevdo, E., Chen, Z., Citro, C., Corrado, G.S., Davis, A., Dean, J., Devin, M., Ghemawat, S.: Tensorflow: large-scale machine learning on heterogeneous distributed systems (2016). arXiv preprint arXiv:1603.04467
17. Kingma, D.P., Ba, J.: Adam: a method for stochastic optimization (2014). arXiv preprint arXiv:1412.6980
18. Ying, R., Bourgeois, D., You, J., Zitnik, M., Leskovec, J.: Gnn explainer: a tool for post-hoc explanation of graph neural networks (2019). arXiv preprint arXiv:1903.03894
19. Park, H., Park, J., Kim, S., Cho, K., Lho, D., Jeong, S., Park, S., Park, G., Sim, B., Kim, S., Kim, Y.: Deep reinforcement learning-based optimal decoupling capacitor design method for silicon interposer-based 2.5-D/3-D ICs. IEEE Trans. Compon. Packag. Manuf. Technol. 10(3), 467–478 (2020)
20. Shi, W., Zhou, Y., Sudhakaran, S.: Power delivery network design and modeling for high bandwidth memory (HBM). In: 2016 IEEE 25th Conference on Electrical Performance of Electronic Packaging And Systems (EPEPS), pp. 3–6. IEEE, Piscataway (2016)
21. Fan, J., Drewniak, J.L., Knighten, J.L., Smith, N.W., Orlandi, A., Van Doren, T.P., Hubing, T.H., DuBroff, R.E.: Quantifying SMT decoupling capacitor placement in DC power-bus design for multilayer PCBs. IEEE Trans. Electromagn. Compatibility 43(4), 588–599 (2001)
22. Kim, K., Hwang, C., Koo, K., Cho, J., Kim, H., Kim, J., Lee, J., Lee, H.D., Park, K.W., Pak, J.S.: Modeling and analysis of a power distribution network in TSV-based 3-D memory IC including P/G TSVs, on-chip decoupling capacitors, and silicon substrate effects. IEEE Trans. Compon. Packag. Manuf. Technol. 2(12), 2057–2070 (2012)
23. Sutton, R.S., Barto, A.G.: Reinforcement Learning: An Introduction. MIT Press, Cambridge (2018)
24. Mnih, V., Kavukcuoglu, K., Silver, D., Rusu, A.A., Veness, J., Bellemare, M.G., Graves, A., Riedmiller, M., Fidjeland, A.K., Ostrovski, G., Petersen, S.: Human-level control through deep reinforcement learning. Nature 518(7540), 529–533 (2015)
25. Koo, K., Luevano, G.R., Wang, T., Özbayat, S., Michalka, T., Drewniak, J.L.: Fast algorithm for minimizing the number of decap in power distribution networks. IEEE Trans. Electromagn. Compatib. 60(3), 725–732 (2017)
26. Lu, Y.C., Nath, S., Khandelwal, V., Lim, S.K.: Rl-sizer: VLSI gate sizing for timing optimization using deep reinforcement learning. In: 2021 58th ACM/IEEE Design Automation Conference (DAC), pp. 733–738. IEEE, Piscataway (2021)
27. Lillicrap, T.P., Hunt, J.J., Pritzel, A., Heess, N., Erez, T., Tassa, Y., Silver, D., Wierstra, D.: Continuous control with deep reinforcement learning (2015). arXiv preprint arXiv:1509.02971
28. Kahng, A.B.: New game, new goal posts: a recent history of timing closure. In: 2015 52nd ACM/EDAC/IEEE Design Automation Conference (DAC), pp. 1–6. IEEE, Piscataway (2015)
29. Stefanidis, A., Mangiras, D., Nicopoulos, C., Dimitrakopoulos, G.: Multi-armed bandits for autonomous timing-driven design optimization. In: 2019 29th International Symposium on Power and Timing Modeling, Optimization and Simulation (PATMOS), pp. 17–22. IEEE, Piscataway (2019)
30. Lattimore, T., Szepesvári, C.: Bandit Algorithms. Cambridge University Press, Cambridge (2020)
31. Flach, G., Fogaça, M., Monteiro, J., Johann, M., Reis, R.: Rsyn: an extensible physical synthesis framework. In: Proceedings of the 2017 ACM on International Symposium on Physical Design, pp. 33–40 (2017)
32. Huang, T.W., Wong, M.D.: OpenTimer: a high-performance timing analysis tool. In: 2015 IEEE/ACM International Conference on Computer-Aided Design (ICCAD), pp. 895–902. IEEE, Piscataway (2015)
33. Swaminathan, M., Han, K.J.: Design and Modeling for 3D ICs and Interposers, vol. 2. World Scientific, Singapore (2013)
34. Park, S.J., Bae, B., Kim, J., Swaminathan, M.: Application of machine learning for optimization of 3-D integrated circuits and systems. IEEE Trans. Very Large Scale Integr. Syst. 25(6), 1856–1865 (2017)

35. Snoek, J., Larochelle, H., Adams, R.P.: Practical Bayesian optimization of machine learning algorithms. In: Advances in Neural Information Processing Systems, vol. 25 (2012)
36. Xie, J., Swaminathan, M.: Electrical-thermal co-simulation of 3D integrated systems with micro-fluidic cooling and Joule heating effects. IEEE Trans. Compon. Packag. Manuf. Technol. **1**(2), 234–246 (2011)
37. Park, S.J., Natu, N., Swaminathan, M.: Analysis, design, and prototyping of temperature resilient clock distribution networks for 3-D ICs. IEEE Trans. Compon. Packag. Manuf. Technol. **5**(11), 1669–1678 (2015)
38. Park, S.J., Yu, H., Swaminathan, M.: Preliminary application of machine-learning techniques for thermal-electrical parameter optimization in 3-D IC. In: 2016 IEEE International Symposium on Electromagnetic Compatibility (EMC), pp. 402–405. IEEE, Piscataway (2016)

Chapter 11
Reinforcement Learning for Routing

Haiguang Liao and Levent Burak Kara

11.1 Introduction

Routing is one of the most challenging and time-consuming steps in EDA physical design. Before routing, a placement step is run to determine the locations for different electronic components. After placement, the routing step determines the paths for the nets on a chip that connect different components or pins. The connecting paths need to be optimized to minimize objectives that are related to the performance of the chips such as the total wirelength of nets or the number of design rule violations. Modern VLSI chip designs usually contain billions of transistors and millions of nets [1], which makes the routing step becoming increasingly more challenging. In order to simplify the routing problem, routing is usually further divided into two steps: **global routing** and **detailed routing**. Firstly, global routing divides the chip into coarse tiles and decides the rough tile-to-tile routes; the purpose of global routing is to allocate the limited routing space in coarse granularity. Then, detailed routing determines the exact routing paths for the design including details like vias (layer-to-layer connections) and tracks for nets. Figure 11.1 shows a schematic of a routing problem.

For a general routing problem, the inputs and outputs can be defined as follows:
Inputs:

- A **placement** solution with fixed locations of components and pins
- A **netlist**, indicating the groupings of components that need to be connected
- A **cost function** to be optimized, such as the total wirelength of routing
- A set of **design rules** for manufacturing, such as the wire/via spacing of each layer

H. Liao (✉) · L. B. Kara
Carnegie Mellon University, Pittsburgh, PA, USA
e-mail: haiguanl@cs.cmu.edu; lkara@cmu.edu

© The Author(s), under exclusive license to Springer Nature Switzerland AG 2022
H. Ren, J. Hu (eds.), *Machine Learning Applications in Electronic Design Automation*, https://doi.org/10.1007/978-3-031-13074-8_11

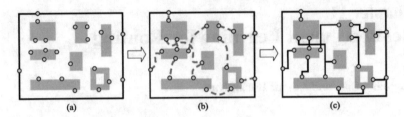

Fig. 11.1 Schematics of routing problem with (**a**) given placement solutions with components location determined, (**b**) global routing, and (**c**) detailed routing [1]

Outputs: wire connection for each net that meets the design rules and optimizes the given objectives.

Most routing problems are formulated as a graph **G(V,E)** where the routing tasks and resources can be represented. A node v in the routing graph can represent a tile in the chip, while an edge $e(u, v)$ represents the boundary of neighboring tiles. Thus, the edge capacity $c(u, v)$ represents the available routing resource of boundaries crossing the edge.

For general-purpose routing (both global routing and detailed routing), there are two basic graph-searching-based routing algorithms: **maze routing** and **A* search routing** [1]. They are frequently applied to general-purpose routing and are also used as the backbone algorithms for more advanced routing methods.

Maze routing (also known as Lee's algorithm) is one of the most widely used routing algorithms for two-pin connection routing, and it is based on breadth-first search (BFS). Maze routing consists of two phases: filling and retracing. The filling phase works analogous to wave propagation: starting from the source node, a BFS is started until the target node is reached. Once the target node is reached, a shortest path is retraced from the target node to the source node. A nice property of maze routing is that the path found is guaranteed to be the shortest path. However, maze routing is slow and memory intensive, especially when the problem is large. One reason that makes maze routing slow is its blind search feature, which means it searches the route region blindly without any prioritized choices.

Considering the shortcoming of blind search for maze routing, a routing algorithm can be improved if it can avoid searching for areas that are not likely to become a routing path. A* routing can overcome this shortcoming by using a function $f(x) = g(x)+h(x)$ to evaluate the cost of a path x. In the cost function, the first term $g(x)$ represents the cost from source node to current node x, and the second term $h(x)$ represents the estimated heuristic cost from current node x to the target node. At each step, the algorithm would select a node with the lowest cost and propagate, which is realized with priority queues. One nice property of A* routing is that if the heuristic cost $h(x)$ is admissible, meaning that it does not overestimate the actual cost from current node to the target node, then it is guaranteed to be optimal. A* has been used extensively in different kinds of EDA routing algorithms [2, 3].

Based on the specific routing algorithm such as maze routing and A* routing, a complete routing algorithm that involves multiple nets (each net consisting of

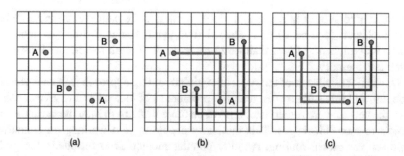

Fig. 11.2 Examples showing the ordering of nets in sequential routing approach. (**a**) Net A and Net B are to be routed; (**b**) routing results by routing net A first and then routing net B; (**c**) routing results by routing net B first and then routing net A [1]

a multitude of pins) can be formed. For global routing, typical complete routing algorithm category includes **sequential global routing** and **concurrent global routing** [1].

For sequential approach, a specific order of nets (or pin pairs within each net) is specified. Then, based on the order, each net is routed sequentially with specific routing algorithm such as A* or maze routing. For sequential approach, the ordering is particularly important to achieve successful results. For instance, Fig. 11.2 gives a simple example demonstrating the effects of net ordering. Given nets A and B, if we order A in front of B and route the two nets with maze routing, then net B has to detour to be connected, as shown in (b). If we choose to prioritize B over A and use maze routing, then we would get the results shown in (c). By comparison, the overall wirelength of case (c) is much smaller than (b) despite using the same routing algorithm. Although net ordering is a critical step, it has been shown by Abel that there is no single net ordering scheme that performs universally better than other in all routing problems and that finding the optimal ordering has proven to be NP-hard [4].

Classical methods to address the net ordering challenge include heuristic net ordering and **rip-up and reroute** [1]. For heuristic net ordering, some popular methods include ordering the nets based on the number of pins in the bounding box of each net and ordering the nets according to the size of their bounding box. However, as mentioned above, these heuristics tend to work for specific kinds of problem and have no guarantees of optimality. The rip-up and reroute method is a more popular approach to address the sequence issues of routing. It usually consists of two major steps: (1) determine the bottleneck (congested) regions and rip up some already routed nets; (2) route the blocked connections and reroute the ripped-up nets. The rip-up and reroute process is usually repeated until a better routing is obtained. However, the iterative nature of rip-up and reroute also means it is a very time-consuming method to improve routing quality. A better negotiation-based rip-up and reroute algorithm called PathFinder was later developed [5], which can balance routability and performance. PathFinder is adopted in more recent academic routers such as Box Router [6], Fast Route [7], and NTHU-Route [3]. Another very useful technique used in sequential global routing is the pattern routing [8], which uses L-shaped (one-bend) or Z-shaped (two-bend) routes to make connections, which is significantly faster than maze router or A* search router, and it also uses much less memory.

One major challenge of the sequential global routing approach is the net ordering problem, and it is still an open research problem as of today. For any given order of nets, the nets that are routed later tend to be more difficult to route because they are subject to more blockages. This motivates people to start working on concurrent approach for global routing. Another popular method is to formulate the global routing as an integer linear programming (ILP) problem. The routing problem is firstly parsed as a routing graph $G(V, E)$, where each node corresponds to a routing tile and each edge corresponds to the boundary between adjacent routing tiles. Given a net to be routed, all of the possible routing solutions can be enumerated, and the design variable x_{ij} can be used to indicate whether certain routing pattern j has been selected for net i. By enumerating the candidate routes for all nets for a routing problem and collecting all the design variables, the routing problem can be formulated as a constrained ILP. However, the ILP problem is NP-complete and the high time complexity makes it difficult to scale.

The abovementioned algorithms are mostly for two-pin net routing, and a better and more natural method to route multi-pin nets is to adopt the Steiner-tree-based approach [9]. Steiner-tree-based routing algorithms have been developed to solve different kinds of routing problems [10–13]. Constructing Steiner tree is a NP-hard problem and most existing algorithms are based on heuristics. For detailed routing, the exact tracks and vias for nets need to be determined while obeying all the constraints. Although the abovementioned generic routing algorithms can still be applied to solve some detailed routing problem. There are some unique problem formulations for detailed routing such as channel routing and full-chip routing. Channel routing is a special case of routing in which wires can only be connected through the routing channels. In order to do channel routing, a routing region is usually decomposed into different routing channels. Channel routing has been proved as a NP-complete problem [14] and has been solved with different algorithms including heuristics [15, 16], linear programming [17], or greedy algorithm [18]. As the ICs scale down to nanometer territory, modern routing considerations also include signal integrity, manufacturability, as well as reliability, which all further complicate the routing problems.

With recent progress in RL in theories, algorithms, as well as applications, it has been applied to come up with better solutions for routing in EDA. Table 11.1 summarizes recent works of RL routing with their respective application domains, methods applied, and their ability to generalize. RL has been applied to solve global routing, detailed routing, as well as Steiner tree constructions with different level of success in terms of performance, generalization, and scalability. For the rest of this chapter, we are going to dive deep into some of these recent works in RL for routing.

Table 11.1 Summary of major RL for routing work

Name	Application	RL methods	Generalization
DQN global routing [19]	Global routing	Deep Q-learning	No
RL Steiner tree [20]	Steiner tree	A2C	Generalize within the same dataset
Attention routing [21]	Detailed routing	REINFORCE	Generalize among placements from the same design
Supervised attention [22]	Detailed routing	Imitation learning	Generalize among placements from the same design
Net order [23]	Detailed routing	A3C	Generalize across different nets
NVCell [24]	Standard cell routing	PPO	Generalize across standard cells of the same technology node

11.2   RL for Global Routing

11.2.1   DQN Global Routing

Recently, RL has been explored as an alternative method to address global routing. In DQN global routing [19], a value-based RL approach deep Q-learning (DQN) is applied for global routing by formulating the problem as a Markov decision process (MDP). The DQN global routing is the first attempt in solving global routing with RL, and the results show that for certain problem types, DQN outperforms the baseline sequential A* search router, which represents a typical heuristic-based router.

11.2.1.1   Problem Formulation

Given the netlist file of a design to be routed, a virtual environment is constructed for the global routing problem. The environment is designed to have the following functions:

- For each net, decompose a multi-pin problem into a set of two pin problems, because the current DQN global router can only handle two-pin routing problems. Feed routing solver with sets of two-pin routing problems stores and merges two-pin solutions to final global routing solution
- Provide reward feedback and observed state to the DQN algorithm. There are positive reward to encourage net connection and negative reward to minimize connection wirelength
- Keep track of the environment information such as edge capacity in real time

Fig. 11.3 Problem
formulation of DQN global
routing [19]

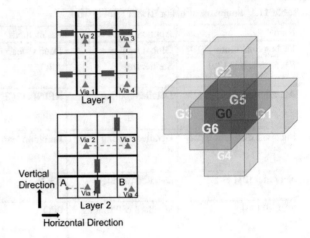

In general, the simulated global routing environment can be compared to a sequential version of a maze game, with edge capacities changing dynamically according to the thus-far routed path. Figure 11.3 shows an example of a simulated two-layer ($4 \times 4 \times 2$) problem environment with an illustrative routing solution for a two-pin problem. Each layer consists of 16 global routing tiles (4×4), with Layer 2 stacked right above Layer 1. Bold edges have zero capacity; therefore the route can only be north-south (referred to as the vertical direction) in Layer 1 and east-west (referred to as the horizontal direction) in Layer 2. Red rectangles represent blocked edges owing to pre-routed wires or the existence of certain components. The two-pin routing problem is to generate a route from pin A to pin B, through different global routing tiles with the shortest wirelength and least overflow (connection that exceeds edge capacity) possible.

At each step, starting from a tile, there are six possible moves, as shown in Fig. 11.3, which are going east (G_1), west (G_3), north (G_5), south (G_6), up (G_2), and down (G_4). In the environment, actions are subject to boundary conditions and capacity constraints; therefore actual possible actions will vary and can be less than 6 depending on the location of the agent. Once a routing step is taken, the capacity of the crossed edge changes accordingly.

11.2.1.2   Method and Results

Deep Q-network (**DQN**) introduced by Mnih et al. [25] in 2015 is viewed as the most important breakthrough in DRL. In DQN, deep neural network is for the first time applied as function approximator for the action value function, and the trained model exhibited performance comparable to human players in 49 Atari games. Before DQN, it is well known that RL is unstable when the action value function is approximated with a nonlinear function like deep neural networks. DQN is able to stabilize the training of action value function approximation with convolutional

Fig. 11.4 Workflow of DQN global routing [19]

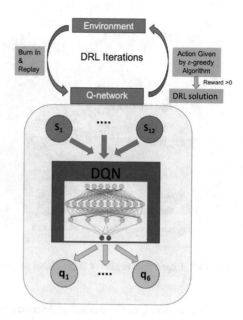

neural networks (CNN) using experience replay and target network. In this work, DQN is applied for the first to solve global routing problem.

In the DQN global routing work, the DQN is used to solve all the two-pin routing problems sequentially. The Q-network of the DQN consists of a multilayer perceptron (MLP) and functions as an agent interacting with the environment as shown in Fig. 11.4. Specifically, for each two-pin problem, the environment provides the network with state information. Then the agent evaluates the Q values of all the potential next states $(q_1, q_2, ..., q_6)$. Finally, based on an ϵ-greedy algorithm, an action is chosen and executed, altering the environment. The agent will have thus taken one "step" in the environment. A reward is calculated according to the new state and the edge capacity information is updated. In parallel, a replay buffer records each transition along the training process. These transitions are used for backpropagation when iteratively updating the weights in the Q-network. Critical components of the DQN models are as follows:

- **State Design**: the state is defined as a 12-dimensional vector. The first three elements are the x,y,z coordinates of the current agent location in the environment. The fourth through the sixth elements encode the distance in the x, y, and z directions from the current agent location to the target pin location. The remaining six dimensions encode the capacity information of all the edges the agent is able to cross. This encoding scheme can be regarded as an admixture of the current state, the navigation, and the local capacity information.
- **Action Space**: the actions are represented with an integer from 0 to 5 corresponding to direction of move from the current state.

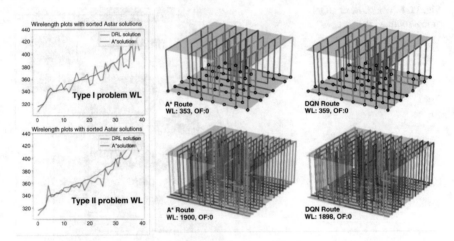

Fig. 11.5 Experiment results of DQN global routing on small-size problems [19]

- **Reward Design**: the reward is defined as a function of the chosen action and the next state $R(a, s')$. In our case,

$$R(a, s') = \begin{cases} 100 & \text{if s' is the target} \\ -1 & \text{otherwise} \end{cases} \tag{11.1}$$

This design forces the agent to learn a path as short as possible since any unfruitful action will cause a decrement in the cumulative reward. Additionally, we limit the maximum number of steps the agent can take when solving each two-pin problem to be less than a maximum threshold T_{max}, depending on the size of the problems to be solved.

- **Replay Buffer**: the replay buffer is an archive of past transitions the agent experiences and is updated during training. A burn-in preprocess is introduced to fill in the replay buffer before training begins, which provides a basic knowledge of the environment to the network. In our case, the burn-in transitions are acquired by A* search. In each training iteration, a batch of transition records are randomly sampled from this replay buffer and used to update the network weights through backpropagation.

Figure 11.5 shows the experiment results of the DQN global routing on small-size global routing problems of different types and their visualization results. In the work, **Type I** problem (also referred to as no-edge-depletion problem) refers to routing problems where edges with positive capacity in the beginning are not fully utilized throughout the routing process, while **Type II** problem (also referred to as partial-edge-depletion problem) refers to routing problems where at least some edges with positive capacity in the beginning are not fully utilized throughout the routing process. The results showed that DQN does not outperform the baseline sequential A* router on Type I problem, while for Type II problem, the DQN router is predominantly better than the sequential A* router. The reason why DQN has similar performance with A* on Type I problem is the Type I problem property makes the sequence of the routing individual net irrelevant to the final routing quality and A* can already route each individual pair optimally. However, for Type II problem, sequential A* will only focus on generating optimal routes for each net in a greedy manner, while missing the global optimality. For DQN router, through the conjoint optimization mechanism, the model is able to take into account all the nets to be routed in a problem conjointly and thus leads to better solution compared with A*. Further explanation on the conjoint optimization mechanism of the DQN router is available in [19].

The DQN global routing method shows some promising preliminary results in terms of solving global routing in a better way with RL compared with heuristics or greedy approaches. However, there are two major limitations that need to be solved to apply this method to larger-scale global routing problems: scalability and generalization. The DQN method is based on step-by-step search in a grid graph, and the search needs to run at each iteration during the training process, which means the method does not scale well when the graph size increases. To make the method scale up to larger-size problems, more efficient state space representation needs to be formulated to realize fast training and inference of the DQN-based routing approach. The second limitation lack of generalization means the current approach needs to train the DQN model every time a new problem needs to be solved and the training procedure can be time-consuming. Enabling the generalization ability of the current problem would make the model solve unseen problems by just running inference or inferencing with only light fine-tuning on unseen problems, which would have significant runtime advantage compared with heuristics or search-based routing algorithms.

11.2.2 Constructing Steiner Tree via Reinforcement Learning

As an alternative method for routing for multiple-pin nets, Steiner tree has been extensively used in global routing. Recently, RL has been applied to construct Steiner tree and demonstrates better performance than previous heuristics [20]. In the work, the authors adopt actor-critic neural network RL model for constructing rectilinear Steiner minimum tree (RSMT). The model is able to learn by itself with RL learning and generate high-quality RSMT much faster than heuristics of similar quality. It is also the first successful attempt to solve RSMT construction with RL method.

11.2.2.1   Problem Formulation

The problem of constructing the RSMT can be defined formally as given a set of nodes V, RSMT is a rectilinear tree $T(U)$ that connects all the points with minimum length, such that U is a superset of V and the newly introduced points in U are the Steiner points. The role of the Steiner points is to reduce the total length of the tree. In this work, the RL model is used to produce an optimal sequence that can be converted to RSMT, and the sequence is called rectilinear edge sequence (RES). It is proved in the work that we can always find an RES such that the tree it represents is an optimal RSMT for the nodes. So the problem is reduced to finding an optimal RES.

11.2.2.2   Methods and Results

Figure 11.6 shows the RL model architecture. The model mainly consists of two components: actor and critic. Both of them take the set of n point coordinates V ($V \in R^{n \times 2}$ as input. The actor works on determining the elements in the RES based on a stochastic policy, and the critic sets a baseline by predicting the expected length of the RSMT found by the current actor; in this way the actor can keep encouraging the actor to improve its performance. The architecture is based on the multi-head attention mechanism.

The model is implemented with PyTorch and trained on a 64-bit Linux machine with Intel Xeon 2.2 GHZ CPUs and an Nvidia Titan V GPU. The model is trained with point sets from degrees 3–50 with Adam optimizer and a learning rate of 2.5×10^{-4} with decay learning rate scheme. After training, the RL model is

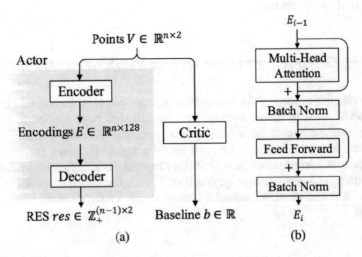

Fig. 11.6 RL-based RSMT construction: (a) simplified structure of the RL architecture; (b) the i_{th} encoding process [20]

applied to construct RSMT on the test sets, and the results are compared with the heuristics Steiner tree construction algorithms including FLUTE and GeoSteiner. The comparative results indicate that the RL model is able to generate similar-quality RSMT as GeoSteiner and FLUTE with significant less time.

11.3 RL for Detailed Routing

Detailed routing is the downstream step of global routing. Guided by the paths obtained in global routing, detailed routing generates the actual paths by assigning actual tracks and vias for nets [9]. Since detailed routing is the final routing stage, the detailed routing solutions also have to meet design rule constraints. The design rule constraints are getting increasingly more complex in recent years, especially in advanced node technologies. Thus, existing heuristic-based detailed routing methods such as channel routing are not good enough to meet the detailed routing design needs in advanced node technologies. In the attention routing work [21], RL is applied to solve a typical detailed routing problem for the first time. The problem is a detailed routing design with track assignment step for industry advanced node technologies. Instead of using RL to solve the detailed routing problem in an end-to-end manner, a solution flow is constructed for the track assignment detailed routing problem first. All the design rules are encoded into the track assignment step, and heuristics algorithm is used to do the track assignment work. Then, the RL only focus on the most critical step in the flow, which is to determine the optimal order of the device pairs to be routed.

Another work [23] also solved detailed routing with RL, focusing on the net ordering steps. An asynchronous RL framework is constructed to optimize the net ordering from multiple designs of the ISPD detailed routing contest. In this section, we will introduce the attention routing and the asynchronous RL detail routing work.

11.3.1 Attention Routing

11.3.1.1 Problem Formulation

Recent works on using attention-model-based REINFORCE—a RL algorithm to solve combinatorial problems—have demonstrated near optimum performance with significant generalization capability compared to existing heuristic-based method. It also outperforms previous RL methods including pointer network (PN) [26] in widely studied problem sets including travelling salesman problem (TSP), orienteering problem (OP), price collecting TSP, etc. In solving these problems, the solution can always be formulated as a sequential decision process. One important reason they tend to be solved reasonably well is that they can be modeled as a Markov decision process (MDP). In the RL for detailed routing work, the attention-

Fig. 11.7 Schematics
showing the track assignment
and routing step of attention
routing and the constraints in
track assignment
constraints [21]

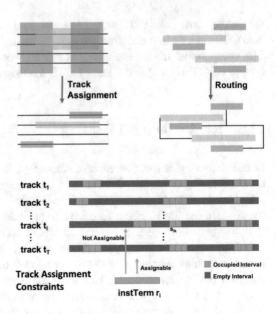

based RL model is used to determine the optimal order of nets, which can also
be considered as a MDP. The similar problem formulation between net ordering
and other combinatorial optimization problem such as TSP is one of the primary
motivations of applying the attention-based RL model to net ordering in detailed
routing.

In attention routing, the problem is solved in two major steps: track assignment
and routing, as shown in Fig. 11.7. To address the design rules in terms of
spacing, a layer of abstraction is introduced, namely, width spacing pattern (WSP).
WSPs define a set of track patterns that consists of different width and spacing
configurations for the metal wires. By restricting the routes on the WSP rows
and tracks, many design rules associated with full custom designs, including those
concerning the spacing, minimum widths, and coloring rules, can be avoided. In
the attention routing work, the routing strategies presented follow the design rule
specifications through the adoption of WSP abstraction. Once the WSP is defined,
the track assignment step works on place the long routes instTerms on tracks defined
by the WSP. The track assignment needs to satisfy the vertical constraints as shown
in Fig. 11.7. A heuristics approach based on bipartite graph matching is used to solve
the track assignment step. After the track assignment step, the routing step works on
sequentially connecting all the pairs of instTerms with RL focusing on determining
the optimal sequence of the instTerm pairs. In formulating the problem, RL is
not applied in an end-to-end fashion to generate the detailed routes from scratch.
Instead, it is only focusing on the most critical net ordering steps. The motivation
here is to reduce the state space to a level that is amenable to the state of the art RL
model's capability.

11.3.1.2 Method and Results

For the track assignment step, which is formulated as a bipartite graph matching and solved with heuristics, two graphs (overlap graph and assignment graph) describing the layout of instTerms are constructed first. With these two graphs, the track assignment problem is reduced to matching the instTerm nodes to the track nodes in the assignment graph while minimizing the matching cost, such that no conflicting instTerm nodes in the overlap graph are matched to the same track. Then a modified bipartite graph matching algorithm is applied to solve the problem. Details about the track assignment step is available at [21]. After the track assignment step, a pin decomposition based on minimum spanning tree is used to decompose each net into a set two-instTerm pairs. After this, the RL-based router works on connecting all the instTerm pairs sequentially.

In order to find an optimized routing solution that minimizes the cost among all possible routing sequences, an attention-based encoder-decoder RL model with rollout baseline is used. We define each problem instance as a graph with n nodes, where each node represents an instTerm pair between instTerm i and instTerm j. Each instTerm pair is in the form

$$(x_{i1}, x_{i2}, y_i, x_{j1}, x_{j2}, y_j, l), i \neq j \tag{11.2}$$

where x_{i1}, x_{i2}, y_i represents the xy-coordinates of instTerm i; similarly x_{j1}, x_{j2}, y_j represents the xy-coordinates of instTerm j; l represents the net index. We define the solution to the routing sequence π as a permutation of the n nodes (instTerm pairs).

In solving the problem, the attention RL model is based on policy gradient method: REINFORCE. In a policy gradient model, a model is used to learn a policy $p(a|s, \theta)$, matching state s of a problem at each time step to a corresponding probability distribution over all actions a by iteratively optimizing the policy model parameters based on training samples. The cost that training process aims to minimize is the expectation of reward r collected after certain policy p has been rolled out for an episode, which can be expressed as $E_p[r(\tau)]$. Following the policy gradient theorem, the gradient of the cost can be expressed as

$$\nabla E_{\pi_\theta}[r(\tau)] = E_{\pi_\theta}\left[r(\tau)\left(\sum_{t=1}^{T} \nabla log p_\theta(a_t|s_t)\right)\right] \tag{11.3}$$

In REINFORCE, the above gradient of cost function is used to train the policy model. However, the training process of REINFORCE tends to be unstable due to the delayed reward mechanism of REINFORCE. Thus, in attention routing, REIN-FORCE with baseline is applied to stabilize REINFORCE training and increase the convergence speed. In attention routing, the formulation of the problem can be described as follows: the solution is defined as a routing sequence $\pi = (\pi_1, ..., \pi_n)$, which is a permutation of the instTerm pairs. The input of the policy model is based on the instTerm pairs physical location information, as shown in Eq. 11.2. At each step, the policy model outputs a probability distribution $p_\theta(\pi_t|s, \pi_{1:t-1})$ over all the instTerm pairs that are likely to be visited at the next time step. In this process, instTerm pairs that are already routed are masked to avoid being chosen repeatedly. Based on this, a problem policy $p(\pi|s)$ is defined as Eq. 11.4, which is simply the product of probability distribution for the n steps. If at each time step, the node with the highest probability is chosen, a solution (path) is said to be given in a deterministic greedy rollout manner:

$$p_\theta(\pi|s) = \prod_{t=1}^{n} p_\theta(\pi_t|s, \pi_{1:t-1}) \qquad (11.4)$$

Based on the problem policy and the policy gradient as shown in Eq. 11.3, the policy model can be trained with backpropagation based on Eq. 11.5. For the attention routing problem, the loss term $L(\pi)$ is simply the total wirelength of the routes. For each instTerm pair, a simplified pattern router is used to make connection fast:

$$\nabla L(\theta|s) = E_{p_\theta}(\pi|s)[(L(\pi)\nabla log p_\theta(\pi|s)] \qquad (11.5)$$

The model architecture is implemented based on attention-based encoder-decoder model which can be considered as an instance of graph attention network [27] as shown in Fig. 11.8. The encoder is similar to the transformer architecture encoder. After a first learned linear projection layer, the major part of encoder consists of N attention layers, with each layer consisting of two sublayers: a multi-head attention (MHA) layer and a fully connected feed-forward (FF) layer. Skip connection and batch normalization (BN) are applied at each sublayer. At the heart of the MHA structure is the attention mechanism which can be summarized as a weighted message passing between the nodes in a graph. The weight of a message value that a node receives from a neighbor depends on the compatibility of its query with the key of the neighbor. For each node, its corresponding key, query, and value are obtained by projecting the node embedding/input by parameter matrices correspondingly, whose weights are automatically learned during training. The decoder consists of only one layer of MHA, and it outputs the probability distribution of nodes to be visited at each time step based on the embeddings from the encoder and the output generated at the previous time steps.

The detailed implementation can be found in Algorithm 11.1. To compare with the RL model results, genetic algorithm (GA) is used as a baseline to optimize the sequence of instTerm pairs, which represents a typical method used in EDA tools to solve similar problems. Although GA works well in solving such kind of combinatorial problems, it is not able to scale up efficiently to larger problems and also not able to learn from previous problems and generalize to new problems.

The methods are tested on two (denoted as small and large) commercial advanced node (sub-16 nm technology) analog design datasets, including comparators, OpAmp, and analog-to-digital converter (ADC). For small problem set, the

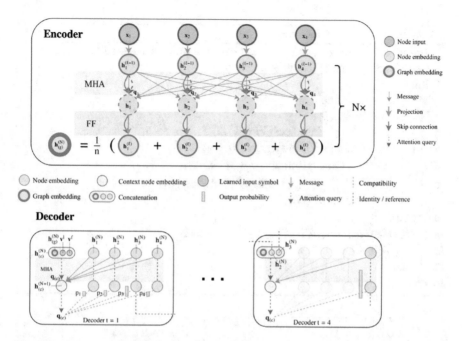

Fig. 11.8 Encoder-decoder model structures used as the policy model in attention routing [21], which is based on graph attention network

number of instTerms for each problem ranges from 10 to 100, and in large problem sets, the number of instTerms for each problem ranges from 100 to 1000. The policy model of the attention routing is trained on training set and works on inferencing unseen test sets; the GA works on both the training and test sets as a comparison. Figure 11.9 shows the result of cost comparison between the attention model and the baseline GA on both training and test set, where the result is ranked in the ascending order of the GA result. The results show that in both training and test sets, the attention model achieves very similar results compared with GA. This indicates that the attention model can generalize well to unseen problems and achieves similar performance to baseline GA. However, in terms of runtime, the inference time of the attention model is more than 100 times faster than the baseline GA, which takes long iterations to converge. The results indicate the superior performance of the attention routing.

As a follow-up work to the attention routing work, a supervised version of the attention routing is developed [22] based upon the existing flow of track assignment detailed routing in the attention routing. The supervised methods leverage labeled data and train the policy model based on minimizing KL divergence between labeled data and the model outputs, as shown in Eq. 11.6. It can be viewed as an imitation learning way of training the policy. Based on the supervised method, the trained

---

Algorithm 11.1: Attention sequencing

Input : Number of epochs E, batch size B, training set T, significance α
Output: Sequence based on best policy

1 Init $\theta, \theta^{BL} \leftarrow \theta$;
2 **for** *epoch=1,...,E* **do**
3   **for** *batch=1,...,B* **do**
4          $t_i \leftarrow$ SampleInstance() $\forall i \in 1, ..., T$;
5          $\pi_i \leftarrow$ SampleRollout(t_i, p_θ) $\forall i \in 1, ..., T$;
6          $\pi_i^{BL} \leftarrow$ GreedyRollout($t_i, p_{\theta BL}$) $\forall i \in 1, ..., T$;
7          $\Delta L \leftarrow \sum_{i=1}^{B} (L(\pi_i) - L(\pi_i^{BL}))\Delta_\theta \log p_\theta(\pi_i)$;
8          $\theta \leftarrow$ Adam($\theta, \Delta L$);
9   **end**
10   **if** *OneSidedPairedTTest($p_\theta, p_{\theta BL}$)* $\leq \alpha$ **then**
11         $\theta^{BL} \leftarrow \theta$;
12   **end**
13 **end**

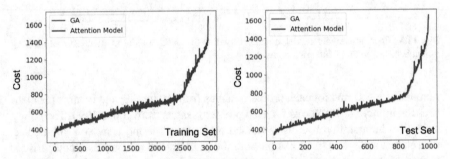

Fig. 11.9 Cost comparison between attention model and GA on training and test set [21], results ranked in ascending order of cost based on GA

model achieved similar performance compared with the model trained with policy gradient. The supervised version provides a useful alternative to train the policy model, where users can leverage past human experts' design to train customized models in sequencing the net (Fig. 11.10):

$$L(s) = D_{KL}(p_{\theta(s)|q(s)}) = p_{\theta(s)}^T(log(p_\theta(s) - log(q(s)))) \tag{11.6}$$

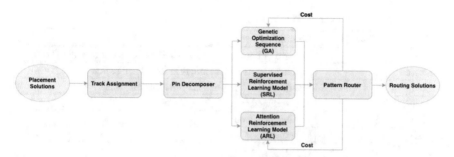

Fig. 11.10 Flowchart showing the flowchart of three different approaches for solving track assignment detailed routing (attention RL, supervised RL, and GA) [22]

11.3.2   Asynchronous RL for Detailed Routing

11.3.2.1   Problem Formulation

A more recent work [23] applied asynchronous RL to solve detailed routing. While also focusing on the net ordering step, the authors formulate the problem such that it can be scaled up to larger problems (ISPD 2018 detailed routing contest benchmarks) and it outperforms the state-of-the-art heuristic-based solver Dr.CU [28].

The problem is solved with a combination of RL net ordering and the rip-up and reroute (RRR) based on Dr.CU approach. So the focus of the work is to use RL to optimize the net ordering of detailed routing. The net ordering problem is defined as:

Problem 1 (Net Ordering)   Given a set of nets N, train a net ordering policy that can generate a ranking score s_i for each net $n_i \in N$ used by a sequential detailed router. The following metrics should be optimized simultaneously: (1) the total wirelength of all nets, (2) the number of the total used vias, and (3) the number of DRC violations [23].

11.3.2.2   Methods and Results

In order to solve the net ordering, the author used asynchronous RL framework to solve the problem. Some basic components are as follows:

- **State**: the state s is a collective feature of all the nets to be ordered. Table 11.2 summarizes the state space representation for each net. The first feature is the size of its routing region; the second feature is its degree, which denotes the number of nets whose routing region overlaps with it; the third feature is the number of times that the net has been routed/rerouted so far. The remaining four features are described in the table.

Table 11.2 Features of each net in work [23]

Feature	Dimension	Description
Size	1	The size of the routing region (half-perimeter of bounding box)
Degree	1	Number of nets with conflicts in its routing region
Count	1	The number of times it has been routed/rerouted
Cost	1	The weighted sum of violations on it
Via	1	Number of via on it
WL	1	Wirelength
LA	16	Layer assignment

- **Action**: the action is a real number vector, with each number corresponding to the ordering score of a net.
- **Reward**: given the ordering score based on the action, the environment (which is the router) will provide feedback. The reward is defined as

$$r = C_{cu} - C_{agent} \tag{11.7}$$

where C_{agent} and C_{cu} represent the cost of a design based on the agent's action and Dr.CU's result. The reward will encourage the RL model to find better solutions compared with Dr.CU's results.

In order to learn more efficiently, asynchronous advantage actor-critic (A3C) method [29] is used so that multiple agents can run in parallel. For A3C, each agent has a local copy of the policy and value network weights. Then it rolls out in its own environment to explore the state space. After the rollout, different agents update the global network in an asynchronous way. In order to adopt A3C, policy network and value network need to be designed. The policy network takes the state s as input and output a probability distribution over the actions. Actions are then sampled from the probability distribution. The value network outputs a value function $V(s)$, which is a gauge of how advantageous certain state is. So, the policy network generates the ordering scores of all nets, while the value network evaluates the state. Figure 11.11 shows the architecture of the policy and value networks. The models are designed to enable the policy model being used across different designs with different number of nets. In order to make the architecture invariant to the number of nets in a design, a net-wise feature encoding is introduced to encode the features of each net independently. The encoded features are then concatenated to be used in the policy and value networks. The major advantage of this architecture is that the policy network can be shared across different designs and nets, since we perform net-wise modeling with the ordering score of each net only depending on its own features.

Once the policy is obtained, the ordering scores are obtained, and Dr.CU flow is used to finish each RRR iteration in order to finish the route. More specifically, the authors schedule all batches at the beginning of an RRR iteration by sorting the nets based on the scores and divide them into batches, such that nets within a batch do not

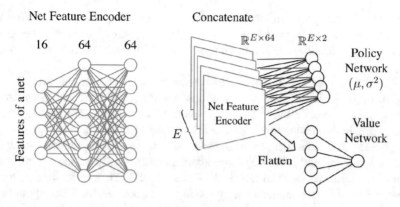

Fig. 11.11 Architecture of the policy and value networks, where E denotes the number of nets. The net feature encoder encodes the features of each net [23]

conflict with each other and can be routed in parallel. If the RRR stopping criteria are not met, the iterations will continue until the maximum number of iterations is reached [23].

In validating the method with experiment, Dr.CU 2.0 is used as the detailed router, and the RL module is implemented with PyTorch. Experiment benchmarks are selected from ISPD 2018 and ISPD 2019 Initial Detailed Routing Contests. The performance are evaluated on both the training and test sets in terms of wirelength, number of vias, DRC violations, total cost, and runtime between the RL framework and baseline Dr.CU. In the training dataset, with similar wirelength and number of vias, the RL model achieved 13% DRC violations reduction compared with the default policy in Dr.CU. In terms of total cost, only small improvements are obtained. This is mainly because the cost is dominated by wirelength due to its large scale according to the definition in the contests. The training set results show that the RL framework is able to learn good policies from the benchmark. On the testing set, the RL model achieved an average of 14% improvement in violations and 0.7% in total cost without compromising wirelength and number of vias. The testing set results indicate the RL model is able to generalize to unseen benchmarks and achieve high-quality results. The future works of this asynchronous RL for detailed routing method include considering the correlations between multiple nets and expanding the state representation to include more features such as macros.

11.4 RL for Standard Cell Routing

Standard cells are the building blocks of digital circuit designs, and modern circuits can be made of millions of standard cells. For specific technology node, IP providers and semiconductor design companies have large teams for designing standard cell libraries, with each of them usually consists of thousands of cells. Lots of the standard cell library design works are done manually today. Thus, automating the standard cell layout design will speed up the design process which enables designers to co-optimize standard cells and chip designs to improve the chip performance [30].

Standard cell design consists of two main steps: placement and routing. The placement step places transistors and pins within a cell, and the downstream routing step connects all the transistor terminals and cell pins. The routing step is challenging in the design flow since feasible routing solutions have to satisfy all the design rule constraints. For advanced technology nodes, the design rules are exponentially growing and becoming more interdependent. Mathematical optimization approaches based on Boolean satisfiability problem (SAT) and mixed-integer linear program (MILP) were applied to solve routing with good results. These approaches depend on the premise that design rule checking (DRC) can be expressed in the explicit math forms. However, it is not guaranteed that all the DRCs especially those on advanced technology node can be expressed efficiently. Even worse, the large number of constraints that need to be created makes these methods difficult to scale up to larger designs. These are some of the reasons why standard cell routing is still done manually today.

The NVCell [24] combined RL with heuristics to provide a first of its kind RL standard cell placement and routing solutions that achieve human level or even better quality designs in advanced node industry standard cell libraries. Although there are applications of RL in both the placement and routing step, we will focus on the routing part of the NVCell in this chapter. In NVCell, genetic algorithm (GA) is applied to create initial routing candidates, and then RL is applied to fix the design rule violations incrementally. A design rule checker sends feedback of the violations to the RL agent so that the agent can learn how to fix them based on the feedback [30].

11.4.1 Background

In recent years, standard cell design has changed drastically when advanced technologies started requiring all routing shapes adhere to a fixed routing grid. This change significantly reduced the set of allowable routing shapes, which leads to an environment more amenable to automation. Figure 11.12 shows the stick diagram for an advanced technology node standard cell with unidirectional metal. There are five layers where nets can be routed: LISD (local interconnect source-drain), LIG (local interconnect gate), M1, M2, and M3. M1 and M3 layers route horizontally, while M2 layer routes vertically. LISD connections essentially allow some vertical routing below M1, and LIG connections allow some minimal horizontal routing. There are many restrictions on how each layer can be routed. The most strict DRC rules are on the M1 layer. On the M1 layer, the routing shapes are on a fixed grid, and cut metal shapes need to be inserted between adjacent routing segments on the same track. The locations of the cut metals can be inferred from the routing assignment

Fig. 11.12 Stick diagram of a standard cell [30]

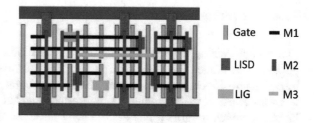

and have to satisfy many DRC constraints. These constraints are particularly difficult to model with mixed-integer linear programming (MILP) constraints. Also, in order to create MILP constraints for this kind of rule, enumerating all possible loops formed by cut metals is needed, increasing the number of constraints exponentially [30].

Different algorithms have previously been proposed to do standard cell routing, which include channel routing, SAT-based routing, and MILP-based routing. For channel routing, the commonly used deterministic methods such as LEA, Dogleg, Greedy, etc. [31] only generate a particular routing solution and fail to address the DRC constraints well. The genetic algorithm (GA)-based channel routing can generate multiple routing candidates, but it might not generate DRC-free routes directly for advanced node technologies [30]. For the SAT-based routing method, it requires DRC checks to prune conflicting routing candidates and often cannot find feasible solutions for complex cells. For the MILP-based routing method, the routing problem is formulated as a mixed-integer linear programming problem, which relies on a MILP solver to solve a large number of constraints and also requires DRC constraints being expressed in analytical forms. Because of this, the supporting of new technology node using MILP solver becomes very challenging. RL has demonstrated great success in different domains and also has been proposed to solve routing problems, including DQN-based global routing [19], attention routing [21], etc. However, these RL routing methods are mostly used in analog and PCB routing, which need to be extended to handle the DRC constraints for advanced technology node. The NVCell demonstrates a novel standard cell routing method that combines RL with GA and can handle complex DRC constraints.

11.4.2 Problem Formulation

The NVCell consists of placement and routing stage. The routing stage depends on the result generated from the placement stage, which is based on combination of simulated annealing and RL. The NVCell addresses the standard cell routing in two steps: a GA-based routing step and a RL-based DRC fixing step. The standard cell routing result is represented as a stick diagram, and a program called Sticks is developed to generate DRC/LVS clean layouts from the stick diagram with grid-based layout [30].

11.4.3   Methods and Results

The NVCell routing focuses on fixing the M1 layer DRC errors, while other DRC errors can be filtered out in the routing step. The M1 layer DRC errors are caused by constraints for cut metal locations. The Sticks program determines the cut metal locations based on the M1 routes. Therefore changing M1 routes modifies cut metal locations, which in turn change the M1 layer DRC. So the environment game for the RL is to incrementally change the routes to minimize the M1 DRCs. The game settings do not remove any existing routes in order to guarantee the routability of the initial route. Some key components for the RL environment are listed below:

- **Internal State**: the internal state of the environment is the multilayer grid space used in the Sticks program.
- **Observation Space**: the observation space *obs* is a tensor [3, H^{M1}, W^{M1}], where H^{M1} and W^{M1} are the M1 grid dimensions. The first dimension of *obs* represents the M1 routes, the second dimension represents the routing mask, and the third dimension represents the DRC information, which records the location of the DRC marks reported by the Sticks program.
- **Action Space**: the action space *act* is a categorical tensor [$H^{M1} \times W^{M1}$], which represents the probability of whether the specific M1 routing grid should be routed.
- **Reward**: there are two kinds of reward in this environment: r_s is a negative reward given at each step, which pushes the agent to finish the task as soon as possible. r_d is a reward associated with the improvement of DRC, which is proportional to the reduction of the number DRC marks reported by the Sticks program. The r_d pushes the agent to reduce DRCs as much as possible.
- **Termination Condition**: there are two termination conditions in the environment: first, when there is no available action for the next step and second when the DRC is clean.

For the RL model, proximal policy optimization (PPO) [32] algorithm is used, which is a policy gradient RL algorithm and has achieved state-of-the-art performance in discrete action space tasks. The PPO includes two policy models: one training model and one rollout model. The training model works on learning parameters, and the rollout model works on collecting data by interacting with the environment. The key feature of PPO is its objective function that guarantees limited divergence between the two (training, rollout) models. Also, the PPO model needs to have two heads: a value head and a policy head. The value head predicts the value of current state, while the policy head generates the probability distribution over actions. These two heads share a common module of model initially, and the output of the shared module is described as embeddings.

The model architecture is shown in Fig. 11.13. The observation is first fed into convolution layers, where each convolution layer produces output activation with the same height and width as input. The output of the last convolution layer is the embedding with a dimension of [512, H^{M1}, W^{M1}]. The embeddings are then

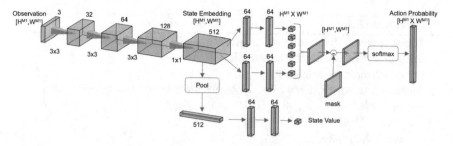

Fig. 11.13 RL model architecture for PPO in NVCell routing part [30]

fed into the value and policy networks. The policy network consists of three fully connected (FC) layers, and each pixel of the embeddings is fed into these FC layers. Then the output of the policy network with a dimension of $[H^{M1}, W^{M1}]$ are connected, masked by action mask to avoid illegal routes. Then the output is fed into a softmax layer to produce the action policy vector. For the value network, the embeddings are pooled together to form a 512-dimension vector. It is then fed into the value network consisting of three FC layers. With this design, the architecture would be independent of the size of the image or number of nets.

The RL mode works in conjunction with GA-based routing in NVCell routing module. Routing segment-based genetic encoding strategy used in channel routing is adopted. It is also extended with a random maze router and support for Stick diagram. The complete routing flow is shown in Algorithm 11.2.

The Stick program is implemented in Perl and the RL environment is implemented in Python based on OpenAI Gym. The GA and maze router is also implemented in Python. The standard cells picked have various complexity and come from an industry standard cell library. The model is trained on an NVIDIA V100 GPU and the training runtime is about 2 h. Figure 11.14 shows the reward curves for multiple runs. The reward curves indicate that in all the runs, models converge fast and stabilize afterward.

Although the model is trained with only one cell with multiple routes, it is able to generalize to unseen cells. Figure 11.15 shows a case of the standard cell stick diagram of initial routes and DRC-fixed routes with RL. The DRCs are able to be cleaned with the RL model. Because the RL model is trained on a large set of random routes, even though it is only trained with one cell, the agent is still able to see enough different routing configurations, which corresponds to many different DRC patterns. Also, it is believed these DRC patterns are invariant between different region of a same standard cell or different cells, which is the main reason why the trained model is transferable to unseen cells.

Before applying RL to solve the DRC fixing, the authors also tried to apply RL to solve routing problem directly by creating routing actions for each net, where the action space is a routing action for each net. The results are promising but not scaling well. The motivation for shifting the focus of RL to DRC fixing is that one should not make RL agent learn the job of a maze router which can be solved

---

Algorithm 11.2: GA-based routing flow in NVCell [30]

   Input   : nets N, net terminals T_n, generations G, population K
   Output: DRC free routing solution R
1 create the terminal pair set P
2 initialize routing solutions $\{R_1, ..., R_k\}$ for the first generation with random routing order
3 **for** *g=1,...,G* **do**
4   **for** *i=1,...,K* **do**
5   select R_{dad} and R_{mom} from $\{R_1, ..., R_K\}$
6 $R' \leftarrow$ crossover(R_{dad}, R_{mom})
7 $R' \leftarrow$ mutate(R')
8 $R'_i \leftarrow$ Maze route unrouted pairs with random order
9   run RL DRC fixer on route R'_i
10 $DRC(R'_i) \leftarrow$ remaining DRCs
11   **if** *$DRC(R'_i) == 0$* **then**
12 | return R'_i
13   **end**
14   **end**
15 $\{R_1, ..., R_k\} \leftarrow$ Top K fitness from $\{R'_1, ..., R'_k, R_1, ..., R_k\}$
16 **end**

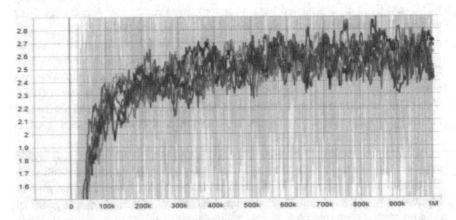

Fig. 11.14 Reward vs. training episodes plot for multiple runs of NVCell routing module RL model [30]

with heuristics like Lee algorithm; instead RL should focus on learning to solve the harder problems such as DRC fixing that heuristics or optimization methods are struggling with. Also, learning DRC fixing is easier to scale to larger cells since the problems are local problem, enabling the RL model to be directly applied to subregions of larger cells.

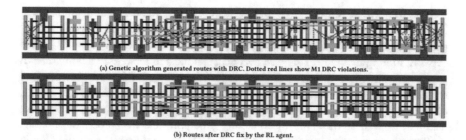

(a) Genetic algorithm generated routes with DRC. Dotted red lines show M1 DRC violations.

(b) Routes after DRC fix by the RL agent.

Fig. 11.15 Stick diagram of standard cell initial routes generated by GA and DRC-fixed routes [30]

11.5   RL for Related Routing Problems

Routing has been a common problem not only in EDA but also in fields such as communication network, traffic control, sensor network, robotics planning, etc. In these fields, the routing problem formulation is usually different from EDA routing. For instance, in routing problems for traffic control, obstacles and other vehicles are often dynamic [33]; thus the routing problem cannot be formulated as a static one as we usually do in EDA routing. However, these non-EDA routing problems also share lots of similarities with EDA routing. In most cases, they can all be formulated as combinatorial optimization just like EDA routing, but with different constraints. This makes lots of the routing solutions in these related problems insightful for solving EDA routing problems.

Recently, there are lots of adoptions of RL in solving routing problems outside EDA field. Some of the works are highlighted here to hopefully provide inspiration on solving EDA routing with RL in novel ways.

11.5.1   RL for Communication Network Routing

In the past decades, RL has been extensively used in communication network routing. In the scope of communication networks, routing problem is defined as selecting paths to send packets from sources to destinations, while meeting QoS (quality of signal) requirements and optimizing networks resources. Typical way of solving communication network routing is to formulate the network as a weighted graph and find paths with minimum cost while satisfying constraints. Many heuristics have been proposed to solve the communication network routing problems, yet these traditional routing techniques are usually based on huge assumptions on the traffic flows and network condition changes. When some of the underlying assumptions are not satisfied, the network performance would significantly deviate from the simulation results. RL has been proposed as an efficient alternative to solve real-world network routing problems [33]. In different RL-based network routing, RL has demonstrated better capabilities of adapting to specific network environment while optimizing resource utilization. Mammeri [33] gives a review of the RL-based network routing methods in the past decades, and most of them use value-based RL approaches and their variations. In the problem formulation, RL model architecture might provide useful insights on RL routing in EDA. Open challenges in RL network routing include slow convergence rate, large state space, collaborative learning, as well as difficulty in efficiently predicting the traffic demands. These challenges are also similar to those in RL for EDA routing and RL's application in other domains.

11.5.2 RL for Path Planning

Path planning has been a central problem in robotics, and the problem setting is similar to two pin routing in EDA while avoiding obstacles generated by pre-routed nets. Recently, value iteration network (VIN) [34] is proposed to solve path planning with RL in better way. In VIN, a neural network-based policy that can efficiently learn to plan is proposed, and the model is able to generalize to a random unseen map after training without an extra planning module. The key innovation of the method is that VIN has a differentiable "planning module" embedded within the network architecture. The intuition for this approach is that the classic value iteration planning algorithm can be represented as a special kind of convolution operation in CNN. Therefore, by embedding such network into the policy model architecture, a model that can learn to plan is constructed. More importantly, the value iteration module is differentiable and thus can be trained together with the whole network through backpropagation. As a result, the VIN is able to achieve state-of-the-art performance in grid world map (as shown in Fig. 11.16) with random obstacles. The performance is significantly better than CNN- or MLP-based architectures without any planning module added in the architectures.

Based on VIN, an improved multiagent version VIN is developed [35], and it focuses on route a fleet or a swarm of vehicles. Specifically, the focus is on the multiagent autonomous mapping problem: for a fleet of vehicles, finding the minimum total cost to map a region subject to traffic constraints. The goal is

Fig. 11.16 Two instances of grid world problem in VIN [34]

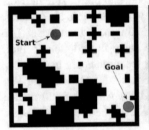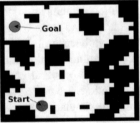

to traverse each road in the city at least certain number of times. The problem setting is very similar to EDA routing problems with multiple nets. The proposed method, multiagent routing value iteration network (MARVIN), is a distributed neural network focusing on coordinating the fleet of vehicles that need to be routed. Specifically, each agent performs the local planning by a single VIN module and exploits interagent communication through a novel learned communication protocol using the attention mechanism. The effectiveness of MARVIN is demonstrated on real road maps extracted from 18 different cities around the world, and realistic traffic flow simulation is used. The MARVIN outperforms the state-of-the-art VRP solvers. Also, it is able to generalize to unseen maps with different number of agents.

The VIN-based methods' success in path planning, especially in multiagent scenario, provides a potentially promising solution in solving EDA routing in generalizable and efficient way with RL. But there are some open challenges to apply VIN-based approach to EDA routing. Firstly, the maps VIN can handle tend to have sparse obstacles, and even that requires curriculum learning to incrementally add more obstacles during training in order to learn to planning successfully on nontrivial maps. For EDA routing, with the great number of nets needed to be routed, the obstacles can be much more challenging than the test cases used to demonstrate VIN. Secondly, the VIN is demonstrated only on relatively small grid graph maps, for instance, 28×28, and the performance get worse as the size of the map increases. It hasn't demonstrated enough scalability to solve EDA routing where the maps are much larger. If these challenges can be addressed, the VIN-based approach can be very promising methods for EDA routing.

11.6 Challenges and Opportunities

One major challenge for applying RL to solve EDA routing is the scalability. For the RL-based routing methods introduced in this chapter, most of them demonstrated efficiency of RL on limited-size problems: DQN global routing works on grid graph less than 100×100 [19], and RL for detailed routing works on design with number of nets less than 1000 [21]. These test cases are still 1 to 2 orders smaller than the typical routing benchmarks comparable to industry design such as ISPD benchmarks. For those that actually use larger benchmarks to demonstrate RL approaches, such as net ordering RL method [23], major assumptions and simplifications are made to make the problem amenable to be solved with RL. The net ordering RL method, for instance, assumes that a single model can be used to all nets and under such assumption it is able to apply to large number of nets. Yet, the simplifications and assumptions come with cost of deteriorating the solution quality. So, it might end up being a tradeoff between model scalability and performance based on the level of assumptions and complexity of models. It is also worth pointing out that scalability has also been an open challenge for RL applications to other domains and problems. This is also an open research topic under active research. For instance, when applying RL to solve travelling salesman problem [36] (a canonical combinatorial optimization problem), the optimality gap of solutions significantly increases (more than 100%) even when the TSP graph size just increases from 20 to 100. In order to address the scalability challenge when applying RL to EDA routing, some possible solutions include simplifying the problem formulation of routing by adding modest assumptions on the problem and applying RL methods that have better scalability ability or higher training efficiency. For instance, leveraging asynchronous and parallel RL method such as A3C would significantly improves the convergence speed of RL in larger problems and thus scale up RL to larger problems [37].

Another major challenge on applying RL for routing is the generalization of RL model, which refers to the RL model's capability of working on unseen problems after training on the training set. A generalizable RL model does not need further training or only very light fine-tuning on unseen problems. Thus, generalizable RL model has the benefit of runtime efficiency, as well as the ability of learning from past designs or expert designs. However, for the existing RL-based routing methods, only some of them have generalization ability across problems that have certain levels of similarity. For instance, NVCell [24] demonstrates good generalization ability of RL model on unseen standard cells from the same standard cell libraries with the same technology node. Attention routing [21] demonstrates generalization among different placement for the same design. It would be part of the future work to further explore the generalization ability for these RL routing methods among more diverse routing problems such as routing problems based on different technology node or different sizes. For some other work, such as DQN global routing [19], it does not demonstrate generalization ability but focus more on iteratively improving the solution quality for one particular problem. In order to improve the generalization ability of RL models in solving routing problems, some improvement on the state representation and model architecture might be useful. For instance, VIN [34] achieved impressive generalization ability by realizing the planning function through part of the model architecture.

Despite the abovementioned open challenges, there are also some promising opportunities of solving EDA routing with RL, considering the recent impressive success of RL in solving similar problems. One particular relevant domain is the RL for combinatorial optimization. In recent years, there have been lots of works in applying RL to solve some of the canonical form combinatorial optimization problems in better ways [38]. The problems include travelling salesman problem (TSP), maximum cut (max-cut) problem, bin packing problem (BPP), etc. These problems are mostly NP-hard and are previously solved with either heuristics or analytical approaches that tend to have limitations in terms of efficiency, scalability, or optimality. The RL-based approaches demonstrate better generalization, smaller optimality gap, as well as time complexities in multiple works [38]. Some EDA routing problem can be formulated as canonical forms of combinatorial optimizations or their slight variations. As a result, leveraging some of the recent success in RL for combinatorial problem by formulating EDA routing problem as canonical form or variations of combinatorial optimization and then leveraging recent progress in RL for combinatorial optimization could be very promising opportunities.

Besides RL-based combinatorial optimization, better and customized RL algorithms as well as some of the abovementioned RL algorithms for solving related routing problems are also potential opportunities to come up with better RL-based routing solutions. RL for performance-optimization-based routing is also another promising direction in RL for routing, especially for critical performance metrics in advanced node technologies such as manufacturability, reliability, and signal integrity.

References

1. Chen, H.-Y., Chang, Y.-W.: Global and detailed routing. In: Electronic Design Automation. Morgan Kaufmann, Los Altos, pp. 687–749 (2009)
2. Clow, G.W.: A global routing algorithm for general cells. In: 21st Design Automation Conference Proceedings. IEEE, Piscataway (1984)
3. Chang, Y.-J., Lee, Y.-T., Wang, T.-C.: NTHU-Route 2.0: a fast and stable global router. In: 2008 IEEE/ACM International Conference on Computer-Aided Design. IEEE, Piscataway (2008)
4. Abel, L.C.: On the ordering of connections for automatic wire routing. IEEE Trans. Comput. **100**(11), 1227–1233 (1972)
5. McMurchie, L., Ebeling, C.: PathFinder: a negotiation-based performance-driven router for FPGAs. In: Reconfigurable Computing, pp. 365–381. Morgan Kaufmann, Los Altos (2008)
6. Cho, M., Pan, D.Z.: BoxRouter: a new global router based on box expansion and progressive ILP. IEEE Trans. Comput.-Aided Design Integr. Circuits Syst. **26**(12), 2130–2143 (2007)
7. Pan, M., Chu, C.: FastRoute 2.0: a high-quality and efficient global router. In: 2007 Asia and South Pacific Design Automation Conference. IEEE, Piscataway (2007)
8. Kastner, R., Bozorgzadeh, E., Sarrafzadeh, M.: Pattern routing: use and theory for increasing predictability and avoiding coupling. IEEE Trans. Comput.-Aided Design Integr. Circuits Syst. **21**(7), 777–790 (2002)
9. Chen, H.Y., Chang, Y.W.: Global and detailed routing. In: Electronic Design Automation, pp. 687–749. Morgan Kaufmann, Los Altos (2009)
10. Tang, H., et al.: A survey on Steiner tree construction and global routing for VLSIdesign. IEEE Access **8**, 68593–68622 (2020)
11. Huang, X., et al.: MLXR: multi-layer obstacle-avoiding X-architecture Steiner tree construction for VLSI routing. Sci. China Inform. Sci. **60**(1), 1–3 (2017)
12. Hwang, F.K., Richards, D.S.: Steiner tree problems. Networks **22**(1), 55–89 (1992)
13. Ajwani, G. Chu, C., Mak, W.-K.: FOARS: FLUTE based obstacle-avoiding rectilinear Steiner tree construction. IEEE Trans. Comput.-Aided Design Integr. Circuits Syst. **30**(2), 194–204 (2011)
14. Szymanski, T.G.: Dogleg channel routing is NP-complete. IEEE Trans. Comput.-Aided Design Integr. Circuits Syst. **4**(1), 31–41 (1985)
15. Lienig, J., Thulasiraman, K.: A genetic algorithm for channel routing in VLSI circuits. Evol. Comput. **1**(4), 293–311 (1993)
16. Chen, Y.K., Liu, M.L.: Three-layer channel routing. IEEE Trans. Comput.-Aided Design Integr. Circuits Syst. **3**(2), 156–163 (1984)
17. Gao, T., & Liu, C. L. (1996). Minimum crosstalk channel routing. IEEE Trans. Comput.-Aided Design Integr. Circuits Syst. **15**(5), 465–474
18. Ho, T-T., Sitharama Iyengar, S., Zheng, S.-Q.: A general greedy channel routing algorithm. IEEE Trans. Comput.-Aided Design Integr. Circuits Syst. **10**(2), 204–211 (1991)
19. Liao, H., Zhang, W., Dong, X., Poczos, B., Shimada, K., Burak Kara, L.: A deep reinforcement learning approach for global routing. J. Mech. Design **142**(6), 061701 (2020)

20. Liu, J., Chen, G., Young, E.F.: REST: constructing rectilinear Steiner minimum tree via reinforcement learning. In: 2021 58th ACM/IEEE Design Automation Conference (DAC). IEEE, Piscataway (2021)
21. Liao, H., Dong, Q., Dong, X., Zhang, W., Zhang, W., Qi, W., Fallon, E., Kara, L.B. (2020). Attention routing: track-assignment detailed routing using attention-based reinforcement learning. In: International Design Engineering Technical Conferences and Computers and Information in Engineering Conference, vol. 84003, p. V11AT11A002. American Society of Mechanical Engineers
22. Liao, H., Dong, Q., Qi, W., Fallon, E., Kara, L. B.: Track-assignment detailed routing using attention-based policy model with supervision. In: Proceedings of the 2020 ACM/IEEE Workshop on Machine Learning for CAD, pp. 105–110 (2020)
23. Qu, T., Lin, Y., Lu, Z., Su, Y., Wei, Y.: Asynchronous reinforcement learning framework for net order exploration in detailed routing. In: 2021 Design, Automation & Test in Europe Conference & Exhibition (DATE), pp. 1815–1820. IEEE, Piscataway (2021)
24. Ren, H., Fojtik, M., Durham, N.C., Khailany, B.: NVCell: Generate Standard Cell Layout in Advanced Technology Nodes with Reinforcement Learning
25. Mnih, V., et al.: Playing atari with deep reinforcement learning (2013). arXiv preprint arXiv:1312.5602
26. Vinyals, O., Fortunato, M., Jaitly, N.: Pointer networks (2015). arXiv preprint arXiv:1506.03134
27. Veličković, P., Cucurull, G., Casanova, A., Romero, A., Lio, P., Bengio, Y.: Graph attention networks (2017). arXiv preprint arXiv:1710.10903
28. Chen, G., Pui, C.W., Li, H., Young, E.F.: Dr. CU: detailed routing by sparse grid graph and minimum-area-captured path search. IEEE Trans. Comput.-Aided Design Integr. Circuits Syst. **39**(9), 1902–1915 (2019)
29. Mnih, V., Badia, A.P., Mirza, M., Graves, A., Lillicrap, T., Harley, T., Silver, D., Kavukcuoglu, K.: Asynchronous methods for deep reinforcement learning. In: International Conference on Machine Learning, pp. 1928–1937. PMLR (2016)
30. Ren, H., Fojtik, M.: Standard cell routing with reinforcement learning and genetic algorithm in advanced technology nodes. In: Proceedings of the 26th Asia and South Pacific Design Automation Conference, pp. 684–689 (2021)
31. Sherwani, N.A.: Algorithms for VLSI physical design automation. Springer, Berlin (2012)
32. Schulman, J., Wolski, F., Dhariwal, P., Radford, A., Klimov, O.: Proximal policy optimization algorithms (2017). arXiv preprint arXiv:1707.06347
33. Mammeri, Z.: Reinforcement learning based routing in networks: review and classification of approaches. IEEE Access **7**, 55916–55950 (2019)
34. Tamar, A., Wu, Y., Thomas, G., Levine, S., Abbeel, P.: Value iteration networks (2016). arXiv preprint arXiv:1602.02867.
35. Sykora, Q., Ren, M., Urtasun, R.: Multi-agent routing value iteration network. In: International Conference on Machine Learning, pp. 9300–9310. PMLR (2020)
36. Kool, W., Van Hoof, H., Welling, M.: Attention, learn to solve routing problems! (2018). arXiv preprint arXiv:1803.08475
37. Frobeen, L.: Asynchronous Methods for Deep Reinforcement Learning (2017)
38. Mazyavkina, N., Sviridov, S., Ivanov, S., Burnaev, E.: Reinforcement learning for combinatorial optimization: a survey. Comput. Oper. Res. **134**, 105400 (2021)

Chapter 12
Machine Learning for Analog Circuit Sizing

Ahmet F. Budak, Shuhan Zhang, Mingjie Liu, Wei Shi, Keren Zhu, and David Z. Pan

12.1 Introduction

Analog sizing is to find correct sizing of devices given a topology for analog integrated circuit (IC) design. The sizing of devices, such as transistors and capacitors, can be critical to the performance, power, and area (PPA). Sizing in manual design usually relies on both design expertise and the feedback from simulators [29]. It usually takes time and efforts to achieve target PPA. Automating analog sizing can therefore reduce labor efforts and accelerate the design cycle.

Efforts to automate analog sizing can be seen decades ago. Conventionally, it can be categorized into *equation-based* and *simulation-based* methods [31]. Equation-based methods embed equations to quantify the PPA and guide the sizing. It is computationally efficient, but the equation might be inaccurate, especially in modern advance technologies. As a result, equation-based methods are questionable to produce satisfying results. Conventional simulation-based methods search the design space with simulators. It leverages simulators to obtain accurate PPA. However, as simulations are expensive, simulation-based methods suffer from efficiency and scalability.

In general, AMS circuit sizing problem can be formulated in two ways. The first way is to formulate it as a constrained optimization problem succinctly as below:

$$\min \ f_0(\mathbf{x})$$
$$s.t. f_i(\mathbf{x}) \le 0 \quad \text{for } i = 1, \ldots, m \tag{12.1}$$

A. F. Budak (✉) · S. Zhang · M. Liu · W. Shi · K. Zhu · D. Z. Pan
The University of Texas at Austin, Austin, TX, USA
e-mail: ahmetfarukbudak@utexas.edu; shuhan.zhang@utexas.edu

© The Author(s), under exclusive license to Springer Nature Switzerland AG 2022

H. Ren, J. Hu (eds.), *Machine Learning Applications in Electronic Design Automation*, https://doi.org/10.1007/978-3-031-13074-8_12

307

where $\mathbf{x} \in \mathbb{D}^d$ is the parameter vector and d is the number of design variables of sizing task. Thus, $\mathbb{D}^d$ is the design space. $f_0(\mathbf{x})$ is the objective performance metric we aim to minimize, and $f_i(\mathbf{x})$ is the ith performance constraint in the design. The second way formulating sizing problem is to transform it into an unconstrained problem by defining a figure of merit (FoM) and optimizing for it. In this way, circuit performance values are lumped into a single equation by using normalization constants, w:

$$\min \text{ FoM} = \sum_{i=0}^{m} w_i \times f_i(\mathbf{x}) \tag{12.2}$$

The literature of analog sizing expands from adapting random black-box optimization methods to state-of-the-art machine learning optimization methods. Initially, random methods, such as evolutionary algorithms, have shown that analog sizing can be formulated as a black-box optimization problem and global-optima can be obtained in an iterative way. Since random methods required excessive number of simulations, more efficient methods using advanced ML techniques are later proposed. Researchers have utilized robust modeling techniques to better interpret simulation results and make guided search of the optimal regions. Further, the advances in hardware technology and computational resources allowed such modeling methods to be included inside the optimization loop. In this line, several Bayesian optimization-based methods are proposed to make the parameter searching more efficiently [23, 48]. Bayesian methods make statistical predictions on unseen designs and query the optimization with highest promising new design. Neural networks are utilized in the evolutionary algorithm for improving the surrogate model. Reinforcement learning (RL) is also applied to the problem for higher efficiency where the sizing problem is interpreted as an environment. In RL setting, the optimization task became the problem of finding correct actions, and a reward expression is engineered in a way that a good design results in a higher reward. Finally, to mitigate the increasing sensitivity to layout parasitics, several parasitic and layout-aware sizing methods are also proposed.

In the rest of this chapter, Sect. 12.2 gives the overview on conventional methods, Sect. 12.3 reviews the Bayesian optimization-based methods on analog sizing, Sect. 12.4 presents the recent development of evolutionary algorithm for analog sizing leveraging deep neural network (DNN)-assisted surrogate model, Sect. 12.5 overviews the sizing algorithms leveraging reinforcement learning, Sect. 12.6 introduces the recent trends in parasitic and layout-aware sizing techniques, and, finally, Sect. 12.7 concludes the chapter.

12.2 Conventional Methods

There are two typical paradigms in conventional sizing framework: the equation-based and the simulation-based. In this section, we briefly review the conventional equation-based algorithms (Sect. 12.2.1) and the simulation-based methods (Sect. 12.2.2). Then we analyze the limitations in them and summarize the recent trends in the field highlighting the applications of ML techniques (Sect. 12.2.3).

12.2.1 Equation-Based Methods

Numerous methods have previously been proposed for automated transistor sizing. Early work [10] generated symbolic AC models and optimized circuit performance using simulated annealing. A symbolic simulator can generate analytical AC models for any non-fixed analog circuit topology. The circuit performance can then be expressed in fully symbolic of mixed numeric-symbolic equations. The model is then passed to the design optimization program using simulated annealing to optimize the design parameters.

Equation-based methods have also been proposed [14], where circuit performance is guaranteed through geometric or constrained programming. The design objectives and constraints have a special form that is posynomial functions of the design variables. As a result, the amplifier design problem can be expressed as a special form of optimization problem called geometric programming, for which very efficient global optimization methods have been developed. As a consequence the globally optimal amplifier designs or globally optimal tradeoffs among competing performance measures such as power, open-loop gain, and bandwidth are achieved. This method, therefore, yields completely automated sizing of globally optimal CMOS amplifiers, directly from specifications. However, the CMOS modeling is restricted and noncompatible with advanced submicron technology nodes, making these methods severely restricted in terms of accuracy when compared with transistor-level SPICE simulation results.

12.2.2 Simulation-Based Methods

As the technology processes became more advanced and circuit topologies got more complex, it became more challenging to bring analytical expressions for the circuit performance specifications. Even single transistor behavior deviates from the classical long-channel approximation due to high device scaling. Therefore the majority of the automation algorithms are simulation-based methods where real circuit simulation evaluations are called during the optimization process. These methods utilize the existed samples and their simulation evaluations to determine new design points to be evaluated. In simulation-based methods, accuracy in specification value is acquired in exchange for embedding costly circuit simulations in the optimization loop.

Simulation-based methods can be further categorized depending on the order of derivative used in the algorithm. First-order optimization methods require expressions for gradients of the specs with respect to design variables. Gradient calculations are either embedded in the simulator and obtained by the chain rule or they can be approximated via numerical methods afterward. In [27], authors defined the sizing problem as a multi-objective nonlinear optimization problem and introduced a solver using a first-order optimization method. The gradient information is used to find a search direction and a sufficiently large step until a stationary point is reached. The algorithm consists of phases where at first the hard constraints are met, second the soft constraints are tuned, and last the objectives are improved.

However, gradient information is not conveniently available for all analysis types which leads to adaptation of zeroth-order black-box optimization algorithms in the field. In [20], the constrained optimization problem is transformed into a fitness function by utilizing augmented Lagrangian method. The resulting fitness value is minimized by a hybrid differential evolution algorithm. Other popular conventional methods in the field are variations and/or combinations of genetic algorithms [18], simulated annealing[1], or particle swarm optimization [41].

12.2.3   Limitations on Conventional Methods

While the conventional approaches demonstrate some success, the development of modern IC design requires more efficient methods in automating analog circuit design. Advance technologies impose challenges on equation-based methods. Modern device characteristics have been complicated and are usually deviated from classical device modeling. Moreover, modern analog design methodology requires more comprehensive metrics in design. As a result, using equations to characterize the circuit functionalities becomes increasingly inaccurate which makes the equation-based method challenging in applying to real-world circuits. Secondly, equations are sometimes design-dependent, and therefore equation-based methods tend to be targeting particular circuit architectures. The lack of generalization also becomes a blockage to the conventional equation-based analog sizing.

Simulation-based methods, on the other hand, face challenges from increasing simulation costs. With more complicated circuit models and simulation procedures, the cost to run a simulation becomes significant for modern analog IC design. Therefore more efficient optimization scheme is needed. Furthermore, conventional simulation-based methods shall better balance between the search for the local optimal and the global optimal. More "intelligent" searching agent is preferred to globally consider different targeting design metrics.

12.3 Bayesian Optimization

Inspired by the equation-based and simulation-based methods, the Bayesian optimization framework has been proposed to fully accelerate the optimization process by combining the equation-based and simulation-based approaches. It is quite suitable for problems that don't have a closed-form expression for the objective function and can only be observed through sampled values. The Bayesian optimization (BO) algorithm is especially efficient in situations when the sampled values are noisy, evaluations are incredibly expensive, or the convexity properties are unknown. Generally, there are two key elements in the Bayesian optimization framework: *probabilistic surrogate model* and *acquisition function*.

Surrogate Model The surrogate model incorporates people's prior belief and provides a posterior distribution with the observed data. The prescribed prior belief is the modeling space of the possible latent function. The posterior distribution means the surrogate model provides not only predictive means but also the corresponding uncertainty estimations. In other words, the surrogate model works as a cheap-to-evaluate substitute for the expensive latent function.

One most commonly used surrogate model in the Bayesian optimization framework is the Gaussian process regression model. Given a d-dimension input design variable $\mathbf{x}$, the unknown objective function is $y = f(\mathbf{x}) + \epsilon$, where ϵ denotes the observation noise $N(0, \sigma_n^2)$. Let us assume the accumulated observations as $D = \{X, \mathbf{y}\}$, where X represents a set of design variables $X = \{\mathbf{x}_1, \mathbf{x}_2, \cdots, \mathbf{x}_N\}$ and $\mathbf{y}$ denotes the corresponding N observations $\mathbf{y} = \{y_1, y_2, \cdots, y_N\}$. By capturing our prior belief about the performances of the unknown objective function with predefined mean function $m(\mathbf{x})$ and kernel function $k(\mathbf{x}_i, \mathbf{x}_j)$, the Gaussian process regression model can provide posterior distribution for an arbitrary location $\mathbf{x}^*$ as follows [28]:

$$\begin{cases} \mu(\mathbf{x}^*) = k(\mathbf{x}^*, X)[K + \sigma_n^2 I]^{-1}\mathbf{y} \\ \sigma^2(\mathbf{x}^*) = k(\mathbf{x}^*, \mathbf{x}^*) - k(\mathbf{x}^*, X)[K + \sigma_n^2 I]^{-1}k(X, \mathbf{x}^*), \end{cases} \tag{12.3}$$

where $\mu(\mathbf{x}^*)$ is the predictive mean, $\sigma(\mathbf{x}^*)$ denotes the uncertainty estimation, $k(\mathbf{x}^*, X) = k^T(X, \mathbf{x}^*)$, and $K = k(X, X)$ is the corresponding covariance matrix. The mean function $m(\mathbf{x})$ can be any function, and the covariance matrix K should be a symmetric positive definite (SPD) matrix. We refer readers to [28] for more details.

Acquisition Function In the Bayesian optimization framework, acquisition function works as a cheap-to-evaluate utility function to guide the sampling decisions. Instead of exploring the design space only with the predictive mean, the acquisition function leverages the uncertainty estimation to explore the unknown area, until they are confidently ruled out as suboptimal. In this way, the acquisition function favors not only the current promising area with high confidence but also the unknown region with large uncertainty estimation. In other words, the acquisition function trades off between the exploration and exploitation based on the posterior beliefs provided by the surrogate model. There are three most widely used acquisition functions.

(1) The Probability of Improvement (PI) Given the current minimum objective function value τ in the dataset, the probability of improvement function tries to measure the probability an arbitrary $\mathbf{x}$ exceeds the current best. The corresponding formulation is as follows [19]:

$$PI(\mathbf{x}) = \Phi(\lambda), \tag{12.4}$$

where $\Phi(\cdot)$ is the CDF of standard normal distribution, $\lambda = (\tau - \xi - \mu(\mathbf{x}))/\sigma(\mathbf{x})$, and ξ is a small positive jitter to encourage exploration.

(2) The Expected Improvement (EI) Compared with PI that only measures the probability of improvement and treats the improvement equally, the expected improvement function tries to measure the amount of improvement upon the current best τ. By maximizing the expected improvement function, the observation $\mathbf{x}$ will not only exceed the current best but also exceed the current best value at the highest magnitude. The corresponding formulation can be expressed as [26]

$$EI(\mathbf{x}) = \sigma(\mathbf{x})(\lambda\Phi(\lambda) + \phi(\lambda)), \tag{12.5}$$

where $\phi(\cdot)$ is the PDF of standard normal distribution.

(3) The Lower Confidence Bound (LCB) Compared with the improvement-based strategies like PI and EI, the lower confidence bound function tries to guide the search from an optimistic perspective. With the carefully designed coefficient β, the cumulative regret is theoretically bounded [38] [39]. Thus, the convergence of the Bayesian optimization algorithm is guaranteed. The corresponding formulation is

$$LCB(\mathbf{x}) = \mu(\mathbf{x}) - \beta\sigma(\mathbf{x}). \tag{12.6}$$

β is the parameter that balances between exploration and exploitation.

Apart from the abovementioned acquisition functions, there are also some other types of acquisition functions, including entropy search (ES) [12], Thompson sampling (TS) [40], predictive entropy search (PES) [13], max-value entropy search (MES) [43], and knowledge gradient (KG) [7, 33]. And it is also possible to explore the state space with a portfolio of acquisition functions [15, 36].

Bayesian optimization framework has demonstrated significant potential in approximating the global optimum with a relatively small number of evaluations [5, 16, 23–25, 45–47]. It gains the efficiency by leveraging both the surrogate model and the acquisition function [35]. The surrogate model works as a simplified representation of the costly simulation process by taking the whole history of optimization into considerations. The informative posterior distribution provided by the surrogate model includes the predictive mean and well-calibrated uncertainty

---

Algorithm 12.1 Bayesian optimization framework

Input: The size of the initial dataset N_{init} and the maximum number of iteration N_{iter}
1: Randomly sample a initial dataset $D_0 = \{X, y\}$
2: **for** $t = 0 \rightarrow N_{iter}$ **do**
3:     Construct a Gaussian process regression model with D_t
4:     $\mathbf{x}_t \leftarrow \text{argmax}_{\mathbf{x}} \alpha(\mathbf{x}; D_t)$
5:     $y_t = f(\mathbf{x}_t)$
6:     $D_{t+1} \leftarrow \{D_t, \{\mathbf{x}_t, y_t\}\}$
7: **end for**
8:
9: **return** Best y recorded after optimization

---

estimation. The acquisition function prioritizes data points in the candidate pool and guides the search by proposing a sequence of promising data points. The overall Bayesian optimization framework is presented in Algorithm 12.1.

Apart from the abovementioned conventional Bayesian optimization algorithm, researchers also leverage the recent advancements in Bayesian optimization algorithm to solve constrained optimization problem and make it work in parallel. Section 12.3.1 reviews Bayesian optimization framework for constrained optimization problem. To fully utilize the hardware resources, Sect. 12.3.2 overviews the advancements in batch Bayesian optimization algorithm.

12.3.1 WEIBO: An Efficient Bayesian Optimization Approach for Automated Optimization of Analog Circuits

Most of the acquisition functions do not handle nonlinear constraints. Expected improvement and predictive entropy search [9, 21] are the acquisition functions that can deal with constraints. According to [9], the advantage of an entropy-based acquisition function is that it can handle the decoupled constraints, i.e., the constraints and the objective functions are evaluated separately, so this approach might be useful if objectives and constraints are obtained from different measures with different test benches. However, the entropy-based acquisition function does not have a closed form. Thus, the weighted expected improvement (wEI) [8, 32] is proposed to deal with the optimization problem with constraints.

For each constraint $u_i = f_i(\mathbf{x})$, a single Gaussian process (GP) model is constructed to approximate the constraint $f_i(\mathbf{x})$. The GP model gives a distribution of u_i as prediction

$$u_i \sim \mathcal{N}(\mu_i, \sigma_i^2), \tag{12.7}$$

where μ_i and σ_i are the corresponding mean and variance, respectively. For the constraint $f_i(\mathbf{x}) < 0$, the probability of the constraint being satisfied can be expressed as

$$PF_i(\mathbf{x}) = \Phi(-\frac{\mu_i}{\sigma_i}), \tag{12.8}$$

where $\Phi(\cdot)$ is the CDF of standard normal distribution.

To deal with the constraints, the improvement function is changed. The improvement is nonzero if and only if the objective function $f_0(\mathbf{x}) < \tau$ and the point $\mathbf{x}$ is feasible:

$$I(y, \tau) = \begin{cases} 0 & y >= \tau \\ \tau - y & y < \tau \text{ and } \forall i \in \{1, \ldots, m\}, f_i < 0 \end{cases} \tag{12.9}$$

Assume that the objective function $f_0(\mathbf{x})$ and all the constraints $f_i(\mathbf{x})$ are mutually independent. The new expected improvement can thus be expressed as

$$\text{wEI}(\mathbf{x}) = \text{EI}(\mathbf{x}) \prod_{i=1}^{m} p(f_i(\mathbf{x}) < 0), \tag{12.10}$$

where m is the number of constraints. The EI function in (12.5) is now weighted by the probability of feasibility of all the constraints. The weighted expected improvement (wEI) in (12.10) is a fully probabilistic acquisition function as it considers the uncertainties of both objective functions and constraints. Note that τ is the current best result. However, it is possible that there are no feasible points found at the beginning state of optimization, so that the wEI in (12.10) is undefined. In such cases, [23] proposes to find the feasible points firstly. The probability of feasibility (PF) is thus used as acquisition function

$$\text{PF}(\mathbf{x}) = \prod_{i=1}^{m} p(f_i(\mathbf{x}) < 0). \tag{12.11}$$

The expression of wEI is based on the assumption that the constraints are independent so that they can be independently modeled and the probabilities could be multiplied. Although the objective functions and constraints are mapped from different specifications of circuits and they may be correlated, [23] demonstrates that the wEI function works well in practice.

The overall framework of weighted expected improvement-based Bayesian optimization algorithm is as follows. Starting from the initial training set, the GP models are built for objective and constraint functions firstly. The hyperparameters of the GP models are determined by maximum likelihood estimation. The acquisition function (EI for unconstrained problem and wEI for constrained problem) is then constructed based on the GP models. The acquisition function is then maximized to find the next data point $\mathbf{x}$. By simulating the circuits, the objective and constraint function values at $\mathbf{x}$ are obtained. The new data point is added to the training set. The GP models are updated, and the iterations continue until the maximum number of simulations is reached or other stop criteria are met.

12.3.2   EasyBO: An Efficient Asynchronous Batch Bayesian Optimization Approach for Analog Circuit Synthesis

Despite that effectiveness and efficiency of the Bayesian optimization framework, the sequential characteristic of the state-of-the-art acquisition function in the Bayesian optimization framework makes it hard to be parallelized. Without parallelism, the hardware is not able to be fully utilized when multicore workstations are available. In order to further reduce the overall time consumption on circuit simulations, efforts have been made to make batch Bayesian optimization possible [16, 24]. By synchronously sampling several points at each iteration and evaluating the performances in parallel, the overall simulation time can be greatly reduced compared to the Bayesian optimization algorithm in sequential mode.

Nevertheless, the main issue with the synchronous batch Bayesian optimization algorithm is that different design parameters can lead to different simulation time consumption. And it is a waste of hardware resources to let workers wait idly for the slowest design in the batch to finish the simulation.

As illustrated in Fig. 12.1, the synchronous batch BO algorithm aims to select a batch of candidate points to evaluate in parallel. The next batch will only be issued when all samples in the previous batch have been evaluated. Due to variability in simulation times for different design parameters, there will always be workers waiting idly for others to finish their jobs in the synchronous batch BO algorithm.

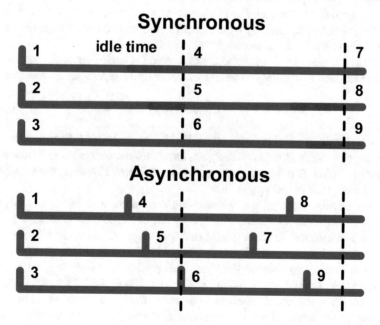

Fig. 12.1 An illustration of asynchronous and synchronous setting when batch size is 3 [48]

Therefore, the synchronous batch BO algorithm is not able to achieve $B\times$ speed-up for batch size B compared to its sequential counterpart. As the batch size increases, the time reduction effect will deteriorate quickly, since more workers will wait idly for others to finish their jobs.

Instead of waiting for the whole batch to finish the evaluation process, [48] propose to asynchronously issue new candidate points whenever a worker becomes available. The key motivation behind the asynchronous batch BO algorithm is to make full use of the hardware resources to reduce the overall time spent on evaluation. Intuitively, the asynchronous batch BO algorithm can process a greater number of evaluations than its synchronous counterpart, in a given period of time. With the increase of the batch size, the time reduction for a given number of simulations will be more significant compared to its synchronous counterpart. And the asynchronous batch BO algorithm is especially suitable for problems where the simulation time of which differs greatly for different design parameters.

However, there are two harsh challenges for the batch BO algorithms to deal with: (1) how to fully leverage our current knowledge about the latent function and select the future query points and (2) how to penalize around the selected locations that are still under evaluation to prevent redundant samples from being chosen in the busy region.

In Bayesian optimization, the acquisition function is carefully designed to balance the exploration and exploitation. For batch Bayesian optimization, it is also desirable for the acquisition function to create the diversity of the query points in a batch. The upper confidence bound as shown in Eq. (12.6) is a direct yet powerful acquisition function. The predictive mean $\mu(\mathbf{x})$ represents the exploitation, while the predictive uncertainty $\sigma(\mathbf{x})$ represents the exploration. The parameter $\mathbf{x}$ is introduced to balance the exploration and exploitation.

In [16], the diversity is introduced to the batch selection by assigning the predictive mean and uncertainty measurement with different weighting parameter w:

$$\alpha_{\text{pBO}}(\mathbf{x}, w) = (1 - w) * \sigma(\mathbf{x}) + w * \sigma(\mathbf{x}). \tag{12.12}$$

For batch size B, B weights $\{w_1, \ldots, w_B\}$ which are uniformly distributed over [0, 1] are selected. With B different weights, the B different acquisition functions are expected to select B different query points for a batch. However, such a strategy does not work well in real applications.

At the starting stage of the optimization procedure, it is important to gather global information about the underlying behavior of the objective function. In other words, the acquisition function should encourage exploration at the initial stage of the optimization process, which means w should be larger to encourage exploration.

After a limited number of iterations, the predictive uncertainty $\sigma(\mathbf{x})$ of the refined model would have a much smaller magnitude than its predictive mean $\mu(\mathbf{x})$. Therefore, the acquisition functions with smaller w would generate almost the same query points, since $(1 - w) * \mu(\mathbf{x})$ dominates the acquisition function in Eq. (12.12). From the previous analysis, we should encourage larger w for the

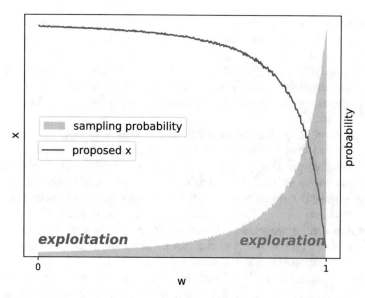

Fig. 12.2 An illustration of the UCB acquisition function with different w over $[0, 1]$. And the sampling probability of our proposed acquisition function with respect to w [48]

acquisition function as shown in (12.12). The uniformly distributed w in [16] is not a good choice.

The corresponding distribution of the selected location **x** with respect to w parameter is shown in Fig. 12.2. It is worth noting that **x** only has small change when w is relatively small and encourages exploitation. And the value of **x** changes quickly when w is relatively large and encourages exploration. Therefore, the sampling density around the region with large w should be increased to maintain the diversity of the selected query points.

EasyBO [48], instead of uniformly sampling the w parameter, proposes a new acquisition function scheme that increases the sampling density as w increases, which can be expressed as

$$\alpha(\mathbf{x}, w) = (1 - w) * \mu(\mathbf{x}) + w * \sigma(\mathbf{x}), \qquad (12.13)$$

where $w = \kappa/(\kappa + 1)$. By randomly sampling κ from the uniform distribution over a proper range $[0, \lambda]$, w tends to approach 1 to encourage exploration at the initial stage and the diversity at the later optimization stage. The corresponding sampling probability of w is shown in Fig. 12.2. It can be observed that w has higher probability near the neighborhood of 1. λ should be carefully designed to prevent the acquisition function from too much exploration during the optimization. In [48], λ is set as 6.0.

An interesting observation is that the predictive uncertainty of GPR model provides a natural penalization for diversity. The uncertainties are lower in the neighborhood of already sampled data points while larger in the unvisited region. Thus, the UCB acquisition function as shown in (12.12) can naturally avoid redundant sampling over the neighborhood of already visited points.

For batch Bayesian optimization, the diversity of query points in one batch can be naturally guaranteed by including query points in the same batch to the training dataset and incorporating the corresponding predictive uncertainty into the acquisition function.

Denote $D = \{X, \mathbf{y}\}$ the observed data points. Denote $\hat{X} = \{\hat{x}_1, \ldots, \hat{x}_{B-1}\}$ the already selected query points in this batch, where B is the batch size. Note the corresponding observations $\mathbf{y} = \{y_1, \ldots, y_{B-1}\}$ are unknown, since the simulations of this batch are not finished yet. Following the same penalization strategy as [5], the underlying objective function follows the same behavior as the current predictive mean. Let $\hat{X} = \{\hat{\mathbf{x}}_1, \ldots, \hat{\mathbf{x}}_{B-1}\}$ denote the query points under evaluations; the observations can be approximated with the predictive mean $\hat{\mathbf{y}} = \{\hat{y}_1, \ldots, \hat{y}_{B-1}\}$ of GPR model with the observed data points $D = \{X, \mathbf{y}\}$ as training data. With the observed data points and the pseudo data points, i.e., $\hat{D} = \{X, \mathbf{y}\} \cup \{\hat{X}, \hat{\mathbf{y}}\}$, the predictive uncertainty of GPR model $\hat{\sigma}(\mathbf{x})$ can be obtained according to (12.3). And this uncertainty estimation $\hat{\sigma}(\mathbf{x})$ can naturally be incorporated into acquisition function to guarantee the diversity of the query points in one batch. The acquisition function with penalization scheme can be expressed as [48]

$$\alpha(\mathbf{x}, w) = (1 - w) * \mu(\mathbf{x}) + w * \hat{\sigma}(\mathbf{x}), \tag{12.14}$$

where $w = \kappa/(\kappa+1)$, and κ is randomly sampled from the uniform distribution over a proper range $[0, \lambda]$. Here, the predictive uncertainty is replaced by the uncertainty estimation $\hat{\sigma}(\mathbf{x})$ to guarantee the exploration as well as diversity simultaneously. We refer readers to [48] for more details.

12.4 Improved Surrogate Model for Evolutionary Algorithm with Neural Networks

Although utilizing Gaussian processes for modeling design space and making predictions for unseen points significantly improved the optimization sample efficiency, it also introduced new challenges to tackle. One major deficiency is the fact that GPs suffer from cubic computational scaling with respect to number of samples used in GP modeling. Another deficiency is the curse of dimensionality since GP inference weakens significantly as design variables are increased in number. Therefore learning-based methods leveraging deep neural networks (DNN) are adapted for higher training efficiency and more modeling capacity.

12.4.1   An Efficient Analog Circuit Sizing Method Based on Machine Learning-Assisted Global Optimization

A successful example of using machine learning (ML) for sizing task is introduced in [2]. The main idea is that a feedforward neural network can be trained to approximate circuit performance specifications. Then the trained neural network is switched to inference mode to predict unseen design points and these predictions used to boost the efficiency of search mechanism. The work introduces how to guide zeroth-order optimization with a neural network. Authors also introduce a novel automated ranking mechanism and a data augmentation method. The resulting algorithm is called ESSAB.

Overall Framework   The overall flow diagram of ESSAB is included in Fig. 12.3. The algorithm is randomly initialized to collect samples for training and optimization iterations follow. In each iteration, based on the ranking of evaluated samples, an elite population is selected for training and design space exploration purposes. Then, the search mechanism is built by an evolutionary strategy. Authors have applied differential evolution (DE) to create potential samples for SPICE evaluation. Once the artificial neural network (ANN) is trained, potential samples are exposed to prescreening where their performance values are approximated by the trained ANN. Then the algorithm is iterated by selecting the best potential candidate and calling real simulation for evaluation.

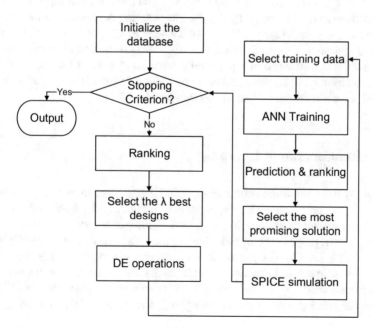

Fig. 12.3   The flow diagram of ESSAB algorithm 12.3

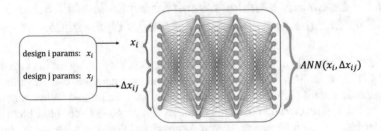

Fig. 12.4 The regression neural network used in ESSAB 12.3

Regression Neural Network A major complication for utilizing a neural network as a simulator proxy is the fact that collecting data for analog circuits is not a cheap process. Most of the time, we are not provided by the data before optimization algorithm is started. Therefore, a good practice is to collect the data and train the network simultaneously. Further, an effective use of data is desired due to costly nature of data collection process. To tackle these issues, a data augmentation method is utilized to process online collected data. Authors have used Cartesian product pairs of collected sample designs to create *pseudo-samples*. By this method, the number of training data is squared and ANN accuracy is improved.

To construct the ANN, design parameters of the circuit are reflected at the input layer, and each node at the output layer corresponds to a performance metric. Input vector of ANN is formed by a design vector x and a corresponding change vector Δx due to the augmentation method used. The resulting ANN answers the question: "If I had a design parameterized by x and changed by Δx, what is the performance of resulting design, $x + \Delta x$?" When this ANN is used for inference purposes, i.e., to predict the performance of a potential sample created by DE steps, the starting point x is assumed to be the parent design, and Δx is calculated as the difference between the child design and the parent design (Fig. 12.4).

12.5 Reinforcement Learning

Nowadays, deep reinforcement learning (DRL) has been extensively applied to many complex problems such as optimization, robotics and AutoML. With a large amount of data and enough exploration, DRL demonstrates superior performances. Moreover, DRL proves to be transferable. As the transistor dimension rapidly scales down, porting existing designs from one technology node to another becomes a common practice. Similarly, design with the layout parasitics can be facilitated by the trained model in the schematic level. Therefore, transferability of DRL can improve the efficiency of automatic sizing in different conditions and stages.

12.5.1 GCN-RL Circuit Designer: Transferable Transistor Sizing with Graph Neural Networks and Reinforcement Learning

Different from conventional optimization approaches, [42] features the capability of transferring the knowledge across different technology nodes and topologies. Given the same circuit topology, the design principle is similar across different technology nodes. A RL agent trained on a technology is able to optimize the circuit quickly on another technology. Moreover, graph convolutional neural network (GCN) is introduced to leverage the circuit topology information. The connection relationship of components in the topology can be learned by GCN. Therefore, the transferability is extended to different topologies. The overview of GCN-designer is shown in Fig. 12.5 and agent architecture is shown in Fig. 12.6.

RL is used to solve complex decision-making problems. In the RL setting, an agent interacts with the environment, getting rewards for the actions it performs in each state. The goal is to maximize the total reward. The unconstrained optimization problem can be formulated as a RL problem. The circuit simulation is wrapped as an environment. FoM can serve as the reward. The agent is trained to design a circuit sizing with the best FoM. The detailed RL problem setting in the GCN-RL designer is the following:

State The index, type, and the electrical properties of each device are encoded into the state vector.

Reward The definition of reward here is similar to that of the FoM mentioned before. It is a weighted sum of the normalized performance metrics.

Action The action vector is the sizing of devices in the circuit. The action space is continuous.

Agent Agent has the actor and critic neural network. The neural network structure is shown in Fig. 12.6. GCN is leveraged to process the circuit topology information during the optimization. Transistor's hidden representations are calculated by the

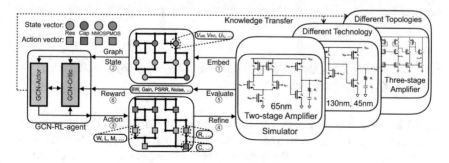

Fig. 12.5 High-level flow of GCN-RL designer [42]

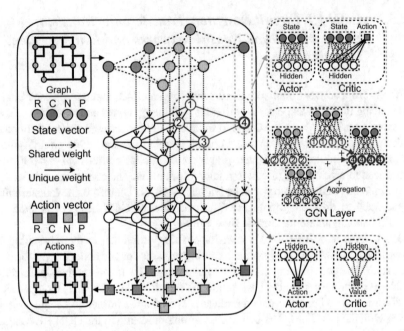

Fig. 12.6 Architecture of GCN-RL agent [42]

GCN layer. For each transistor, feature vectors from its neighbors are aggregated together. One node can receive information from farther and farther nodes by stacking multiple layers. The actor's first layer is a FC layer shared among all components. The critic's first layer is a shared FC layer with a component-specific encoder to encode different actions. The actor's last layer has a component-specific decode to decode the hidden activations to different actions. The critic's last layer is a shared FC layer with the predicted reward value as the outputs.

In the experiments, the GCN-RL designer is capable of transferring the knowledge between the different technologies and similar topologies. Experiments are conducted on 45 nm, 65 nm, 130 nm, 250 nm CMOS technologies and two-stage/three-stage TIA. With the GCN-RL agent, fewer steps are needed to reach high FoM given an agent trained on another technology or a similar circuit topology.

12.5.2 AutoCkt: Deep Reinforcement Learning of Analog Circuit Designs

GCN-RL and other optimization algorithms view circuit sizing as an optimization problem in one-step horizon. The circuit performance will be obtained once the sizing parameters are given. The optimal performance and corresponding parameters are only what we desired. On the contrary, [34] trains a RL agent to have the knowledge about the entire design space. Consequently, the agent is able to choose the next best action at a certain point in design.

Suppose there are N parameters (sizing) to tune for optimizing M target design specifications. Therefore, parameter space is $x \in Z^N$ and the design specification space is $y \in R^M$.

State The state vector contains the target circuit specifications, current circuit performances, and the current parameters.

Reward The reward of each trajectory is the accumulation of the reward of each step which is

$$R = \begin{cases} r, & r < -0.01 \\ 10 + r, & r \geq 0.01 \end{cases} \tag{12.15}$$

$$r = \sum_{i=1}^{M-T} \min \left\{ \frac{o_{pt\_i} - o^*_{pt\_i}}{o_{pt\_i} + o^*_{pt_i}}, 0 \right\} - \sum_{j=1}^{T} \epsilon \frac{o_{pt\_j} - o^*_{pt\_j}}{o_{pt\_j} + o^*_{pt\_j}} \tag{12.16}$$

where o_{pt} represents hard constraint design specifications and $o_t h$ represents design specifications which should be minimized. The reward is a measure of how close the current circuit performance is to the target.

Action The set of available actions contains three discrete actions: increment, decrement, or retain the same parameter values.

Agent The RL agent is implemented by 3 layers with 50 neurons each.

The training and deployment of AutoCkt is shown in Fig. 12.7. The agent learns to design a circuit through many trajectories. Agent either reaches the target or terminates the trajectory due to hitting the maximum number of steps. After training, the agent is deployed to an unseen target. The more effective the training is, the higher success rate agent can achieve in the deployment.

Fifty target specifications are randomly sampled from the design specification space before training starts. The trajectories generated during the training is visualized in Fig. 12.8. The maximum number of total steps is H. The trajectory will end early if the target is reached.

After training, another set of randomly chosen target specifications will be used to test the agent. The success rate of achieving these target specifications is used to measure the agent's capability of designing circuit given an arbitrary target. The trained agent can also transfer the knowledge gained on the training in the schematic level to the PEX level. Since the similar tradeoffs between parameters and corresponding circuit performance hold in the schematic and PEX simulation, the agent can still hit the target in the PEX condition after it learns how to take best actions to move toward the goal.

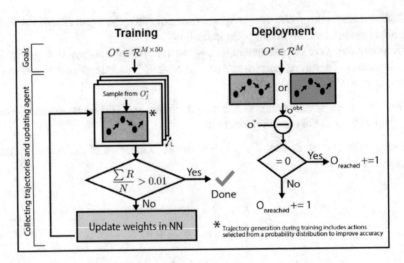

Fig. 12.7 Training and deployment of AutoCkt [34]

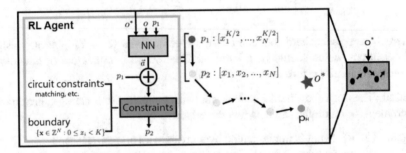

Fig. 12.8 Trajectory generation of AutoCkt [34]

12.5.3 DNN-Opt: An RL-Inspired Optimization for Analog Circuit Sizing Using Deep Neural Networks

DNN-Opt [3] is a deep neural network black-box optimization algorithm that is constructed on an RL structure. It borrows its neural network structure from a continuous action space actor-critic algorithm introduced in the RL community. Then this structure is tailored with significant modifications for sizing problem.

In DNN-Opt, the neural network performance metric approximation, introduced in [2], is utilized as the critic-network. Critic-network is trained by the data created via *pseudo-sample* generation mechanism. Further, actions are determined by the actor-network to determine the next sample point. In that perspective, the state representation of the environment is the vector of design variables, and actions are change vector for the design variables. The core architecture of DNN-Opt is shown in Fig. 12.9.

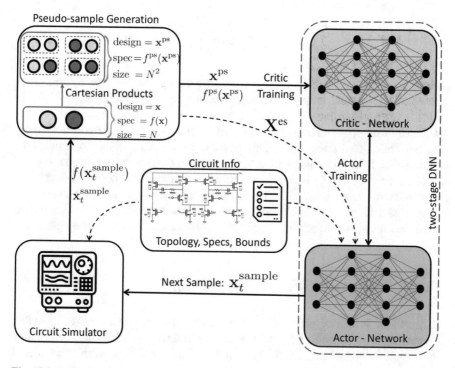

Fig. 12.9 DNN-Opt framework [3]

Now we explain how critic- and actor-networks are trained and used to better understand the way DNN-Opt iterates the optimization.

Critic-Network Since critic-network is the regression network for circuit performance metrics, it is trained by the mean squared error expressions obtained by the difference between network predictions and real simulation outputs. For a data batch of size N_b, the loss expression for critic-network, Q, parameterized by θ^Q is given as follows:

$$
L\left(\theta^Q\right) = \frac{1}{N_b(m+1)} \sum_{k=1}^{N_b} \sum_{l=1}^{m+1} \left(Q(\mathbf{x}_k, \Delta\mathbf{x}_k)^l - f(\mathbf{x}_k + \Delta\mathbf{x}_k)^l \right)^2
$$

where $Q(\mathbf{x}_k, \Delta\mathbf{x}_k)^l$ is the critic-network's approximation for k^{th} sample's l^{th} performance and $f(\mathbf{x}_k + \Delta\mathbf{x}_k)^l$ is the SPICE simulated value for the same design-performance pair.

Actor-Network Once the critic training is done, actor is trained based on the fixed critic. The analogy here is that critic-network behaves as a surrogate model for the design space and actor-network conducts a search within this model to find potential *good* designs. The training of actor-network has two main elements stated as follows:

$$L\left(\theta^{\mu}\right) = \frac{1}{N_b} \sum_{k=1}^{N_b} \left(g\left[Q(\mathbf{x}_k, \mu(\mathbf{x}_k \mid \theta^{\mu}))\right] + \|\lambda * \text{viol}_k\|_2\right) \tag{12.17}$$

where $\mu(\mathbf{x}_k \mid \theta^{\mu})$ is the proposed parameter change vector $\Delta\mathbf{x}_k$ by the actor-network. The second term includes penalization for actor proposals resulting in outside restricted parameter region. The restricted region is determined by the statistics of a population based on the "elite" solutions. This regularization forces the population to converge in a finite horizon. In this way, DNN-Opt mimics the convergence behavior of population-based methods (such as evolutionary).

Once the next sample is determined by actor-network's proposal, this sample is sent to circuit simulator, and its real evaluation is obtained. Then the population is updated with this evaluation and a new iteration loop starts. The optimization continues until either a satisfactory design is met or simulation budget is consumed.

12.5.4 Discussion: RL for Analog Sizing

RL methods in analog sizing differ from each other in several ways. Typically, researchers have to decide how to define the environment, what are the actions, how the reward is engineered, and the type of RL algorithm to maximize accumulated return.

In terms of the state representation, GCN-RL uses device-specific properties; AutoCkt uses design parameters, current performance and target performance; and DNN-Opt only uses design variables. They also differ in the action space; GCN-RL and DNN-Opt utilize a continuous action space and interpret design parameter values as the actions for the agent. However, AutoCkt has a discrete action space where the actions are to determine increment, decrement, or retain the same value of design parameters. In terms of the reward expression, GCN-RL and DNN-Opt have similar approach to use an FoM to build reward and return. On the other hand, AutoCkt's reward has a slightly different form, while it still depends on the target specifications. Finally, we see they all differ in their engineering capacity. GCN-RL and DNN-Opt collects all their samples in the same episode, but AutoCkt separates the training and deployment phases. GCN-RL proposes methods to transfer the learning from between topologies and technology nodes. AutoCkt uses its learned patterns to expand the optimization into PEX level. DNN-Opt has pseudo-samples and tailored a population-based convergence scheme which results in high sample efficiency.

12.6   Parasitic and Layout-Aware Sizing

12.6.1   BagNet: Layout-Aware Circuit Optimization

BagNet [11] is a layout-aware evolutionary algorithm-based circuit optimizer boosted with DNN. It contains three main components: (1) an evolutionary core engine to generate offsprings; (2) a layout generation tool, the Berkeley Analog Generator (BAG) [4]; and (3) a DNN model acting as an oracle to two designs. The layout generation tool enables the framework to consider parasitics from layouts, and the oracle allows the framework to be more efficient. The efficiency boosting from DNN is especially important as the layout generation and post-layout simulation is more costly compared to schematic simulation.

Figure 12.10 shows the high-level architecture of the framework. The evolutionary algorithm generates the next generation of offsprings based on the current population. The DNN-based oracle compares the generated candidates and acts as a discriminator to remove those of lower scores. The generated children after the discrimination stage are then implemented in layouts using BAG and being

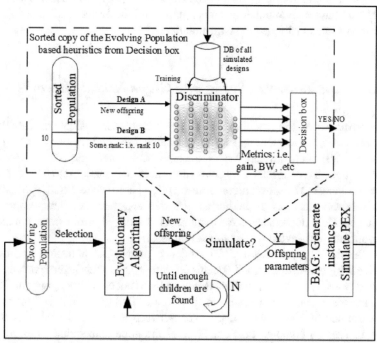

Measured children appended to both evolving population and database

Fig. 12.10 High-level architecture of BagNet [4]

simulated to obtain the metrics. The simulation results are then used to fine-tune the oracle and guide the next generation in the evolutionary algorithm core.

Evolutionary Algorithm The BagNet framework uses the current population and performs some evolutionary operations to get the next generation of the population. Cross-entropy and some canonical $\mu + \sigma$ evolutionary strategies are used in the process.

Layout Generation BAG is a procedural-based analog layout generation tool. For a design, the designer can codify the procedures for drawing the layout in a parameterized way. Using BAG allows the layout generation being automated given different parameters or sizing in the schematic design. The BagNet can therefore obtain the layouts for a new sizing without additional human interference.

DNN-Based Oracle The DNN oracle is a fully connected network that takes input of the parameters from two designs, D_A and D_B. The oracle predicts which design performs better in each individual metric. In other words, without loss of generality, if there are N metrics, the output of the oracle is an N-dimensional vector, and each element means the probability that D_A is preferred to D_B for that metric. There is a subtle constraint on the network. The network is constrained to predict complementary probabilities for inputs $[DA, DB]$ vs. $[DB, DA]$. Each sub-DNN's layer should have even number of hidden units, and the corresponding weight and bias matrices are constrained to be symmetric.

12.6.2 Parasitic-Aware Sizing with Graph Neural Networks

Recent advancements in machine learning have enabled statistical and data-driven approaches to accurately estimate layout parasitics directly from circuit schematics. ParaGraph [30] proposes the use of a graph neural network (GNN) model to predict net parasitic capacitance by converting circuit schematics into heterogeneous graphs. MLParest [37] trains random forest models based on extracted features from circuit schematic netlists to predict an effective resistance and lumped parasitic capacitance. Both approaches have demonstrated reduced errors when comparing pre-layout and post-layout circuit simulation results.

ParaGraph [30] proposes a GNN model to predict net parasitics and device parameters by converting circuit schematics into graphs with heterogeneous edge and node types. Each device and net is mapped into nodes within the graph. Nodes are connected in the graph based on the circuit topology, while different edges are determined by the connected pin type between the net and device as shown in Fig. 12.12a. To compute the next layer of the node embeddings, the previous layer node embeddings of the target node neighbors are fed into different weight matrices based on the edge type, aggregated with self-attention, concatenated with the target nodes embedding, and fed into a final linear layer to generate the target's next-layer embedding. Details of the paragraph node embedding updates are shown

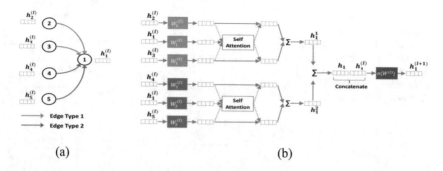

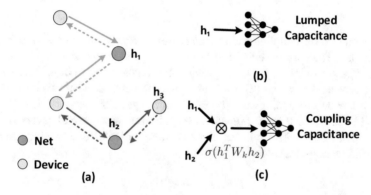

Fig. 12.11 ParaGraph node embedding update in [30]. (**a**) Node 1 has four input edges with two different edge types. Each node has its own embedding on layer l. (**b**) Compute graph of a ParaGraph embedding layer. Each input node embedding is fed into different weight metrics based on their input edge types, aggregated with self attention, then concatenated with target node's layer l embedding before fed into a weight matrix to generate embedding on layer $l + 1$

Fig. 12.12 Improved parasitic prediction. (**a**) ParaGraph node embedding with heterogeneous graphs. (**b**) Lumped capacitance prediction. (**c**) Coupling prediction with bilinear layer

in Fig. 12.11. To predict a parasitic value, the target node embedding is fed into a neural network with fully connected layers for regression in Fig. 12.12b.

Paragraph's capabilities can be extended by adding predictions to the model for parasitic resistance and coupling capacitance. As shown in Fig. 12.13, layout extraction creates a distributed RC network with multiple intermediate nodes for a single net in the schematic. We only model the effective resistance of each net to the connected transistor gates where the effective resistance is obtained with DC simulations as shown in Fig. 12.13c. The final net parasitic model is shown in Fig. 12.13d with only the effective resistance of gate connections, lumped capacitance, and coupling capacitance. Compared with [37], where the resistance is uniform and coupling is disregarded, the extended ParaGraph parasitic model has coupling capacitance and nonuniform effective resistance modeled.

Resistance and coupling regression tasks are formulated as similar tasks to link prediction in social network analysis. Where net lumped capacitance only involves

Fig. 12.13 Parasitic
modeling. (**a**) Circuit
schematic. (**b**) Distributed RC
network from extraction. (**c**)
Measurement of effective
resistance for labeling. (**d**)
Proposed parasitic model

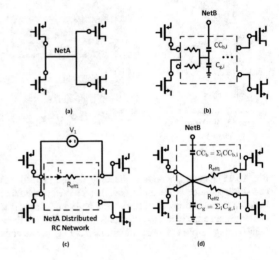

a single graph node, coupling and resistance need information from both source and
target nodes. The entire prediction process is mapped in two stages: (1) the node
embeddings are obtained by aggregating local neighborhood graph features with
graph neural networks and (2) the final regression network where both the source
and target node embeddings are the input. The regression network demonstrated
in Fig. 12.12c consists of a bilinear layer followed with several layers of fully
connected networks. The bilinear layer is more suitable for modeling pairwise
feature interactions, calculated as follows:

$$f(h_1, h_2) = h_1^T W_k h_2, \tag{12.18}$$

where h_1, h_2 are the input node embeddings of the source and target and W_k are k
trainable weight matrices to generate an output vector of size k.

The parasitic-aware sizing framework is depicted in Fig. 12.14. During each
optimization iteration, the design parameters are updated with Bayesian optimization. The parasitic prediction engine (ParaGraph) predicts post-layout parasitics for
the circuit based on the circuit topology and design parameters, which are back-
annotated into the schematic design. The circuit FOM is obtained with SPICE
simulations and added to the training set for the surrogate model. The neural
network models the objectives and constraints and is trained every iteration with
information from both the design variables and parasitic information in ParaGraph.
Finally the acquisition function is maximized using the surrogate model to obtain
the next design parameters.

The key differences between the proposed parasitic-aware transistor sizing
framework and prior work is the use of graph neural network parasitic prediction
during SPICE simulations and surrogate modeling. Parasitic prediction removes
the need for in-the-loop layout generation when automated layout tools are limited
and unstable. Furthermore, including parasitic information in the surrogate model

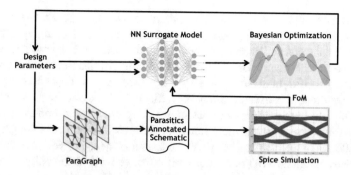

Fig. 12.14 Parasitic-aware analog sizing framework

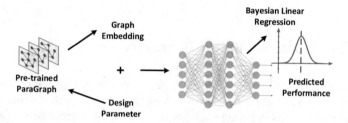

Fig. 12.15 Performance model with parasitic graph embedding

greatly improves the accuracy and increases the convergence for Bayesian optimization.

The work of [22] further proposes an improved performance surrogate model using graph embeddings from the pre-trained parasitic graph neural network as additional parasitic information. Since both device sizing and critical parasitics can affect the circuit performance, we naturally want to include parasitic-related information into the performance surrogate model. We leverage the pre-trained ParaGraph GNN model described to encode latent information of the circuit parasitics with graph embeddings. The parasitic GNN model is pre-trained with an abundant amount of data, including circuits with different topologies and sizing, while the performance model is targeted to a specific circuit.

Figure 12.15 illustrates the proposed performance model with parasitic graph embedding. Given a specific design, the circuit graph is fixed. The node attributes input to the ParaGraph GNN model are related to the design parameters. The node embedding matrix $H = \{h_1 \cdots h_n\} \in \mathbb{R}^{n \times d}$ is obtained from the GNN by concatenating all n node embeddings of dimension d. The graph embedding g is then obtained by aggregating the node embedding matrix H, which includes information from both the design and parasitic prediction. Finally, the design parameter and graph embedding is fed into several layers of fully connected neural networks for performance prediction. The last fully connected layer could be replaced with BLR to model uncertainty.

To estimate the predictive mean and predictive uncertainty, the results of stochastic forward passes through the model were retained in training mode, and their statistical mean and variance were computed. During training with dropout, the neurons are stochastically "dropped out" by sampling from a Bernoulli distribution. With modern GPU hardware, the forward passes can be done concurrently in batches, resulting in constant running time. Since gradients for the predicted uncertainty with dropout are difficult to obtain, we use particle swarm optimization [17] (PSO) to maximize the acquisition function. We limit the number of budget queries (i.e., 10,000) and select the top-k (10) candidates for batched Bayesian optimization. In practice, we found PSO is a simple yet effective method to explore the design space more effectively since swarms of particles converge to different local maxima, and the selected candidates within a batch still maintain differences.

12.7   Conclusion and Future Directions

As we introduced throughout the chapter, there has been a considerable effort on automating analog circuit sizing. How sizing problem is approached by time evolves in parallel to the trends in engineering practices. It is getting more data driven, and improvements in the learning community have a big impact as EDA being one of its major application areas.

Recent works in analog sizing show the increasing interest toward utilizing deep neural networks via supervised or reinforcement learning algorithms. With more capacity, such methods could handle more design variables which is the case with larger circuits. Many learning-based methods also allow transfer learning to improve solution quality across different tasks. Learning transfer is particularly important for sizing since many times designs are carried over to different technology process. One complication EDA researchers have to cope is how such algorithms are adapted for sizing problem. For supervised-learning methods, neural network engineering and data handling is crucial. For RL methods, how sizing problem is transformed into MDP space and what state representation is used to describe the design will be a major determinant of success of the algorithm.

One way to improve sizing algorithms is to integrate it with other phases of the analog design cycle. Although the majority of the sizing methods fixes the topology inside the problem prerequisites, several recent efforts proposed ML-based solutions for topology generation. In [6], authors have proposed an upper-confidence-bound-tree-based RL algorithm to solve topology generation problem for power converters. The topology is interpreted as a graph, and a state representation is created by using a component set, a port set, and an adjacency matrix specifying the connections between the ports. The action space consists of two phases: first is to determine the device type for a selected component and second is to make decisions on the port connections. In each state, optimal actions are found based on the parameters of the Monte Carlo tree search algorithm and the reward engineering. Another topology generation work where the circuit is represented as a graph is introduced in [44].

They used geometric properties and the relative placement of nodes in the edge embedding due to the nature of the distributed circuits. A graph neural network is trained to model electromagnetic properties of distributed circuits, and an inverse optimization method is proposed to generate the topology matching with the desired specifications.

Finally, we realize that industry adaptations of sizing tools have not been widely accepted yet despite the large academic literature. This is mainly caused by the fact that level of abstraction made by the academia falls short from capturing the industrial design closure. In general, industrial circuits are larger and have more design variables that need to be tuned, and simulation costs are also larger. Those require sizing automation tools to have more modeling capacity and to be more efficient. Further, analog IC design process is an iterative process with various stages. Designers have to consider PVT corners, layout effects, and variation/yield in order to finalize the sizing. In fact, most of the time, these phases have considerable effects on one another, and they have to be considered jointly. There are already steps taken toward extending the awareness of sizing with other elements such as PVT and layout effects. We believe, as sizing community introduce more robust and more comprehensive tools, we will witness more shift in the way analog sizing is done. This chapter only considers the front-end analog circuit sizing techniques. For the back-end design automation, we refer the interested readers to "Machine Learning for Analog Layout" chapter for more details.

References

1. Alpaydin, G., Balkir, S., Dundar, G.: An evolutionary approach to automatic synthesis of high-performance analog integrated circuits. IEEE Trans. Evol. Comput. **7**(3), 240–252 (2003)
2. Budak, A., Gandara, M., Shi, W., Pan, D., Sun, N., Liu, B.: An efficient analog circuit sizing method based on machine learning assisted global optimization. IEEE Trans. Comput.-Aided Design Integr. Circuits Syst. pp. 1–1 (2021). https://doi.org/10.1109/TCAD.2021.3081405
3. Budak, A.F., Bhansali, P., Liu, B., Sun, N., Pan, D.Z., Kashyap, C.V.: DNN-Opt an RL inspired optimization for analog circuit sizing using deep neural networks. In: Proceedings of the 58th ACM/EDAC/IEEE Design Automation Conference, DAC '21 (2021)
4. Chang, E., Han, J., Bae, W., Wang, Z., Narevsky, N., NikoliC, B., Alon, E.: BAG2: a process-portable framework for generator-based AMS circuit design. In: IEEE Custom Integrated Circuits Conference (CICC), pp. 1–8 (2018)
5. Desautels, T., Krause, A., Burdick, J.W.: Parallelizing exploration-exploitation tradeoffs in gaussian process bandit optimization. J. Mach. Learn. Res. **15**(1), 3873–3923 (2014)
6. Fan, S., Cao, N., Zhang, S., Li, J., Guo, X., Zhang, X.: From specification to topology: automatic power converter design via reinforcement learning. In: 2021 IEEE/ACM International Conference On Computer Aided Design (ICCAD), pp. 1–9 (2021)
7. Frazier, P., Powell, W., Dayanik, S.: The knowledge-gradient policy for correlated normal beliefs. INFORMS J. Comput. **21**(4), 599–613 (2009)
8. Gardner, J.R., Kusner, M.J., Xu, Z., Weinberger, K.Q., Cunningham, J.P.: Bayesian optimization with inequality constraints. In: International Conference on Machine Learning (ICML), pp. II-937–II-945 (2014)
9. Gelbart, M.A., Snoek, J., Adams, R.P.: Bayesian optimization with unknown constraints (2014). arXiv preprint arXiv:1403.5607

10. Gielen, G., Walscharts, H., Sansen, W.: Analog circuit design optimization based on symbolic simulation and simulated annealing. IEEE J. Solid-State Circuits **25**(3), 707–713 (1990). https://doi.org/10.1109/4.102664
11. Hakhamaneshi, K., Werblun, N., Abbeel, P., Stojanović, V.: Bagnet: berkeley analog generator with layout optimizer boosted with deep neural networks. In: IEEE/ACM International Conference on Computer-Aided Design (ICCAD) (2019). https://doi.org/10.1109/ICCAD45719.2019.8942062
12. Hennig, P., Schuler, C.J.: Entropy search for information-efficient global optimization. J. Mach. Learn. Res. **13**(1), 1809–1837 (2012)
13. Hernández-Lobato, J.M., Hoffman, M.W., Ghahramani, Z.: Predictive entropy search for efficient global optimization of black-box functions. In: Advances in Neural Information Processing Systems, pp. 918–926 (2014)
14. Hershenson, M., Boyd, S., Lee, T.: Optimal design of a CMOS op-amp via geometric programming. IEEE Trans. Comput.-Aided Design Integr. Circuits Syst. **20**(1), 1–21 (2001). https://doi.org/10.1109/43.905671
15. Hoffman, M.D., Brochu, E., de Freitas, N.: Portfolio allocation for Bayesian optimization. In: UAI, pp. 327–336. Citeseer (2011)
16. Hu, H., Li, P., Huang, J.Z.: Parallelizable Bayesian optimization for analog and mixed-signal rare failure detection with high coverage. In: Proceedings of the International Conference on Computer-Aided Design, pp. 1–8 (2018)
17. Kennedy, J., Eberhart, R.: Particle swarm optimization. In: Proceedings of ICNN'95— International Conference on Neural Networks, vol. 4, pp. 1942–1948 (1995). https://doi.org/10.1109/ICNN.1995.488968
18. Koza, J., Bennett, F., Andre, D., Keane, M., Dunlap, F.: Automated synthesis of analog electrical circuits by means of genetic programming. IEEE Trans. Evol. Comput. **1**(2), 109–128 (1997). https://doi.org/10.1109/4235.687879
19. Kushner, H.J.: A new method of locating the maximum point of an arbitrary multipeak curve in the presence of noise (1964)
20. Liu, B., Wang, Y., Yu, Z., Liu, L., Li, M., Wang, Z., Lu, J., Fernández, F.V.: Analog circuit optimization system based on hybrid evolutionary algorithms. Integr. VLSI J. **42**(2), 137–148 (2009). https://doi.org/10.1016/j.vlsi.2008.04.003
21. Liu, B., Zhao, D., Reynaert, P., Gielen, G.G.E.: Gaspad: a general and efficient mm-wave integrated circuit synthesis method based on surrogate model assisted evolutionary algorithm. IEEE Trans. Comput.-Aided Design Integr. Circuits Syst. (2014). https://doi.org/10.1109/TCAD.2013.2284109
22. Liu, M., Turner, W.J., Kokai, G.F., Khailany, B., Pan, D.Z., Ren, H.: Parasitic-aware analog circuit sizing with graph neural networks and Bayesian optimization. In: 2021 Design, Automation Test in Europe Conference Exhibition (DATE), pp. 1372–1377 (2021). https://doi.org/10.23919/DATE51398.2021.9474253
23. Lyu, W., Xue, P., Yang, F., Yan, C., Hong, Z., Zeng, X., Zhou, D.: An efficient Bayesian optimization approach for automated optimization of analog circuits. IEEE Trans. Circuits Syst. I: Regul. Pap. **65**(6), 1954–1967 (2017)
24. Lyu, W., Yang, F., Yan, C., Zhou, D., Zeng, X.: Batch Bayesian optimization via multi-objective acquisition ensemble for automated analog circuit design. In: International Conference on Machine Learning, pp. 3312–3320 (2018)
25. Lyu, W., Yang, F., Yan, C., Zhou, D., Zeng, X.: Multi-objective Bayesian optimization for analog/RF circuit synthesis. In: Proceedings of the 55th Annual Design Automation Conference, pp. 1–6. ACM, New York (2018)
26. Mockus, J., Tiesis, V., Zilinskas, A.: The application of Bayesian methods for seeking the extremum. Towards Global Optim. **2**(117–129), 2 (1978)
27. Nye, W., Riley, D., Sangiovanni-Vincentelli, A., Tits, A.: Delight.spice: an optimization-based system for the design of integrated circuits. IEEE Trans. Comput.-Aided Design Integr. Circuits Syst. **7**(4), 501–519 (1988). https://doi.org/10.1109/43.3185
28. Rasmussen, C.E., Williams, C.K.I.: Gaussian Processes for Machine Learning (Adaptive Computation and Machine Learning. The MIT Press, Cambridge (2005)

29. Razavi, B.: Design of Analog CMOS Integrated Circuits, 1 edn. McGraw-Hill, New York (2001)
30. Ren, H., Kokai, G.F., Turner, W.J., Ku, T.S.: ParaGraph: layout parasitics and device parameter prediction using graph neural networks. In: ACM/IEEE Design Automation Conference (DAC) (2020)
31. Rutenbar, R.: Analog design automation: Where are we? Where are we going? In: IEEE Custom Integrated Circuits Conference (CICC) (1993). https://doi.org/10.1109/CICC.1993.590704
32. Schonlau, M., Welch, W.J., Jones, D.R.: Global versus local search in constrained optimization of computer models. Lecture Notes-Monograph Series, pp. 11–25 (1998)
33. Scott, W., Frazier, P., Powell, W.: The correlated knowledge gradient for simulation optimization of continuous parameters using gaussian process regression. SIAM J. Optim. **21**(3), 996–1026 (2011)
34. Settaluri, K., Haj-Ali, A., Huang, Q., Hakhamaneshi, K., Nikolić, B.: Autockt: deep reinforcement learning of analog circuit designs. In: IEEE/ACM Proceedings Design, Automation and Test in Europe (DATE) (2020)
35. Shahriari, B., Swersky, K., Wang, Z., Adams, R.P., De Freitas, N.: Taking the human out of the loop: a review of Bayesian optimization. Proc. IEEE **104**(1), 148–175 (2015)
36. Shahriari, B., Wang, Z., Hoffman, M.W., Bouchard-Côté, A., de Freitas, N.: An entropy search portfolio for Bayesian optimization (2014). arXiv preprint arXiv:1406.4625
37. Shook, B., Bhansali, P., Kashyap, C., Amin, C., Joshi, S.: MLParest: machine leaning based parasitic estimation for custom circuit design. In: ACM/IEEE Design Automation Conference (DAC) (2020)
38. Srinivas, N., Krause, A., Kakade, S.M., Seeger, M.: Gaussian process optimization in the bandit setting: no regret and experimental design (2009). arXiv preprint arXiv:0912.3995
39. Srinivas, N., Krause, A., Kakade, S.M., Seeger, M.W.: Information-theoretic regret bounds for gaussian process optimization in the bandit setting. IEEE Trans. Inform. Theory **58**(5), 3250–3265 (2012)
40. Thompson, W.R.: On the likelihood that one unknown probability exceeds another in view of the evidence of two samples. Biometrika **25**(3/4), 285–294 (1933)
41. Vural, R.A., Yildirim, T.: Swarm intelligence based sizing methodology for CMOS operational amplifier. In: 2011 IEEE 12th International Symposium on Computational Intelligence and Informatics (CINTI), pp. 525–528 (2011)
42. Wang, H., Wang, K., Yang, J., Shen, L., Sun, N., Lee, H., Han, S.: GCN-RL circuit designer: transferable transistor sizing with graph neural networks and reinforcement learning. In: ACM/IEEE Design Automation Conference (DAC) (2020)
43. Wang, Z., Jegelka, S.: Max-value entropy search for efficient Bayesian optimization (2017). arXiv preprint arXiv:1703.01968
44. Zhang, G., He, H., Katabi, D.: Circuit-GNN: graph neural networks for distributed circuit design. In: Chaudhuri, K., Salakhutdinov, R. (eds.) Proceedings of the 36th International Conference on Machine Learning. Proceedings of Machine Learning Research, vol. 97, pp. 7364–7373. PMLR (2019)
45. Zhang, S., Lyu, W., Yang, F., Yan, C., Zhou, D., Zeng, X.: Bayesian optimization approach for analog circuit synthesis using neural network. In: 2019 Design, Automation & Test in Europe Conference & Exhibition (DATE), pp. 1463–1468. IEEE, Piscataway (2019)
46. Zhang, S., Lyu, W., Yang, F., Yan, C., Zhou, D., Zeng, X., Hu, X.: An efficient multi-fidelity bayesian optimization approach for analog circuit synthesis. In: Proceedings of the 56th Annual Design Automation Conference 2019, p. 64. ACM, New York (2019)
47. Zhang, S., Yang, F., Zhou, D., Zeng, X.: Bayesian methods for the yield optimization of analog and sram circuits. In: 2020 25th Asia and South Pacific Design Automation Conference (ASP-DAC), pp. 440–445. IEEE, Piscataway (2020)
48. Zhang, S., Yang, F., Zhou, D., Zeng, X.: An efficient asynchronous batch Bayesian optimization approach for analog circuit synthesis. In: 2020 57th ACM/IEEE Design Automation Conference (DAC), pp. 1–6. IEEE, Piscataway (2020)

Part III
Machine Learning Applications in Various Design Domains

Chapter 13
The Interplay of Online and Offline Machine Learning for Design Flow Tuning

Matthew M. Ziegler, Jihye Kwon, Hung-Yi Liu, and Luca P. Carloni

13.1 Introduction

The complexity and schedules of modern industrial chip design coupled with the market pressure for optimal PPA (performance, power, area) often present conflicting goals. Increasing chip complexity and the pressure for faster time to market calls for raising the level of abstraction and streamlining productivity, which may lead to sacrificing PPA. In contrast, improving PPA calls for investing more time and effort to explore design points and understand nuances.

Fortunately, within the scope of digital VLSI design, computer-aided design (CAD) tools for logic synthesis, placement, and routing have evolved to a level of sophistication where balancing these competing objectives may be feasible. But, while the automation capabilities of the tools are astounding, the usage complexity of the tools makes achieving near-optimal PPA far from automatic.

VLSI CAD tools capable of delivering industrial quality chip designs typically present the user, i.e., the designer, with upward of a thousand tool parameters, settings, and options. Consequently, a primary task of the designer is to determine a suitable scenario of tool settings. But, given the sheer number of the possible scenarios (parameter configurations) and the expensive compute cost to evaluate a

M. M. Ziegler (✉)
IBM T. J. Watson Research Center, Yorktown Heights, NY, USA
e-mail: zieglerm@us.ibm.com

J. Kwon · L. P. Carloni
Department of Computer Science, Columbia University, New York, NY, USA
e-mail: jihyekwon@cs.columbia.edu; luca@cs.columbia.edu

H.-Y. Liu
Cadence Design Systems, San Jose, CA, USA
e-mail: hungyil@cadence.com

© The Author(s), under exclusive license to Springer Nature Switzerland AG 2022
H. Ren, J. Hu (eds.), *Machine Learning Applications in Electronic Design Automation*, https://doi.org/10.1007/978-3-031-13074-8_13

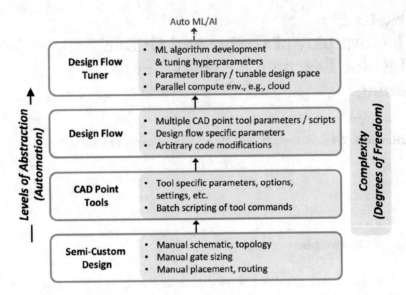

Fig. 13.1 Levels of VLSI design automation and abstractions

scenario (often a day or more of runtime), this task can be challenging even for highly skilled designers and daunting for novice designers.

Fortunately (again), several research efforts are now focusing on tuning CAD tool parameters and, more broadly, tuning design flows. Figure 13.1 illustrates levels of automation and abstraction in modern digital design. Higher levels of abstraction automate many of the lower-level complexities, but often introduce new complexities.

The "design flow tuner" abstraction aims to automate the tuning of the design flow. We distinguish between CAD point tool parameter tuning and design flow tuning in that the latter involves the tuning of multiple CAD point tools. Furthermore, design flow tuning may also include design flow-specific parameters and other designer modifications to the flow scripts that can be automated. Design flow tuning offers the potential for improving productivity and PPA, but new challenges arise.

Most recently proposed approaches for CAD tool parameter and design flow tuning employ some form of machine learning to navigate the vast design space. Tasks involved in the implementation of a tuner include selecting an appropriate machine learning algorithm as well as tuning the algorithm hyperparameters. Another common task is defining a parameter library or tunable design space. In addition, it is often necessary to parallelize the scenario evaluation, which may require a cloud or grid computing environment.

While there are some commonalities that apply to most design flow tuning implementations, in this chapter we focus on two distinct approaches, illustrated in Table 13.1, which we refer to as "online" and "offline" machine learning, respectively. The key characteristic of an online approach is learning without a

Table 13.1 An "online" and "offline" machine learning comparison

Approach	Description, Examples, Challenges
Online	*Learning without a prior dataset, as data becomes available, typically iterative*
	Examples: Bayesian Optimization, Active Learning, Reinforcement Learning
	Keys Challenges: compute cost of parallel trials, iterative runtime latency
Offline	*Batch Learning, i.e., training from an existing dataset, typically followed by inference*
	Examples: Neural Networks, Recommender Systems, Transfer Learning
	Keys Challenges: dataset curation, training runtime

previously available dataset. Data is presented to the online system as it becomes available, with the system learning incrementally. In a design flow tuner application, the data is typically a sample point from running the design flow, or a subset of design flow steps, for a specific design and specific scenario of design flow settings. In contrast, an offline approach batch learns from an existing dataset, where the dataset is typically composed of many design flow samples from various designs and scenarios.

Considering Table 13.1, we present the main thesis of the chapter: design flow tuner implementations do not need to employ only one of these approaches; instead, we postulate that the most effective strategy might be employing both online and offline machine learning simultaneously. To illustrate this holistic approach, we describe how our SynTunSys (STS) online learning system and our offline recommender system work together to achieve better PPA than either system alone.

The following section provides a background in design flow tuning for online and offline machine learning approaches. Section 13.3 focuses on online design flow tuning and reviews prior work across the application domains of high-level synthesis (HLS), field-programmable gate array (FPGA) synthesis and place-and-route, and VLSI logic synthesis and physical design (LSPD). The online industrial case study VLSI design flow tuning system (STS) is also detailed. An analogous treatment of offline design flow tuning is given in Sect. 13.4. A central theme of this chapter is that the design flow tuner abstraction can be further enhanced by considering the interplay of online and offline systems. Section 13.5 reviews prior work on systems with both online and offline components and details the case study system. We then present experimental results from the case study system supporting the benefits of these forms of interplay (Sect. 13.6). Finally, Sect. 13.7 presents potential research directions that explore extending the design tuner abstraction to more adaptable, collaborative, multi-macro, and generalized systems.

13.2 Background

The objective of design flow tuning is to optimize a selected set of design metrics for a specific design by searching the design flow parameter space. While the high-level

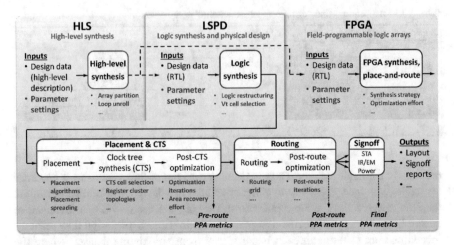

Fig. 13.2 Advanced CAD flows for HLS (high-level synthesis), LSPD (logic synthesis and physical design), and FPGAs (field-programmable gate arrays)

goal of design flow tuning may be straightforward, efficiently implementing a design flow tuning system can be challenging. Large parameter spaces and expensive computing requirements can make manual tuning of design flows nearly intractable, even for experienced human designers. Fortunately, machine learning approaches can help to navigate the parameter space, and cloud computing environments can provide the necessary computing resources.

Machine-learning-based approaches have been proposed for a variety of CAD tool parameter and design flow tuning problems illustrated in Fig. 13.2. The execution of each CAD tool (represented as a white rounded square) is guided or controlled by the parameter settings (in the purple text below white squares). The key prerequisite is that the design flow must be automated to allow numerous trials and evaluations to be run by a higher-level program. The transition from human-in-the-loop design flows, such as semi-custom design, to automated CAD-oriented design flows results in an executable application that can be tuned by machine learning in a distributed computing environment. The CAD tools and design flows in Fig. 13.2 can all be automated in this way. The figure highlights three areas of active design flow tuning research: high-level synthesis (HLS), logic synthesis and physical design (LSPD), and FPGA synthesis and physical implementation (FPGA). To the first order, the tuning objectives for these design flows are similar. However, differences in complexity of parameter space and computing requirements among the applications imply different tuning solutions may be needed.

LSPD design flows typically have the highest complexity among the design flows in Fig. 13.2 and thus present a more complicated tuning problem. As such, LSPD design flow tuning will be the primary focus of this chapter, although we also provide a peripheral discussion on both HLS and FPGA design flow tuning literature. LSPD flows typically consist of a "pipeline" of flow steps, with each flow step calling a CAD point tool. Flow scripts, i.e., code that controls the execution and interaction between flow steps, add another layer of complexity to LSPD flows, as well as more potential tuning options. Thus, LSPD flow tuning is a superset of tuning the individual CAD tools at each flow step as well as the flow scripts. Furthermore, while CAD point tools typically have a predefined (finite) set of parameters, the flow scripts may be modified in an immeasurable number of ways. For example, a human designer can modify, add, or delete nearly any code snippet in the flow scripts when attempting to improve the design. The resulting challenges for LSPD design flow tuning are based on the large parameter space and expensive computing requirements, i.e., long runtimes and high memory usage. To tackle the LSPD design flow tuning challenges, we discuss two broad categories of approaches for design flow tuning: "online" and "offline." We believe the differentiation of online and offline approaches is fundamental, affecting both the overall machine learning algorithm and how and when compute effort is allocated. Table 13.1 describes the machine learning algorithms and general properties of these two approaches. Offline approaches spend computational effort in an upfront training phase with the goal of reducing design-specific computations. While the training phase may not provide design-specific improvements in the near term, the long-term compute effort over multiple designs may be reduced. In contrast, online approaches allocate compute effort to a specific design for near-term results. However, tuning subsequent designs may incur the same compute overhead, unless the learning can be passed to future designs. While we describe the distinct characteristics of online and offline machine learning, design flow tuning solutions can employ hybrid approaches that blend these approaches.

To delineate online and offline approaches more precisely, we define **online design flow tuning** as *Given a new design, e.g., design data for a single macro, and no prior trained model (i.e., model trained from a dataset or an reinforcement learning policy), the approach can achieve "adequate" results within the required latency (iteration count) and parallel compute availability. The compute costs of the solution must allow tuning multiple designs within an overall chip project or product schedule.*

We also use the following three criteria as requirements to classify an approach as an online design flow tuning solution:

- The capability of tuning from the scope of single design instance, i.e., without using data from other design instances.
- No preexisting dataset OR trained model.
- Tuning latency and computing requirements for the solution to complete within feasible limits.

In contrast, solutions that do not meet the online definition and criteria are generally considered offline. Furthermore, we define a **hybrid online/offline design flow**

tuning system as *a system with an online subsystem and an offline subsystem where the two subsystems interplay, but also have the ability to function independently.*

13.2.1 Online Design Flow Tuning Implications

Design flow tuning for LSPD presents particular challenges for online machine learning. LSPD runtimes are considerably longer than HLS, logic synthesis-only, or FPGA parameter tuning runtimes. Furthermore, the additional design flow steps provide a larger set of tunable parameters, i.e., a larger design space. But, LSPD tuning provides higher accuracy and more optimization opportunities via the additional parameters. The following are key LSPD implications, but may also be relevant to other applications that have similar computing requirements.

Implication 1: Long runtimes $\longrightarrow$ parallel trials.

Running a single scenario of an industrial chip design flow to build a design of even moderate size and difficulty may take a day or more of runtime. The high compute cost rules out strictly sequential optimization approaches and requires parallel computing.

Implication 2: Large parameter space $\longrightarrow$ iterative refinement.

The large number of parameters, often in the 1000s, requires an iterative approach, because it is clearly unfeasible to enumerate and run all parameter combinations in parallel. Secondly, some form of design space reduction will most likely be needed, e.g., thoughtfully selecting a subset of parameters for tuning.

Implication 3: Overall tuning latency constraints $\longrightarrow$ few iterations.

Industrial design schedules often have frequent and periodic input changes to LSPD, e.g., weekly logic update cycles, requiring a tuner to reach a near-optimal solution in a low number of iterations.

Implication 4: Distributed compute $\longrightarrow$ tolerance of trial failures.

Processing many parallel trials requires a distributed computing environment, e.g., a cloud or grid computing cluster. In particular, applications having high memory requirements, as it is typically the case for LSPD jobs, will need distributed computing resources. Within a distributed computing environment, the tuning algorithm should not assume precise control over machine resources or job scheduling. In practice, distributed computing environments are subject to hardware failures, varying machine loads that may slow execution, inherently slower hardware on certain nodes, and scheduling traffic that may delay the execution of a job. The result is that trials may see varying runtimes and in some case may not complete, e.g., in the case of a hardware failure.

The implications described above will most likely lead to an online design flow tuning compute framework as illustrated in Fig. 13.3. Furthermore, this fault-tolerant, parallel, and iterative compute framework is generally applicable for tuning any application with long runtimes, high memory footprint, and a large parameter space.

Fig. 13.3 Generalized online design flow tuning compute framework

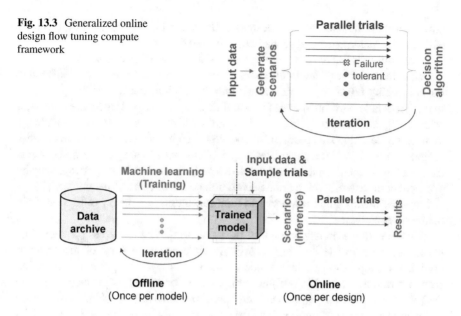

Fig. 13.4 Generalized offline-learning-based design flow tuning framework

13.2.2 Offline Design Flow Tuning Implications

Offline approaches for design flow tuning aim at learning from prior data for different designs to expedite and enhance the tuning task of new designs. Since different designs exhibit diverse characteristics, a major challenge for offline machine learning lies in extracting useful information from a large and sparse data archive that contains results from different designs. The following are key implications for offline-learning-based tuning approaches.

Implication 1: Numerous designs $\longrightarrow$ tuning results collection.

Advanced computer systems consist of numerous heterogeneous components. A complex chip is partitioned into many blocks of distinct functions that are synthesized separately. Furthermore, in industrial settings, the logic for the same block of design may be periodically updated throughout the research and development cycles. Each of those design instances is a target for the design flow tuning, usually with the same design flow. Hence, the tuning results for such prior designs can be exploited by machine learning approaches, for the design flow tuning of a new design. The collection of those results is shown as *Data archive* in Fig. 13.4.

Implication 2: Diverse designs $\longrightarrow$ hidden information extraction.

Whereas online design flow tuning targets a single design at a time, offline approaches usually train a model that can be exploited for many new designs. As shown in Fig. 13.4, the training is performed offline using prior tuning results, and the *Trained model* is used for online inference. Due to the diversity of prior designs as well as of new designs, the machine learning system needs to extract hidden information to characterize those designs so that it can effectively learn the different effects of the parameters and scenarios on diverse designs.

Implication 3: Large parameter space $\longrightarrow$ dimensionality reduction.

When the number of the tuning parameters is in the 1000s, it may be infeasible for any tuning effort in practice to explore a sufficient portion of the parameter space, especially with fast-paced industrial design schedules. Hence, the dataset of prior tuning results often turns out to be sparse, which poses a significant challenge to machine learning. This problem can be alleviated by reducing the dimensionality of the feature space. For instance, the principal component analysis method projects data in a high-dimensional space down to a lower-dimensional subspace spanned by principal components, which are linear combinations of original parameters [1].

Implication 4: Online resource and time constraints $\longrightarrow$ heavy offline and light online computations.

To cope with the limited amount of time and computing resources for the online phase, offline machine learning algorithms often consume a lot of resources to optimize a single model. Such models can be employed to generate promising scenarios online, after lightweight sample trials or model adjustment for new designs. This way, an offline-learning-based framework (e.g., Fig. 13.4) spends much less time and resources online than an online-only framework (e.g., Fig. 13.3). Since a trained model is used for many new designs, the offline training time can be amortized across the designs so that it becomes small or negligible for each design.

The aforementioned offline-learning-based system is also generally applicable for tuning the parameters of other applications with numerous cases, diverse examples, a large parameter space, and online resource constraints.

13.2.3 LSPD Application-Specific Considerations

While the implications above may generalize for tuning other applications with similar characteristics, the following considerations are specific to LSPD flows and CAD tools. Considering these application-specific characteristics may help in developing a more efficient tuning solution:

Consideration 1: LSPD CAD tools often have many *categorical parameters*, i.e., parameters that can take on values having no underlying innate ordering. Categorical parameters may present challenges to some optimization approaches, e.g., Bayesian optimization. Although some techniques have been proposed to handle categorical parameters [2], the choice of the parameter types may guide the selection of optimization approaches.

Consideration 2: LSPD CAD tool *default parameter settings are not random.* Tool developers often spend considerable effort tuning the parameters to choose default settings that are expected to perform well on average. While the LSPD default settings are almost certainly nonoptimal for a specific design, the presence of well-intended default parameters may be useful information.

Consideration 3: LSPD CAD tool *parameters may affect some metrics more than others.* With multiple objectives, such as timing, power, routing congestion, etc., some parameters are directly focused on specific objectives. Although modifying such a directed parameter will most likely affect all other metric values in some way, knowing that some parameters are intended to affect specific metrics can be useful.

13.2.4 Design Flow Tuner Development: CAD Tool Vendor, Open-Source, or In-House

The complexity of chip design often leads to design flows and CAD tool abstractions. One consideration for design flow tuning is the ownership at these levels of abstractions. Typically CAD tools are provided by commercial CAD tool vendors or in-house CAD tool organizations. On the other hand, the design flow is often owned and customized by the design team, or in-house methodology team, focusing on a specific chip.

Recently, CAD tool vendors have announced entries into the design flow tuning space. Synopsys has introduced DSO.ai [3] and Cadence has released Cerebrus [4], both of which employ reinforcement learning techniques. In contrast, Jung et al. describe an open-source design flow tuner [5]. Another option is for the in-house design methodology team to develop the design flow tuner, which is the case for the case study tuner in this chapter.

At this time, it is unclear which development approach will be more common for design flow tuning in the long term. Our speculation is that a successful design flow tuner will need to be flexible to support various design flows and adapt to the changing needs of the design team running the tuner. As the design team or methodology owners enhance and modify the design flow, ideally the design flow tuner will also adapt accordingly.

13.2.5 Case Study Background: STS Terminology and Design Space

In order to convey practical implementation decisions for design flow tuning, we use our industrial tuning system called SynTunSys as a case study throughout this chapter. SynTunSys [6, 7] (synthesis tuning system), often abbreviated as STS, utilizes both online and offline machine learning approaches in a complimentary manner. An early version of STS was first used in production for optimizing the IBM POWER7+ processor [8, 9]. Over the years, STS has been continually enhanced in numerous ways, e.g., with the addition of a Bayesian-inspired decision engine [10] and a recommender system [11].

Fig. 13.5 STS terminology

- *Parameter:* a fine-grained option for a CAD point tool (e.g., logic synthesis, place-and-route) or a design flow

- *Primitive:* a Boolean valued option grouping one or more native synthesis parameters settings

- *Primitive Library:* a library of primitives that is formed using prior expert knowledge to reduce the design space

- *Scenario:* a combination of one or more primitives

- *Trial:* an evaluation of a scenario for a particular design

- *QoR (quality of results):* one or more metric values reflecting a design's quality, analogous to PPA

- *Cost:* a numerical evaluation of the QoR of a scenario, after a trial

- *Cost Function:* an algorithm for generating a cost value

Table 13.2 An example of primitive names and primitive descriptions

Primitive name	Primitive description
restruct_a	Logic restructuring to reduce area
restruct_t	Logic restructuring to improve timing
rvt_lvt10	Native RVT, allow 10% LVT
rvt_lvt50	Native RVT, allow 50% LVT
vtr_he	High-effort VT recovery
area_he	High-effort area reduction
wireopt_t	Wire optimization for timing
wireopt_c	Wire optimization for congestion

An early version of STS was an online design flow tuning system, given that a training dataset did not exist at that time. In fact, many of the synthesized macros (i.e., independently synthesized blocks of a large chip) on the POWER7+ chip were semi-custom designs in the predecessor chip (POWER7). Often, a downside of the online approach is the intensive computation it requires compared to an offline approach that employs a pre-trained model. Thus, STS saves meta-data for all LSPD runs as a future dataset.

The STS design space, i.e., the space of tunable parameters, is a reduced version of the inherent ~1000 multi-valued CAD tool parameter space.[1] Figure 13.5 summarizes STS key terms.

Primitives: Rather than directly processing ~1000 multi-valued parameters, STS takes as input a smaller set of Boolean parameters called *primitives*. A primitive is a higher-level construct containing one or more CAD tool parameters set to specific values. The *Primitive Library* is the set of all available primitives (currently containing ~500 primitives). Table 13.2 shows an example of a small primitive library.

---

[1] We describe only primitives using CAD tool parameters for brevity, although primitives may consist of design flow parameters outside of the CAD tool and include snippets of code to modify the design flow.

Fig. 13.6 Illustration of the interaction of parameters, primitives, and scenarios. Primitives consist of one or more parameters. Scenarios consist of one or more primitives

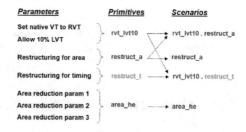

Scenarios: STS creates scenarios consisting of one or more primitives as Fig. 13.6 shows. The construction of quality scenarios is not trivial, as some primitives are complementary and others are noncomplementary. The optimal scenarios often are macro-specific with respect to a set of weighted design objectives, while default CAD tool settings require balancing quality of results (QoR)[2] across multiple macros.

Cost Function: The STS cost function conveys the optimization goals. It converts multiple design metrics into a single cost value, allowing cost ranking of scenarios. Examples of available metrics include multiple timing metrics,[3] power consumption, congestion scores, utilized area, electrical violations, runtime, etc. The selected metrics are assigned weights to signify their relative importance. The overall cost function is then a "normalized weighted sum" of the ℓ selected metrics, expressed by Eq. 13.1 where $Norm(Q_i)$ is the normalized i-th metric Q_i across the results of all scenarios in a STS run for a single design:

$$Cost = \sum_{i=1}^{\ell} W_i \cdot Norm(Q_i) \tag{13.1}$$

The STS design space is further reduced for a specific tuning run. A subset of primitives are selected based on the cost function metrics and historical primitive performance. Typically ~50 primitives are selected, although this number can vary based on the user's compute quota, macro size, and desired exploration effort.[4] To reduce trial latency, we often tune over a subset of design flow steps. We typically use metrics just prior to the routing step, although the tuning flow steps are user configurable.

---

[2] We consider QoR and PPA synonymous terms, although we use QoR in the STS context since we often tune numerous metrics, beyond performance, power, and area.

[3] Typical timing metrics of interest are (i) internal slack, e.g., latch-to-latch slack (L2L), (ii) worst negative slack (WNS), and (iii) total negative slack (TNS).

[4] The primitive library and historical primitive performance are two ways the online system improves "offline"; however, a new online system can still be effective without these evolving improvements.

In the next two sections, we provide more detailed considerations of online and offline design flow tuning solutions, including a brief survey of prior work for both approaches and details of our case study system.

13.3 Online Design Flow Tuning Approaches

This section presents a variety of online machine-learning-based approaches for design flow tuning, with an emphasis on the STS Online System case study.

13.3.1 Approaches for HLS Flows

The two main inputs to an HLS flow are the high-level description and a scenario that consists of the parameter settings. Given an input description, online approaches attempt to generate optimal scenarios after running the HLS flow for that description with a number of sample scenarios. Liu et al. perform active learning via transductive experimental design to sample representative and hard-to-predict scenarios [12]. The HLS tool is repeatedly executed with the sampled scenarios whose results are fed to train (random forest) models for predicting the area and latency. Meng et al. also perform active learning to sample unobserved scenarios and train random forests that predict the QoR such as application throughput and area utilization [13]. Instead of sampling the scenarios that are expected to improve the model prediction accuracy, they iteratively eliminate scenarios that are highly likely to be nonoptimal from the pool of all candidate scenarios.

13.3.2 Approaches for FPGA Flows

Xu et al. propose a reinforcement learning approach to allocate resources for tuning the parameters of an FPGA compilation flow (from an RTL description to a configuration bitstream) [14]. The search space, which is defined by the (discretized) FPGA parameters, grows exponentially with respect to the number of parameters; hence, they dynamically partition the search space into subspaces to be explored in parallel on multiple machines. The partitions are made without prior knowledge and based on the information gains, measured after the initial FPGA flow runs for the given input description with a number of sample scenarios. Then, they apply the multiarmed bandit technique, which is a class of reinforcement learning that chooses a sequence of actions to maximize the total payoff obtained by performing those actions. In this case, the actions correspond to allocating computing resources to specific subspaces so that those subspaces are explored by OpenTuner, an open-source autotuning framework [15] that invokes the FPGA flows. The payoff of selecting a subspace is defined as the average QoR of samples in that subspace, plus an additional term that balances exploitation and exploration.

13.3.3   Approaches for LSPD Flows

13.3.3.1   Bayesian Optimization

Ma et al. perform a Bayesian optimization in the LSPD flow tuning to minimize the QoR values [16]. In each iteration, an acquisition function samples a configuration, and the LSPD flow is executed with that configuration to output the QoR values, which are used to update a surrogate model. Given a configuration, the surrogate model based on Gaussian process regression outputs a predicted QoR value (a single QoR value or a weighted sum of multiple QoR values) along with the uncertainty of the prediction. The acquisition function exploits the updated surrogate model to sample a new configuration in the next iteration.

Ziegler et al. employ a Bayesian-inspired algorithm for the STS Online System [6, 7, 10]. Section 13.3.4 provides more details on the STS Online System that is used as a case study throughout this chapter.

13.3.3.2   Reinforcement Learning

Reinforcement learning is another technique that has potential for tuning LSPD design flows. Reinforcement learning is a fairly broad category of machine learning techniques with many variants. Based on the definition of online design flow tuning given in Sect. 13.2, some reinforcement learning approaches may be considered online, while others may be considered offline. For example, Hosny et al. target the logic synthesis step with an approach where the reinforcement learning agent performs actions of adding primitive transformations to scenarios. The logic synthesis tool is repeatedly initialized and executed with the goal of maximizing the reward that increases when the area decreases. This approach is better classified as online because it does not require an existing model or dataset, does not appear to require excessive runtime, and has a scope of a single design. On the other hand, in Sect. 13.4.3.3 we review an offline reinforcement learning approach for physical design tuning.

Furthermore, the previously mentioned vendor CAD tool design flow tuning solutions (DSO.ai [3] and Cerebrus [4]) are also reported to be based on reinforcement learning; however, as the inner workings of these systems are not published as of this writing, we will not attempt to classify them as online or offline.

13.3.4 Case Study: STS Online System

The online component of the case study system, i.e., the STS Online System, is a scalable industrial LSPD flow tuning system presented by Ziegler et al. [6, 7, 10]. The first descriptions and product impact of STS were given in [9] and [17].

The STS Online System has been enhanced over time and is currently a Bayesian-inspired algorithm, fitting the needs described in Sect. 13.2.1. The STS Online System includes the STS Online-Learning Algorithm (SOLAR). SOLAR begins by launching parallel trials for an initial iteration ($i = 0$) consisting of one trial for each given primitive, i.e., a one-hot sensitivity test. This sensitivity test is analogous to a Bayesian prior. For subsequent iterations, SOLAR generates a given number of scenarios (k) that are run as parallel trials (i.e., maximized utilization of computing resource) and dynamically adapts to the k scenarios in subsequent iterations that are more likely to return lower costs (i.e., adaptive exploration).

Figure 13.7a illustrates the main idea of SOLAR. Following the sensitivity test ($i = 0$) on the given primitives, SOLAR *estimates* the cost of unevaluated composite scenarios by taking the average cost of the contributing scenarios as a cost predictor. For instance, to estimate the cost of a scenario that comprises three primitives (**b**, **j**, **d** in Fig. 13.7a), SOLAR calculates the average cost of the three contributing scenarios, each comprising of just one of the three primitives. SOLAR generates candidate scenarios by "looking ahead" using a combination order $O > 1$, which allows combining up to O prior scenarios for cost estimation ($O = 3$ in Fig. 13.7a).

SOLAR then selects the top-k composite scenarios with the lowest estimated costs from the candidate scenarios to form a *potential* set and then submits k parallel LSPD jobs with the selected scenarios. Since the size k of the potential set is constrained, with a combination order greater than 1, SOLAR can filter out non-promising scenarios early in the tuning loop and instead allocate the tuning resource budget to the more promising scenarios. This estimation-selection-submission process repeats for every tuning iteration until an exit criterion is met (e.g., max iteration i is reached).

Furthermore, SOLAR leverages the iterative process to continuously refine its cost-estimation accuracy on *noncomplementary* combinations. Specifically, at any iteration i, whenever a composite scenario, say **restruct_t + area_he**, was predicted good (i.e., low timing and area costs) and run as a trial, but the result turns out to be not good (mediocre weighted cost because of conflicting underlying optimization mechanisms), then the algorithm can *learn* the actual effectiveness of combining scenarios **restruct_t** and **area_he**. Therefore, at any future iterations, it will demote any composite scenario that involves **restruct_t + area_he**. In summary, SOLAR uses cost estimation to avoid non-promising scenarios and refines its estimation after learning actual LSPD results.

Moreover, to better estimate the cost based on *nontrivial* contributing scenarios (i.e., scenarios comprising more than one primitive), SOLAR includes a fine-grained cost estimation (see Fig. 13.7b). For instance, given two scenarios, **s1 = (b + d)** and **s2 = (b + j + f)**, the algorithm not only regards the average cost of **s1** and **s2** as

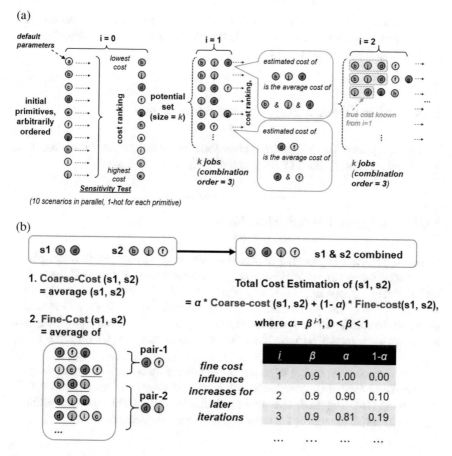

Fig. 13.7 Illustration of the online algorithm (SOLAR). (**a**) Parallel and iterative refinement for SOLAR. (**b**) Cost-estimation process for SOLAR

the *coarse-cost*, but also considers a *fine-cost*. The fine-cost aims to learn the *inter-primitive* interaction (such as **d** + **j**), in contrast to the coarse-cost that targets the inter-scenario interaction (such as **s1** + **s2**). To this end, SOLAR calculates the fine-cost of **s1** and **s2** by using the average cost of their *reference scenarios*, which are the scenarios that have been run in the previous iterations. Each of these includes a pair of primitives, such that one primitive (e.g., **d**) comes from **s1** and the other (e.g., **j** or **f**) from **s2**. See pair-1 and pair-2 in Fig. 13.7b for illustration, where the scenarios listed to the left of pair-1 and pair-2 are the reference scenarios for **s1** and **s2**. Hence, the fine-cost of **s1** and **s2** is the average cost of these reference scenarios.

Overall, SOLAR's cost-estimation function is a dynamic weighted sum of the coarse- and fine-cost with a changing weighting factor α for the coarse-cost and a factor $(1 - \alpha)$ for the fine-cost, where $0 \leq \alpha \leq 1$. The value of α is computed with the formula $\alpha = \beta^{(i-1)}$ where $0 < \beta < 1$ and i is the current iteration number. This formula favors the fine-cost component in the later tuning iterations, where there are more reference scenarios available.

13.4 Offline Design Flow Tuning Approaches

This section presents various offline approaches for the design flow tuning problem, with an elaboration on the STS Recommender System case study.

13.4.1 Approaches for HLS Flows

Offline approaches exploit the design flow tuning results of previous designs to train a model that can be applied online for a new design. Kwon et al. propose a transfer learning approach for tuning parameters of HLS flows. Transfer learning aims to improve the performance of a target task by reusing the knowledge extracted from different but related source tasks [18]. They train an input-dependent neural network model on prior data (source tasks) and transfer only the input-independent part to a new model for new designs (a target task). Wang et al. train predictive models for tuning hyperparameters of the meta-heuristic methods (e.g., the simulated annealing, genetic algorithm, and ant colony optimization) for HLS flow tuning [19]. The predictive models are trained offline and exploited online to tune the hyperparameters of the meta-heuristics for new designs.

13.4.2 Approaches for FPGA Flows

Agnesina et al. propose a model stacking approach for tuning the parameters of an FPGA place-and-route flow for FPGA-based logic emulation [20]. For hard-to-compile netlists, they first train multiple base models (a random forest, extra trees, gradient-boosted trees, and an artificial neural network) that each takes in the extracted features of the input netlist and predicts the winning probability for each FPGA flow scenario. In this problem, a scenario wins if it achieves the shortest FPGA compilation time. Then, they train a stacker (a logistic regressor) that collects the output of the base models and makes a final prediction. The models are trained using a large-scale commercial FPGA compilation database.

13.4.3 Approaches for LSPD Flows

13.4.3.1 Parameter Importance

Xie et al. propose a feature-importance-based approach for physical design flows that exploits offline a large amount of prior data to determine the importance of each parameter [21]. A parameter's importance is defined as the variance in the QoR values obtained by varying the value of that parameter. Then, for a new design, all candidate parameter settings are clustered based on the parameter importance in order to improve the sampling efficiency during the online learning with tree-based prediction models.

13.4.3.2 Transfer Learning

Davis et al. propose a transfer learning approach for QoR prediction in RTL-to-GDS flows [22]. A neural network model is trained offline for predicting the QoR values from the flow parameters using a prior dataset. For new designs, only the last layers of the trained model are re-trained online with a small number of the design flow runs.

13.4.3.3 Reinforcement Learning

Mirhoseini et al. propose a reinforcement learning approach for the chip floorplanning step in the placement stage of physical design [23]. Their approach focuses on generalizing the problem so that a reinforcement learning policy network trained with a large set of designs can effectively guide the placement of new designs. For this, they first train a graph neural network that embeds the netlist information and predicts the reward defined as a linear combination of the approximate wirelength, congestion, and density. This neural network (except the last prediction layer) is used as the encoder of the reinforcement learning policy network. While the trained policy network may also be fine-tuned for new designs, we consider this an offline approach as the policy is trained using multiple designs and requires significant computing resources.

13.4.3.4 Recommender Systems

Kwon et al. propose a recommender system approach for tuning industrial LSPD flows [11]. This system is referred to as the STS Recommender System and is detailed in Sect. 13.4.4. Recommender systems predict the affinity between each *user* and *items*, such as movies, music, or restaurants, to make personalized recommendations [24, 25]. In the STS Recommender System, the LSPD flow *scenarios* correspond to the *items* and the *macros* to the *users*.

13.4.4 Case Study: STS Offline System

The STS Recommender System consists of two modules: (1) the offline learning module and (2) the scenario recommendation module. The offline learning module trains a QoR prediction model using the collaborative filtering approach with the data archive. The scenario recommendation module takes as input the trained model, the target macro (the macro name for a legacy macro or sample LSPD results for a new macro), the QoR cost function or weight vector, and the number of scenario recommendations to generate. The module makes an inference using the given model and finally returns the number of requested scenarios.

Critical challenges for the recommender system approach include (1) the sparsity of the training data with respect to the huge search space and (2) the complex nature of the problem, e.g., macro-specific or designer-specific parameters that the recommender system cannot address. In some cases, the performance can be improved by combining the machine-generated scenarios with a design expert's input scenario. The collaboration between the recommender system and the design experts could lead to a QoR that is not achievable by either of them working solely.

System Model and Problem Statement To discuss this approach from a mathematical perspective, let us use the following notations. $\mathcal{M}$ is a set of d macros: $\mathcal{M} = \{m_1, \cdots, m_d\}$, where m_i is a symbolic representation, e.g., the macro's name or index. Let $\mathcal{P}$ be a set of n binary meta-parameters, also called primitives: $\mathcal{P} = \{p_1, \cdots, p_n\}$. A scenario s is a subset of $\mathcal{P}$, i.e., selected primitives that are set to *True*, while others are set to *False*. Then, function QoR maps a (*macro, scenario*)—pair to normalized QoR scores in ℓ metrics, presented as a real-valued ℓ-dimensional vector:

$$QoR(m, s) = (q_1, \cdots, q_\ell) \in [0, 1]^\ell, \ m \in \mathcal{M}, \ s \subseteq \mathcal{P}. \tag{13.2}$$

The data archive is assumed to contain QoR results from prior LSPD flow runs. For each macro m, let $S(m)$ represent the set of all scenarios that were applied during the LSPD flow tuning for this macro. Then, an archive $\mathcal{A}$ contains the $(m, s, QoR(m, s))$—tuples for all macros $m \in \mathcal{M}$ and scenarios $s \in S(m)$.

The target problem is to find a prediction model F that approximates the QoR function, where F also maps a (*macro, scenario*)—pair to an ℓ-dimensional vector:

$$\underset{F}{\arg\min} \sum_{m \in \mathcal{M}, s \subseteq \mathcal{P}} \|QoR(m, s) - F(m, s)\|^2. \tag{13.3}$$

In this problem, the goal is to minimize the sum of L^2 distances between $QoR(m, s)$ and $F(m, s)$ for all macros m and scenarios s. However, for scenarios $s \notin S(m)$, the golden $QoR(m, s)$ values are unknown. Thus, the aim is to minimize the distances between $QoR(m, s)$ and $F(m, s)$ only for the scenarios recorded in the archive $\mathcal{A}$, which acts as the training data for machine learning.

A critical challenge in solving the above problem is the lack of input features to describe the macros m and scenarios s. A full specification of a macro is a collection of the RTL description, scripts, constraints, and libraries that are neither easily available nor quantifiable.

Architecture of the Prediction Model To address the aforementioned challenge, the STS Recommender System exploits a collaborative filtering approach, which is widely used for recommender systems [24]. For instance, a movie recommender system recommends a new movie to a user based on this user's rating for other movies and all other users' ratings. Let matrix A represent the movie ratings by all users, where A_{ij} represents user i's rating on movie j. Then, without further information, the system can learn latent features of each user and each movie, by factorizing matrix A into a user matrix B and the transpose of a movie matrix C, i.e., $A = B \cdot C^{Tr}$. This factorization can be approximately done even when some elements of A are missing. After B and C are learned, a missing rating A_{ij} can be predicted as the (i, j)—element of $B \cdot C^{Tr}$ [24].

The architecture of the prediction model in the STS Recommender System is motivated by the above approach for movie recommender systems, but it differs in addressing the following additional challenges:

C1. Unlike the movies, the observed scenarios are very sparse. For example, a macro tuned with SOLAR in Sect. 13.3 may only observe 150–250 scenarios, out of about 2^{250} possible scenarios. Moreover, the best scenarios for one macro are rarely observed while tuning the LSPD flows for other macros.

C2. While a movie rating prediction model outputs a single value for each $(user, movie)$–pair, the QoR prediction model outputs a vector with ℓ elements for each $(macro, scenario)$–pair.

To cope with **C1**, the prediction model factorizes the QoR information into a macro matrix, a primitive matrix (instead of a scenario matrix), and the part that relates the latent information for $(macro, primitive)$—pairs to a QoR vector. On the other hand, **C2** indicates that ℓ individual models could be needed to predict each of the ℓ QoR metrics. Instead, the STS Recommender System constructs one holistic model that predicts all ℓ metrics. This model can be described with a $(macro, primitive, metric)$–tensor, in analogy to a $(user, movie)$–matrix. A tensor is a generalization of a vector or a matrix, which can also be represented as a multidimensional array. With this approach, the number of variables describing the model is greatly reduced.[5]

---

[5] The term *variables* is often referred to as *parameters* or *weights* in other machine learning applications and recommender systems. Here, we refer to them as *variables* to avoid the confusion with the design flow *parameters*.

The prediction model F describes the relationship between $(macro, scenario)$—pairs and their ℓ-dimensional QoR vectors by (1) tensor decomposition and (2) regression with a neural network. Let T be a tensor whose (i, j, k)—element T_{ijk} represents an intermediate value of the k-th QoR metric for the (macro m_i, primitive p_j)–pair. These intermediate values, which are unknown at first, propagate through a neural network G that predicts the final QoR for a scenario. Thus, the intermediate values, as well as other variables of G, can be adjusted with the backward propagation of errors. Tensor T containing the intermediate values can be decomposed into factor matrices containing latent features.

Specifically, the tensor T has the shape of $|\mathcal{M}| \times (|\mathcal{P}| + 1) \times \ell$, where $|\mathcal{M}|$ is the number of archived macros and $|\mathcal{P}| + 1$ is the number of primitives, plus one special primitive that is always $True$. ℓ is the number of QoR metrics. By CP decomposition [26],[6] T is decomposed into the macro matrix M, primitive matrix P, QoR metric matrix Q, and a super-diagonal tensor of shape $h \times h \times h$, where h indicates the dimension of the latent features. The factor matrices M, P, and Q have dimensions of $|\mathcal{M}| \times h$, $(|\mathcal{P}| + 1) \times h$, and $\ell \times h$, respectively. Then, an (i, j, k)—element of tensor T can be computed as

$$T_{ijk} = \sum_{\alpha=1}^{h} M_{i\alpha} \cdot P_{j\alpha} \cdot Q_{k\alpha}. \tag{13.4}$$

Given an input vector taken from tensor T, a fully connected single-layer neural network G predicts the final QoR vector. The input vector that corresponds to a (macro m_i, scenario s)—pair is defined as follows. For any primitive p_j, the vector $T_{ij:} = (T_{ij1}, \cdots, T_{ij\ell})$ represents intermediate QoR values for the (m_i, p_j)–pair. For notational simplicity, let $s(p)$ be 1 when the primitive p is in the scenario s, and 0 otherwise. The input vector is the concatenation of vectors $s(p_1) \cdot T_{i1:}, \cdots, s(p_n) \cdot T_{in:}$. Then, QoR for an (m_i, s)—pair can be predicted by $G(s(p_1) \cdot T_{i1:}, \cdots, s(p_n) \cdot T_{in:})$.

To summarize, the prediction model F is constructed as follows:

$$F(m_i, s) = G(s(p_1) \cdot T_{i1:}, \cdots, s(p_n) \cdot T_{in:}) \tag{13.5}$$

Since T can be described by the latent feature matrices M, P, and Q, the model F can be written as $F(m_i, s; M, P, Q, G)$. Here, F has two types of input: (1) the original input (macro m, scenario s) to the QoR prediction model and (2) F's variables M, P, Q, and G.

---

[6] CP and Tucker are two widely used methods for tensor decomposition. The acronym CP stands for (1) CANDECOMP (canonical decomposition)/PARAFAC (parallel factor analysis) or for (2) canonical polyadic (decomposition). Tucker is a generalization of CP, where the core tensor is not super-diagonal and contains hidden features [27]. With CP, it is possible to explicitly represent the latent information for each macro separately.

Offline Learning Module The offline learning module trains model F using the data archive. The data is split into the training set $\mathcal{A}$ and the validation set $\mathcal{B}$. With $\mathcal{A}$, we apply the stochastic gradient descent method to learn model F's variables:

$$\underset{F}{\arg\min} \sum_{(m,s,QoR(m,s)) \in \mathcal{A}} ||QoR(m, s) - F(m, s)||^2 + r(F) \qquad (13.6)$$

where $r(F)$ is a regularization term added to prevent or mitigate overfitting. A trained model F is evaluated on $\mathcal{B}$ in terms of the validation error computed on $\mathcal{B}$. The offline learning module returns model F with the smallest validation error.

Scenario Recommendation Module Given model $F = F(m_i, s; M, P, Q, G)$, a cost function or metric weights w, number t of scenarios, and target macro m, the scenario recommendation module returns t best scenarios (in terms of the cost function or weights w) for m according to F.

For a legacy macro $m_i \in \mathcal{M}$, the $QoR(m_i, s)$ for any scenario s can be predicted by computing $F(m_i, s)$. *Making an inference using this model F takes much less time than applying an LSPD flow (e.g., a few minutes vs. a number of hours).*

For a new macro m^*, the scenario recommendation module first collects a number of sample LSPD results for this macro. In the case of a legacy macro $m_i \in \mathcal{M}$, the i-th row of the macro matrix M contains the latent features for the macro. Similarly, let μ denote the (unknown) latent feature vector for m^*. Then, $QoR(m^*, s)$ for a scenario s can be estimated by $F(m_1, s; \mu; P, Q, G)$. In this model, only the values of μ are unknown, which can be learned using the model F and a set of sample LSPD results $\mathcal{C}$:

$$\underset{\mu}{\arg\min} \sum_{(m^*,s,QoR(m^*,s)) \in \mathcal{C}} ||QoR(m^*, s) - F(m_1, s; \mu; P, Q, G)||^2 + r(F).$$

$$(13.7)$$

After μ is learned by the gradient descent method, the model F can be again used to make inferences for the new macro m^*.

13.5 The Interplay of Online and Offline Machine Learning

While the previous two sections describe online and offline design flow tuning systems separately, the interplay of these two systems presents opportunities for improving tuning results and efficiency beyond what each system can achieve independently. Although it is typical for offline systems to employ some form of online refinement, we believe a more holistic interaction of online and offline design flow tuning systems is a promising direction. In this section we review and describe hybrid methods that blend online and offline machine learning approaches.

13.5.1 Approaches for FPGA Flows

Kapre et al. propose a hybrid online/offline approach for FPGA flows such that the online learning can be carried out immediately without prior data or models [28]. For the goal of accelerating timing closure, the online Bayesian learning component learns to predict the total negative slack of the implementation from the settings of the Boolean parameters of the target FPGA flow. The offline component performs principal component analysis to reduce the number of parameters to be tuned in order to improve the online machine learning performance.

13.5.2 Approaches for LSPD Flows

Liang et al. propose a multistage approach for LSPD flows that consists of the online parameter tuning and the offline warm-up [29]. For each stage of an LSPD flow (e.g., logic synthesis, placement, clock tree synthesis, routing, and optimizations), an ant colony optimization engine suggests the settings of the parameters for that individual stage, while cooperating across all stages to evaluate the settings (by running the LSPD flows). This online flow tuning may be enhanced via the offline warm-up that initializes the online engines and reduces the number of parameters to be tuned by grouping the frequently interacting ones and selecting only the important ones determined from the archived data.

Ziegler et al. propose the Hybrid STS System [30]. This system is based on the interactions of the STS Online and Offline Systems, described in Sects. 13.3.4 and 13.4.4, respectively. The following subsection provides more details on the Hybrid STS System.

13.5.3 Case Study: Hybrid STS System

Figure 13.8 shows the Hybrid STS System [30]. The system consists of three main components, (1) an online learning system, (2) a data archive, and (3) a recommender system. The aim of the overall system is to create a holistic approach that can cover all phases of the design cycle and availability of suitable training dataset. For example, if little or no suitable training data is available, STS can emphasize the online learning system. However, as the data archive contains suitable training data, the system can utilize the recommender system. The system also inherently supports hybrid or mixed-mode online and offline approaches. For example, the scenarios originating from the recommender system can be mixed with scenarios from the online component and iteratively refined. Furthermore, the system can take scenarios from additional sources of prior knowledge. For example, in an industrial setting, there are often legacy designs that could provide specific

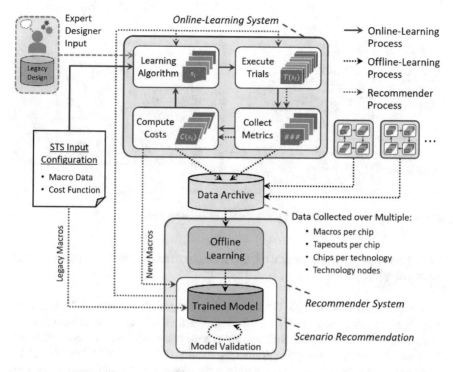

Fig. 13.8 Overall STS system diagram

scenarios that would be beneficial to the iterative refinement process. The legacy design scenarios may not exist in the STS data archive, as they could be generated using ad hoc methods. Expert designers can also directly supply input to the online system.

The STS Online System (SOLAR algorithm) is the vehicle for evaluating, combining, and refining parameter configurations from various sources. Figure 13.9 shows a more explicit illustration of three "generators" of scenarios $\{S\}$ and STS primitives $\{P\}$ for the online system.

Primitive Library The *Primitive Library*, introduced in Sect. 13.2.5, contains primitives to the online system based on the cost function provided by the user. Primitives are selected using prior empirical evidence of strong performance for the given cost function. In the case of a new primitive or little historical performance data, primitives can be selected based on the expected performance for the cost function.[7] The selection of primitives is not macro specific, so while the *Primitive Library* provides a less customized source of tunable parameters, it provides

---

[7] A primitive is typically categorized based on the expected metrics it will affect, which follows **Consideration 3** in Sect. 13.2.1.

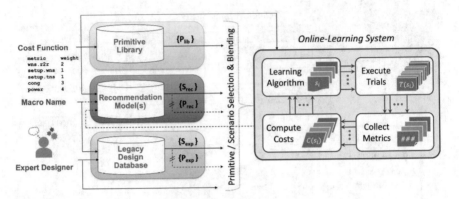

Fig. 13.9 Sources of scenarios and primitives for online refinement

parameters that generally work well, and it can be used in absence of a prior dataset or expert input.

Recommendation Model(s) One or more "recommendation models" trained via the recommender system, as described in Sect. 13.4, may also generate input scenarios to the STS Online System. More than one model may be created based on either varying training hyperparameters, training datasets, or other variables. The recommended scenarios are customized for the specific macro and cost function, as the model takes macro specific information and the cost function as input. In the case of a legacy macro, the recommendation model takes the macro name and cost function as input. For a new macro, one iteration of the SOLAR algorithm is run using primitives from the *Primitive Library* generator, and the results are provided to the model along with the cost function. The recommendation model returns complete scenarios, represented as $\{S_{rec}\}$ in Fig. 13.9; however, the scenarios may be optionally "diced" into the primitives that constitute the scenario. For example, using an illustrative scenario from Fig. 13.6, if the scenario **rvt_lvt10.restruct_a** is returned in $\{S_{rec}\}$ by a model, it could be converted into the primitives **rvt_lvt10** and **restruct_a** as part of $\{P_{rec}\}$. To the first order, using recommended scenarios should speed up the tuning latency, as more complex, yet coarser, tunable parameters are given to the STS Online System, while using primitives may require more iterations, but allow a finer grain parameter search.

Legacy Design or Expert Designer Other generators of tunable parameters can come from outside the STS components. One such generator is a *legacy design*, i.e., a prior version of the macro or a similar macro. Quite often industrial chip design projects archive the final version of design data used for tapeout. The archived design data typically includes the parameter settings used for the final implementation. These final implementation parameter settings may be valuable prior knowledge and useful for newer versions of the macro, even if the rationale for using the parameter settings may not have been documented. In our experience, we can often access the legacy scenario, $\{S_{exp}\}$, using just the macro name. While this scenario may not be assembled from primitives, it may be mapped to a set of primitives, $\{P_{exp}\}$, plus a residual set of parameters. The residual parameters can then be grouped as a new primitive.

Another source of tunable parameter input is directly from an expert designer. In this case, the expert designer can create complete scenarios or primitives using any empirical evidence or intuition. This interaction with an expert designer can be static, e.g., user-defined primitives are provided at the beginning of the STS online tuning process, or dynamic, in which case the designer may modify conditions of an STS run in flight, e.g., modifying settings for the next iteration.

Since the STS Online System takes only a limited number of tunable parameters, a selection and blending algorithm is needed to choose among the various input sources. As mentioned in Sect. 13.2.5, ~50 primitives are often selected for average complexity macros in our design flow. In practice, we use a set of heuristic-based selection and blending algorithms, as there are numerous potential algorithm implementations.[8] One approach to a selection and blending algorithm could be similar to ensemble learning methods and in particular *stacking*, where the goal is to boost prediction accuracy by blending predictions of multiple machine learning models [31, 32]. While a thorough investigation into approaches for a selection and blending algorithm is beyond the scope of this chapter, we empirically compare a number of approaches in Sect. 13.6.2.

13.6   STS Experimental Results

This section highlights results from our STS case study system. We discuss the impact of STS on the design of IBM products as well as the potential improvements from the overall hybrid online/offline system.

13.6.1   STS Product Impact

The first impact of STS on an industrial product was demonstrated during the design of the IBM POWER7+ [8] and POWER8 [33] processors. STS was employed to close timing for difficult macros and reduce the power of targeted macros. A prior STS dataset was not available at the beginning of these chip design projects, and thus the STS Online System was used for optimization. Chronicles of results for these two processors can be found in [9, 17].

---

[8] While, in concept, empirically comparing multiple selection and blending algorithms is feasible, the practical compute effort would be quite expensive.

Table 13.3 Average STS improvement over the best known prior solution based on post-route timing and power analysis

STS pass (chip release)	Internal slack	Total negative slack	Total power
Improvement %	(%)	(%)	(%)
Pass 1	60%	36%	7%
Pass 2	24%	2%	3%
Sum of 200 macros	(ps)	(ps)	(arb. units)
Pre-pass 1	−1929	−2,150,385	17,770
Post-pass 1	−765	−1,370,731	16,508
Sum of 25 macros	(ps)	(ps)	(arb. units)
Pre-pass 2	−260	−185,087	3472
Post-pass 2	−198	−180,896	3379

Subsequently, STS was widely adopted for the design of nearly all macros for the IBM z13 22nm server processor [34]. The processor underwent two chip releases (tapeouts) over a multi-year design cycle. The chip consists of a few hundred macros that average around 30K gates in size, with larger macros in the 300K gate range. The STS Online System was also used for this chip, as the STS archive was not yet populated. During the z13 design, a systematic performance evaluation of the STS Online System was conducted. Results were tracked for approximately 200 macros from the processor core. The "pass 1" rows of Table 13.3 report the average improvements achieved by STS over the best solution previously achieved by the macro owners for the first chip release. Note that these results are based on the routed macro timing and power analysis; in most cases the best known prior solutions included manual parameter tuning by the macro owner. The first pass of STS resulted in a 36% improvement in TNS (total negative slack), a 60% improvement in L2L (latch-to-latch, i.e., internal slack), and a 7% power reduction. The actual values of the metrics, summed across all the macros, underscore that the changes in the absolute numbers were significant, e.g., for pass 1, ∼780,000 picoseconds of TNS was saved across ∼200 macros.

A second STS tuning run was performed during the second chip release for a subset of timing-critical and high-power macros. These second-pass tuning runs build off the prior tuning run results to further search the design space. Based on data from 25 macros (Table 13.3, "pass 2"), we see considerable improvements after the second pass of STS.

The success of STS during the z13 server design led to STS being used for all subsequent server processors to date. More STS results for z13 and later IBM server chips are provided in [6, 10, 35].

13.6.2 STS Hybrid Online/Offline System Results

Although only the STS Online System was employed for the earlier IBM product deployments, the use of the online system during these projects led to a populated STS archive sufficient for the STS Offline System. The resulting STS archive was a by-product of online tuning by the macro owners during product development and did not require explicit effort. Alternatively, a dedicated effort could be undertaken to create a dataset, which could lead to a useful dataset in less time, but at a higher compute and human resource cost.

In this subsection we provide a brief example of the results for the Hybrid STS System based on a floating-point pipeline macro with approximately 63K gates from an IBM server product. Additional results for the offline and hybrid systems are available in [11]. For this comparison we provide results based on the three generators shown in Fig. 13.9 individually, as well as a few blended cases. While we are not aiming for a comprehensive study of the possible algorithms for selecting and blending the input parameters from multiple generators, we look to convey a few of the possible approaches and compare them to the generators in isolation. For this experiment we define the following methods that use only one of the generators:

- *STS Online*: Only primitives from the *Primitive Library*. This method is common when there is no prior dataset or legacy designs. This method uses five iterations of the SOLAR algorithm.
- *STS Offline*: Only scenarios from a recommendation model. The recommended system was trained with data in STS archive. The 30 recommended scenarios are run in parallel with no additional SOLAR iterations.
- *Expert Designer*: Only the legacy scenario is run, without SOLAR iterations. While for our experiment only one trial is run, the amount of design effort required to develop the scenario during the legacy chip design project is unknown.

Next, we define two blended generator methods for the experiment:

- *STS Online + Expert Designer*: The *Expert Designer* scenario is combined with each of the scenarios generated during the *STS Online* run. This method uses five iterations of the SOLAR algorithm.
- *STS Offline + Expert Designer*: The *Expert Designer* scenario combined with each of the 30 scenarios generated by the recommender system, i.e., the *STS Offline* run.

Table 13.4 shows the experimental results for each method. The table includes measures of compute effort in terms of relative tuning latency (SOLAR iterations) and total number of trial evaluations. The cost calculation employs the cost function from Eq. 13.1 and compares the lowest cost (best) scenario from each experiment. The smaller table to the right provides the individual metric weights for cost calculation.

From the results in Table 13.4, we see the two hybrid methods outperforming the individual generators. While the *STS Online* approach results in a competitive QoR values, the compute effort is quite high. For a similar compute cost, the *STS Online + Expert Designer* method produces generally better QoR. Likewise, the *STS Offline + Expert Designer* can achieve competitive QoR with significantly less compute effort. Overall, these results confirm the advantages to using hybrid online/offline strategies for design flow tuning.

13.7 Future Directions

In this section we outline some avenues of future research for design flow tuning systems. In particular, we consider directions that explore extending the design tuner abstraction to more adaptable, collaborative, multi-macro, and generalized systems.

13.7.1 Adaptable Online and Offline Systems

There are many ways in which the online and offline components of the system work together and dynamically. The approach we propose consists of dynamically adapting the percentage of trials originating from the online learning algorithm (Stream 1) and the recommender system (Stream 2). The adaption can occur over the course of the cumulative STS tuning runs of system, i.e., over the lifetime of the system, and across the iterations of a single tuning run. Figure 13.10 provides an example of four phases where the system adapts the percentage of trials allocated to Streams 1 and 2. The illustration assumes the system begins without a suitable dataset. Thus, in Phase (a) all trials are allocated to Stream 1 for both tuning the current designs and accumulating data for the dataset. As the dataset is building over multiple STS runs, the offline recommender system is being trained and tested periodically. Once the model validation provides sufficient test accuracy, the system moves into Phase (b), where STS runs begin to allocate an increasing percent of trials to Stream 2. During Phase (b) a quality comparison occurs between Stream 1

Table 13.4 Experimental results from blending parameters from multiple generators

Experiment	Compute Effort		Cost	QoR						Metric	Cost Weight
	SOLAR iterations	Total trials		WNS (ps)	R2R (ps)	TNS (ps)	Cong (a.u.)	Power (a.u.)			
STS Online + Expert Designer	5	164	1.65	-12	-11.8	-2409	88.5	44.4		WNS	1
STS Offline + Expert Designer	1	30	1.68	-4	1.4	-137	88.7	47.6		R2R	2
STS Online	5	173	1.85	-11	-11.3	-1751	88.7	45.0		TNS	1
Expert Designer	NA	many	3.98	-18	-18.3	-5145	89.3	46.6		Cong	3
STS Offline	1	30	8.48	-16	-16.2	-5255	103.2	51.0		Power	4
default parameters	1	1	11.00	-25	-25.2	-9120	104.6	52.5			

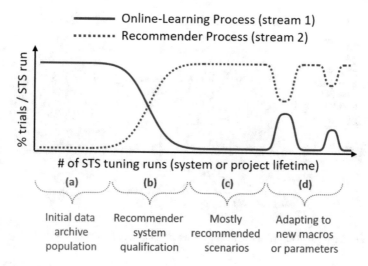

Fig. 13.10 Adaptive online and offline modules

and 2 scenarios, and if the Stream 2 scenarios perform adequately, they become the primary source of trials, and the system enters Phase (c). But, as Phase (d) illustrates, as the system encounters either new macros for which the Stream 2 scenarios are not successful or if new primitives arise, e.g., a new LSPD CAD tool version having new parameters, the system can shift back into a mode that allocates an increasing number of Stream 1 trials to incrementally update the dataset with sample points for the new primitive and/or make optimization progress on the new macros. While there are a number of situations where adaption could occur at varying complexity, we believe some form of dynamic interplay between the online and offline components will be needed over the system lifetime.

13.7.2   Enhanced Online Tuning Compute Frameworks

One direction of research is enhancing the online compute framework illustrated in Fig. 13.3. The parallel, iterative, and fault-tolerant characteristics of the compute framework provide the basic support for the online tuning of applications with high memory requirements, long runtimes, and many parameters, such as LSPD flows. While the compute framework in Fig. 13.3 represents the minimal requirements, there are a number of techniques that may improve the compute throughput and the quality of results, albeit with added algorithmic complexity.

One possible improvement is allowing asynchronous parallel trial submission, i.e., relaxing the need for all trials to complete within an iteration before submitting additional trials. One example of such an algorithm is asynchronous batch Bayesian optimization [37]. Within the context of CAD, similar approaches have been applied

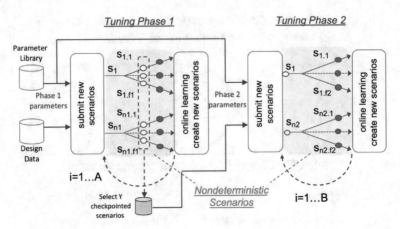

Fig. 13.11 Example of an enhanced online compute framework [36]

to analog circuit tuning [38]. Online design flow tuning may also benefit from asynchronous parallel approaches.

Additional efficiency enhancements to the online compute framework reduce the need for running all the tuning trial design flow steps. For one approach, referred to as early scenario pruning [39] or early stopping [29], trials with non-promising metrics in the earlier steps of a design flow can be terminated prior to completing the last tuning flow step, saving computing resources. A second approach reuses run data of prior trials for which parameter settings are common during earlier flow steps. This approach is referred to as checkpointing and forking [40] or jump starting [29].

Yet another compute framework modification is based on running multiple passes of the design flow. In this approach a subset of design flow steps are tuned and trials are checkpointed. The most promising trials are restarted and further tuned during later design flow steps. This approach can be combined with running multiple forked trials that vary due to inherent CAD tool nondeterminism or by explicitly injecting variation into the forked trials. Figure 13.11 illustrates these approaches, referred to as multiphase positive deterministic tuning [36].

While the approaches mentioned in this subsection have been initially explored to some degree, there are most likely many more compute framework enhancements that have yet to be discovered. It seems likely that future generations of design flow tuners will combine multiple enhancements for advancements in efficiency and quality of results.

13.7.3 HLS and LSPD Tuning System Collaboration

Although the scope of this chapter is LSPD, HLS poses an analogous tuning problem, and perhaps even a larger body of research exists in the HLS domain [41].

Table 13.5 Comparison of HLS and LSPD tuning application properties

Property	HLS	LSPD
Number of parameters	10 s	100 s or 1000 s
Runtime	Minutes, hours	Hours, days
Type of parameters	Numerical, categorical	Categorical, numerical
Parameter commonality	Design specific	Common to most designs
Optimization constraint	Behavioral equivalency	Cycle-accurate
Typical objective	Pareto-optimal space exploration	Optimized single operating point

Table 13.5 shows a first-order comparison between HLS and LSPD tuning applications. LSPD tuning is typically more compute intensive, as the number of parameters and runtime per trial is higher than HLS. Thus, LSPD tuners may require more parallelism and allow fewer iterations. Both applications have numerical and categorical parameter types, although HLS often has relatively more numerical parameters than categorical parameters, while the opposite is true for LSPD. One challenge in HLS tuning is that many parameters are design specific [18], while most parameters can be tuned for all LSPD designs. On the other hand, LSPD is more tightly constrained and typically targets maintaining cycle accuracy, while HLS has more freedom for changing logic and only needs to maintain behavioral equivalency. More broadly, the overall objective of HLS tuning is often architectural exploration and determining a multi-objective Pareto frontier. On the other hand, LSPD optimization often occurs after locking in a specific operating point, so optimization may be more amenable to single objective optimization, e.g., the multi-metric weighted cost function as presented for STS.

Although differences between the two tuning problems exist, we believe a generalized online/offline tuning framework could be applicable for both HLS and LSPD tuning. To account for inherent application differences, we believe the HLS and LSPD tuner instances could be customized by setting the tuner framework hyperparameters, e.g., a higher number of parallel submissions for LSPD, versus less parallelism and more iterations for HLS. If such tuning systems for both HLS and LSPD were developed, we believe there is an opportunity for the two systems to collaborate in various ways, as illustrated in Fig. 13.12. For example, Path (a) in the figure shows the resulting RTL from the HLS tuner being passed as input to the LSPD tuner. However, for the interaction we envision, multiple versions of the RTL can be passed to the LSPD tuner, i.e., the RTL can be considered a tuning primitive. Further, new versions of RTL can be passed over time as the HLS tuner performs iterative optimization. A second type of interaction involves determining if correlations exist between HLS primitives and LSPD primitives, i.e., Path (b) in the figure. For highly correlated cases, we believe there is an opportunity to further

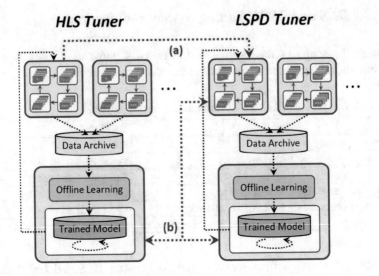

Fig. 13.12 HLS and LSPD tuner collaboration

enhance the recommended scenarios for the LSPD. For brevity, we mention here two possibilities for collaboration between the systems, but more complex interactions are possible.

13.7.4 Human Designer and Tuning System Collaboration

Another type of interaction is between the design flow tuning system and the human designer. Modern designers must often be a "jack-of-all-trades" to balance various design tasks to deliver a functionally correct and appropriately optimized design

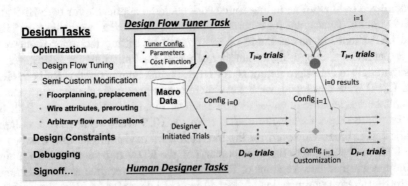

Fig. 13.13 Designer and tuning system interaction

within the allotted project schedule. The left side of Fig. 13.13 shows a number of design tasks, including optimization, setting designs constraints, reviewing and debugging tool run logs, and ultimately meeting high-accuracy signoff goals. Fortunately, a busy human designer can offload the design flow tuning task to the design flow tuner. But, while design flow tuning can often be effective, difficult and/or high-priority designs may require additional optimization effort. Employing semi-custom design techniques within the LSPD flow is one way the human designer can achieve results beyond what the design flow can natively deliver. These semi-custom approaches allow the human design to override CAD tool decisions for critical portions of the design. In practice, the precise usage of semi-custom techniques, such as instance preplacement, setting wire attributes, or prerouting, provides targeted guidance to CAD tools for closing challenging designs [17].

Figure 13.13 shows one effective way to combine the strengths of a design flow tuner and a human designer. The design flow tuner strength lies in the tireless optimization of design parameters, typically based on parallel and iterative refinement using machine learning. On the other hand, the human designer may have insight based on a deeper understanding of the design problems and/or from design experience collected over the course of a career. Using the respective strengths of each, the tuner and designer can attempt to optimize the design in parallel from two different perspectives. For example, as the design flow tuner runs iterative refinement, the human designer can develop and test hypothesis using intuitive design insight. When the human designer finds an approach that improves the design, the solution can be combined with the tuner trials on the next automated tuner iteration. Likewise, as the tuner makes progress on the design, the human design can incorporate the results of each iteration into the next set of hypotheses. Thus, the design flow tuner and human designer act as collaborators, approaching the problem in parallel, from different perspectives, and merging solutions as progress is being made.

Another method of interaction between the human designer and tuner is for the designer to modify the tuner hyperparameters during a tuning run in flight, i.e., "dynamically steering" the tuner. The long runtimes of LSPD trials make this dynamic interaction more feasible and in some cases more of a necessity. In addition to providing customized input to the tuner, as described above, the designer may reevaluate the tuner goals, e.g., if the tuner is able to close high-priority metrics, such as timing, the designer could steer the tuner to focus more on secondary metrics, like power reduction. For high-priority designs, this type of interaction may lead to a tuner being run continuously during the course of the chip design project with incremental guidance from a human designer.

13.7.5   Multi-Macro and Hierarchical Tuning Challenges

The scope of design flow tuning throughout this chapter thus far has focused on achieving QoR improvements for a single macro; however, large design projects

Fig. 13.14 Architecture of
the STS Scheduler (STSS)

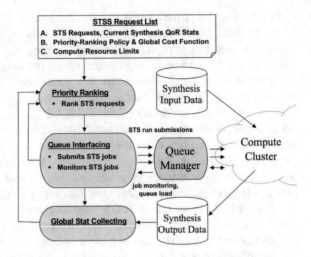

often consist of many macros that are concurrently designed by multiple human designers. Furthermore, limited computing resources, even in an industrial setting, require ROI considerations when investing effort into design flow tuning. Considering the interplay between multiple currently designed macros and the available computing resources holistically allows maximizing the QoR improvement for an entire chip. Hierarchical design also imposes another form of multi-macro interplay for design flow tuning. For example, tuning parent macros may require significant runtime during which child macro instances may change.

We believe design flow tuning research considering multiple macros and hierarchical design dependencies is another exciting direction. Some initial work has been presented on multi-macro tuning and CAD tool scheduling. Ziegler et al. present the *SynTunSys Scheduler*, a.k.a., *STSS*, a system that manages multiple STS runs for multiple macros [10]. This system works at a higher level of abstraction that considers the ROI of STS runs at the project level. Figure 13.14 shows a diagram of the components and processes of STSS. The general goal of STSS is to take a list of STS-run requests and optimally determine the order in which to submit them to the queue manager, given resource limits. Agrawal et al. also address the broader topic of scheduling compute jobs for IC design using mixed interlinear programs [42].

13.7.6 Generalized Code Tuning

One of our overall goals is to enable a design flow tuner to utilize the same breadth of techniques that are at the disposal of human designers. Stated more ambitiously, rather than automating the tuning of CAD tool parameters, we are aiming to automate the tuning of *designer actions*. Human designers ultimately implement their optimization trials by modifying an instance of a design flow for

Fig. 13.15 Evolution to generalized code tuning

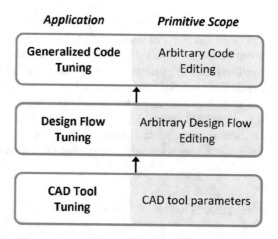

a specific macro and scenario. Setting CAD tool parameters is just a subset of possible modifications to a design flow instance. Designers often have considerable more degrees of freedom to modify flow scripts by adding or deleting code, files, flow steps, etc. In fact, to fully account for any designer action, the tuning system must provide the means for any arbitrary modification to the files in the design flow instance, yielding an essentially unlimited design space. To provide a formal mechanism for designers to express design actions that may require arbitrary code modification, we define a tuning primitive to be a snippet of code that can modify one or more files of a design flow instance. The primitives may also take arguments to allow a single primitive to have broader use, e.g., primitive arguments could be values for CAD tool parameters or variables used in conditional statements in the code snippet.

A design flow is typically a relatively complex code base, consisting of multiple files, written in multiple programming languages, that invoke multiple computing processes. Once we have broadened the definition of design flow tuning to include arbitrary code modification, a framework for tuning any code base can be established, as Fig. 13.15 illustrates. Thus, one interesting direction is exploring how such a framework can be applied more broadly to code optimization, in tandem with machine learning.

13.8   Conclusion

Design flow tuning has emerged as a new design abstraction and a powerful automation approach for modern VLSI design optimization. In this chapter we have described both online and offline machine learning techniques for design flow tuning, as well the interplay between these two approaches. We argue that the most effective systems will consider online and offline machine learning holistically. This chapter has also provided an overview of recent design flow tuning research, spanning the application domains of high-level synthesis (HLS), field-programmable gate array (FPGA) synthesis and place-and-route, and VLSI logic synthesis and physical design (LSPD). Our central focus has been LSPD, the most complex design flow tuning domain of the three. Throughout the chapter, IBM's SynTunSys (STS) design flow tuning system has been used as a case study. STS consists of a parallel and iterative online tuning system that works in tandem with an offline trained recommender system. STS has been a primary optimization tool for numerous IBM server chips, yielding considerable improvements in timing, power, and routing congestion. Finally, we speculated on a number of future directions of design flow tuning research, suggesting the topic has a promising future.

References

1. Abdi, H., Williams, L.J.: Principal component analysis. Wiley Interdiscip. Rev. Comput. Stat. **2**(4), 433–459 (2010)
2. Garrido-Merchán, E.C., Hernández-Lobato, D.: Dealing with categorical and integer-valued variables in Bayesian optimization with Gaussian processes. Neurocomputing **380**, 20–35 (2020)
3. Dso.ai: Ai-driven design applications. https://www.synopsys.com/implementation-and-signoff/ml-ai-design/dso-ai.html. Accessed 11 Aug 2021
4. Cerebrus intelligent chip design. https://www.cadence.com/en_US/home/tools/digital-design-and-signoff/soc-implementation-and-floorplanning/cerebrus-intelligent-chip-explorer.html. Accessed 11 Aug 2021
5. Jung, J., Kahng, A.B., Kim, S., Varadarajan, R.: METRICS2.1 and flow tuning in the IEEE CEDA robust design flow and OpenROAD. In: IEEE/ACM International Conference on Computer-Aided Design (ICCAD) (2021)
6. Ziegler, M.M., Liu, H.Y., Gristede, G., Owens, B., Nigaglioni, R., Carloni, L.P.: A synthesis-parameter tuning system for autonomous design-space exploration. In: Design, Automation & Test in Europe Conference & Exhibition (DATE), pp. 1148–1151 (2016)
7. Ziegler, M.M., Liu, H.Y., Gristede, G., Owens, B., Nigaglioni, R., Kwon, J., Carloni, L.P.: SynTunSys: a synthesis parameter autotuning system for optimizing high-performance processors. In: Machine Learning in VLSI Computer-Aided Design, pp. 539–570. Springer, Berlin (2019)
8. Taylor, S.: POWER7+: IBM's next generation POWER microprocessor. In: Hot Chips 24 (2012)
9. Ziegler, M.M., Gristede, G.D., Zyuban, V.V.: Power reduction by aggressive synthesis design space exploration. In: International Symposium on Low Power Electronics and Design (ISLPED), pp. 421–426 (2013)
10. Ziegler, M.M., Liu, H.Y., Carloni, L.P.: Scalable auto-tuning of synthesis parameters for optimizing high-performance processors. In: International Symposium on Low Power Electronics and Design (ISPLED), pp. 180–185 (2016)
11. Kwon, J., Ziegler, M.M., Carloni, L.P.: A learning-based recommender system for autotuning design flows of industrial high-performance processors. In: ACM/IEEE Design Automation Conference (DAC) (2019)
12. Liu, H.Y., Carloni, L.P.: On learning-based methods for design-space exploration with high-level synthesis. In: Design Automation Conference (DAC) (2013)
13. Meng, P., Althoff, A., Gautier, Q., Kastner, R.: Adaptive threshold non-Pareto elimination: rethinking machine learning for system level design space exploration on FPGAs. In: Design, Automation & Test in Europe Conference & Exhibition (DATE), pp. 918–923 (2016)

14. Xu, C., Liu, G., Zhao, R., Yang, S., Luo, G., Zhang, Z.: A parallel bandit-based approach for autotuning FPGA compilation. In: International Symposium on Field-Programmable Gate Arrays (FPGA), pp. 157–166 (2017)

15. Ansel, J., Kamil, S., Veeramachaneni, K., Ragan-Kelley, J., Bosboom, J., O'Reilly, U.M., Amarasinghe, S.: Opentuner: an extensible framework for program autotuning. In: Proceedings of the 23rd International Conference on Parallel Architectures and Compilation, pp. 303–316 (2014)

16. Ma, Y., Yu, Z., Yu, B.: CAD tool design space exploration via Bayesian optimization. In: Workshop on Machine Learning for CAD (MLCAD) (2019)

17. Ziegler, M.M., Puri, R., Philhower, B., Franch, R., Luk, W., Leenstra, J., Verwegen, P., Fricke, N., Gristede, G., Fluhr, E., Zyuban, V.: Power8 design methodology innovations for improving productivity and reducing power. In: Custom Integrated Circuits Conference (CICC) (2014)

18. Kwon, J., Carloni, L.P.: Transfer learning for design-space exploration with high-level synthesis. In: Workshop on Machine Learning for CAD (MLCAD), pp. 163–168 (2020)

19. Wang, Z., Schafer, B.C.: Machine learning to set meta-heuristic specific parameters for high-level synthesis design space exploration. In: ACM/EDAC/IEEE Design Automation Conference (DAC) (2020)

20. Agnesina, A., Lim, S.K., Lepercq, E., Cid, J.E.D.: Improving FPGA-based logic emulation systems through machine learning. ACM Trans. Des. Autom. Electron. Syst. **25**(5), 1–20 (2020)

21. Xie, Z., Fang, G.Q., Huang, Y.H., Ren, H., Zhang, Y., Khailany, B., Fang, S.Y., Hu, J., Chen, Y., Barboza, E.C.: FIST: A feature-importance sampling and tree-based method for automatic design flow parameter tuning. In: Asia and South Pacific Design Automation Conference (ASP-DAC), pp. 19–25 (2020)

22. Davis, R., Franzon, P., Francisco, L., Huggins, B., Jain, R.: Fast and accurate PPA modeling with transfer learning. In: IEEE/ACM International Conference on Computer-Aided Design (ICCAD) (2021)

23. Mirhoseini, A., Goldie, A., Yazgan, M., Jiang, J.W., Songhori, E., Wang, S., Lee, Y.J., Johnson, E., Pathak, O., Nazi, A., et al.: A graph placement methodology for fast chip design. Nature **594**(7862), 207–212 (2021)

24. Koren, Y., Bell, R., Volinsky, C.: Matrix factorization techniques for recommender systems. IEEE Comput. **42**(8) (2009)

25. Zhang, F., Yuan, N.J., Zheng, K., Lian, D., Xie, X., Rui, Y.: Exploiting dining preference for restaurant recommendation. In: International Conference on World Wide Web (2016)

26. Sidiropoulos, N.D., De Lathauwer, L., Fu, X., Huang, K., Papalexakis, E.E., Faloutsos, C.: Tensor decomposition for signal processing and machine learning. IEEE Trans. Sig. Proces. **65**(13), 3551–3582 (2017)

27. Kolda, T.G., Bader, B.W.: Tensor decompositions and applications. SIAM Rev. **51**(3), 455–500 (2009)

28. Kapre, N., Chandrashekaran, B., Ng, H., Teo, K.: Driving timing convergence of FPGA designs through machine learning and cloud computing. In: International Symposium on Field-Programmable Custom Computing Machines, pp. 119–126. IEEE, Piscataway (2015)

29. Liang, R., Jung, J., Xiang, H., Reddy, L., Lvov, A., Hu, J., Nam, G.J.: Flowtuner: a multi-stage EDA flow tuner exploiting parameter knowledge transfer. In: IEEE/ACM International Conference on Computer-Aided Design (ICCAD) (2021)

30. Ziegler, M.M., Kwon, J., Liu, H.Y., Carloni, L.P.: Online and offline machine learning for industrial design flow tuning: (Invited-ICCAD special session paper). In: IEEE/ACM International Conference on Computer-Aided Design (ICCAD) (2021)

31. Wolpert, D.H.: Stacked generalization. Neural Netw. **5**(2), 241–259 (1992)

32. Sill, J., Takacs, G., Mackey, L., Lin, D.: Feature-weighted linear stacking. arXiv:0911.0460, 2009

33. Fluhr, E.J., Friedrich, J., Dreps, D., Zyuban, V., Still, G., Gonzalez, C., Hall, A., Hogenmiller, D., Malgioglio, F., Nett, R., Paredes, J., Pille, J., Plass, D., Puri, R., Restle, P., Shan, D., Stawiasz, K., Deniz, Z.T., Wendel, D., Ziegler, M.: 5.1 power8^{tm}: A 12-core server-class

processor in 22nm soi with 7.6tb/s off-chip bandwidth. In: International Solid-State Circuits Conference (ISSCC), pp. 96–97 (2014)

34. Warnock, J., Curran, B., Badar, J., Fredeman, G., Plass, D., Chan, Y., Carey, S., Salem, G., Schroeder, F., Malgioglio, F., Mayer, G., Berry, C., Wood, M., Chan, Y.H., Mayo, M., Isakson, J., Nagarajan, C., Werner, T., Sigal, L., Nigaglioni, R., Cichanowski, M., Zitz, J., Ziegler, M., Bronson, T., Strevig, G., Dreps, D., Puri, R., Malone, D., Wendel, D., Mak, P.K., Blake, M.: 22nm next-generation IBM system z microprocessor. In: International Solid-State Circuits Conference (ISSCC), pp. 1–3 (2015)

35. Ziegler, M.M., Bertran, R., Buyuktosunoglu, A., Bose, P.: Machine learning techniques for taming the complexity of modern hardware design. IBM J. Res. Develop. **61**(4/5), 13:1–13:14 (2017)

36. Ziegler, M.M., Reddy, L.N., Franch, R.L.: Design flow parameter optimization with multi-phase positive nondeterministic tuning. In: Proceedings of the 2022 International Symposium on Physical Design (2022)

37. Kandasamy, K., Krishnamurthy, A., Schneider, J.G., Póczos, B.: Parallelised bayesian optimisation via thompson sampling. In: nternational Conference on Artificial Intelligence and Statistics (AISTATS) (2018)

38. Zhang, S., Yang, F., Zhou, D., Zeng, X.: An efficient asynchronous batch bayesian optimization approach for analog circuit synthesis. In: ACM/EDAC/IEEE Design Automation Conference (DAC) (2020)

39. Anwar, M., Saha, S., Ziegler, M.M., Reddy, L.: Early scenario pruning for efficient design space exploration in physical synthesis. In: International Conference on VLSI Design (VLSID), pp. 116–121 (2016)

40. Ziegler, M.M., Gristede, G.D.: Synthesis tuning system for VLSI design optimization. U.S. Patent 9910949, 2018-3-6

41. Schafer, B.C., Wang, Z.: High-level synthesis design space exploration: past, present, and future. IEEE Trans. Comput. Aided Des. Integr. Circuits Syst. **39**(10), 2628–2639 (2019)

42. Agrawal, P., Broxterman, M., Chatterjee, B., Cuevas, P., Hayashi, K.H., Kahng, A.B., Myana, P.K., Nath, S.: Optimal scheduling and allocation for ic design management and cost reduction. ACM Trans. Des. Autom. Electron. Syst. **22**(4), 293–306 (2017)

Chapter 14
Machine Learning in the Service of Hardware Functional Verification

Raviv Gal and Avi Ziv

14.1 Introduction

Hardware verification is the process of ensuring that the implementation matches the specification [85]. In the hardware design domain verification is broken into many types and aspects, such as performance verification, power verification, and more. Functional verification ensures that the functionality of the implementation matches its functional specification; it is the most important and labor-intensive type of verification in the entire hardware design cycle. Some market estimations claim that functional verification costs reach 50–70% of the overall design development effort [23].

Functional verification is generally part of the front end of the design process. Its goal is to confirm that the implementation at the *register transfer level* (RTL) matches the specification. Other verification processes, such as equivalence checking [59] and manufacturing testing [32], ensure that the RTL implementation is equivalent to the manufacturing information sent to the fab and that the fabricated chips match this information.

Functional verification originated with the introduction of *hardware description languages* (HDLs) and logic simulators in the 1980s. These offered design teams a relatively fast way to simulate the design. They also made it possible to add a verification code that replaced the manual injection of stimuli and the checking of results. As the complexity of the RTL increased, following Moore's law, the requirements for the verification increased as well. Human written tests were no longer sufficient, and automation was called in to close the gap. First came automatic checking methods, which slowly reduced the need for the manual inspection of

R. Gal · A. Ziv (✉)
IBM Research, Haifa, Israel
e-mail: RAVIVG@il.ibm.com; aziv@il.ibm.com

© The Author(s), under exclusive license to Springer Nature Switzerland AG 2022
H. Ren, J. Hu (eds.), *Machine Learning Applications in Electronic Design
Automation*, https://doi.org/10.1007/978-3-031-13074-8_14

wave forms to determine whether or not a test-case passed. This opened the door for the simulation of more scenarios and the introduction of randomness into the test-cases. Over the years, many new technologies were added to the arsenal of verification teams, including formal verification [14] and tools that address specific areas in the verification. The development of new stimuli generation techniques, such as constrained random generation [8], reduced the need for directed tests; these techniques also allowed random stimuli generators to generate many different random test-cases from *test-templates*. These templates hold the specification for the tests given to the random stimuli generator, which can range from generic to very specific. The ability to generate and simulate large numbers of test-cases increased the automation in the verification process, but it also meant that it became much harder for the verification teams to track how well the design under verification (DUV) is exercised and whether all aspects of it are thoroughly tested. The solution for this problem was the adaptation and development of coverage techniques [64].

Modern verification is a highly automated process in which an endless number of verification jobs are continuously being executed in verification farms [73]. The process involves many tools and subsystems that handle tasks such as the documentation and management of the verification plan, track changes in the design and the verification environment, schedule verification jobs, generate stimuli, simulate the DUV, collect coverage data, and more. These verification tools produce a large amount of data that is essential for understanding the state and progress of the verification process. For example, simulation traces are needed for efficient debug. Coverage analysis is used to monitor the state of the verification process and identify areas that need more attention [5]. The ability of verification teams to successfully verify large and complex designs greatly depends on their ability to extract useful information from the data produced by the verification process; they can then use this information to steer the verification process and its components. This can be done either by using the information for decision support or by automatically feeding the information back to the verification process to affect its course.

Data science [11] in general and machine learning [74] specifically are fast-growing disciplines in computer science that deal with extracting patterns and information out of datasets. Machine learning uses algorithms that improve based on experience or that learn from their data. In general, machine learning provides one of three types of analyses: descriptive analyses that describes what happened, predictive analyses that predict what will happen, and prescriptive analyses that provide foresight on how we can improve the future [67]. For example, identifying changes in bug discovery trends uses descriptive analysis, predicting the files in which bugs will be found during the next week uses predictive analysis, and creating test-templates that hit a previously uncovered coverage event uses prescriptive analysis.

Data science and machine learning have already been incorporated into many verification tools. In [80] and its follow-on work, Vasudevan et al. combine the data mining of simulation trace data to identify candidate assertions. This was done using formal verification techniques that filter out spurious assertions, thereby yielding high-quality assertions that can be used in the verification process. For stimuli generation, Wu et al. [86] use decision trees to explore the search space of a constraint problem and discover favorable regions where the probability of finding satisfying assignments is higher. In [20], Fine et al. use Bayesian networks to learn the initial conditions that improve the success rate of system-level and processor core-level stimuli generators.

Machine learning (ML) and formal verification (FV) seem like totally unrelated or even contradicting disciplines. On the one hand, the goal of formal verification is to prove the correctness of a given design. Machine learning, on the other hand, generalizes and extracts properties from observed data and is based on statistics and probabilities. Still, connecting these two disciplines can benefit both. Specifically, machine learning techniques can improve many formal verification techniques and algorithms [2]. The high complexity of most formal verification problems means that their solution techniques is based on heuristics that can be tuned to specific instances or instance families of the problem. Machine learning methods can be used to learn from the characteristics and previous activations of an FV algorithm and provide information that helps improve its behavior in future activations.

A prominent example for the use of machine learning techniques in formal verification can be found in SAT solving [7]. The best-known SAT solving technique is the Davis-Putnam-Loveland-Logemann algorithm (DPLL) [18]. Simply stated, in each step the algorithm selects a branching variable and a value to assign to the variable. It then simplifies the formula and repeats the step until the formula is satisfied or a conflict is detected. If a conflict is detected, the algorithm backtracks to its last unexplored branch. Over the years, many machine learning-based heuristics for selecting the branching variable, the assigned value, and the branch to backtrack to or restart from have been proposed.

There are many features that characterize SAT problem instances. Some of these features are static, like the number of clauses and variables in the instance [60], while others are dynamic and specific for a certain algorithm, such as the number of assigned variables and search depth statistics [40]. These features can be used to learn the expected runtime of a SAT instance and help select the best algorithm [88] or best restart policy [40].

A different machine learning technique, namely, reinforcement learning (RL), can be used for selecting the right branching variable. At each step, the algorithm selects the variable (or variable value pair) that maximizes the expected reward, and the expected reward is updated based on the progress of the algorithm. For example, in [51], Lagoudakis and Littman use the expected runtime as the reward. The actions in their system are seven variable selection rules, and their algorithm updates the reward by penalizing actions that lead to backtracks. In [54], Liang et al. propose a branching heuristic called conflict history-based branching heuristic (CHB) that uses a stateless reinforcement learning with a reward function that is updated online by detected conflicts.

This chapter takes a fresh look at the use of data science and machine learning in verification. It shows how these technologies can be integrated into the verification process in a holistic manner and become an integral part of the verification process backbone. This means that data analytics and machine learning tools have access to all the verification data on the one side and can produce information that impacts the entire verification process or, at least, large parts of it. This information can be used in human decision-making or to automatically point important activities in the right direction.

The first step in adding machine learning to the verification process backbone is to provide the analysis tools with access to the entire verification data. This step, which is often overlooked, presents several challenges. First, there is the need to connect data coming from tools and data sources that speak different languages and use different terminologies. Performing analysis on data coming from different sources requires translation of the data into a common language to relate data items coming from various sources to each other. The second challenge is efficient access to the data. Analysis tools often require access to large quantities of data using complex queries. Production tools, on the other hand, are optimized for operation work, which means that the data handling is optimized for simple insert, modify, and delete operations.

In Sect. 14.2, we present the *Verification Cockpit* (VC). The Verification Cockpit is a platform designed to serve as the lifecycle management tool [83] for the hardware verification of high-end IBM systems. The Verification Cockpit uses a big data warehouse [65] to store the vast amount of data collected by the individual data sources for reporting and analysis. The data warehouse is fed with the data from the various data sources using *Extract, Transform, and Load* (ETL) processes [65]. It organizes the data in a star schema, which is efficient for data retrieval, and creates the required connectivity between the data items from different sources. The Verification Cockpit provides several interfaces to its data. These include direct query access to the data warehouse and to the REST API [84]. This REST API itself provides access to the data warehouse and to other data sources maintained by the Verification Cockpit; it also exposes the connections and relations between data sources that feed the Verification Cockpit.

In addition to interfacing with analysis tools, the Verification Cockpit includes a built-in report engine that provides a first level of descriptive reports, some of which are described in the section. These reports provide the verification team with informative views into the state and progress of the verification process. However, in many cases, deeper analysis can purify this information and make it easier to use. In Sects. 14.3 and 14.4, we highlight two aspects of the verification process that can benefit from such analysis and describe several analysis techniques that target these areas.

Coverage closure is the process of advancing coverage goals in general and coverage levels specifically [64]. Because coverage is one of the main measures for the quality of the verification process, finding and closing large coverage holes is one of the most important tasks faced by the verification team. Section 14.3 provides several analysis techniques related to coverage closure. First, in Sect. 14.3.1, we provide an overview of several descriptive analysis techniques that are used to identify areas and coverage events that need attention. In the rest of the section, we address one of the holy grails of verification, namely, automatic techniques for creating test-cases or test-templates that hit specific coverage events. *Coverage directed generation* (CDG) is a general term that is used for many methods and techniques developed to achieve this goal. Sections 14.3.2–14.3.5 describe CDG in more detail and review several CDG techniques based on data analytics and machine learning.

Another important area is the identification of areas or aspects in the design that are less stable and thus need more attention. In Sect. 14.4, we describe two such techniques. First, we show how trend analysis can be used to detect anomalies and trend changes in the bug discovery data, as these are good indicators for changes in the stability of the DUV or verification process. To complement this, we also present a second technique that attempts to detect changes in the stability at the file level before they are exposed. This technique, which is based on a machine learning engine, uses data from the source code change logs and defect records to identify at-risk files; these are files in which there is a high probability that a bug will be found in the next few days. The verification team can use the information provided by both analysis techniques to take preventive and corrective actions to fix the problem and avoid reduced stability.

We summarize the chapter with a discussion on the importance of data analytics and machine learning to the verification process and the unique challenges that verification brings to these domains. The importance of data analytics and machine learning to the verification process cannot be overstated due to the vast amount of data produced by the various verification tools and the need for concise useful insight to better manage and operate the process. The practical benefits of data analytics and machine learning presented in this chapter have brought them to center stage in the verification domain and led to their integration in many state-of-the-art verification environments.

14.2 The Verification Cockpit

For data science to serve as the backbone of the entire verification process, we need access to the verification data. As any data scientist knows, having the right data available is a gating requirement for any data-driven analytics. In many practical cases, most of the challenge may cover handling the data from different data sources, including cleaning, transforming, and extracting the meaningful features. In this section, we describe the *Verification Cockpit* (VC) [4], a consolidated platform for the planning, tracking, analysis, and optimization of a large-scale verification project. VC is a platform designed to serve as the lifecycle management tool [67] for the hardware verification of IBM high-end systems. We start by describing the different aspects of the modern verification flow, the data it produces, and the challenges it introduces. We then describe the VC architecture and the initial steps in using the data for descriptive analytics (dashboards).

The modern verification process involves many tools and subsystems that handle tasks such as the documentation and management of the verification plan, tracking changes in the design and the verification environment, scheduling verification jobs, generating stimuli, simulating the design under verification (DUV), collecting coverage data, and more.

These verification tools, which we sometimes refer to as data sources, produce a large amount of data that is essential for understanding the state and progress of the verification process. For example, simulation traces are needed for efficient debugging. Another example is coverage data, which is used to monitor the state of the verification process and identify areas that need more attention [5]. These examples use data from a single tool or data source. In many cases, cross-referencing data from several sources can provide additional insight into the verification process. For example, the correlation between coverage, bugs found, and new features in the DUV can help the verification team assess the quality of new features and identify holes in the verification and coverage plans.

While keeping the description general, we use IBM Verification Cockpit, described in [4], as our main source of examples for the implementation and usage of analytics.

14.2.1 Motivation and Goals

A typical modern verification environment comprises many tools originating from many vendors. Often, a tool suite from one vendor provides most of the needs, but such suites are generally augmented by tools from other vendors and in-house solutions. The major components of the verification environment can be seen in Fig. 14.1.

The *Verification Plan* holds the features that need to be verified, the verification means (simulation, special checker, formal verification, etc.), and the metrics (coverage or other). The *Work Plan* breaks down the features into tasks. The *Test Submission* system keeps statistics, such as how many times we ran each test-template, how many times it failed, the time and memory it took to generate and simulate, and more. The *Failure Tracking* system provides the data required for the triage, debug, and rerun of failures (error message, number of failures, test-template, etc.). The *Coverage Tracking* tool holds the definition and the status of the coverage events and models. The *Bug Tracking* tool provides the status of and information regarding the bugs found in the verification process. Finally, the *Design* and the *Test Bench* data can be provided by the version control tool (owner, number of rows, changes, etc.).

While most of the data described is structured, some tools produce unstructured data, such as test logs and definition documents. We focus here on the structured data, as it covers many aspects of the data sources and provides a lot of essential information.

To allow extensive use of data analytics in the hardware functional domain, we need a platform that can connect the various tools involved in the verification process and is capable of handling the vast amount of data created by these tools. There are many inherit challenges in developing such a platform. With so many

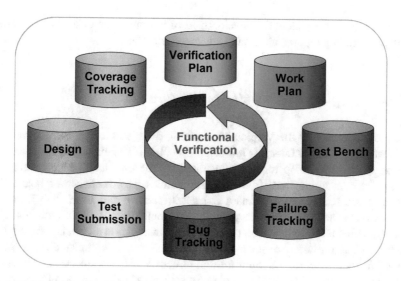

Fig. 14.1 Modern verification environment components

tools involved in the verification process, each with its own data repository, its own API, and own vocabulary (e.g., naming the same objects differently in different tools), we need the platform to connect the independent data sources and translate the data to a common vocabulary. While each tool is optimized for operational work (e.g., data handling is optimized for insert-update-delete), the platform itself should be optimized for data analytics, where data retrieval and manipulation is the main task. Finally, the platform needs to handle the vast amount of data created in the verification process, which includes billions of cycles, millions of tests per day, hundreds of thousands of coverage events, and more. The granularity at which data is stored must balance accuracy and performance.

Having an open architecture is a fundamental requirement for such a platform. The openness of the architecture should exist for both the connection of new verification tools as data sources and the analytics engines as consumers of the data. The platform also needs to provide a centralized data model that defines the relationship between data items coming from the different data sources. For example, the data model needs to ensure that all data sources use the same names for the hardware units. Alternatively, if different names are used, the name used by one tool should be mapped to the name used by the other tools.

To summarize, the fundamental requirements for the VC are the following:

- Open architecture to allow the connection of any tool;
- Centralized *data model* defining the relationship between data items coming from different data sources;
- Means for connecting data items from different data sources;
- Ability to handle large amounts of structural data;

- Open interface to allow the development of a large range of analytics over the data, including reports, alerts, and advanced analytics.

14.2.2 Verification Cockpit Architecture

The basic approach of the Verification Cockpit, described in Fig. 14.2, stems from the concept of *Product Lifecycle Management* (PLM). The interconnection between the different verification tools is illustrated by the circular line connecting the tools. The converging lines from the data sources into the data warehouse (i.e., Big Data tool) represent a standard *Extract Transform and Load* (ETL) process. This extracts the data from the data source, transforms it to the coherent data model, and loads it into the target database. Data analytics at all levels—descriptive, predictive, and prescriptive—over the data are then used to optimize directives over the verification process. The optimization directives may be reports and alerts that recommend actions. They can also directly activate actions, such as modifying the test submission policy to maximize the coverage rate of a feature in the DUV.

We define a coherent *data model* for the data warehouse based on the star schema methodology [65]. There is a fundamental difference between the operational database usually managed by each tool and the star schema model used in the data warehouse. While the operational database is optimized for insert, update, and delete operations, the star schema model is optimized for data analytics, where the common operation would be data retrieval. A partial example of the VC star schema is presented in Fig. 14.3. The star schema is built from dimensions and metrics. While the metrics usually hold numeric values that can be aggregated, the dimensions define the "aggregate by" options. There are common dimensions such as the date and the project name. There are also tool-specific dimensions such as the test-template name for the test submission data and the coverage model name

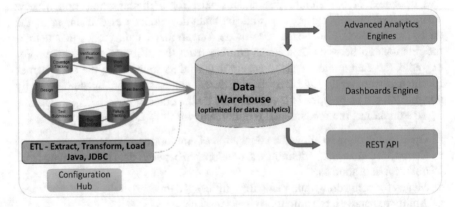

Fig. 14.2 Verification Cockpit architecture

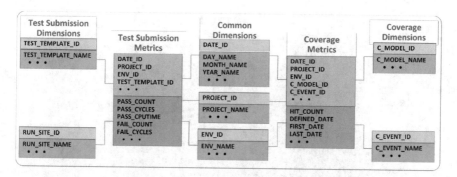

Fig. 14.3 Star schema data model for test submission and coverage data (partial)

for the coverage data. The VC maintains separate metric tables per data source, with the relevant dimensions, to allow many rich queries. The common dimensions are the essential components needed to cross-reference data queries. The metric tables, which can hold millions of rows, usually hold numeric values. Hence, they are thin tables in which queries can be handled very quickly using modern relational databases that are optimized for data analytics, such as IBM DB2, BLU version [45].

Clearly, as the tools were developed independently, their data does not follow this model directly. This is a known issue with data warehousing and is handled in the *Transform* step of the ETL process. In this process, we select the data we want to load into the data warehouse and transform it to the VC common data model. This includes, for example, data aggregation and renaming. To allow the process admin to control the data naming, we developed the *configuration hub*, a tool to define and control the naming translations needed for the ETLs.

The data warehouse holds several terabytes of compressed data collected over several years. It holds over 200 tables (metrics and dimension), where the big metric tables can hold several millions of lines. VC processes several gigabytes of data daily. Therefore, it performs a routine cleanup to control the database size. The most detailed data (per test per day) is kept for a few weeks, while aggregated data is kept forever and not cleaned. This allows teams to compare usage between projects over the years. Storage is just one factor for the cleanup. The other is that aggregated data allows for faster computation over large periods of time, as is done when projects are compared.

Some of the ETLs run on a daily basis, while other runs on an hourly basis. Runtime can vary from seconds to a few minutes and to hours. The ETLs run in a non-blocking mode, meaning the data is always available for any user.

There are three consumers of the data loaded into the data warehouse, as illustrated in Fig. 14.2. The *Dashboards Engine* provides a large inventory of descriptive analytics. It should further allow any user to create her own dashboards. Tools such as IBM Cognos Analytics [44] can be used. Examples that show the capabilities of descriptive reports are provided next. The *Rest API* allows access to the raw data for any new consumer. Its main goal is to encapsulate the complex structure of the data model from the user and allow clear and easy query capabilities.

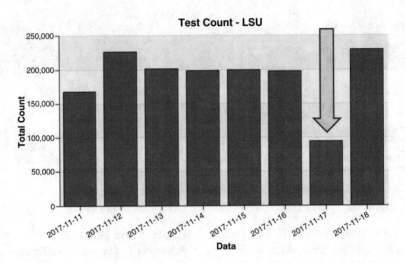

Fig. 14.4 Report presenting daily test runs

The *Advance Analytics Engines* represent the more sophisticated applications that use ML and other techniques to provide insights and corrective recommendations based on the data stored. These include engines for Template Aware Coverage (TAC) and Coverage directed generation (CDG), as described in Sect. 14.3.

14.2.3 Descriptive Analytics

The descriptive analytics engine that provides visualization of the data is an important layer for any data-driven process. For data scientists, it allows them to explore and visualize the data. When it comes to the program manager, the verification leader, or the verification engineer, it can also provide essential information that serves as a basis for immediate action. The Verification Cockpit, described in the previous section, brings data from different data sources into a unified data model and data warehouse; it also provides the infrastructure for dashboards containing single-source or multi-source reports. The following are a few examples of such reports and views and the information to which the dashboard can point.

The verification engineer, or the verification team lead, can use a dashboard to display daily statistics for the test submissions and failure-tracking tools. Figure 14.4 shows the daily number of tests run for a specific verification environment. We can see a drop in the seventh day. The reason for such a drop may be a known issue in the simulation farm, but it can also indicate an unknown problem. Connecting such an event to an automatic alarm can alert the team and cut down the time lost due to the problem.

Rank	Menu	List	CQ Per M Tests	HDWB				Tracker			ClearQuest		
				Total Count	Pass Rate	Fail Rate	zFail Rate	Avg Bone Pile	Sim Errors	CQs	Opened	Answered	Closed
1	lsu_dev	lsu_uldnkis	4,065.041	246	44.715%	46.748%	8.537%	12	12	1	1	1	1
2	ls_misc	lsu_stressExceptions	2.615	764,963	99.932%	0.024%	0.043%	14	4	1	2	1	1
3	ls_main	ls_gold_vec_rev	0.880	3,409,594	99.966%	0.012%	0.002%	13	14	2	3	2	2
4	ls_main	ls_gold	0.307	9,785,930	99.989%	0.005%	0.006%	23	15	1	3	2	1
5	ls_core	ls_core_vec_rev_directed	0.000	219	0.000%	100.000%	0.000%	0	0	0	0	0	0
5	ls_main	ls_gold_mvcrl	0.000	1,297,118	99.996%	0.000%	0.004%	20	4	1	0	0	0
5	ls_misc	lsu_clkGatingOnVsOff	0.000	12,615	99.849%	0.055%	0.095%	2	2	1	0	0	0
5	ls_misc	lsu_disables	0.000	346,978	99.973%	0.002%	0.025%	2	4	1	0	0	0
5	ls_misc	lsu_hw_prefetch	0.000	1,004,005	99.990%	0.005%	0.005%	6	9	1	0	0	0
5	ls_misc	lsu_setpAggr	0.000	1,034,860	99.963%	0.011%	0.026%	6	8	1	0	0	0

Fig. 14.5 Report presenting test-suite quality, sorted by defects per millions of test-runs

The previous view is based on a single data source. VC also allows us to use multiple data sources for descriptive analytics. A common practice in verification is to collect coverage data for all the tests that pass. A simple report that compares the number of passing tests with the number of tests reported to the coverage engine should show relatively similar numbers (small gaps due to communication latency are acceptable). A gap in this report, showing that not all passing tests reported their coverage, indicates a problem in the verification environment where simulation cycles are not contributing to coverage.

A more complex report is presented in Fig. 14.5. It uses information coming from the test submission data, the failure tracking data, and the defects handling tool, in order to grade test-suite quality (it can also grade test-templates). The report looks at different metrics that may indicate the quality of the test-suite. The first column shows the defects found by this suite, per million runs. The next statistics provide the number of test-instances generated from this suite, along with the number of tests passed, failed, and z-fail (i.e., failed on "cycle 0"). The next set of columns shows the average bone pile (i.e., number of failures for this suite while waiting for debug) and the number of different error messages. The last set of columns presents information from the defects tool, about the number of defects that were opened while debugging failures from this suite. Sorting the list by the first column shows us the most effective test-suites in terms of defects per run. The first line is a debug test-suite, showing a very high rate. The second row, highlighted in green, represents a very effective test-suite that is running fewer tests than the other suites. The team can use this information to change the balance of work between the different test-suites, according to their bug-finding effectiveness.

14.3   Coverage Closure

Coverage closure is the process of advancing coverage goals in general and, more specifically, coverage levels [64]. Since coverage is one of the main quality indicators of the verification process and the *design-under-verification* (DUV), coverage status is an important criterion for many project milestones, such as tape-outs. As a result, the verification team can spend significant time and effort on coverage closure.

To achieve coverage closure, the verification team needs to analyze the whole picture, identify features and functions in the DUV or aspects in the verification plan that need more attention, and steer the verification process toward these areas. In terms of coverage, it means identifying coverage events that are unhit or lightly hit (i.e., not hit enough). After the relevant events are identified, the verification team can create new test-cases or test-templates (i.e., the input to a random stimuli generator) that hit the identified events or modify existing ones to improve the probability of hitting these events.

Both aspects of the coverage closure are time-consuming and require expertise from the verification team. Simply viewing the status of all coverage events is not sufficient for identifying areas that need more attention because of the large number of coverage events that exist in the DUV. Instead, the verification team needs to know how to query the coverage database and which questions to ask it to extract such information. Descriptive analyses, ranging from visualization techniques [4, 72] to complex analysis algorithms [5], can help the team navigate the coverage data and automatically find large holes in the coverage picture. In Sect. 14.3.1, we provide an overview of several descriptive analysis techniques used to identify areas and coverage events that need attention.

To change the coverage picture, the verification team needs to steer or direct the verification toward specific risky areas in the design. In coverage terms, this means directing the verification toward specific coverage events or sets of closely related events. Writing or modifying test-cases or test-templates to hit specific coverage events requires expertise and a deep understanding of the DUV. First, the verification team needs to analyze the conditions for hitting the desired events. Then, they need to understand how to propagate these conditions to the inputs of the DUV. Finally, the conditions for the inputs need to be translated into a test-case or a test-template. This process can be time-consuming because it often requires iterative improvements to the resulting test-case. As a result, automatic techniques for creating test-cases or test-templates that hit specific coverage events are considered one of the holy grails of verification. *Coverage directed generation* (CDG) is a general term that is used for many methods and techniques developed to achieve this goal. Sections 14.3.2–14.3.5 describe CDG in more detail and review several CDG techniques, with a focus on data-driven techniques.

14.3.1 Descriptive Coverage Analysis

The goal of descriptive coverage analysis, as its name suggests, is to describe the coverage picture. In other words, the goal is to extract simple, concise, and useful information from the coverage data and present it to users. Descriptive coverage analysis begins with simple data visualization techniques, such as the ones shown in Sect. 14.2. For example, showing the coverage status of events in different environments, such as levels in the hierarchy or based on the simulation platform, can help the verification team distinguish between global problems and problems in a specific environment [72]. Trend plots can show how coverage is progressing over time and help detect when the verification effort runs out of steam. Moreover, the data for these plots can be used to predict future levels of coverage [41] and identify aged-out events that have not been hit for a long time [9].

In many cases, descriptive coverage analysis exploits the structure of coverage models to improve the quality of the analysis and help users adapt the analysis and resulting reports to their needs. In this section, we focus on cross-product coverage [36, 64]. A cross-product functional coverage model is based upon a schema comprising the following components: a semantic description (story) of what needs to be covered, a list of the attributes mentioned in the story, a set of all the possible values for each attribute, and an optional set of partitions for each attribute. A simple example of cross-product functional coverage is all the possible combinations of requests sent to a memory subsystem and the responses associated with them. In this case, a coverage event is composed of the pair of attributes, `<request, response>`, where `request` is any of the possible requests that can be sent to the memory (e.g., `memory read`, `memory write`, `I/O read`, `I/O write`) and `response` is one of the possible responses (e.g., `ack`, `retry`, `reject`) that can be received.

Cross-product coverage provides a simple way to define views that allow users to concentrate on the specific aspects of the model they are interested in while ignoring the areas not of interest. These views provide useful coverage reports that address the needs of the users. The views into the coverage data of cross-product coverage models are based on three operations: projection, selection, and grouping. With projection, the coverage data is projected onto a subset of the cross-product attributes. This allows users, for example, to examine the distribution of `responses` in the coverage model described above and to find out if one of the `responses` was not hit at all. The second operation selects specific events, for example, only events with good responses. Selection allows users to ignore parts of the coverage model that are currently not in the main focus of the verification effort or to filter out events based on their coverage data. The third operation allows coverage events to be grouped together. For example, we can group together `read` and `write` responses for every request. Grouping provides a more coarse, more abstract, view of the coverage data.

While coverage views provide a means for finding areas that need more attention, finding such areas stipulates an understanding of which views to examine. This, in turn, often requires time and expertise. Hole analysis [5, 50] is a technique for automatically detecting and reporting large coverage holes, which often indicate areas that need more attention. The main idea of hole analysis is to group together sets of uncovered events that are closely related based on the schema definition, thus allowing the coverage tool to provide shorter and more meaningful coverage reports to the user.

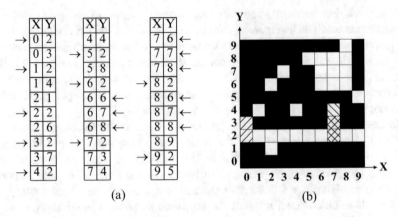

Fig. 14.6 Example of coverage hole. (**a**) List of uncovered events. (**b**) Hole visualization

To illustrate the importance of hole analysis, consider a model with just two integer attributes, X and Y, each capable of taking on values between 0 and 9. Figure 14.6a shows the individual uncovered events in the model after 70% of the events have been covered. The two meaningful holes that exist in the coverage data, indicated by arrows in the figure, are not immediately obvious. One hole occurs whenever Y equals 2, and a second hole exists when both attributes have values 6, 7, or 8. This, however, is readily seen in Fig. 14.6b, which shows the coverage of the model in a 2D graph and marks uncovered events as white squares. Such graphs provide a convenient way to present holes that are clustered along ordered values in models with a small number of attributes. The challenge for automatic holes analysis is to discover more complex holes in arbitrarily large models and to present these holes in such a way that their root cause may be more easily discerned.

Meaningful holes can be automatically discovered between uncovered events that are quantitatively similar. A simple metric for similarity is the rectilinear distance between two events. This corresponds to the number of attributes for which the two differ. The distance will be one for events that differ in only one attribute, two for holes that differ in two attributes, and so on, up to the number of attributes in the coverage space. We aggregate together any two holes whose rectilinear distance is one. Thus, the two uncovered events <0,2> and <0,3> from Fig. 14.6a can be aggregated into the single hole <0,{2,3}> and similarly for other such events.

Rectilinear distances can also be computed on aggregated holes. Again, the distance between any two holes is equal to the number of differing attributes, but now the comparison is done for aggregated sets as well as atomic values. The process can be applied iteratively until no more new aggregated holes are discovered. Figure 14.7 shows how the five events, <0,2>, <0,3>, <7,2>, <7,3>, and <7,4>, can be aggregated together until only the holes <{0,7},{2,3}> and <7,{2,3,4}> remain. This technique is similar to Karnaugh binary mapping [48] and is useful for much the same reasons. The aggregated holes of Fig. 14.7 are marked by left and right diagonal lines in Fig. 14.6b.

Fig. 14.7 Aggregated holes calculation

$$
\begin{array}{ccc}
\begin{array}{l}
< 0,2 > \\
< 0,3 > \\
< 7,2 > \\
< 7,3 > \\
< 7,4 >
\end{array}
&
\Longrightarrow
\begin{array}{l}
< \{0,7\},2 > \\
< \{0,7\},3 > \\
< 0,\{2,3\} > \\
< 7,\{2,3,4\} >
\end{array}
&
\Longrightarrow
\begin{array}{l}
< \{0,7\},\{2,3\} > \\
\\
< 7,\{2,3,4\} >
\end{array}
\end{array}
$$

Of particular interest are those cases where all the possible values of one or more attributes have not been covered for some specific values of the other attributes. An example of this is the value 2 for the Y attribute in Fig. 14.6. We call such a hole a projected hole and mark it $< *; 2 >$, where the $*$ sign indicates a wild card.

A coverage model of n attributes can be viewed as a set of coverage events or points in an n-dimensional space. A projected subspace of dimension 1 contains events that lie on a line parallel to the axis in the hyperspace. In other words, a projected hole of dimension 1 corresponds to a set of events whose attributes all have fixed values, except a single attribute that has a wild card. A projected subspace of dimension 2 describes events that lie on a plane parallel to the axis, and so on. A projected subspace of a higher dimension subsumes all contained subspaces of a lower dimension. For example, the subspace $p = < *; *; x_3; \ldots x_n >$ of dimension 2 contains all the events described by the subspace $q = < *; x_2; x_3; \ldots x_n >$. In such a case, we say that p is the ancestor of q and that q is the descendant of p in dimension 2. We denote by S_v the set of dimensions of subspace v that do not have a value (i.e., their value is $*$). s_v is the smallest member in S_v. For example, for the subspace p above, $S_p = \{1; 2\}$ and $s_p = 1$. In the algorithm below, descendants $(v; s)$ denotes all the direct descendants of subspace v in dimension s. That is, it refers to all the subspaces that are created by replacing the $*$ in the sth dimension of v with a specific value for that dimension. Descendants $(v; S_v)$ denotes all direct descendants of v in all possible dimensions.

Holes of higher dimension are, in general, larger and more meaningful than the holes of lower dimension they subsume. Therefore, it is desirable to report holes of the highest possible dimension and leave the subsumed holes out of the report.

The skeleton of the projected hole algorithm is shown in Algorithm 14.1. The algorithm is based on the double traversal of all the possible subspaces in the coverage space, first in increasing and then in decreasing order of dimensions. In the first phase, the algorithm marks all the subspaces that are covered. That is, at least one event in the subspace is covered. This is done by marking all the ancestors of a covered space, starting with the covered events themselves (subspaces of dimension 0). During the second phase, the algorithm traverses all the subspaces in decreasing order of dimensions. For each subspace, if it is not marked as covered or as a subsumed hole, the algorithm reports it as a hole and recursively marks all its descendants because they are subsumed by this hole.

The complexity of the algorithm described here is exponential to the number of dimensions (attributes) in the coverage model. The performance of the algorithm can be significantly improved by applying pruning techniques, both during the construction of the projected subspaces graph and while traversing the graph to report the holes.

14.3.1.1 Automatic Discovery of Coverage Models Structure

Earlier we showed how the inherent structure of coverage models can be used to improve the analysis of the coverage space. Specifically, we showed how hole analysis can find large uncovered holes in cross-product coverage models. In many cases, coverage models are defined as a set of individual coverage events without an explicit structure. In such cases, a manual search for large coverage holes is needed. This search involves scanning a list of uncovered events to find sets of closely related events. This is a tedious and time-consuming task.

Machine learning techniques have proven effective in many classification and clustering problems. Therefore, applying such techniques can improve the efficiency and quality of finding coverage holes and reduce the manual effort involved in this task. Gal et al. in [28] describe a clustering-based technique that finds large coverage holes in verification environments, such as cross-product, in which coverage models do not have an explicit structure. The proposed technique exploits the relations between the names of the coverage events. For example, event `reg_msr_data_read` is close to event `reg_pcr_data_write`, but not to event `instr_add_flush`.

To find large coverage holes, this technique combines classic clustering with domain-specific optimizations and uses it to map individually defined coverage events to cross-product spaces. Then, it uses cross-product hole analysis techniques, such as the ones described earlier in the section, to find large coverage holes in these spaces. The analysis comprises three main steps. In the first step, it clusters all the coverage events and maps the clusters into cross-product spaces. The second step applies domain-specific optimization to improve the cross-products created, by cleaning imperfections left by the clustering algorithm. Finally, in the third step, a standard cross-product hole analysis is performed, and the holes are reported to the user. The following is a brief description of the first two steps. More details and an elaborated description of the algorithms can be found in [28].

The first step in the analysis is to cluster all the coverage events based on their names. This clustering begins with breaking each coverage event into words. In the

---

Algorithm 14.1 Algorithm for projected holes

---

procedure HOLEPROJECTION
    // Mark covered subspaces
    for $i = 0$ to $n - 1$ **do**
        for all subspace v with dimension i **do**
            if v is marked covered **then**
                Mark all direct ancestors of v as covered
    //Mark and report holes
    for $i = n$ downto 0 **do**
        for all subspace v with dimension i **do**
            if v is not marked (as hole or covered) **then**
                Report v as hole
                recursively mark all descendants of v as subsumed holes

---

example here, the names of coverage events are in the form of `w1_w2_...._wn`, where each `wi` is a word. That said, breaking events names into words can work with any other naming convention, such as a capital first letter (i.e., camel case). The words of all the events are used as the features for the clustering algorithm. This technique is common to many document clustering techniques [68]. The analysis technique uses the K-means algorithm [47] with a non-negative matrix factorization algorithm (NMF) [52]. The one major difference between the clustering deployed here and standard document classification is the addition of the index of each word to the word. The reason is that events sharing the same word in the same location are much more likely to be related than events that share the same word in different locations. That is, the event `reg_msr_read` is more likely to be related to the event `reg_pcr_write` than to the event `set_data_reg_to_0`. Practically, each word in the event name is used with two indices: a positive index from the start of the name and a negative index from the end. This helps relate events that share the last word, before last word, and so on.

The next step, after the clusters are formed, is to extract the cross-product structure out of each cluster. Figure 14.8a–c provides an example of this step. It begins by creating, for each possible location, the set of all words that appear in that location with the number of times each word appears. For example, the cluster of the events in Fig. 14.8a yields the ten location sets in Fig. 14.8b. Note that the number of sets is twice the length of the longest events. The location sets are used to find *anchor locations*. Anchors are locations that have a single word in them, and that word appears in all the events in the cluster. In the example, location 1 with the set {`reg` (6)} and location −2 with the set {`data` (6)} are anchors. Location 5 with the set {`rmw` (2)} is not an anchor because the word `rmw` does not appear in all the events. The underlined locations in Fig. 14.8b mark the anchors in the example cluster.

After the anchors are identified, all the words between them are considered the dimensions or attributes of the cross-product space. This includes the words before the first anchor, if it is not in location 1, and the words after the last anchor, if it is not in the last location. The dimensions of the cross-product in the example are shown in Fig. 14.8c. In the general case, dimensions may include several words and may have a different number of words in each event. For example, the first dimension in the example includes two words for the second and last events and one word for the other events. This dimension semantically corresponds to two real dimensions. In the next subsection, we show how to reveal these real dimensions.

After forming the initial cross-products, the technique improves their quality by increasing their size and increasing the number of dimensions in the clusters. The technique also improves their density, which is defined as the ratio between the number of events in the cross-product and the size of the cross-product space. The improvement process can be divided into two main steps. The first step adds events to the cross-products and combines similar cross-products to overcome deficiencies of the clustering algorithm described in the previous section. The second step utilizes domain knowledge on cross-product coverage to apply heuristics that increase the number of dimensions and improve the density of the models.

```
reg_msr_data_read          reg_msr_data_write
reg_msr_atomic_data_rmw    reg_pcr_data_read
reg_pcr_data_write         reg_ir_atomic_data_rmw
```

(a)

```
1: {reg(6)}
2: {msr(3), pcr(2), ir(1)}
3: {atomic(2), data(4)}
4: {data(2), read(2), write(2)}
5: {rmw(2)}
-1: {read(2), rmw(2), write(2)}
-2: {data(6)}
-3: {atomic(2), msr(2), pcr(2)}
-4: {ir(1), msr(1), reg(4)}
-5: {reg(2)}
```

(b)

```
1: loc. 2 - -3 {msr, msr_atomic, pcr, ir_atomic}
2: loc. -1     {read, rmw, write}
```

(c)

```
1: loc. 2     {msr, pcr, ir}
2: loc. 3 - -3 {atomic, φ}
3: loc. -1     {read, rmw, write}
```

(d)

```
1: loc. 2 {msr, pcr, ir}
2: loc. -1 {read, rmw, write}
```

(e)

Fig. 14.8 Extracting cross-product example. (**a**) Events in the cluster. (**b**) Location sets. (**c**) Initial cross-product dimensions. (**d**) Cross-product dimensions after breaking the first dimension. (**e**) Cross-product dimensions after removing redundant second dimension

The clustering and mapping to cross-product algorithm leaves behind some dirt in the form of outlier events: orphan events that should belong to a cross-product but are not clustered with it and several clusters that should belong to the same cross-product. The reason for this dirt is that the distance measure used by the clustering algorithm does not exactly match the distance needed for the cross-products. The cleaning step comprises two sub-steps: adding orphan events to cross-products and combining similar cross-products. To find orphan events, the algorithm compares all the events in the unit to the pattern of each cross-product. If an event matches the pattern exactly or almost exactly, it is added to that cross-product. An event almost matches a pattern if it matches all the anchors and differs in at most one dimension. In this case, the dimension value of the event is added to the cross-product dimension. For example, when event reg_ir_data_read, which is not part of the initial cluster in Fig. 14.8a, is compared to its cross-product pattern, it matches the two anchors and the second dimension, but not the first dimension. In this case, the event is added to the cross-product, and the first dimension is updated to include the value ir.

To combine close clusters, the algorithm compares the patterns of each pair of clusters. If the two patterns differ in just one location, either an anchor or a dimension, it combines the two clusters into a larger cluster. In this comparison, patterns are equal in a dimension if they have the same values in the dimension or if the values in one pattern are a subset of the values in the second dimension. Note that if two patterns that differ in a single anchor are combined, the anchor becomes an attribute in the combined pattern.

The second part of the improvements is specific to cross-product coverage models. These heuristics were developed to improve the quality of the cross-products after examining the results of many real-life models. The heuristics improve the quality of the cross-product by adding more dimensions to the cross-products on the one hand and by removing redundant dimensions on the other. The resulting cross-product can have more dimensions or be denser (or both). The process repeats two steps of breaking dimensions and removing redundant dimensions until convergence is reached for each cross-product.

Dimensions in the cross-products found in the previous steps can span multiple words. For example, the first dimension in Fig. 14.8c is from location 2 to location -3 in the events, and it spans 1–2 words. In such cases, breaking the long attribute into two or more smaller attributes can improve the quality of the cross-product. In the example, the first attribute comprises two separate semantic meanings. The first word in the attribute (location 2) is the name of a register (ir, msr, pcr), and the second word is an indicator of whether or not the data access is atomic. Understanding the semantic of each word is hard. However, if breaking a dimension into two (or more) dimensions results in a dense cross-product, then, in most cases, this break is semantically correct and improves the quality of the cross-product.

For each single breaking point, the heuristics check the ratio between the number of values in the dimension and the size of the cross-product of the broken dimension. If the ratio is high enough ($\geq 50\%$ in the current implementation), the algorithm performs the break. In the example, the number of values in the first dimension is three. When breaking it into two attributes (location 2 with values {ir, msr, pcr} and locations 3 to -3 with values {atomic, ϕ}), the size of the original dimension equals 50% of the product of the sizes of the new attributes ($3 \cdot 2 = 6$), so the algorithm performs the break. The resulting dimensions appear in Fig. 14.8d.

Removing redundant dimensions from the cross-product reduces the size of the cross-product space, thus improving its density. A dimension d is redundant if a projection of the n dimensional space into the $n - 1$ space that does not include d leaves any point in the new space with at most one event associated with it. The implementation uses a simpler form of a pair-wise redundancy: dimension d_1 is redundant with respect to dimension d_2; if considering the value pairs for all the events, the values in dimension d_1 partition the values in dimension d_2. In the cross-product in Fig. 14.8d and the events in Fig. 14.8a, the pairs of values for dimensions 2 and 3 are (atomic, rmw), (ϕ, read), (ϕ, write). Therefore, the second dimension partitions the third dimension into (atomic, {rmw}), (ϕ, {read, write}), which makes it redundant. Figure 14.8e shows the dimensions after removing the redundant dimension. In cases where the locations of dimensions i and j are adjacent, instead of removing the redundant dimension d_1, it is combined with d_2. This leads to the same number of dimensions and space size as if dimension d_1 is removed.

After completing all the improvements, a final cleanup removes cross-products that are either too large (e.g., more than one million events) or too sparse (e.g., less than 10% density). Cross-products that have just one dimension are also removed. Therefore, the final set of cross-products does not cover all the events in the unit where the algorithm started. Usage results show that in a typical unit, the analysis technique is able to map 25%–60% of the events to cross-products and that more than 60% of the cross-product identified have more than 100 events.

14.3.2 Coverage Directed Generation (CDG)

Coverage-Directed Generation (CDG) [79] is a generic name used for a multitude of techniques that create test-cases or test-templates for hitting targeted events; these are usually, uncovered events or lightly hit events. CDG has long been on the wish list of verification teams and the target of a vast amount of research. To understand the challenges faced by CDG, we first take a closer look at the goals it is trying to achieve. Figure 14.9a shows the normal flow of coverage in the verification process. The inputs to the flow are either test-templates that are fed to the verification environment and the stimuli generator in it or test-cases that are fed directly to the DUV. When the DUV is executed (e.g., simulated, emulated), it produces coverage data regarding the coverage events that are hit during the execution.

CDG systems, as shown in Fig. 14.9b, need to work in the other direction. That is, the input to the system is the desired coverage and the output is either a test-template or a test that, after execution on the DUV and the verification environment, achieves the desired coverage. Constructing a system that can convert any desired coverage to test-cases or test-templates[1] is difficult because the system needs to provide the inverse of the normal complex flow mapping from tests to coverage.

Over the years, many CDG systems have been proposed that either approximate the inverse mapping or calculate it in special cases. Generally speaking, these systems can be classified into two types: systems that use models to directly calculate the inverse mapping from coverage to tests and systems that use the

---

[1] To avoid tedious repetitions, we use the general term tests from here on, unless the distinction is important.

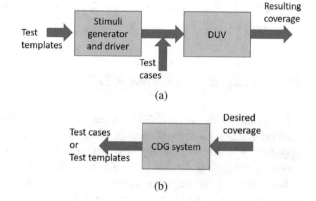

Fig. 14.9 Normal verification and CDG flows. (a) Normal verification flow. (b) CDG flow

normal verification flow to search for the inverse mapping. The first class of systems that directly calculate the inverse mapping can be divided into two subclasses: systems that use existing built-in or external models to calculate the inverse mapping and systems that infer such models from observations of instances of the forward simulation mapping. A different classification of CDG systems [21] calls the first subclass of systems with existing models direct CDG or model-based CDG. The second subclass of systems that use inference and the second class of systems that use the normal verification flow are referred to as data-driven CDG.

14.3.3   Model-Based CDG

Model-based CDG does not use data analytics or machine learning techniques, and therefore it is outside the scope of this chapter. For completeness, a brief description and some references are given below. The idea behind model-based CDG is to build a model of the DUV, or parts of it, and use this model to generate tests that hit desired coverage events [79]. The model in model-based CDG needs to be simple to allow easy maintenance. Therefore, most model-based CDG systems model only the DUV without the verification environment. This means that such systems usually generate test-cases, not test-templates. If the model is accurate, this is a big advantage for model-based CDG because the generated tests are guaranteed to hit the desired events. On the other hand, an inaccurate model can be almost useless, and constructing and maintaining a model can be a big challenge.

The main method for model-based CDG is to build an abstract finite-state machine (FSM) of the DUV and use formal verification tools such as model checkers to traverse the FSM from an initial state to a state corresponding to the target coverage event [79]. Several systems that use this technique have been proposed (e.g., [12, 46, 57, 79]). These systems differ in the type of DUV they model, the type of model used, and the method to convert the abstract test-case extracted from the abstract model to a concrete test-case. The main drawback of

FSM-based CDG systems is that they fail to scale due to the limitations of formal methods.

14.3.4   Direct Data-Driven CDG

To avoid the model-based CDG need for an accurate model of the DUV and dealing with problems in obtaining and maintaining it, many CDG systems use a data-driven approach. Instead of modeling the behavior of the design, the CDG system directly models the mapping from coverage to tests. In direct data-driven CDG, the CDG system observes the mapping from tests to coverage during simulation and learns from the observations about the inverse mapping from coverage to tests. After the system is trained, it can be queried about the best tests for reaching the desired coverage.

One of the inherent problems of data-driven CDG systems that does not exist in model-based system is the lack of positive evidence for hitting uncovered events. This lack of evidence makes it much harder for the CDG system to learn how to hit such events. One possible solution is to replace the query for tests that best hit the desired coverage with a query for tests that best hit the events that are near the desired coverage (a.k.a approximated target events) [29]. The idea, which mimics the work of verification experts, is that by improving the probability of hitting these neighbors, we exercise the relevant area in the DUV. This, in turn, increases the probability of hitting the target event itself. The validity of using approximated targets is backed up by the usage results.

There are many possible ways to automatically find the neighbors of a coverage event. For example, Wagner et al. [82] used the natural order of buffer utilization to learn how to fill a buffer. Fine and Ziv [22] exploited the structure of a cross-product coverage model. Gal et al. [27] used formal methods to find a set of neighbor events with positive and negative information regarding the probability of a test hitting the target event. All these methods are used in CDG systems, such as AS-CDG [29], and demonstrate their effectiveness in many cases.

Next, we describe several types of direct data-driven CDG systems starting from a simple system that harvests test-cases through a system that works on existing test-templates to improve coverage. From there we move on to a system that creates new test-templates on its quest to hit uncovered events.

14.3.4.1   Test-Case Harvesting

Test-case harvesting is the simplest data-driven CDG system. Simply stated, a test-case harvesting system observes the simulated test-cases and their resulting coverage and harvests (i.e., collects and stores) the last test-case that hits a desired event. When the verification team wants to hit an event, they query the system about the test-case that last hit and re-simulate the test-case. This CDG system is not a machine learning system because it does not learn from its observations and cannot infer any new knowledge about unseen tests. Instead, this is a memorization system that simply memorizes some of the data it observes and provides the data in response to queries.

Test-case harvesting for coverage provides only limited benefits to the verification team. Simulating the same test-case on the DUV without changing any of it yields the same behavior from the DUV and thus provides no new information to the verification team.[2] Any change to either the test-case or the DUV can completely change the DUV behavior and lead to the desired coverage events not being covered anymore. For this reason, the use of test-case harvesting for coverage is usually limited to the most difficult coverage events, where the verification team is not sure it can re-hit the events using directed random stimuli, even with the best help of CDG.

The opposite approach is used by Roy et al. in [70]. Instead of filtering tests, they use machine learning techniques to identify a small subset of coverage events that can predict the coverage of the entire set of events. The method uses k-means clustering [68] to find representative modules in the design and then a deep neural network (DNN) [33] to predict the coverage of the rest of the modules based on the coverage of the representative modules. This allows the system to reduce the overhead of coverage collection with a minimal impact on the accuracy of the coverage results.

14.3.4.2   Template-Aware Coverage

Most data-driven CDG systems work at the test-template level instead of test-cases because of the limited benefits in reusing test-cases. That is, as a response to queries, the systems produce test-templates instead of test-cases. These test-templates are fed to the stimuli generator that is part of the verification environment. Random stimuli generators are designed to generate many significantly different test-cases from the same test-template. Therefore, the same test-template can be used over and over, generating different new test-cases and exploring different paths in the DUV. In other words, CDG systems operate at the test-template level to capture the relationships between coverage and test-templates in a less precise manner than test-case based systems. They combine this knowledge with the power of the random stimuli generator to generate stimuli that improve the possibility of hitting the desired events. While this type of CDG systems cannot produce test-cases that guarantee hitting the desired events, as done by model-based CDG or test-case harvesting, they are much simpler than model-based CDG systems and require a less accurate model. This also makes them less sensitive to changes in the DUV.

---

[2] We assume that simulation is deterministic. Verification tools work very hard to maintain this assumption because of the difficulty in debugging non-deterministic simulations.

Template-aware coverage (TAC) [24] is a simple memorization data-driven CDG system that operates at the test-template level. In many aspects, TAC is the equivalent of test-case harvesting but works at the test-template level. At the heart of TAC lies the hit matrix that maintains first-order statistics on the coverage of each event by each test-template. The statistics approximate the probability of hitting the event with a test instance generated from the test-template. Formally, the definition of the hit matrix is the following [17]. Let $\mathbf{T} = (t_i)$ denote the test-template vector of size $|\mathbf{T}|$, and let $\mathbf{E} = (e_j)$ denote the coverage events vector of size $|\mathbf{E}|$. The hit probability matrix P_{hit} is defined as

$$P_{hit} = \left. \begin{pmatrix} \cdots \cdots \cdots \cdots \\ \cdots p_{i,j} \cdots \\ \cdots \cdots \cdots \cdots \end{pmatrix} \right\} \textit{ test-templates} \qquad (14.1)$$

$$\underbrace{\phantom{\begin{pmatrix} \cdots \cdots \cdots \end{pmatrix}}}_{\textit{Coverage Events}}$$

where $p_{i,j}$ is the probability that a test-instance generated from test-template t_i will hit event e_j. We calculate the hit probability matrix using first-order statistics of the actual coverage collected. Specifically, let w_i denote the number of test instances generated from test-template t_i in a given time frame, and let $hit_{i,j}$ denote the number of test instances generated from test-template t_i that hit event e_j at least once in that time frame. Then

$$p_{i,j} = \frac{hit_{i,j}}{w_i}.$$

Using the P_{hit} matrix, TAC can provide useful information to users regarding the relations between coverage and test-templates. For example, it can analyze whether a test-template achieves its coverage goal and hits all the events it should. For CDG purposes, the simplest query TAC can answer is to identify the test-template t_{best}^j that best hits a given coverage event e_j. The best test-template is simply the one with the highest value of $p_{i,j}$, or $t_{best}^j = \arg\max_i p_{i,j}$.

TAC can perform more complicated analyses and answer more difficult CDG queries. A commonly used CDG function of TAC is to calculate a regression suite with certain coverage properties. Given the hit probability matrix P_{hit} and a test execution policy $\mathbf{TP} = (w_i)$, where w_i is the number of test instances that are generated from template t_i, the probability of covering event e_j is

$$P_j = 1 - \prod_i (1 - p_{i,j})^{w_i}. \qquad (14.2)$$

Using this equation, we can formulate the problem of finding a test policy $TP = (w_i)$ that minimizes the number of simulation runs, with probability Ψ of covering each previously covered event as [17]:

$$\min_{TP} \sum w_i$$

$$\texttt{s.t.} \quad \forall j \quad P_j = 1 - \prod_i (1 - p_{i,j})^{w_i} \geq \Psi$$

$$\forall i \; w_i \geq 0, \quad w_i \in \mathbb{N}$$

This is a nonlinear integer programming problem. It can be relaxed to real numbers, which requires discretization of the solution. We can also approximate it as a linear programming (*LP*) problem by taking the log of P_{hit} [17].

Note that regression suites can be calculated on a subset of the coverage events and test-templates. One common use is to calculate such regression suites for events that are hard-to-hit or lightly covered (e.g., events that are hit less than a certain number of times in the last 2 weeks). Such regression suites are used daily as part of the overall IBM regression. These suites not only cause a significant reduction in the number of hard-to-hit events, but they also help hit many events that have never been hit before and discover new bugs [24].

14.3.4.3 Learning the Mapping

So far, we discussed data-driven CDG systems that memorize the mapping between existing tests and coverage and use the existing tests to achieve some coverage goals. The next step is to construct CDG systems that can learn from similar observations and infer insight about the behavior of unseen test-templates and the best overall test-templates to achieve the desired coverage.

Building such a CDG system raises several challenges. First, the flow of the verification process, as shown in Fig. 14.9a, is from test-templates to coverage. However, the flow of CDG is in the other direction, from coverage to test-templates, as shown in Fig. 14.9b. One possible solution is to use a machine learning engine in the CDG reverse direction. That is, use the coverage as the features and the test-templates as the labels. To the best of our knowledge, there are no publications regarding such CDG systems. A second approach is to train the machine learning engine in the normal flow direction but query it from its output to its input. Some machine learning techniques, such as Bayesian networks [62], naturally fit this type of diagnostic query. Later in the section, we describe such a system. Another possible solution is to use techniques similar to *generative adversary networks* (GAN) [34] that can manipulate and slightly modify existing input to change the label of the input. We are not aware of such techniques being used for CDG. A third solution is to search the input space using the trained learning engine. We discuss this solution in detail in Sect. 14.3.5.

A second challenge in building a learning CDG system is the need to handle previously uncovered coverage events. The machine learning literature covers several techniques for dealing with one-sided classification, when there are no samples for one of the possible labels. However, these techniques are used in the

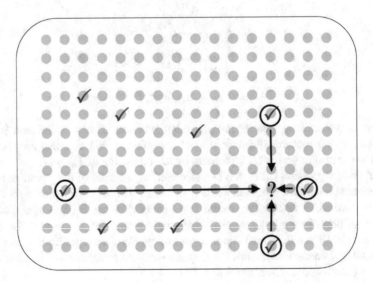

Fig. 14.10 Using the coverage model structure to learn how to reach an uncovered event in two-dimensional space

forward direction and cannot handle the large labels space of CDG, making them impractical for our needs. Earlier, we described a solution that approximately targets neighbor events. Another solution is to use the structure of the coverage model in the structure or architecture of the machine learning engine. Consider the two-dimensional coverage space in Fig. 14.10. The covered events in the space are marked with checkmarks, and the desired event is marked by a question mark (?). The CDG system can try to learn the condition for hitting each dimension separately and independently and then combine the two results to infer how to reach the target event.

A CDG system that breaks cross-product coverage into its attributes and uses inductive logic to extract rules for controlling each of the attributes is described in [43]. A CDG system based on Bayesian networks [22] is described next.

A Bayesian network is a graphical representation of the joint probability distribution for a set of variables [62]. Simply stated, a Bayesian network is a directed acyclic graph whose nodes are random variables and whose edges indicate direct influence from the edge source to its destination. The nodes of the Bayesian network contain the conditional distribution of the random variable, represented by the node given it parents. It is easy to show that if direct influence exists, and if and only if there is an edge in the network, the Bayesian network compactly captures the joint probability of all its random variables. During the training of a Bayesian network, the conditional probabilities in its nodes are updated based on the observations in the training set.

To illustrate the use of Bayesian networks for CDG, we use a simplified a high-level model of the dispatch unit and the two arithmetic pipes of NorthStar, a

Fig. 14.11 Abstract pipeline model: two pipes with three stages

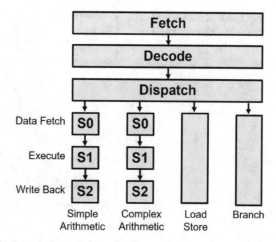

superscalar in-order processor. Each of the pipes comprises three stages: data fetch, execution, and write-back. Of the two pipes, the simple pipe handles only simple instructions (e.g., *add*), and the complex pipe handles both complex (e.g., *mul*) and simple instructions. A model of the processor is shown in Fig. 14.11.

The random generator uses four parameters to generate programs for the processor. OP determines the instruction type. SR and TR are used to select the source and target registers, and CR determines whether the condition register is used. The coverage model comprises five attributes: two for the type of the instructions in the data fetch stages of the complex and simple pipes, C0 and S0; two indicators whether the execution stages are occupied C1 and S1; and an indicator whether the instruction in the execution stage of the simple pipe is using the condition register S1CR.

Figure 14.12 shows the structure of the Bayesian network for the NorthStar example. The figure shows nodes for the generator parameters in the top box, nodes for the coverage attributes in the bottom box, and two additional hidden nodes, H1 and H2. The edges in the network indicate direct influence from their source to the target. For example, the type of generated instructions controlled by OP directly influences the types of instructions in each of the pipe stages C0, S0, C1, and S1. Hidden nodes are variables for which we do not have any physical evidence (observations) for their interactions. These nodes are added to the Bayesian network structure primarily to reflect expert domain knowledge regarding hidden causes and functionalities that impose some structure on the interaction between the interface (observed) nodes. For example, the hidden node H1 in Fig. 14.12 represents the domain knowledge that source and target registers together are used to create data hazards.

The ability and need to use different Bayesian network structures for each CDG instance introduce both big advantages and disadvantage. It is an advantage because

Fig. 14.12 Bayesian network
for the NorthStar example

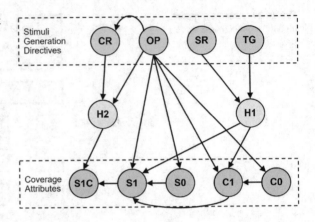

it allows the easy incorporation of domain expert knowledge into the CDG system. This domain knowledge can make the job of the CDG system easier and be the difference between success and failure. It is a disadvantage because it makes the construction of the CDG system more difficult and it requires some knowledge in Bayesian networks that verification teams may not have. Attempts to learn the structure of Bayesian networks [6] yielded results that were inferior to manual construction of the networks structure.

In a recent paper [81], Vasudevan et al. propose a new CDG system, called Design2Vec, that incorporates the design source code into the machine learning model. This approach is able to combine the advantages of model-based CDG presented in Sect. 14.3.3 with the advantages of machine learning techniques that learn the relation between the test-templates and coverage. In a nutshell, Design2Vec converts the control and data flow graph (CDFG) of the RTL code of the design into a graph neural network (GNN) [42]. Design2Vec uses branch coverage as its coverage goal, where branches are expressed as local transitions in the control graph of the design. The GNN model is trained using test-templates and the coverage they achieve. After training it can predict the probability of a given test-template hitting a given coverage event. Gradient-based search techniques can be used on the GNN model to find the test-template with the highest probability of hitting a given coverage event. Experimental results show that Design2Vec is able to hit hard-to-hit events even when these events are not presented to the GNN during training.

14.3.5 Search-Based CDG

Search-based CDG is a different approach for data-driven CDG. It replaces the reverse flow of direct CDG from coverage to tests with a search for tests that achieve the desired coverage goals. The search methods are based on exploring the tests space and the mapping from them to coverage until elements in the space that achieve the coverage goals are found. Because the tests to coverage mapping are unknown, the search for the tests that achieve the coverage goals requires the mapping to be sampled at individual points. We present two type of techniques, one that is based on fast evaluation of the test to coverage mapping and a second that is based on *derivative free optimization* (DFO) techniques [16].

14.3.5.1 Fast Evaluation of the Test to Coverage Mapping

Simulation of the DUV is an expensive operation that can require a significant amount of time and compute resources. In many cases, the cost of generating test-cases out of test-templates is much lower than the cost of simulation. Here, the generation can be separated from the rest of the verification environment and take place before the simulation begins. If a verification environment that fulfills these conditions is augmented with a component that evaluates the test-case to coverage mapping for a given test, it can filter out test-cases that do not contribute to the coverage goal. Filtering out test-cases that do not contribute to coverage can reduce the number of simulations executed without affecting the resulting coverage.

While this technique appears similar to the test-case harvesting presented in Sect. 14.3.4.1, there are important differences between the two. Test-case harvesting collects test-cases with desired coverage after they were simulated on the DUV. This makes their usability after the harvest limited, as discussed in 14.3.4.1. When test-cases are harvested after passing through the fast evaluation, they are not yet simulated on the DUV. Therefore, they can be used at least once.

Transaction-level models (TLM) and other high-level models of a DUV often exist. These models are used to build and tune the verification environment to help create high-quality test-templates. Although these models can simulate very quickly, they are not often used for test-case harvesting. There are several reasons for this. First, in many cases, the high-level model does not use the same interface as the DUV, and translation of high-level tests-cases to test-cases for the DUV is difficult. Second, the desired coverage is not always implemented or can be implemented in the high-level model. Third, test-case harvesting using fast evaluation can tolerate many false-positive errors and ends up harvesting test-cases that do not contribute to coverage. On the other hand, it can barely tolerate false-negative errors and misses test-cases that hit very hard-to-hit events. Because it is difficult to control the accuracy direction of high-level executable models, they are not popular for test-cases harvesting.

A different approach uses machine learning models to assess the value of stimuli and use only stimuli that are predicted to be useful. For example, Guzey et al. [39] propose an unsupervised support vector analysis [75] with kernel function that measures the similarity between test-cases. This kernel can be used to reject test-cases that are too close to test-cases that have already been simulated. It can also be used to cluster large sets of test-cases into n clusters and select n test-cases at or near the cluster center to maximize the coverage of n test-cases. A similar approach is presented by Blackmore et al. in [10]. The paper presents a test-selection technique that favors test-cases that are believed to be different from previous test-cases. Specifically, the paper uses autoencoder to detect anomalies in test-cases and select test-cases with high anomalies for simulation. Gou et al. [37] use a one-class SVM classifier to distinguish between stimuli that do not contribute to coverage (the class with data) and stimuli that do contribute (the class without data). Their system classifies the stimuli before the stimuli is simulated on a cycle-by-cycle basis, and if the stimuli is rejected, a new stimuli is generated instead.

The methods for fast evaluation of the test-case to coverage mapping can identify test-cases that are doing something different in terms of coverage in the machine learning-based systems. In the fast execution systems, they can even determine which events each test-case covers. But, they cannot target previously uncovered events, as is done by the direct CDG system in Sect. 14.3.4.3. Next, we describe another search-based CDG approach based on optimization techniques that does target specific events.

14.3.5.2   DFO-Based CDG

Finding the test-template that best hits a coverage event or a set of coverage events can be viewed as an optimization problem whose goal is to find the test-template that maximizes the probability of hitting the target events. Because the mapping from test-template is unknown, analytical optimization techniques cannot be used. Instead, optimization methods that do not rely on the ability to calculate derivatives of the mapping, known as derivative-free optimization (DFO), are used. DFO methods are based on sampling the mapping at several points (test-templates) and taking actions based on the sampled values.

The mapping from test-templates to coverage is probabilistic in nature. This is because random stimuli generation can lead to different coverage results when simulating different test-cases generated from the same test-template. This probabilistic nature of the mapping can be viewed as dynamic noise, which the optimization technique must be able to handle. There are several DFO techniques that can handle such noise, including implicit filtering (IF) [49], Bayesian optimization (BO) [58], simulated annealing (SA) [1], genetic algorithms (GA) [78], and particle swarm (PS) [15]. While these methods greatly differ in the way they select the test-templates to simulate at every iteration of the algorithm, all are based on the same principle of repetitive sampling of the mapping at each iteration of the algorithm. In practical use of DFO-based CDG, to reduce the level of noise, each test-template is simulated n times, and the statistics on the average number of test-cases generated from the test-template that hit an event is used as an approximation to the probability of the test-template hitting the event.

In [77], Sokorac proposes the use of a simple genetic algorithm that maximizes the score of tests, where the score is a combination of the overall number of events the test hits (the volume of the test) and the rarity of the events (its breadth) for toggle pair coverage model. To avoid convergence on a single area of the design, the proposed system uses a clustering algorithm to separate the tests to clusters that hit similar events and apply the genetic algorithm to each cluster separately.

AS-CDG [29] is a CDG system that utilizes several of the techniques and solutions described in this section. First, it uses approximated target events to replace the desired coverage and overcome the lack of evidence problem of data-driven CDG. Next it uses TAC to find the best existing test-template for hitting the approximated target. This existing test-template is used to identify the parameters within the test-template that are most relevant to the desired coverage. After the starting test-template is found, a search for the optimal setting of the parameters begins. This search starts with random sampling of the parameters space. The random sampling finds a good starting point for the IF algorithm that finds the best setting for the parameters that maximizes the probability of the desired coverage.

AS-CDG is being used in the verification of high-end processor systems in IBM. In almost all cases, each step in the CDG flow contributes to the coverage by improving the number of hits for lightly hit events and by hitting uncovered events. The few cases where the flow failed to provide the desired results occurred because the events were unhittable or because the verification environment lacked the required capabilities.

One of the weaknesses of the implicit filtering algorithm used in AS-CDG is that it does not use the history of simulated test-templates it already explored to improve its search. In [71], Roy et al. use Bayesian optimization (BO) as their DFO algorithm for their search-based CDG. BO uses the results of previous simulations to improve its belief about the best test-template. The paper shows that BO can find the SystemVerilog test-templates that maximizes the probability of hitting coverage events in small blocks. While the BO algorithm can speed up the search for the optimal test-template, it does not scale well with the number of parameters in the test-template. In the next section, we present another method that utilizes machine learning to speed up and improve the performance of "stateless" DFOs.

14.3.6   CDG for Large Sets of Unrelated Events

The DFO-based CDG techniques presented in Sect. 14.3.5.2 have been efficient in covering single coverage events or small sets of coverage events, as experimental [31] and deployment [29] results show. But using DFO-based CDG for covering larger sets of unrelated events may not be the most efficient method. The reason for this is that most DFO techniques mentioned earlier, except for Bayesian optimization, do not keep track and use the results of past simulations. As a result, the search for the best test-template for a new desired coverage cannot take advantage of previous searches, except, perhaps, in selecting the starting point for the search.[3]

---

[3] The exception is when a search for one event accidentally hit another, unrelated event with a high enough probability, making the search for the event unnecessary.

Machine learning techniques for CDG, such as the ones presented in Sect. 14.3.4.3, can handle large sets of coverage events. But, they are not being used in practice because of their disadvantages and their need for large training sets.

Next, we present a method in which the power of machine learning can be harnessed to improve the performance of DFO techniques and help them cover large sets of unrelated events. In each iteration of a DFO algorithm, the algorithm randomly selects k test-templates according to some conditions. It then simulates the selected test-templates and acts according to the simulation results. For example, in the implicit filtering algorithm, the algorithm selects k random directions and uses test-templates in a given distance from the current center as the test-templates to simulate. After simulation, if the best test-template is better than the center, it moves the center. Otherwise, it takes other actions [31, 49]. Now, assume we have an oracle in the form of a model that can predict the probability of any test-template hitting any coverage events in a very short time. The oracle can improve DFO techniques in the following way. If the oracle is very accurate, we can simply ask it for the best test-template. Even if the oracle cannot directly answer such a query, we can apply a DFO algorithm to the oracle itself to find the template.

When the oracle is not that accurate, but it is accurate in the sense of a weak learner in ensemble methods [89] (i.e., when comparing between two test-templates, the oracle is correct with probability $0.5 + \epsilon$), it can still help the DFO select better random test-templates. At each iteration of the DFO algorithm, instead of selecting k random test-templates according to the conditions, the DFO selects a much larger number (e.g., $100 \cdot k$). The DFO algorithm then uses the oracle to rank these test-templates and select the best k for simulation. It is easy to see that if the weak learner condition holds, the progress of the modified DFO will be faster on average than that of the original DFO.

One possible implementation for the oracle is a machine learning model that is trained on all previous simulations of the CDG run. Here, the oracle is equivalent to the machine learning-based test-case filtering described in Sect. 14.3.5.1 at the test-template level. But unlike test-case filtering that requires very high recall, these accuracy requirements are weaker because of the randomization of the stimuli generator. Because we cannot be sure that the weak learner condition is upheld, we let the oracle select only part (say half) of the K test-templates and choose the rest randomly. This guarantees that even if the oracle is completely wrong, the DFO would still converge to the optimal solution, but at a slightly slower pace.

The ability to improve the accuracy of the machine learning model by continuously retraining it with larger and larger training sets not only improves the performance of an individual DFO activation. It can also improve the performance of the DFO based on knowledge gained from previous DFO activations on different events. Therefore, a simple schema that uses DFO with a machine learning model for each event or small set of related events and continuously retrains the model can improve the efficiency of CDG for large sets of unrelated events. In [26], a slightly different schema is used. There, instead of running the DFO for each event, or set of related events, until it converges or a good enough test-template is found, each instance of the DFO runs for only a small number of iterations before moving to the

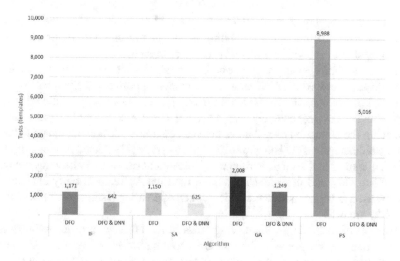

Fig. 14.13 Comparing DFOs performance, without DNN and with DNN

next event. This BFS-like schema performs better (i.e., requires fewer simulations until test-templates that nicely cover all events are found) because it provides a more uniform training for the machine learning model. A more detailed description of the schema that includes a description of various methods for selecting the next targets can be found in [26]. Figure 14.13, taken from [30], compares the number of test-templates needed to cover all the events in the NorthStar pipeline coverage model described in Sect. 14.3.4.3. The figure compares the number of test-templates needed when only DFO is used to the number of test-templates needed when a DNN is added as the machine learning model. The comparison is done for four types of DFOs: implicit filtering (IF), simulated annealing (SA), genetic algorithm (GA), and particle swarm (PS). The figure shows that adding the DNN reduces the number of test-templates needed and thus the number of simulations needed, by a factor of 1.6 (for SA) to 1.9 (for GA). The figure also shows that the simpler IF and SA algorithms significantly outperform the more complex GA and PS algorithms, both with and without the DNN. We believe that this method, using lightly trained machine learning model to boost DFO-based optimization, can also be used for other problems in general and specifically in the EDA domain.

14.4   Risk Analysis

During the execution of the verification process, there is a continuous need to react and fix areas in the design that are less reliable or less stable than others. An area may be less stable simply because it undergoes a more intensive development, due to design flaws, or even because of changes in the surrounding areas. In most cases, it will be an interplay of these factors. If we could accurately reflect the relative or, better yet, the absolute current stability levels, then design and verification teams would be able to act with the correct priority or urgency.

Accurate stability analysis can be an important decision-support tool in the hands of the design and verification teams. It provides opportunities for many types of actions. For example, the design team can target unstable areas by holding deep dive design reviews or peer code reviews, deferring a release, rescheduling plans, and re-balancing the allocation of resources between continued development and debug or other stabilization tasks. The verification team can target selected design areas with directed testing and formal verification, implement additional checking methods, add debug aids, and use refined coverage metrics. Accurate stability analysis can also be used to trigger automatic actions, such as creating testing policies that target the risky areas [24].

One set of methods for identifying areas in the DUV with lower stability is to observe and analyze some of the measures coming out of the verification process. For example, the fail rate of simulations targeting a given area and the number of bugs opened against it are good indicators for its stability. The visualization of such measures over time (e.g., Fig. 14.15) or the use of time series analysis [13] can be used to identify changes in the stability. These methods are efficient at identifying unstable areas and changes in the stability, but they can only detect such changes after they are exposed. This can cause unnecessary delays in the initiation of corrective actions to address the instability. Another disadvantage of these methods is the need for a deeper analysis to understand whether the changes in the measures are caused by changes in the stability. For example, if the job submitter stops sending simulation jobs from a particular test-template after enough failing test-cases generated are collected, then the fail rate for the target area of the test-template may improve without any change in its stability.

The main source for instability and changes in the stability of areas is the source code of the DUV (and its verification environment) and changes made in that code. Analysis, understanding the source code, and determining the changes it goes through can help identify unstable areas in the DUV. Moreover, this type of analysis can help detect such areas before they are exposed by the verification process. There are two main approaches for stability, or at-risk, analysis: static and dynamic. The static approach is based on analyzing the complexity of the code [91]. The complexity of the code is assessed using static measures, such as the length of functions, properties of the call graph, and more. The problem is that this approach does not consider the level of changes in the code. The dynamic approach is based on the history of changes to the code. Here, sources of information that can help analyze the stability and risk for areas in the design are version control logs, defects records, and the connection between the two. This helps identify the areas of the design (e.g., source files) in which defects were found and fixed or new features were added.

In this section, we provide two examples of techniques for identifying at-risk areas in the design. The first is based on the time-series trend analysis of verification measures. Specifically, we show how the use of the Shewhart control charts [76] on bug discovery rate information helps identify changes in the stability of large areas in the DUV (e.g., units). The second example describes a system that identifies at-risk files in hardware designs, where the risk of a file is defined as the probability that a bug will be found in the file in the course of the next D days [25]. The approach uses a simple, machine learning-based, regression system that receives as input data from various sources, including the version control logs and the defect tracking records. For each file in the system, it produces a grade for its risk, as defined above.

14.4.1 Trend Analysis of Bug Discovery Rate

Time series and trend analysis have been used in the hardware verification and software testing domains for many years [55, 56]. The most common measure used in such analysis is the bug discovery rate. However, other measures such as coverage progress [41] have also been used. Many analysis techniques focus on predictive analysis. These techniques try to fit the desired time-series data to a given parameterized model by finding the values of the parameters in the model that best fit the data. If the fit is good enough, the model can be used to predict future behavior. For example, if the bug discovery rate data fits a given model, then the model can predict when the discovery rate will become low enough to meet the tape-out criteria or what the discovery rate will be at the planned tape-out date.

Descriptive analysis of the data often uses simpler models and simpler analysis techniques to describe the current state as reflected by the data. For example, the Shewhart control charts [76] developed by Walter Shewhart at the AT&T Bell Laboratories in the 1920s examine the behavior of the desired time-series and compare it to the expected behavior (mean) and lower and upper control limits. The analysis identifies three types of behaviors: data above the upper, and below the lower, control limit, indicating an anomaly in the data; a trend of x consecutive increasing or decreasing observations, which indicates a change in trend; and a sequence of y consecutive observations above or below the mean, which also indicates a trend. Figure 14.14 shows an example of a Shewhart control chart for a given time-series. The green line presents the mean of the sampled data, while the two red lines are the upper and lower control limits. The red marks indicate positive and negatives anomalies. The blue marks indicate a positive trend of seven consecutive measures above the mean.

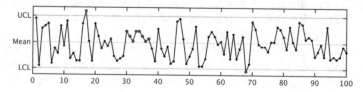

Fig. 14.14 Example of Shewhart control chart

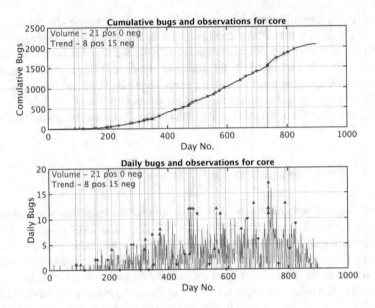

Fig. 14.15 Trend analysis results for high-end processor core

Applying Shewhart control charts to bug discovery raises many challenges. First, most models for bug discovery data are nonlinear [55]. Therefore, the constant behavior assumed by Shewhart control charts does not hold. Our experience shows that the bug discovery rate is affected by two main factors: the ability of the verification team to find bugs and their ability to handle bugs. The ability to find bugs depends on the maturity of the DUV and the number of bugs left. This means it changes throughout the lifetime of the project and behaves like known concave or S-shaped models [55]. The ability to handle bugs, on the other hand, depends mostly on the capabilities of the team, which do not change over time. We observed that throughout most of the project, the dominant factor in determining the bug discovery rate is the ability to handle bugs, not to find them. This is true except for the early stages of the project, when the team is not fully engaged in finding and fixing bugs, and toward the end of the project, when bugs are much harder to find. Figure 14.15 shows the cumulative number of bugs found in the processor core of a high-end processor from the beginning of the project until tape-out and the number of bugs found daily. While it is hard to see it from the daily plot, the cumulative plot shows that between days 300 and 750, the growth in the total number of bugs found is almost linear. Note that there is another linear growth period between days 0 and 200, but this interval in the early stages of the project is of less interest.

Even when the total number of bugs grows linearly, the daily discovery rate of bugs is not constant. There is less verification activity during weekends, so the average number of bugs found during the weekend is lower than the number of bugs found during weekdays. This behavior can affect the Shewhart control charts, which depend on consecutive days of similar behavior. A simple solution to this issue is to replace the daily measures with a sliding window of the last week. Daily measures are still used, but every measure used in the analysis contains five weekdays and two weekend days. This sliding window delays the detection of trend changes and reduces the sensitivity of anomaly detection; however, it increases the precision of trend change detection. Other calendar events, such as holiday seasons, are harder to handle because of their length and irregularity. The analysis method presented here does not consider such periods. Indeed the analysis results show the expected negative trends during holiday season (e.g., days 420–450 in Fig. 14.15) followed by a positive trend when the season ends.

A third challenge in applying Shewhart control charts to bug discovery is the fact that the parameters for the analysis (i.e., the mean and lower and upper control limits) are not known and cannot be predicted from previous or similar designs. To address this challenge, the analysis method estimates the parameters from the data itself. Specifically, given a period of linear growth in the total number of bugs, the mean for the Shewhart control charts is the average number of daily bugs found during the period when the lower and upper control limits are placed three standard deviations away from it, similar to considering the sliding window effect. In all cases where the analysis was used, the standard deviation of the daily bugs found was large compared to the mean; hence, the lower control limit becomes negative and thus meaningless.

Algorithm 14.2 summarizes the analysis technique based on Shewhart control charts. The input to the algorithm is a time-series of the daily bug discovery data. To overcome the weekend challenge, the algorithm first replaces each daily value with the average value of the 7 days preceding that day (including the date itself). Next the algorithm searches for an interval where the total number of bugs grows almost linearly. It does so by calculating a linear regressor for all possible intervals with a length greater than 50 days that end on the current date. It then selects the interval with the minimal mean squared error (MSE). If the minimal MSE is not small enough, the algorithm reports that it cannot perform the analysis and exits.

Algorithm 14.2 Shewhart control charts for bug discovery data

---

procedure TRENDANALYSISFORBUGS(D)   ▷ D is a time-series of daily bug discovery data
  $SW \leftarrow SlidingWindow(D)$
  $Cum \leftarrow cumsum(D)$ ▷ Total number of bugs found
  $len \leftarrow length(D)$
  // *Find long linear interval*
  for $i \leftarrow 0$ **to** $len - 50$
  $e[i] \leftarrow MSE(LinearRegressor(Cum[i : len]))$
  $best \leftarrow \arg\max_i(e[i])$ ▷ *best* is starting point of best interval
  // *Calculate parameters for Shewhart control charts and apply it to interval*
  $mean \leftarrow avg(SW[best : len])$
  $ucl \leftarrow mean + 3 \cdot std(SW[best : len])$
  $lcl \leftarrow mean - 3 \cdot std(SW[best : len])$
  Apply $ShewhartControlCharts(SW[best : len], mean, ucl, lcl)$
  Report results for last day

Otherwise, the algorithm calculates the mean and lower and upper control limits for the selected interval, performs the Shewhart control charts analysis, and reports the results of the last day.

Figure 14.15 shows the analysis results for the core of a high-end processor from the beginning of the project until the first tape-out. The figure shows the results on top of the daily data (bottom plot) and the cumulative number of bugs (top plot). The red triangles mark the days when an anomaly was detected, and the blue triangles mark the beginning of changes in the trend, where the triangle points in the direction of the trend. The number of events marked by the analysis (44) is not large for the $2\frac{1}{2}$ years' duration of the project, so the analysis does not overburden the verification team with reports. When validating the reported events with the verification team leaders, they confirmed the reported events, but they were aware of almost all of them before they were reported. This illustrates one of the challenges of such analysis techniques, namely, the need to reduce the reporting of information that the verification team already has.

14.4.2 Source Code Repository Analysis

The relation between source code changes that appear in version control logs and defects is the target of vast amount of research, mostly in the software engineering community. For example, Lewis and Ou [53] claim that simply ranking files by the number of times they have been changed with a bug-fixing commits will find the hot spots in a code base. This method is simple, fast, and easy to communicate to others. Zimmermann et al. [90] describe a tool that learns association rules between files and predicts further changes in files or functions. In [69], Rolfsnes et al. improved these results by analyzing more partial cases. They were also able to provide many interesting insights for the system characteristics examined.

Similar techniques have also been proposed in the hardware verification domain. The PinDown tool by Verifyter [63] mines the source code repository and performs a smart search on the list of changes to identify the commit (or set of commits) that cause regression tests to fail. Parizi et al. [61] used a system that predicts bugs in hardware designs. Their system relies solely on the activity of source code revision to calculate the entropy of the changes and predict bugs based on this method. Guo et al. [38] added more features to the analysis. These features are based on the code complexity (e.g., number of lines and number of decision points) and its changes (e.g., number of changes and number of bug fixes). Efendioglu et al. [19] applied similar techniques, but instead of using them at the RT level, they applied them to high-level models in SystemC. The system in [25] uses a different target. Instead of identifying files that contain bugs, it looks for files at risk, that is, files in which a bug will be found within the next D days. This is an important measure because it predicts the near future of the verification process and allows the design and verification teams to take preventive measures, such as code reviews or mitigating measures, such as directing more tests to the files at risk. Next, we review the system designed to identify files at risk in the IBM hardware design and verification processes.

Designing a learning system that predicts the risk level for components in a system under development involves many design-point decisions. The first decision involved the granularity level of the analysis. Here, the choices ranged from the sub-file level (e.g., entity or function level) all the way to the module level (a top level of few entities implementing a higher-level functionality). We decided to work at the file level because it provides a natural interface to the users. Moreover, in the design methodology at IBM, most files contain a single entity, so the sub-file level would not be much different from the file level; the module level is too coarse, and every module is always at risk.

The second decision involves the source data for the analysis. The source code itself can be a repository for information about the complexity of the code [91]. This information is more or less static. The at-risk analysis is interested in the dynamics of code changes; therefore, it was not included in the source data. Instead, the system is based on the history of changes because it better fits the evolving nature of a system under development and it implicitly considers the maturity of the code. Another data source that is used comes from the bug tracking system, because areas in the code where bugs were found recently are most likely to have more bugs until they converge. Finally, because the system tries to predict the existence of bugs and their discovery, it uses statistics from the batch job submission tool.

Once the data sources have been identified, the next steps are to connect these data sources and decide which features to extract from the data sources. Connecting the data sources is needed for when we map bugs described in the bugs tracking system to commits and files. It is also used to extract the number of executed and failed test-cases in a given period. Connecting bugs to commits is a tricky challenge because in many cases, there is no strict rule or methodology to enforce documenting this information. The system identifies defects handled in a commit in two ways: the ID of the defect appears in the commit message, or the ID appears in the changed text of a changed file. These two methods rely on the *proper* action of the committer and involve free text parsing. Hence it is not guaranteed that they detect all defects related to a file change. Nevertheless, experimental results show that the system can predict quite well based on this method. To extract the number of test-cases executed for each file, the system uses a very coarse measure of the number of test-cases executed in the environment of the unit in which the file is included. While a finer mapping that maps test-templates to files via coverage events exists, this measure is too ambiguous and imprecise to be used in the direction from test-templates to files. Later in this section, we describe how this mapping is used in the other direction, from files to test-templates.

The system uses hundreds of features per file. Some of the features provide general information about the file. Other features are related to the version control activities of the file, bug fixes in the file, and verification activities of the file. These features are time related. Examples of general information features used in the system include the size of the file (number of lines) and the file owner. Version control features include the number of commits and the number of lines added, changed, or deleted. Features related to bug fixes include the number of defects and high-severity defects connected to the file and the number of defects connected to the neighbors of the file. The neighbors of a file are files that often change along with it. The system uses another learning phase to learn the association rules between files, meaning the probability that if $file_i$ is changed, then $file_j$ is changed as well. Features related to the verification effort include the number of test-cases run on the unit containing the file and the failing rate of these test-cases.

Generally, a risk analysis system can operate in one of two modes. In the first mode, the analysis is done after every commit or push operation on all the files involved in the commit. In this mode, the goal of the system is to report to the committer on committed files at-risk or other files that are at risk because they are not part of the commit. This type of system fits CI/CD development methodologies because it identifies the risk of every commit. But such a system requires very high accuracy because users tend to ignore systems that provide many false alarms or missed detections. In the second mode, the analysis is done periodically on all the files in the design, where a period can be either time-based (e.g., weekly) or release-based. The system uses the periodic analysis because when the system was developed, it was a better fit for the IBM development methodology.

Because the system tracks the history of changes, the next decision involved how much history to use. On the one hand, recent history is the most relevant to current predictions. On the other hand, using only recent history may lead to insufficient data for training the system. In addition, the recent history may not be complete because bugs that were found recently may not yet be debugged, and therefore, the file that caused them is not yet known. To address the issues of shorter and longer histories, the system uses separate features that are related to the long- and short-term history. Specifically, VC uses a history of 12 weeks in its training, and the periodic features of files use 12 short periods of a week (7 days) and 3 long periods of a month (30 days). Therefore, if the sets of periodic features for a file contain p features, there are $p \cdot (12 + 3)$ features overall, and the features matrix at time t, X_t is the following:

$$
X_t = \begin{pmatrix}
\vdots & & \vdots & \vdots & & \vdots \\
features\ at & & features\ at & features\ at & & features\ at \\
week_{t-1} & \cdots & week_{t-12} & month_{t-1} & \cdots & month_{t-3} \\
\vdots & & \vdots & \vdots & & \vdots
\end{pmatrix}
$$

where each column in the matrix above represents p columns in the actual matrix.

The labels vector y_t contains the binary indicators whether each file is at risk at time t. That is, $y_t[i] = 1$ if and only if a bug is found in $file_i$ in the D days following t. The goal of the system is to use the model to predict y_t using the current features matrix X_t. To achieve this goal, the system is trained on the data of the 2 weeks preceding t, as shown in Eq. (14.3).

$$\begin{pmatrix} X_{t-1} \\ X_{t-2} \end{pmatrix} \begin{pmatrix} y_{t-1} \\ y_{t-2} \end{pmatrix} \qquad (14.3)$$

The final decision involved the output of the system. The simplest approach was to use a classifier as the learning engine and use its output as the output of the system. More complex solutions use regressors as the learning engine. The regressor grades the files according to the probability of being at risk. Given these grades, there are several methods for deciding which files are at risk and which are not. The first methods use a constant-grade threshold or report a constant number of files. Experimental results showed that with these methods, it is very difficult to find the correct constant because it is changing both with the design and over time. The method used in the system is to dynamically change the number of risky files identified by the system. Specifically, the system uses a threshold based on last week, *number of actual files at risk / number of files changed*, which performs better than a default value or the last week's *number of actual files at risk*.

The system uses the XGBoost library [87] as the machine learning engine. The XGBoost regressor is trained to maximize the **F1** score, targeting both precision and recall. XGBoost is an implementation of gradient boosted decision trees designed for speed and performance. It is an ensemble of weak learners, where each weak learner is a decision tree. The boosting in XGBoost is optimized using the gradient descent algorithm. The system uses the XGBoost classic machine learning model, instead of the deep neural network mode. This is because XGBoost can provide explainable results in the form of a ranked list of influencing features, which makes the system more appealing to users. The output of the XGBoost regressor is a grade for each file's risk level. The grades are used to create an ordered list of the relative risks. To provide the list of files at risk, we use the dynamic threshold described above.

The system was evaluated on data from two high-end processor designs in IBM. For both projects, the evaluation included using the system on the data of the entire project and on the data of individual units in the project. The evaluation focused on two important measures: the positive precision, defined as the probability of having a defect when we predict one, and positive recall, defined as the probability of predicting a defect when there is one. High values for these measures ensure that the design and verification teams who respond to the reports generated by the system do not waste time on falsely identified files at risk, and do not miss files at risk that are not identified by the system.

Figure 14.16 shows the positive precision and recall for the entire project, over 45 weeks. The blue dashed line is the positive precision on a weekly run, and the orange dashed line is the recall. The bold lines are polynomial approximations of these lines. For the entire project time, we reached roughly 80% precision and 60% recall. These numbers are not low, but they are not extremely high. Therefore, while the system can detect at-risk files and provide information to the design and verification teams, it suffers from false alarms and misdetections.

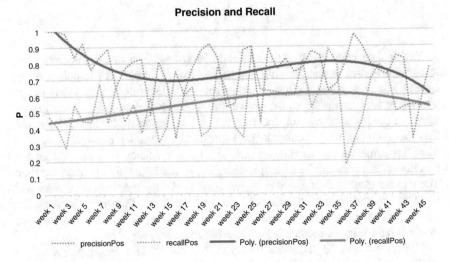

Fig. 14.16 Precision and recall for the entire processor over 45 weeks

The precision and recall of the system are not sufficient to judge its quality and usefulness. This requires two more factors. First, the performance of the system needs to be compared to trivial alternatives. In addition, the actual use-cases of the system and their benefits need to be evaluated. There are two trivial alternatives to this system. The first ranks the risk level of a file based on the number of changes it went through during the previous period [53]. The second puts every file in which a bug was found in the previous period at risk. Analysis of the evaluation data and results shows that more than 20% of the positive predictions made by the system do not come out of one of these trivial sets and that many of these nontrivial positive predictions are correct.

The system can be used in two ways. First, it can provide valuable information to support decisions made by the design and the verification teams. This requires not only reporting correctly but also introducing the experts to *something they do not know*. Given the levels of expertise in many verification teams, this is a very high bar to clear. The reason is that out of the at-risk files correctly reported by the system, most of the files are already known to the team; sifting out the known files is beyond the capabilities of the system.

The second way to provide benefit is to feed the information from the system directly back to the verification process. This direct feedback does not allow all the actions available when at-risk files are reported to the design and verification teams (e.g., code reviews or use of formal verification). However, it is less sensitive to previously known information and mistakes in the identification of at-risk files. One possible automatic action is to explore the mapping between test-templates and files and slightly shift the simulation effort toward the at-risk files by increasing the number of test-cases generated from templates associated with the at-risk files during regression runs. This mapping from files to test-templates begins with finding the coverage events that are related to the file. This can be done using static analysis of the design and verification environment code. Once the coverage events related to the at-risk files are identified, coverage analysis techniques, such as TAC (see Sect. 14.3.4.2 and [24]), can map coverage events to test-templates and design regression suites that maximize the coverage of the coverage events related to the file.

14.5 Summary

Data analytics and machine learning provide powerful weapons for analyzing vast amounts of data and extracting useful information alongside insights. Verification environments utilize many tools producing a lot of data that is important not just for the tools themselves but also for the entire verification process. Therefore, harnessing the power of data analytics and machine learning for the verification process can improve its quality and ease the life of the verification teams.

This chapter showed how such harnessing happens. It begins by connecting all the relevant verification tools and data sources to the Verification Cockpit. The Verification Cockpit is a platform that collects and stores all the relevant verification data. It also provides internal reporting and analysis engines alongside efficient data retrieval interfaces that connect to external analysis tools and engines. Several such analysis engines, like the ones presented in this chapter, are already connected to the Verification Cockpit.

Despite the power of data analytics and machine learning, and their ability to assist in the verification effort, their use is still not widespread, and they do not address some important aspects of the verification process. Triage is one such example. Triage in the hardware development domain is a pre-debug process that handles a set of failures (e.g., from a nightly regression run) and establishes a set of traits for each failure to simplify and increase the efficiency of its handling. The most common and important triage activities include clustering failures with the same root cause and identifying failures with a known bug record. Even though these activities are naturally mapped to unsupervised and supervised machine learning tasks, and despite the research in this area (e.g., [3, 35, 66]), machine learning-based triage systems are still not available.

There are several reasons why data analytics and machine learning cannot efficiently address the triage problem. Some of these reasons are specific to the triage problem, while others arise from the verification domain. Among the specific reasons, we briefly mention the difficulty in obtaining labeled data and the rapid changes a design goes through during development, which makes even its recent history less relevant.

One issue that verification raises for data analytics and machine learning is the type of user of these technologies. Verification engineers are usually experts in their domain and as such have a higher level of expectations from decision support systems. For example, they do not tolerate systems that provide trivial information.

Therefore, the at-risk file identification system described in Sect. 14.4.2, which finds many trivial at-risk files, is not widely used in a decision support role. One possible solution is to use automatic feedback given to the system instead of providing the information to the users. This solution, of course, is not always feasible. When it is feasible, it can work only if the system is highly accurate or the cost of errors is not high, which is true for the at-risk files system but not for triage.

Another challenge with verification is its chaotic nature. For example, random stimuli generators try to put as much noise as possible in generated test-cases while preserving the test-case requirements, to ensure that every test-case explores new paths in the DUV and performs useful verification work. While machine learning techniques are designed to generalize and remove noise from the data, some of the verification data may be too noisy for data analytics and machine learning techniques.

This does not mean that we should lose hope for new solutions that utilize data analytics and machine learning in the verification process. It just means that we need to better adapt existing techniques to the challenges of verification, develop new data analytics and machine learning techniques that fit verification, adapt the verification process to data analytics and machine learning, and find ways to combine them with the existing solutions and the domain expert knowledge.

References

1. Aarts, E., Korst, J., Michiels, W.: Simulated annealing. In: Burke, E.K., Kendall, G. (eds.) Search Methodologies: Introductory Tutorials in Optimization and Decision Support Techniques, pp. 187–210. Springer US, Boston, MA (2005). https://doi.org/10.1007/0-387-28356-0_7
2. Amrani, M., Lúcio, L., Bibal, A.: ML + FV = ♥? A survey on the application of machine learning to formal verification. CoRR abs/1806.03600 (2018). https://arxiv.org/abs/1806.03600
3. Angell, R., Oztalay, B., DeOrio, A.: A topological approach to hardware bug triage. In: 2015 16th International Workshop on Microprocessor and SOC Test and Verification (MTV), pp. 20–25 (2015). https://doi.org/10.1109/MTV.2015.10
4. Arar, M., Behm, M., Boni, O., Gal, R., Goldin, A., Ilyaev, M., Kermany, E., Reysa, J., Saleh, B., Schubert, K.D., Shurek, G., Ziv, A.: The verification cockpit—creating the dream playground for data analytics over the verification process. In: Proceedings of the 11th Haifa Verification Conference, pp. 104–119 (2015)
5. Azatchi, H., Fournier, L., Marcus, E., Ur, S., Ziv, A., Zohar, K.: Advanced analysis techniques for cross-product coverage. IEEE Trans. Comput. **55**(11), 1367–1379 (2006)
6. Baras, D., Fine, S., Fournier, L., Geiger, D., Ziv, A.: Automatic boosting of cross-product coverage using Bayesian networks. Int. J. Softw. Tools Technol. Transfer **13**(3), 247–261 (2011)
7. Biere, A., Heule, M., van Maaren, H., Walsch, T.: Handbook of Satisfiability. IOS Press, New York (2008)
8. Bin, E., Emek, R., Shurek, G., Ziv, A.: Using a constraint satisfaction formulation and solution techniques for random test program generation. IBM Syst. J. **41**(3), 386–402 (2002)
9. Birnbaum, A., Fournier, L., Mittermaier, S., Ziv, A.: Reverse coverage analysis. In: Eder, K., Lourenço, J., Shehory, O. (eds.) Hardware and Software: Verification and Testing, pp. 190–202. Springer, Berlin (2012)

10. Blackmore, T., Hodson, R., Schaal, S.: Novelty-driven verification: Using machine learning to identify novel stimuli and close coverage. In: Proceedings of the design and verification conference and exhibition US (DVCon) (2021)
11. Blum, A., Hopcroft, J., Kannan, R.: Foundations of Data Science. Cambridge University Press, Cambridge (2020)
12. Campenhout, D.V., Mudge, T., Hayes, J.P.: High-level test generation for design verification of pipelined microprocessors. In: Proceedings of the 36th Design Automation Conference, pp. 185–188 (1999)
13. Chatfield, C., Xing, H.: The Analysis of Time Series: An Introduction with R. CRC Press, New York (2019)
14. Clarke, E.M., Grumberg, O., Peled, D.A.: Model Checking. MIT-Press, New York (1999)
15. Clerc, M.: Particle Swarm Optimization. ISTE. Wiley, New York (2010). https://books.google.co.il/books?id=Slee72idZ8EC
16. Conn, A., Scheinberg, K., Vicente, L.: Introduction to Derivative-Free Optimization. SIAM, Philadelphia (2009)
17. Copty, S., Fine, S., Ur, S., Yom-Tov, E., Ziv, A.: A probabilistic alternative to regression suites. Theor. Comput. Sci. **404**(3), 219–234 (2008)
18. Davis, M., Putnam, H.: A computing procedure for quantification theory. J. ACM **7**(3), 201–215 (1960). https://doi.org/10.1145/321033.321034
19. Efendioglu, M., Sen, A., Koroglu, Y.: Bug prediction of systemC models using machine learning. IEEE Trans. Comput. Aided Des. Integr. Circuits Syst. **38**(3), 419–429 (2019). https://doi.org/10.1109/TCAD.2018.2878193
20. Fine, S., Freund, A., Jaeger, I., Mansour, Y., Naveh, Y., Ziv, A.: Harnessing machine learning to improve the success rate of stimuli generation. IEEE Trans. Comput. **55**(11), 1344–1355 (2006)
21. Fine, S., Fournier, L., Ziv, A.: Using Bayesian networks and virtual coverage to hit hard-to-reach events. Int. J. Softw. Tools Technol. Transfer **11**(4), 291–305 (2009)
22. Fine, S., Ziv, A.: Coverage directed test generation for functional verification using Bayesian networks. In: Proceedings of the 40th Design Automation Conference, pp. 286–291 (2003)
23. Foster, H.: The 2020 Wilson research group functional verification study part 8 IC/ASIC resource trends. https://blogs.sw.siemens.com/verificationhorizons/2021/01/06/part-8-the-2020-wilson-research-group-functional-verification-study
24. Gal, R., Kermany, E., Saleh, B., A.Ziv, Behm, M.L., Hickerson, B.G.: Template aware coverage: Taking coverage analysis to the next level. In: Proceedings of the 54th Design Automation Conference, pp. 36:1–36:6 (2017)
25. Gal, R., Shurek, G., Simchoni, G., Ziv, A.: Risk analysis based on design version control data. In: 2019 ACM/IEEE 1st Workshop on Machine Learning for CAD (MLCAD), pp. 1–6 (2019). https://doi.org/10.1109/MLCAD48534.2019.9142105
26. Gal, R., Haber, E., Ziv, A.: Using dnns and smart sampling for coverage closure acceleration. In: Proceedings of the 2020 ACM/IEEE Workshop on Machine Learning for CAD, pp. 15–20 (2020). https://doi.org/10.1145/3380446.3430627
27. Gal, R., Kermany, H., Ivrii, A., Nevo, Z., Ziv, A.: Late breaking results: Friends—finding related interesting events via neighbor detection. In: 2020 57th ACM/IEEE Design Automation Conference (DAC), pp. 1–2 (2020). https://doi.org/10.1109/DAC18072.2020.9218685
28. Gal, R., Simchoni, G., Ziv, A.: Using machine learning clustering to find large coverage holes. In: Proceedings of the 2020 ACM/IEEE Workshop on Machine Learning for CAD, MLCAD '20, pp. 139–144. Association for Computing Machinery, New York (2020). https://doi.org/10.1145/3380446.3430621
29. Gal, R., Haber, E., Ibraheem, W., Irwin, B., Nevo, Z., Ziv, A.: Automatic scalable system for the coverage-directed generation (CDG) problem. In: Proceedings of the Design, Automation and Test in Europe Conference (2021)
30. Gal, R., Haber, E., Irwin, B., Mouallem, M., Saleh, B., Ziv, A.: Using deep neural networks and derivative free optimization to accelerate coverage closure. In: Proceedings of the 2021 ACM/IEEE Workshop on Machine Learning for CAD (2021)

31. Gal, R., Haber, E., Irwin, B., Saleh, B., Ziv, A.: How to catch a lion in the desert: on the solution of the coverage directed generation (CDG) problem. Optim. Eng. **22**(1), 217–245 (2021). https://doi.org/10.1007/s11081-020-09507-w

32. Geng, H.: Semiconductor Manufacturing Handbook. McGraw-Hill Education, New York (2017)

33. Goodfellow, I., Bengio, Y., Courville, A.: Deep Learning. MIT Press, New York (2016). http://www.deeplearningbook.org

34. Goodfellow, I., Bengio, Y., Courville, A.: Deep Learning. MIT Press, New York (2016)

35. Goyal, A., Sardana, N.: Empirical analysis of ensemble machine learning techniques for bug triaging. In: 2019 Twelfth International Conference on Contemporary Computing (IC3), pp. 1–6 (2019). https://doi.org/10.1109/IC3.2019.8844876

36. Grinwald, R., Harel, E., Orgad, M., Ur, S., Ziv, A.: User defined coverage—a tool supported methodology for design verification. In: Proceedings of the 35th Design Automation Conference, pp. 158–165 (1998)

37. Guo, Q., Chen, T., Shen, H., Chen, Y., Hu, W.: On-the-fly reduction of stimuli for functional verification. In: 2010 19th IEEE Asian Test Symposium, pp. 448–454 (2010). https://doi.org/10.1109/ATS.2010.82

38. Guo, Q., Chen, T., Chen, Y., Wang, R., Chen, H., Hu, W., Chen, G.: Pre-silicon bug forecast. IEEE Trans. Comput. Aided Des. Integr. Circuits Syst. **33**(3), 451–463 (2014). https://doi.org/10.1109/TCAD.2013.2288688

39. Guzey, O., Wang, L.C., Levitt, J.R., Foster, H.: Increasing the efficiency of simulation-based functional verification through unsupervised support vector analysis. IEEE Trans. Comput. Aided Des. Integr. Circuits Syst. **29**(1), 138–148 (2010). https://doi.org/10.1109/TCAD.2009.2034347

40. Haim, S., Walsh, T.: Restart strategy selection using machine learning techniques. In: Kullmann, O. (ed.) Theory and Applications of Satisfiability Testing—SAT 2009, pp. 312–325. Springer, Berlin (2009)

41. Hajjar, A., Chen, T., Munn, I., Andrews, A., Bjorkman, M.: High quality behavioral verification using statistical stopping criteria. In: Proceedings of the 2001 Design, Automation and Test in Europe Conference, pp. 411–418 (2001)

42. Hamilton, W.L.: Graph Representation Learning. Morgan & Claypool Publishers, Los Altos (2020)

43. Hsiou-Wen, H., Eder, K.: Test directive generation for functional coverage closure using inductive logic programming. In: Proceedings of the High-Level Design Validation and Test Workshop, pp. 11–18 (2006)

44. Ibm cognos analytics. https://www.ibm.com/products/cognos-analytics

45. IBM DB2 with BLU Acceleration. https://www.redbooks.ibm.com/abstracts/tips1204.html

46. Iwashita, H., Kowatari, S., Nakata, T., Hirose, F.: Automatic test program generation for pipelined processors. In: Proceedings of the International Conference on Computer Aided Design, pp. 580–583 (1994)

47. James, G., Witten, D., Hastie, T., Tibshirani, R.: An Introduction to Statistical Learning with Applications in R. Springer Text in Statistics. Springer, Berlin (2013)

48. Karnaugh, M.: The map method for synthesis of combinational logic circuits. Trans. Am. Inst. Electr. Eng. **72**(9), 593–599 (1953)

49. Kelley, C.: Implicit Filtering. SIAM, Philadelphia (2011)

50. Lachish, O., Marcus, E., Ur, S., Ziv, A.: Hole analysis for functional coverage data. In: Proceedings of the 39th Design Automation Conference, pp. 807–812 (2002)

51. Lagoudakis, M.G., Littman, M.L.: Learning to select branching rules in the DPLL procedure for satisfiability. Electron Notes Discrete Math. **9**, 344–359 (2001). https://doi.org/10.1016/S1571-0653(04)00332-4. https://www.sciencedirect.com/science/article/pii/S1571065304003324. LICS 2001 Workshop on Theory and Applications of Satisfiability Testing (SAT 2001)

52. Lee, D.D., Seung, H.S.: Algorithms for non-negative matrix factorization. In: Leen, T.K., Dietterich, T.G., Tresp, V. (eds.) Advances in Neural Information Processing Systems, vol. 13, pp. 556–562 (2001)
53. Lewis, C., Ou, R.: Bug prediction at Google (2011). http://google-engtools.blogspot.sg/2011/12/bug-prediction-at-google.html. [Online; accessed 25-Oct-2019]
54. Liang, J., Ganesh, V., Poupart, P., Czarnecki, K.: Exponential recency weighted average branching heuristic for sat solvers. Proceedings of the AAAI Conference on Artificial Intelligence 30(1) (2016). https://ojs.aaai.org/index.php/AAAI/article/view/10439
55. Lyu, M.: The Handbook of Software Reliability Engineering. McGraw Hill, New York (1996)
56. Malka, Y., Ziv, A.: Design reliability—estimation through statistical analysis of bug discovery data. In: Proceedings of the 35th Design Automation Conference, pp. 644–649 (1998)
57. Mishra, P., Dutt, N.: Automatic functional test program generation for pipelined processors using model checking. In: Seventh Annual IEEE International Workshop on High-Level Design Validation and Test, pp. 99–103 (2002)
58. Mockus, J.: Bayesian Approach to Global Optimization: Theory and Applications. Springer, Berlin (1989)
59. Molitor, P., Mohnke, J.: Equivalence Checking of Digital Circuits: Fundamentals, Principles, Methods. Kluwer Academic Publishers, Dordrecht (2004)
60. Nudelman, E., Leyton-Brown, K., Hoos, H.H., Devkar, A., Shoham, Y.: Understanding random SAT: Beyond the clauses-to-variables ratio. In: Wallace, M. (ed.) Principles and Practice of Constraint Programming—CP 2004, pp. 438–452. Springer, Berlin (2004)
61. Parizy, M., Takayama, K., Kanazawa, Y.: Software defect prediction for LSI designs. In: 2014 IEEE International Conference on Software Maintenance and Evolution, pp. 565–568 (2014). https://doi.org/10.1109/ICSME.2014.96
62. Pearl, J.: Probabilistic Reasoning in Intelligent Systems: Network of Plausible Inference. Morgan Kaufmann, Los Altos (1988)
63. PinDown—an Automatic Debugger of Regression Failures. https://verifyter.com/technology/debug. [Online; accessed 25-Oct-2021]
64. Piziali, A.: Functional Verification Coverage Measurement and Analysis. Springer, Berlin (2004)
65. Ponniah, P.: Data Warehousing Fundamentals for IT Professionals, 2nd edn. Wiley, Hoboken (2010)
66. Poulos, Z., Veneris, A.: Exemplar-based failure triage for regression design debugging. In: 2015 16th Latin-American Test Symposium (LATS), pp. 1–6 (2015). https://doi.org/10.1109/LATW.2015.7102521
67. Pyne, S., Rao, B.P., Rao, S.: Big Data Analytics Methods and Applications. Springer, Berlin (2016)
68. Reddy, C.K., Aggarwal, C.C.: Data Clustering. Chapman and Hall/CRC, New York (2016)
69. Rolfsnes, T., Alesio, S.D., Behjati, R., Moonen, L., Binkley, D.W.: Generalizing the analysis of evolutionary coupling for software change impact analysis. In: IEEE 23rd International Conference on Software Analysis, Evolution, and Reengineering (SANER), pp. 201–212 (2016)
70. Roy, R., Duvedi, C., Godil, S., Williams, M.: Deep predictive coverage collection. In: Proceedings of the Design and Verification Conference and Exhibition US (DVCon) (2018)
71. Roy, R., Benipal, M., Godil, S.: Dynamically optimized test generation using machine learning. In: Proceedings of the Design and Verification Conference and Exhibition US (DVCon) (2021)
72. Schubert, K.D., Roesner, W., Ludden, J.M., Jackson, J., Buchert, J., Paruthi, V., Behm, M., Ziv, A., Schumann, J., Meissner, C., Koesters, J., Hsu, J., Brock, B.: Functional verification of the IBM POWER7 microprocessor and POWER7 multiprocessor systems. IBM J. Res. Dev. 55(3), 308–324 (2011)
73. Schubert, K., et al.: Solutions to IBM POWER8 verification challenges. IBM J. Res. Dev. 59(1), 1–17 (2015)
74. Shalev-Shwartz, S., Ben-David, S.: Understanding Machine Learning: From Theory to Algorithms. Cambridge University, Cambridge (2014)

75. Shawe-Taylor, J., Cristianini, N.: Kernel methods: An overview. In: Kernel Methods for Pattern Analysis, pp. 25–44. Cambridge University, Cambridge (2004)
76. Shewhart, W.: Statistical Method from the Viewpoint of Quality Control. In: Dover Books on Mathematics. Dover Publications, New York (2012)
77. Sokorac, S.: Optimizing random test constraints using machine learning algorithms. In: Proceedings of the Design and Verification Conference and Exhibition US (DVCon) (2017)
78. Thompson, M.P., Hamann, J.D., Sessions, J.: Selection and penalty strategies for genetic algorithms designed to solve spatial forest planning problems. International Journal of Forestry Research **2009**, 1–14 (2009). https://doi.org/10.1155/2009/527392
79. Ur, S., Yadin, Y.: Micro-architecture coverage directed generation of test programs. In: Proceedings of the 36th Design Automation Conference, pp. 175–180 (1999)
80. Vasudevan, S., Sheridan, D., Patel, S., Tcheng, D., Tuohy, B., Johnson, D.: Goldmine: Automatic assertion generation using data mining and static analysis. In: 2010 Design, Automation Test in Europe Conference Exhibition (DATE 2010), pp. 626–629 (2010). https://doi.org/10.1109/DATE.2010.5457129
81. Vasudevan, S., Jiang, W., Bieber, D., Singh, R., Hamid Shojaei, Ho, R., Sutton, C.: Learning semantic representations to verify hardware designs. In: Advances in Neural Information Processing Systems 34 (NeurIPS 2021) (2021)
82. Wagner, I., Bertacco, V., Austin, T.: Microprocessor verification via feedback-adjusted Markov models. IEEE Trans. Comput. Aided Des. Integr. Circuits Syst. **26**(6), 1126–1138 (2007)
83. Wikipedia: Product lifecycle—Wikipedia, the free encyclopedia. https://en.wikipedia.org/wiki/Product_lifecycle. [Online; accessed 25-Oct-2021]
84. Wikipedia: Representational state transfer—Wikipedia, the free encyclopedia. https://en.wikipedia.org/wiki/Representational_state_transfer. [Online; accessed 25-Oct-2021]
85. Wile, B., Goss, J.C., Roesner, W.: Comprehensive Functional Verification—The Complete Industry Cycle. Elsevier, Amsterdam (2005)
86. Wu, B.H., Huang, C.Y.: A robust constraint solving framework for multiple constraint sets in constrained random verification. In: 2013 50th ACM/EDAC/IEEE Design Automation Conference (DAC), pp. 1–7 (2013)
87. XGBoost Documentation. https://xgboost.readthedocs.io/en/latest/. [Online; accessed 25-Oct-2021]
88. Xu, L., Hutter, F., Hoos, H.H., Leyton-Brown, K.: SATzilla: Portfolio-based algorithm selection for SAT. J. Artif. Intell. Res. **32**(3), 201–215 (2008). https://doi.org/10.1613/jair.2490
89. Zhou, Z.H.: Ensemble Methods: Foundations and Algorithms. CRC Press, New York (2012)
90. Zimmermann, T., Zeller, A., Weissgerber, P., Diehl, S.: Mining version histories to guide software changes. IEEE Trans. Softw. Eng. **31**(6), 429–445 (2005)
91. Zuse, H.: Software Complexity—Measures and Methods. Walter de Gruyter, Berlin (2019)

Chapter 15
Machine Learning for Mask Synthesis and Verification

Haoyu Yang, Yibo Lin, and Bei Yu

15.1 Introduction

Moore's law has guided fast and continuous development of VLSI design and manufacturing technologies, which tend to enable the scaling of design feature size to integrate more components into circuit chips. However, the significant gap between circuit feature size and lithography systems has brought great manufacturing challenges.

A classical lithography system consists of mainly five stages that include *source*, *condenser lens*, *mask*, *objective lens*, and *wafer*, as shown in Fig. 15.1. The source stage ejects ultraviolet light beams toward the condenser lens which collects light beams that can go toward the mask stage for further imaging. The remaining light beams that can pass through the mask stage are supposed to leave expected circuit patterns on the wafer stage. As the manufacturing feature size enters single-digit nanometer era, diffraction is inevitable when a light beam enters the mask stage. The objective lens tries to collect diffraction information as much as possible for better transferred image quality. Because of the limited size of the objective lens, higher-order diffraction patterns will be discarded when forming the image on the wafer that results in a lower pattern fidelity [1]. Typically, to ensure the mask image can be transferred onto the wafer as accurate as possible, at least the zero and ± 1st

H. Yang (✉)
nVIDIA Corp., Austin, TX, USA
e-mail: haoyuy@nvidia.com

Y. Lin
Peking University, Beijing, China
e-mail: yibolin@pku.edu.cn

B. Yu
The Chinese University of Hong Kong, Hong Kong, China
e-mail: byu@cse.cuhk.edu.hk

© The Author(s), under exclusive license to Springer Nature Switzerland AG 2022
H. Ren, J. Hu (eds.), *Machine Learning Applications in Electronic Design Automation*, https://doi.org/10.1007/978-3-031-13074-8_15

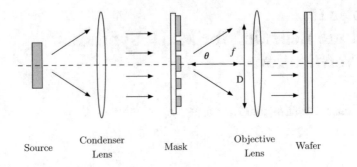

Fig. 15.1 An example of a classical lithography system

diffraction order should be captured by the objective lens. Accordingly, the smallest design pitch can be defined as Eq. (15.1),

$$\frac{1}{p} \propto \frac{NA}{\lambda}, \tag{15.1}$$

where p denotes design pitch, λ is the wavelength of the light source, and NA is the numerical aperture of the objective lens which determines how much information can be collected by the objective lens and is given by

$$NA = n \sin \theta_{\max} = \frac{D}{2f}, \tag{15.2}$$

where n is the index of refraction of the medium, $\theta_{\max}$ is the largest half-angle of the diffraction light that can be collected by the objective lens, D denotes the diameter of physical aperture seen in front of the objective lens, and f represents the focal length [2].

Although research has pushed higher NA design of lithography systems, diffraction information loss still causes mismatches between printed patterns on a wafer and the patterns in the design, which is well known as the *lithography proximity effect*. A mainstream solution is called resolution enhancement technique (RET) that includes multiple patterning lithography (MPL) [3–6], sub-resolution assist feature (SRAF) [7–9] insertion, and optical proximity correction (OPC) [10]. MPL attempts split designs into multiple masks to achieve higher resolution, while SRAF and OPC aim to compensate for the diffraction information loss in the lithography procedure. A lithography system is also subject to process condition variations such as focus and dose, which are likely to deviate from optimal settings. Figure 15.2 illustrates the different effects of process variations with focus and dose variations resulting in image distortion and contour deviation, respectively. It should be noted that high-quality RETs also make designs robust to process variations.

Physics and process limitations have therefore posed various challenges on VLSI design for manufacturing. This chapter focuses on four representative and

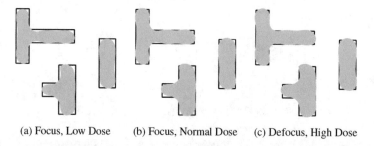

(a) Focus, Low Dose (b) Focus, Normal Dose (c) Defocus, High Dose

Fig. 15.2 Lithography contour versus design target under process variations: (**a**) low dose at focus produces small contours; (**b**) normal dose at focus produces regular contours; (**c**) high dose at defocus produces larger contours

Fig. 15.3 Lithography simulation with optical and resist model

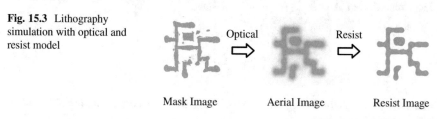

Mask Image Aerial Image Resist Image

critical DFM problems that cover lithography modeling (Sect. 15.2), layout hotspot detection (Sect. 15.3), mask optimization (Sect. 15.4), and layout pattern generation (Sect. 15.5). We will present recent progress and attempts using emerging machine learning solutions to tackle these challenges.

15.2 Lithography Modeling

Lithography simulation is critical in modern DFM flows which enable reliable mask optimization and layout verification (Fig. 15.3). Optical modeling and resist modeling are two major steps in the lithography simulation procedure. Optical modeling maps a mask image to light intensity (aerial image) that is projected on a silicon wafer. Resist modeling deals with the interaction between light intensity and resist materials and determines the final shape formed on the silicon wafer.

15.2.1 Physics in Lithography Modeling

15.2.1.1 Optical Modeling

Singular value decomposition model is the most popular optical lithography approximation model, which has been widely adopted in mask printability estimation and optimization [11–14]. This approximation starts from the Hopkins diffraction model

[15] which is given by

$$I(m, n) = \tilde{s}^H A \tilde{s}, \quad m, n = 1, 2, \ldots, N, \tag{15.3}$$

where

$$\tilde{s}_i = \tilde{M}(p, q) \exp[i2\pi(pm + qn)], \quad i = 1, 2, \ldots, N^2, \tag{15.4}$$

where $\tilde{M} = \mathcal{F}(M)$ is the mask in Fourier space, $\tilde{s}_i$ is the ith element of $\tilde{s}$,[1] $p = i$ mod N, and $q = \lceil \frac{i}{n} \rceil$. $A \in \mathbb{C}^{N^2 \times N^2}$ contains the information of the transmission cross-coefficients that are optical-related parameters. Taking the singular value decomposition of $A = \sum_{k=1}^{N^2} \alpha_k v_k v_k^H$, we have

$$I(m, n) = \sum_{k=1}^{N^2} \alpha_k |\tilde{s}^H v_k|^2, \tag{15.5}$$

where $v_k \in \mathbb{C}^{N^2}$ is the kth eigenvector of A and α_k is the corresponding eigenvalue. We can therefore define

$$\tilde{h}_k(p, q) = \begin{bmatrix} v_{k,1} & v_{k,N+1} & \cdots & v_{k,N(N-1)+1} \\ v_{k,2} & v_{k,N+1} & \cdots & v_{k,N(N-1)+2} \\ \cdots & \cdots & \cdots & \cdots \\ v_{k,N} & v_{k,2N} & \cdots & v_{k,N^2} \end{bmatrix}. \tag{15.6}$$

Let $h_k(m, n) = \mathcal{F}^{-1}(\tilde{h}_k(p, q))$, and connecting Eqs. (15.3)–(15.5), we have

$$I(m, n) = \sum_{k=1}^{N^2} \alpha_k |h_k(m, n) \otimes M(m, n)|^2, \tag{15.7}$$

where $\otimes$ denotes the convolution operation and h_k's are usually called the lithography kernels. Given a mask $M(m, n)$ in real domain, we can calculate its aerial image $I(m, n)$ (light intensity projected on resist materials) via Eq. (15.7).

15.2.1.2 Resist Modeling

Resist models basically aim to perform thresholding on aerial images and obtain the final resist contour. Constant threshold resist (CTR) model is well accepted in

---

[1] Here s itself is meaningless, and we simply use $\tilde{s}$ to indicate the term is related to frequency domain.

Table 15.1 Machine learning solutions for lithography modeling

Model	Framework	Keywords
End-to-end	[17]	Conditional GAN; Single Contact Simulation; CNN Alignment
	[18]	UNet++; Multi Contact Simulation; Multi-Channel Input; Perceptual Loss
Optical	[19]	Conditional GAN; Thin→Thick Mask Modeling
Resist	[16]	Aerial Image→Resist Threshold; CNN
	[20]	Concentrated Circle Sampling; Multi-layer Perceptron; Resist Height Prediction
	[21]	Aerial Image→Resist Threshold;ResNet;Active Sampling

literature study for its simplicity [16]. Given an aerial image I, the degree of resist chemical reaction D is given by

$$D = I(m, n) \otimes G, \tag{15.8}$$

where G is a Gaussian kernel to simulate chemical reactions. And the final resist shape Z is defined as

$$Z(m, n) = \begin{cases} 1, & \text{if } D(m, n) > D_{th}, \\ 0, & \text{otherwise.} \end{cases} \tag{15.9}$$

The strong assumption in CTR makes the model less reliable when facing complicated designs. Variable-threshold resist model (VTR) is then proposed to execute more accurate simulation by assigning local patterns with different thresholds. This is determined by

$$D_{th} = k_1 I_{max} + k_2 I_{min} + k_3 s, \tag{15.10}$$

where I_{max}, I_{min}, and s are max aerial image intensity, min aerial image intensity, and the slope of aerial image profile, respectively.

15.2.2 Machine Learning Solutions for Lithogrpahy Modeling

Physics shows that rigorous lithography simulations either are computationally expensive or suffer performance drop. Therefore, various machine learning frameworks are proposed to meet both runtime and accuracy requirements. As summarized in Table 15.1, prior arts can be categorized into three aspects: end-to-end models, optical models, and resist models.

End-to-end models target to complete optical and resist simulation as a whole, i.e., take the input of a mask and output its corresponding resist patterns. End-to-end modes are fast for rough lithography estimation, however, lacks the details of light intensity information. LithoGAN [17] is a very early attempt to use conditional generative adversarial networks (cGAN) for end-to-end modeling. The major component of LithoGAN is a standard cGAN generator, which takes the input of a mask with the target shape located in the center of the mask. cGAN can then generate the post-lithography contour of the target shape. The misalignment between the predicted contour and design location will question the reliability of such a framework. Therefore, a CNN forward path is introduced as an assist component to output target shape coordinates and hence fix the alignment issue for the generated contours. Very recently, deep lithography simulator (DLS) [18] is proposed as supporting neural networks for full mask optimization. DLS is still a cGAN-backboned structure. Thanks to its UNet [22] backbone and residual feature layers, DLS is able to perform lithography contour prediction on multiple shapes within a $4\,\mu m^2$ tile. To tackle the alignment issue, DLS splits the input mask image into three-channel tensors that include SRAF, OPC, and design patterns, respectively. Apart from traditional cGAN training objectives, DLS also introduces perceptual loss for high-level feature matching, yielding better generation performance. Instead of directly measuring the differences between the generated contours and ground-truth contours, the perceptual loss compares ground-truth images and generated images based on high-level representations from pre-trained convolutional layers.

Optical models deal with aerial image generation. A representative work is TEMPO [19], a framework that takes inputs of a mask image and a series of 2D aerial images at different resist heights. The TEMPO will be trained toward generating aerial images under thick mask consumption through an encoder-decoder structure following root mean square error.

Resist models try to estimate the interaction between light beams and resist materials and hence obtain the final resist patterns. Several ML solutions are proposed to achieve fast and accurate resist modeling. The work of [16] starts to investigate standard CNN-based resist threshold prediction framework. It accepts aerial images as input and generates proper thresholds for the corresponding patterns. Another work [20] deals with resist modeling in a more detailed way. Instead of obtaining a resist threshold for the given pattern, it tries to predict the remaining resist height on a given location after the etching process. Given a layout design, the technique of [20] extracts a series of layout features in terms of a position of interest. These features will then be fed into a multi-layer perceptron and output predicted resist height. Although these learning-based frameworks have demonstrated their effectiveness on resist modeling, achieving high accuracy with CNNs requires a large amount of training data, which is often not easy to get at the early stage of process development. Lin et al. [21] formulate a data-efficient learning problem leveraging the data from old technology nodes to assist model training at the target node. In this formulation, they construct CNN models taking an aerial image as input and outputting the slicing thresholds at the boundary of the target pattern.

15.2.3 Case Studies

In this section, we will detail two representative works [18] and [21] for machine learning-based lithography modeling.

15.2.3.1 Deep Lithography Simulator [18]

The key contributions of the deep lithography simulator [18] are the DCGAN-HD generator, multi-scale discriminator, and perceptual loss.

DCGANHD Generator Left part of Fig. 15.4 shows the architecture of a high-resolution generator, which completes mask-to-wafer mapping from design and SRAF pattern groups. Previous work [12] and [17] adopt traditional UNet [22] for lithography-related tasks where input features are downsampled multiple times. With the decreasing feature resolution, it is easier for a network to gather high-level features such as context features, while low-level information such as the position of each shape becomes harder to collect. However, in lithography physics, low-level information matters more than in the common computer vision tasks. For example, the shape and relative distance of design or SRAF patterns must remain unchanged after the deep mask optimization or deep lithography process. The number and relative distance of via patterns in an input layout have a crucial influence on the result. The features of OPC datasets determine the vital importance of the low-level features. UNet++ [23] is hence proposed for better feature extraction by assembling multiple UNets that have different numbers of downsampling operations. It redesigns the skip pathways to bridge the semantic gap between the encoder and decoder feature maps, contributing to the more accurate low-level feature extraction. The dense skip connections on UNet++ skip pathways improve gradient flow in high-resolution tasks. Although UNet++ has a better performance than UNet, it is not qualified to be the generator of DCGAN-HD.

Fig. 15.4 DCGAN-HD generator [18]. Both deep mask generator (DMG) and deep lithography simulator (DLS) in the paper share the same architecture

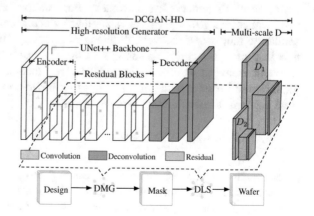

For further improvement, the UNet++ backbone is manipulated with the guidelines suggested in DCGAN [24].

Multi-Scale Discriminator The high-resolution input also imposes a critical challenge to the discriminator design. A simple discriminator that only has three convolutional layers with LeakyReLU and Dropout is presented. Since the patterns in OPC datasets have simple and homogeneous distribution, a deeper discriminator has a higher risk of overfitting. Therefore, we simplify the discriminator by reducing the depth of the neural network. Meanwhile, a dropout layer is attached after each convolutional layer. We use 3×3 convolution kernels in the generator for parameter-saving purposes and 4×4 kernels in the discriminator to increase receptive fields. However, during training, we find that the simple discriminator fails to distinguish between the real and the synthesized images when more via patterns occur in a window. Because when the number of vias reaches 5 or 6 in a window, the via patterns have a larger impact on each other, and the features become more complicated. Inspired by Wang et al. in pix2pixHD [25], we design multi-scale discriminators. Different from pix2pixHD [25] which uses three discriminators, our design uses two discriminators that have an identical network structure but operate at different image scales, which are named $D1$, $D2$, as shown in the right part of Fig. 15.4. Specifically, the discriminators $D1$, $D2$ are trained to differentiate real and synthesized images at the two different scales, 1024×1024 and 512×512, respectively, which helps the training of the high-resolution model easier. In our tasks, the multi-scale design also shows its strengths in flexibility. For example, when the training set has only one via in a window, we can use only $D1$ to avoid overfitting and reduce the training time.

Perceptual Loss Instead of using per-pixel loss such as L_1 loss or L_2 loss, DSL adopts the perceptual loss which has been proven successful in style transfer [26], image super-resolution, and high-resolution image synthesis [25]. A per-pixel loss function is used as a metric for understanding differences between input and output on a pixel level. While the function is valuable for understanding interpolation on a pixel level, the process has drawbacks. For example, as stated in [26], consider two identical images offset from each other by one pixel; despite their perceptual similarity, they would be very different as measured by per-pixel losses. More than that, previous work [24] shows L_2 Loss will cause blur on the output image. Different from per-pixel loss, perceptual loss function in Eq. (15.11) compares ground-truth image x with generated image $\hat{x}$ based on high-level representations from pre-trained convolutional neural networks Φ.

$$\mathcal{L}_{L_P}^{G,\Phi}(x, \hat{x}) = \mathcal{L}_{L_1}(\Phi(x), \Phi(\hat{x})) = \mathbb{E}_{x,\hat{x}} \left[\|\Phi(x) - \Phi(\hat{x})\|_1 \right]. \tag{15.11}$$

DLS Pipeline Figure 15.5 shows the training process of our deep lithography simulator. As a customized design of cGAN, DLS is trained in an alternative scheme using paired mask image x and wafer image y obtained from Mentor Calibre.

Fig. 15.5 The training details of DLS, where the input images are mask-wafer pairs [18]

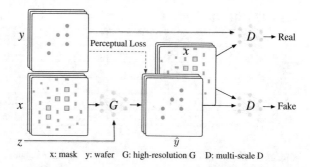

x: mask   y: wafer   G: high-resolution G   D: multi-scale D

z indicates randomly initialized images. The objectives of DLS include training the generator G that produces fake wafer images $G(x, z)$ by learning the feature distribution from $x–y$ pairs and training the discriminators D_1, D_2 to identify the paired $(x, G(x, z))$ as fake. This motivates the design of DLS loss function. The first part of the loss function comes from vanilla GAN that allows the generator and the discriminator interacting with each other in an adversarial way:

$$\mathcal{L}_{cGAN}(G, D) = \mathbb{E}_{x,y}[\log D(x, y)] + \mathbb{E}_{x,z}[\log(1 - D(x, G(x, z)))]. \quad (15.12)$$

Combined with multi-scale discriminators, Eq. (15.12) can be modified as:

$$\sum_{k=1,2} \mathcal{L}_{cGAN}(G_{DLS}, D_{DLS_k}) = \sum_{k=1,2} \mathbb{E}_{x,y}[\log D_{DLS_k}(x, y)]$$

$$+ \mathbb{E}_{x,z}[\log(1 - D_{DLS_k}(x, G_{DLS}(x, z)))], \quad (15.13)$$

where D_{DLS_k} is the kth discriminator of DLS. In DLS design, the perceptual loss is added to the objective, we denote $\hat{y}$ as $G(x, z)$, and loss network Φ is a pre-trained VGG19 on ImageNet. The perceptual loss is given by:

$$\mathcal{L}_{LP}^{G_{DLS},\Phi}(y, \hat{y}) = \sum_{j=1...5} \mathcal{L}_{L_1}(\phi_j(y), \phi_j(\hat{y}))$$

$$= \sum_{j=1...5} \mathbb{E}_{y,\hat{y}}\left[\|\phi_j(y) - \phi_j(\hat{y})\|_1\right], \quad (15.14)$$

where ϕ_j is the feature representation on j-th layer of the pre-trained VGG19 Φ. By combining Eqs. (15.13) and (15.14):

$$\mathcal{L}_{DLS} = \sum_{k=1,2} \mathcal{L}_{cGAN}(G_{DLS}, D_{DLS_k}) + \lambda_0 \mathcal{L}_{LP}^{G_{DLS},\Phi}(y, \hat{y}). \quad (15.15)$$

Experiments Since the DLS model is based on the cGAN framework, we set up an ablation experiment to illustrate the advantages of our generator and discriminators. The results shown in Table 15.2 are the average of six groups of the validation

Table 15.2 Results of DLS [18]

Generator	Discriminator	Loss	mIoU (%)	pixAcc (%)
UNet (cGAN)	D (cGAN)	L_1	94.16	97.12
UNet++	D (cGAN)	L_1	93.98	96.74
G (Our)	D (cGAN)	L_1	96.23	97.50
G (Our)	D (Our)	L_1	97.63	98.76
G (Our)	D (Our)	Perceptual	**98.68**	**99.50**

Fig. 15.6 The overall flow of resist modeling [21]

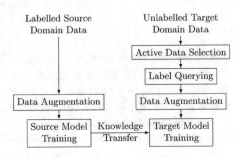

set. Firstly, cGAN (used in LithoGAN) provides a baseline mIoU of 94.16% which is far away from practical application. Then, UNet++ is used to replace the UNet generator in cGAN for better performance. However, the original UNet++ is not qualified to be a generator of a cGAN, and the mIoU is reduced to 93.98% (as shown in Table 15.2).

Following DCGAN, we made some amendments in UNet++, and high-resolution generator is adopted in our DLS model. After applying our high-resolution generator, mIoU is improved to 97.63%, which outperforms UNet and UNet++ generators by a large margin when using the same discriminator. The huge gain in mIoU implies that our developed high-resolution generator is a strong candidate for DLS. Next are the newly designed multi-scale discriminators. Results in Table 15.2 show that mIoU is further boosted to 97.63%.

Lastly, we replace the L_1 loss with the perceptual loss, and the mIoU reaches 98.68%. Additionally, DLS can handle multiple vias in a single clip, which overcomes the limitation of LithoGAN [17].

15.2.3.2   Resist Modeling with Transfer Learning and Active Data Selection [21]

The motivation of learning-based resit modeling is to leverage the simulation speed and accuracy. This comes with dataset efficiency and model design. Lin et al. [21] tackle the resist modeling problem with transfer learning and active data selection. The overall flow is shown in Fig. 15.6.

Fig. 15.7 Transfer learning scheme [21]

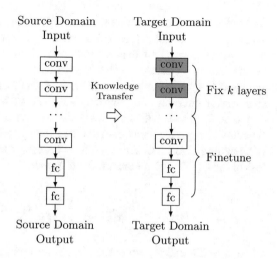

Transfer Learning Since the lithography configurations evolve from one generation to another with the advancement of technology nodes, there is plenty of historical data available for the old generation. If the lithography configurations have no fundamental changes, the knowledge learned from the historical data may still be applicable to the new configuration, which can eventually help reduce the amount of new data required. Transfer learning aims at adapting the knowledge learned from data in source domains to a target domain. The transferred knowledge will benefit the learning in the target domain with faster convergence and better generalization. Suppose the data in the source domain has a distribution P_s and that in the target domain has a distribution P_t. The underlying assumption of transfer learning lies in the common factors that need to be captured for learning the variations of P_s and P_t, so that the knowledge for P_s is also useful for P_t. An intuitive example is that learning to recognize cats and dogs in the source task helps the recognition of ants and wasps in the target task, especially when the source task has a significantly larger dataset than that of the target task. The reason comes from the low-level notions of edges, shapes, etc., shared by many visual categories. In resist modeling, different lithography configurations can be viewed as separate tasks with different distributions.

Typical transfer learning scheme for neural networks fixes the first several layers of the model trained for another domain and fine-tune the successive layers with data from the target domain. The first several layers usually extract general features, which are considered to be similar between the source and the target domains, while the successive layers are classifiers or regressors that need to be adjusted. Figure 15.7 shows an example of the transfer learning scheme. We first train a model with source domain data and then use the source domain model as the starting point for the training of the target domain. During the training for the target domain, the first k layers are fixed, while the rest of the layers are fine-tuned. For simplicity, we denote this scheme as TF_k, shortened from "Transfer and Fix," where k is the parameter for the number of fixed layers.

Active Data Selection Although transfer learning is potentially able to improve the accuracy of the target dataset using knowledge from a source dataset, the selection of representative target data samples may further improve the accuracy. Let D be the unlabeled dataset in the target domain and s be the set of selected data samples for label querying, where $|s| \le k$ and k is the maximum number of data samples for querying. For any $(x_i, y_i) \in D$, x_i is the feature, e.g., aerial image, and y_i is the label, e.g., threshold, where y_i is unknown for D. Consider a loss function $l(x_i, y_i; w)$ parameterized over the hypothesis class (w), e.g., parameters of a learning algorithm. The objective of active learning is to minimize the average loss of dataset D with a model trained from s,

$$\min_{s:|s| \le k, s \in D} \frac{1}{n} \sum_{i=1}^{n} l(x_i, y_i; w_s), \tag{15.16}$$

where $n = |D|$ and w_s represents the parameters of a model trained from s.

Experiments Table 15.3 presents the accuracy metrics, i.e., relative threshold RMS error (ϵ_r^{th}) and CD RMS error (ϵ^{CD}), for learning N7$_b$ from various source domain datasets. Since we consider the data efficiency of different learning schemes, we focus on the small training dataset for N7$_b$, from 1% to 20%. Situations such as no source domain data (∅), only source domain data from N10 (N10$^{50\%}$), only source domain data from N7$_a$ (N7$_a^{50\%}$), and combined source domain datasets are examined. The fidelity between relative threshold RMS error and CD RMS error is very consistent, so they share almost the same trends. Transfer learning with any source domain dataset enables an average improvement of 22% to 38% from that without knowledge transfer. In small training datasets of N7$_b$, ResNet also achieves around 10% better performance on average than CNN in the transfer learning scheme. At 1% of N7$_b$, combined source domain datasets have better performance compared with that of N10$^{50\%}$ only, but the benefits vanish with the increase of the N7$_b$ dataset (Table 15.4).

15.3 Layout Hotspot Detection

Design-process weak points also known as hotspots cause systematic yield loss in semiconductor manufacturing. One of the main goals of DFM is to detect such hotspots. Traditionally, searching for hotspots heavily relies on lithography and manufacturing simulation which are accurate yet time-consuming. Since a layout design can be treated as an image, popular machine learning techniques in computer vision can be naturally applied for manufacturability hotspot detection. In this section, we will introduce how dedicated classification and object detection techniques enable fast and accurate hotspot detection frameworks (Fig. 15.8).

Table 15.3 Relative threshold RMS Error and CD RMS Error for $N7_b$ with different source domain datasets

Source datasets	Ø		$N10^{50\%}$				$N7_a^{50\%}$				$N10^{50\%} + N7_a^{5\%}$		$N10^{50\%} + N7_a^{10\%}$	
Neural networks	CNN		CNN TF$_0$		ResNet TF$_0$		CNN TF$_0$		ResNet TF$_0$		ResNet TF$_0$		ResNet TF$_0$	
	ϵ_r^{th} (10^{-2})	ϵ^{CD}	ϵ_r^{th} (10^{-2})	ϵ^{CD}	ϵ_r^{th} (10^{-2})	ϵ^{CD}	ϵ_r^{th} (10^{-2})	ϵ^{CD}	ϵ_r^{th} (10^{-2})	ϵ^{CD}	ϵ_r^{th} (10^{-2})	ϵ^{CD}	ϵ_r^{th} (10^{-2})	ϵ^{CD}
$N7_b$ 1%	4.31	4.67	2.23	2.36	1.88	2.00	1.55	1.65	1.36	1.45	1.70	1.83	1.63	1.75
5%	2.57	2.74	1.79	1.94	1.57	1.71	1.60	1.73	1.40	1.52	1.66	1.81	1.63	1.77
10%	1.95	2.10	1.73	1.87	1.52	1.65	1.54	1.67	1.39	1.50	1.54	1.66	1.56	1.69
15%	1.83	1.98	1.60	1.76	1.42	1.56	1.49	1.62	1.32	1.44	1.43	1.56	1.45	1.58
20%	1.67	1.81	1.56	1.70	1.36	1.48	1.47	1.59	1.31	1.42	1.38	1.51	1.41	1.54
Ratio	1.00	1.00	0.78	0.79	0.68	0.69	0.69	0.70	0.62	0.62	0.69	0.69	0.69	0.70

Table 15.4 Machine learning solutions for hotspot detection

Model	Framework	Keywords
Traditional ML	[28]	Layout Density; Decision Tree; AdaBoost
	[29]	Concentrated Circle Sampling; Naive Bayes; SmoothBoost
	[30]	Critical Feature Extraction; Multi-kernel SVM
Deep learning	[31]	Feature Tensor Extraction; CNN; Batch Biased Learning
	[32]	ResNet; Binary Neural Networks
	[33]	Multi-task Learning; Transformer
	[34]	Metric Learning; Transformer
	[35]	Object Detection; Mask R-CNN; Hotspot Non-Maximum Suppression
	[36]	Multi-task Learning; CNN

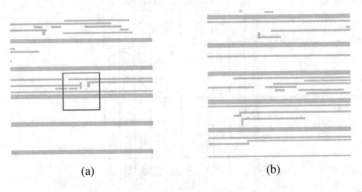

(a)   (b)

Fig. 15.8 Layout design examples that contain hotspots and hotspot-free (Layout source [27]). (**a**) Hotspot Pattern. (**b**) Non-Hotspot Pattern

15.3.1 Machine Learning for Layout Hotspot Detection

Research of machine learning-based hotspot detectors dates back to decades ago before the exploding of deep learning. This section will introduce the recent progress of hotspot detection research from traditional machine learning to emerging deep learning techniques.

Analyzing the layout to reduce lithography hotspots in the early stage is a necessary step in semiconductor manufacturing. Common traditional machine learning techniques for hotspot detection include SVM [30, 37], decision tree [28], and Bayes method [29]. [28, 29] are also two representation contributions using ensemble learning. Feature engineering plays an important role in legacy machine learning solutions, which convert original designs into complete and discriminating layout representations. Effective layout features include layout density [28], Concentrated Circle Sampling [29], Critical Dimension [30], and Tangent Space [38]. Layout feature engineering and learning model together contribute to the research of traditional machine learning on hotspot detection.

Exploding of deep learning techniques has made feature engineering trivial, where, mostly, layout images can be easily converted to discriminative feature space via trained convolutional neural networks. M. Shin and J.-H. Lee [39] and Yang et al. [40] are the two earliest works using CNNs for lithography hotspot detection, where standard VGG-like CNN structures are employed for automatic feature learning and label prediction.

Following the pioneering CNN solution on hotspot detection, [31] presented a feature tensor extraction method to perform representative layout features and got remarkable speedup with the deep neural network. A batch-biased learning training algorithm is also proposed to achieve a better trade-off between hotspot detection accuracy and false-positive penalty. More advanced layout feature-friendly neural network architectures are investigated to enhance hotspot detection performance. Jiang et al. [32], for the first time, brings the binary neural network to hotspot detection tasks, which achieves state-of-the-art results with much smaller neural models and faster inference time. Geng et al. [34] starts investigating the popular transformer structure that incorporates spatial and channel attention in CNN layers. Such methods have been demonstrated efficient in feature learning. Chen et al. [36] considers the availability of data in hotspot detection tasks. When there is a limited number of labeled data, the proposed multi-task learning in [36] learns regular classification cross-entropy loss and an unsupervised contrastive loss to handle the drawbacks of pseudo-label training flow.

Conventional classification-based solutions have potential challenges due to the increased design complexity. To tackle this problem, Chen et al. proposed a region-based hotspot detection framework to detect multiple hotspots on large scales [35]. The framework takes a full-/large-scale layout design as the input and performs classification and regression at the same time to localize the hotspot regions. They report a $45\times$ speedup compared to the classification-based deep learning model. A transformer encoder-based single-stage detection network in [33] can capture the long-range dependencies between polygons and makes the region-based detector more robust.

15.3.2 Case Studies

In this section, we will introduce two representative hotspot detection solutions based on classification and object detection.

15.3.2.1 Detecting Lithography Hotspots with Feature Tensor Generation and Batch-Biased Learning [31]

Feature Tensor Generation The feature tensor extraction method provides a lower-scale representation of the original clips while keeping the spatial information of the clips. After feature tensor extraction, each layout image $\mathbf{I}$ is converted into a

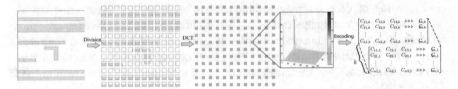

Fig. 15.9 Feature Tensor generation example ($n = 12$). The original clip (1200×1200 nm^2) is divided into 12×12 blocks, and each block is converted to a 100×100 image representing a 100×100 nm^2 subregion of the original clip. Feature tensor is then obtained by encoding on first k DCT coefficients of each block

hyper-image (image with a customized number of channels) **F** with the following properties: (1) size of each channel is much smaller than **I**, and (2) an approximation of **I** can be recovered from **F**.

Spectral analysis of mask patterns for wafer clustering was recently explored in literature and achieved good clustering performance. Inspired by these research, we express the subregion as a finite combination of different frequency components. The high sparsity of the discrete cosine transform (DCT) makes it preferable over other frequency representations in terms of spectral feature extraction, and it is consistent with the expected properties of the feature tensor. To sum up, the process of feature tensor generation contains the following steps.

Step 1: Divide each layout clip into $n \times n$ subregions, and then obtain feature representations of all subregions for multi-level perceptions of layout clips.

Step 2: Convert each subregion of the layout clip $\mathbf{I}_{i,j}$ ($i, j = 0, 1, \ldots, n-1$) into a frequency domain:

$$\mathbf{D}_{i,j}(m, n) = \sum_{x=0}^{B} \sum_{y=0}^{B} \mathbf{I}_{i,j}(x, y) \cos\left[\frac{\pi}{B}\left(x + \frac{1}{2}\right)m\right] \cos\left[\frac{\pi}{B}\left(y + \frac{1}{2}\right)n\right],$$

where $B = \frac{N}{n}$ is subregion size and (x, y) and (m, n) are original layout image and frequency domain indexes, respectively. Particularly, the upper-left side of DCT coefficients in each block corresponds to low-frequency components that contain high-density information, as depicted in Fig. 15.9.

Step 3: Flatten $\mathbf{D}_{i,j}$'s into vectors in zigzag form [41] with the larger index being higher frequency coefficients as follows:

$$\mathbf{C}_{i,j}^* = [\mathbf{D}_{i,j}(0, 0), \mathbf{D}_{i,j}(0, 1), \mathbf{D}_{i,j}(1, 0), \ldots, \mathbf{D}_{i,j}(B, B)]^\mathsf{T}. \tag{15.17}$$

Step 4: Pick the first $k \ll B \times B$ elements of each $\mathbf{C}_{i,j}^*$,

$$\mathbf{C}_{i,j} = \mathbf{C}_{i,j}^*[: k], \tag{15.18}$$

and combine $\mathbf{C}_{i,j}$, i, $j \in \{0, 1, \ldots, n-1\}$ with their spatial relationships unchanged. Finally, the feature tensor is given as follows:

$$
\mathbf{F} = \begin{bmatrix} \mathbf{C}_{11} & \mathbf{C}_{12} & \mathbf{C}_{13} & \ldots & \mathbf{C}_{1n} \\ \mathbf{C}_{21} & \mathbf{C}_{22} & \mathbf{C}_{23} & \ldots & \mathbf{C}_{2n} \\ \vdots & \vdots & \vdots & \ddots & \vdots \\ \mathbf{C}_{n1} & \mathbf{C}_{n2} & \mathbf{C}_{n3} & \ldots & \mathbf{C}_{nn} \end{bmatrix}, \tag{15.19}
$$

where $\mathbf{F} \in \mathbb{R}^{n \times n \times k}$. By reversing the above procedure, an original clip can be recovered from an extracted feature tensor.

The nature of discrete cosine transform ensures that high-frequency coefficients are near zero. As shown in Fig. 15.9, large responses only present at the entries with smaller indexes, i.e., low-frequency regions. Therefore, most information is kept even when a large number of elements in $\mathbf{C}_{i,j}^{*}$ are dropped. The feature tensor also has the following advantages when applied in neural networks: (1) highly compatible with the data packet transference in convolutional neural networks and (2) forward propagation time is significantly reduced when compared with using an original layout image as input, because the scale of the neural network is reduced with the smaller input size.

Batch-biased Learning Intuitively, a too confident classifier is not necessary to give a good prediction performance and, on the contrary, may induce more training pressure or even overfitting problems. Therefore, we propose a new learning algorithm to dynamically adjust the learning taget for different instances. We define a bias function as follows:

$$
\epsilon(l) = \begin{cases} \frac{1}{1+\exp(\beta l)}, & \text{if } l \leq 0.3, \\ 0, & \text{if } l > 0.3, \end{cases} \tag{15.20}
$$

where l is the training loss of the current instance or batch in terms of the unbiased ground truth and β is a manually determined hyperparameter that controls how much the bias is affected by the loss. Because the training loss of the instance at the decision boundary is $-\log 0.5 \approx 0.3$, we set the bias function to take effect when $l \leq 0.3$. With the bias function, we can train the neural network in a single-round mini-batch gradient descent and obtain a better model performance. Because $\epsilon(l)$ is fixed within each training step, no additional computing effort is required for back-propagation.

The training procedure is summarized as Algorithm 15.1, where β is a hyperparameter defined in Eq. (15.20). We initialize the neural network (line 1) and update the weight until meeting the convergence condition (lines 2–16). Within each iteration, we first sample the same amount of hotspot and non-hotspot instances to make sure the training procedure is balanced (lines 3–4); we then calculate

Algorithm 15.1 Batch Biased-learning

Require: Learning rate λ, learning rate decay factor α, learning rate decay step k, hotspot label
  $\mathbf{y}_h^*$, non-hotspot label $\mathbf{y}_n^*$, β;
Ensure: Neural network parameters $\mathbf{W}$;
1: Initialize parameters $\mathbf{W}$, $\mathbf{y}_h^* \leftarrow [0, 1]$;
2: **while not** stop condition **do**
3:　　Sample m non-hotspot instances $\{\mathbf{N}_1, \mathbf{N}_2, \ldots, \mathbf{N}_m\}$;
4:　　Sample m hotspot instances $\{\mathbf{H}_1, \mathbf{H}_2, \ldots, \mathbf{H}_m\}$;
5:　　Calculate average loss of non-hotspot samples l_n with ground truth $[1, 0]$;
6:　　$\mathbf{y}_n^* \leftarrow [1 - \epsilon(l_n), \epsilon(l_n)]$;
7:　　**for** $i \leftarrow 1, 2, \ldots, m$ **do**
8:　　　　$\mathcal{G}_{h,i} \leftarrow \text{backprop}(\mathbf{H}_i)$;
9:　　　　$\mathcal{G}_{n,i} \leftarrow \text{backprop}(\mathbf{N}_i)$;
10:　　**end for**
11:　　Calculate gradient $\bar{\mathcal{G}} \leftarrow \frac{1}{2m} \sum_{i=1}^{m} (\mathcal{G}_{h,i} + \mathcal{G}_{n,i})$;
12:　　Update weight $\mathbf{W} \leftarrow \mathbf{W} - \lambda \bar{\mathcal{G}}$;
13:　　**if** $j \mod k = 0$ **then**
14:　　　　$\lambda \leftarrow \alpha\lambda, j \leftarrow 0$;
15:　　**end if**
16: **end while**

Table 15.5 Performance comparison with state-of-the-art hotspot detectors on target layouts

Benchmarks	SPIE'15[28]		ICCAD'16[29]		SOCC'17[43]		SPIE'17[42]		BBL+AUG	
	Accu (%)	FA#	Accu (%)	FA#	Accu (%)	FA#	Accu (%)	FA#	Accu (%)	FA#
ICCAD	84.20	2919	97.70	4497	96.90	**1960**	97.70	2703	**98.40**	3535
Industry0	93.63	**30**	96.07	1148	97.77	100	97.55	85	**99.36**	387
Average	88.92	1475	96.89	2823	97.34	**1030**	97.63	1394	**98.88**	1961
Ratio	0.899	0.752	0.980	1.439	0.984	**0.525**	0.987	0.711	**1.000**	1.000

the average loss of non-hotspot instances to obtain the bias level and the biased ground truth (lines 5–6); the gradients of the hotspot and non-hotspot instances are calculated separately (lines 8–9); the rest of the steps are the normal weight update through back-propagation and learning rate decay (lines 11–15).

Experiments We compare the hotspot detection results with four state-of-the-art hotspot detectors in Table 15.5. Here "Accu (%)" denotes the hotspot prediction accuracy, and "FA#" represents the number of false positives. "SPIE'15 [28]" is a traditional machine learning-based hotspot detector that applies the density-based layout features and the AdaBoost–DecisionTree model. "ICCAD'16 [29]" takes the lithographic properties into account during feature extraction and adopts the more robust Smooth Boosting algorithm. "SPIE'17 [42]" is another deep learning solution for hotspot detection that takes the original layout image as input and contains more than 20 layers. "SOCC'17 [43]" employs deep neural networks that replace

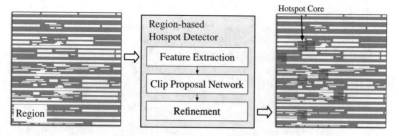

Fig. 15.10 The proposed region-based hotspot detection flow

all pooling layers with strided convolution layers and contain the same number of layers as SPIE'17.

Overall, the framework performs better than traditional machine learning techniques (SPIE'15 [28] and ICCAD'16 [29]) with at least a 2% advantage for the detection accuracy (98.88% of BBL v.s. 96.89% of ICCAD'16) on target layouts. Traditional machine learning models are effective for the benchmarks with regular patterns (ICCAD and Industry0) with the highest detection accuracy of 97.7% on the ICCAD and 96.07% on the Industry0 achieved by Zhang et al. [29].

15.3.2.2 Faster Region-Based Hotspot Detection [35]

The proposed region-based hotspot detection (R-HSD) neural network, as illustrated in Fig. 15.10, is composed of three steps: (1) feature extraction, (2) clip proposal network, and (3) refinement. In this section, we will discuss each step in detail. At first glance, R-HSD problem is similar to the object detection problem, which is a hot topic in the computer vision domain recently. In object detection problems, objects with different shapes, types, and patterns are the target to be detected. However, as we will discuss, there is a gap between hotspot detection and object detection, e.g., the hotspot pattern features are quite different from the objects in real scenes; thus, typical strategies and frameworks utilized in object detection cannot be applied here directly.

Inception-Based Feature Extraction According to the recent progress of deep learning-based research in the computer vision region, a deeper neural network can give a much more robust feature expression and get higher accuracy compared to a shallow structure as it increases the model complexity. However, it also brings sacrifice on the speed at both the inference stage and training stage. Another point we need to be concerned about is that the feature expression of the layout pattern is monotonous, while the feature space of layout patterns is still in low dimension after we transform it by the encoder-decoder structure. According to these issues, we propose an inception-based structure. The following three points are the main rules of our design:

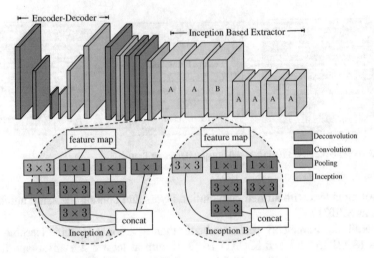

Fig. 15.11 Tensor structure of feature extractor

- Increase the number of filters in width at each stage. For each stage, multiple filters do the convolution operation with different convolution kernel sizes and then concatenate them in channel direction as feature fusion.
- Prune the depth of the output channel for each stage.
- Downsample the feature map size in height and width direction.

With the above rules, the inception structure [44] can take a good balance between accuracy and time. The blobs shown in Fig. 15.11 are what we apply in our framework. We construct module A with the operation stride one and four branches. The aim of module A is to extract multiple features without downsampling the feature map. The operation stride of each layer in module B is two. This setting makes the output feature map half of the input and further reduces the operations. We only use one Module B here, because the feature map size should not be too small, while the low dimension of feature expression in final layers may bring negative effects to the final result.

The 1×1 convolution kernel with low channel numbers brings the dimension reduction which controls the number of the parameters and operations. The multiple branches bring more abundant feature expressions, which give the network ability to do the kernel selection with no operation penalty.

Clip Proposal Networks Given the extracted features, a clip proposal network is developed here to detect potential hotspot clips. For both feature maps and convolutional filters, the tensor structures of the clip proposal network are illustrated in Fig. 15.12. Per preliminary experiments, clips with a single aspect ratio and scale (e.g., square equal to the ground truth) may lead to poor performance. Therefore, for each pixel in the feature map, a group of 12 clips with different aspect ratios is generated. The network is split into two branches: one is for classification and

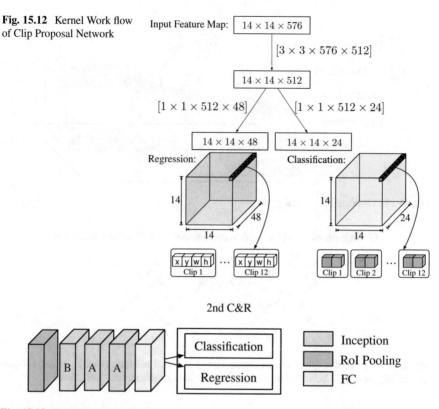

Fig. 15.12 Kernel Work flow of Clip Proposal Network

Fig. 15.13 Tensor structure of Refinement

the other is for regression. In the classification branch, for each clip, a probability of hotspot and a probability of non-hotspot are calculated through the softmax function. In the regression branch, the location and the shape of each clip are determined by a vector $[x, y, w, h]$.

Refinement Network   After the prediction of the first classification and regression in the clip proposal network stage, we get a rough prediction on hotspot localization and filtered region which is classified as non-hotspot. While the greedy method of clip filtering cannot guarantee all the reserved clips are classified correctly, the false alarm may be too high. To bring a robust detection with a lower false alarm, we further construct refinement stage in the whole neural network, which includes a Region of Interests (RoI) pooling layer, three inception modules, as well as another classification and regression. The structure of refinement is shown in Fig. 15.13.

The coordinates of each clip are the actual location from the original input image. We scale down the coordinates to conform with the spatial extent of the last feature map before the refinement. In traditional image processing, the most common ways to resize the image are cropping and warping. The crop method cuts the pattern

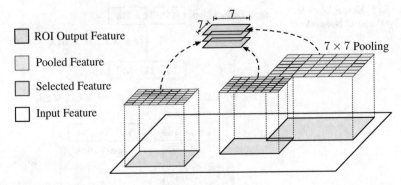

Fig. 15.14 Visualized 7×7 RoI pooling

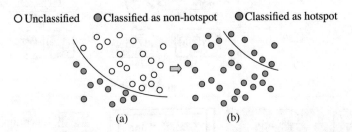

Fig. 15.15 (**a**) The first hotspot classification in clip proposal network; (**b**) the labelled hotspots are fed into the second hotspot classification in refinement stage to reduce false alarm

boundary to fix the target size which leads to information loss. The warping will reshape the size which changes the shape of origin features. Here we apply Region of Interests (RoI) pooling to transform the selected feature region $h \times w$ to a fixed spatial size of $H \times W$ (H and W are the hyperparameters, and we use 7×7 in this work). For each pooled feature region $\lfloor h/H \times w/W \rfloor$, the max-pooling is applied independently. The RoI pooling transforms clips with different sizes into a fixed size which reserves the whole feature information and makes further hotspot classification and regression feasible. Figure 15.14 gives an example of RoI pooling operations.

Besides classification and regression in clip proposal network, here additional classification and regression are designed to fine-tune the clip location and give a more reliable classification result. At this stage, most non-hotspot clips have been removed; thus, the two stages of hotspot classification can efficiently reduce false alarms. Figure 15.15 illustrates the flow of the two-stage hotspot classification.

Loss Function We design a multi-task loss function called classification and regression (C&R) to calibrate our model. As shown in Figs. 15.12 and 15.13, C&R is applied both in clip proposal network stage and refinement stage. The input tensors of the 1st C&R are boxes in Fig. 15.12. W, H, and C are width, height, and channel, respectively. The probability score of the hotspot, non-hotspot, and prediction of clip coordinates is grouped in the channel direction. Here x and y are the coordinates of the hotspot, which means the center of the clip area. w and h are the width and height of the clip. In the 2nd C&R, the tensor flow of the classification and regression are the same as [45] using fully connected layers.

In the task of region-based hotspot detection, h_i is the predicted probability of clip i being a hotspot, and h'_i is the ground truth of clip i. $l_i = \{l_x, l_y, l_w, l_h\} \in \mathbb{R}^4$ and $l'_i = \{l'_x, l'_y, l'_w, l'_h\} \in \mathbb{R}^4$ are assigned as coordinates of clips with index i representing the encoded coordinates of the prediction and ground truth, respectively. The encoded coordinates can be expressed as:

$$
\begin{aligned}
l_x &= (x - x_g)/w_g, & l_y &= (y - y_g)/h_g, \\
l'_x &= (x' - x_g)/w_g, & l'_y &= (y' - y_g)/w_g, \\
l_w &= \log(w/w_g), & l_h &= \log(h/h_g), \\
l'_w &= \log(w'/w_g), & l'_h &= \log(h'/h_g),
\end{aligned}
\tag{15.21}
$$

Variables x, x_g, and x' are for the prediction of clip, g-clip, and the ground-truth clip, respectively (same as the y, w, and h).

The classification and regression loss function for clips can be expressed as:

$$
L_{C\&R}(h_i, l_i) = \alpha_{loc} \sum_i h'_i l_{loc}(l_i, l'_i) + \sum_i l_{hotspot}\left(h_i, h'_i\right)
$$
$$
+ \frac{1}{2}\beta \left(\|\boldsymbol{T}_{loc}\|_2^2 + \|\boldsymbol{T}_{hotspot}\|_2^2 \right),
\tag{15.22}
$$

where β is a hyperparameter which controls the regularization strength. α_{loc} is the hyperparameter which controls the balance between two tasks. $\boldsymbol{T}_{loc}$ and $\boldsymbol{T}_{hotspot}$ are the weights of the neural network. For elements $l_i[j]$ and $l'_i[j]$ ($j \in [1, 4)$) in l_i, l'_i, respectively, l_{loc} can be expressed as

$$
l_{loc}\left(l_i[j], l'_i[j]\right) = \begin{cases} \dfrac{1}{2}\left(l_i[j] - l'_i[j]\right)^2, & \text{if } |l_i[j] - l'_i[j]| < 1, \\ |l_i[j] - l'_i[j]| - 0.5, & \text{otherwise,} \end{cases}
\tag{15.23}
$$

which is a robust L_1 loss used to avoid the exploding gradients problem at training stage. $l_{hotspot}$ is the cross-entropy loss which is calculated as:

$$
l_{hotspot}\left(h_i, h'_i\right) = -\left(h_i \log h'_i + h'_i \log h_i\right).
\tag{15.24}
$$

Experiments We list the detailed result comparison in Table 15.6. Column "Bench" lists three benchmarks used in our experiments. Columns "Acc," "FA,"

Table 15.6 Comparison with state of the art

Bench	TCAD'18 [31]			Faster R-CNN [46]			SSD [47]			Ours		
	Acc (%)	FA	Time (s)	Acc (%)	FA	Time (s)	Acc (%)	FA	Time (s)	Acc (%)	FA	Time (s)
Case2	77.78	48	60.0	1.8	3	1.0	71.9	519	1.0	93.02	17	2.0
Case3	91.20	263	265.0	57.1	74	11.0	57.4	1730	3.0	94.5	34	10.0
Case4	100.00	511	428.0	6.9	69	8.0	77.8	275	2.0	100.00	201	6.0
Average	89.66	274.0	251.0	21.9	48.7	6.67	69.0	841.3	2.0	95.8	84	6.0
Ratio	1.00	1.00	1.00	0.24	0.18	0.03	0.87	3.07	0.01	**1.07**	**0.31**	0.02

"Time" denote hotspot detection accuracy, false alarm count, and detection runtime, respectively. Column "TCAD'18" lists the result of a deep learning-based hotspot detector proposed in [31] that adopts frequency-domain feature extraction and biased learning strategy. We also implement two baseline frameworks that employ Faster R-CNN [46] and SSD [47], respectively, which are two classic techniques matching our region-based hotspot detection objectives well. The corresponding results are listed in columns "Faster R-CNN" [46] and "SSD" [47]. The results show that our framework gets better hotspot detection accuracy on average with 6.14% improvement with ∼ 200 less false alarm penalty compared to [31]. Especially, our framework behaves much better on Case2 with 93.02% detection accuracy compared to 77.78%, 1.8%, and 71.9% for [31], Faster R-CNN, and SSD, respectively. The advantage of the proposed two-stage classification and regression flow can also be seen here that [31] achieves similar hotspot detection accuracy compared to our framework but has extremely large false alarms that will introduce additional. It should be noted that the detection runtime is much faster than [31], thanks to the region-based detection scheme. We can also observe that although Faster R-CNN and SSD are originally designed for large region object detection, they perform poorly on hotspot detection tasks which reflects the effectiveness and efficiency of our customized framework.

15.4 Mask Optimization

RETs try to minimize the error when transferring a design onto silicon wafers. Mainstream solutions vary from lithography source configuration [48, 49], mask pattern optimization [8, 11, 50, 51], to multiple patterning lithography [5, 52]. Among the above, mask optimization is one of the most critical stages in sign-off flows. It tweaks the features or contexts of design layout patterns to circumvent side effects from the lithography proximity effect, which requires frequent interaction with lithography simulation engines, resulting in significant computation overhead. Also, mask optimization recipes need to be carefully crafted for good result convergence.

15.4.1 Machine Learning for Mask Optimization

Mask optimization involves constrained optimization and mask printability querying. This requires an extensive understanding of lithography physics. The exploding of machine learning and deep neural networks brings opportunities to learn lithography behavior either implicitly or explicitly. We will introduce recent research from the perspective of SRAF and OPC (as in Table 15.7).

SRAF Insertion Sub-resolution assist feature (SRAF) insertion is one representative strategy among numerous RET techniques. Without printing SRAF patterns

Table 15.7 Machine learning solutions for mask optimization

Task	Framework	Kewords
SRAF	[8]	Concentrated Area Sampling; Decision Tree; Logistic Regression
	[7]	Concentrated Area Sampling; Dictionary Learning; SVM
	GAN-SRAF [51]	Conditional GAN; Heat Map Encoding
OPC	[55]	Concentrated Area Sampling; EPE Prediction; XGBoost
	[53]	Concentrated Area Sampling; Bayes Model; Markov Chain Monte Carlo
	GAN-OPC [54]	Conditional GAN; ILT-guided Training; UNet
	DAMO [18]	Conditional GAN; UNet++; Perceptual Loss
	DevelSet [14]	Conditional GAN; LevelSet; Signed Distance Field

themselves, the small SRAF patterns can transfer light to the positions of target patterns, and therefore SRAFs are able to improve the robustness of the target patterns under different lithographic variations. Xu et al. [8] is one of the earliest works using machine learning for SRAF insertion. Given a layout location, the machine learning model is trained to identify whether the location should include an SRAF or not. Concentrated circle area sampling is investigated for feature extraction and applied on legacy machine learning models. Geng et al. [7] improves [8] from the perspective of feature engineering. A supervised dictionary learning is proposed to further optimize the concentrated circle area sampling features. However, legacy machine learning approaches still suffer runtime and accuracy issues. Therefore, GAN-SRAF [51] for the first time brings conditional generative adversarial networks for SRAF insertion. A novel heat map encoding is proposed to address the problem that neural networks are not talented to generate sharp edges.

OPC Early attempts using machine learning for OPC regard the entire optimization process as a black box. Matsunawa et al. [53] builds up a Bayes model to directly predict the edge correction level in model-based OPC that can reduce the overall OPC iterations by a significant amount. Since model-based OPC limits the solution space for mask optimization, [14, 54] tend to generate initial solutions at the format of coverage image and level set field, respectively. These initial solutions can then be fed into the legacy inverse lithography technique (ILT) engine for faster and better optimization performance.

Machine learning-based OPC solutions all deal with the overhead of lithography simulation. Instead of reducing iterations and lithography simulator query count, [55] works on per-iteration lithography query time. It builds a machine learning-based lithography estimator that predicts EPE at certain OPC control points and guides edge segment movement. DAMO [18] is also a framework developed for fast OPC iterations using a deep lithography simulator (DLS). The DLS is able to backpropagate lithography error gradients back to another neural network for mask generation.

Fig. 15.16 The proposed
SRAF insertion flow

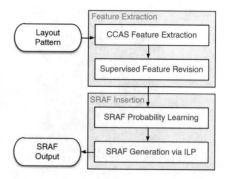

15.4.2   Case Studies

We will discuss details of one legacy machine learning solution for SRAF insertion
[7] and the pioneer work using GAN for mask optimization [54].

15.4.2.1   SRAF Insertion with Supervised Dictionary Learning [7]

The overall flow of SRAF insertion is shown in Fig. 15.16, which consists of
two stages: feature extraction and SRAF insertion. In the feature extraction stage,
after feature extraction via concentric circle area sampling (CCAS), we propose
supervised feature revision, namely, mapping features into a discriminative low-
dimension space. Through dictionary training, our dictionary consists of atoms
which are representatives of original features. The original features are sparsely
encoded over a well-trained dictionary and described as combinations of atoms. Due
to space transformation, the new features (i.e., sparse codes) are more abstract and
discriminative with little important information loss for classification. Therefore,
proposed supervised feature revision is expected to avoid overfitting of a machine
learning model. In the second stage, based on the predictions inferred by the learning
model, SRAF insertion can be treated as a mathematical optimization problem
accompanied by taking design rules into consideration.

Supervised Feature Revision With considering concentric propagation of
diffracted light from mask patterns, the recently proposed CCAS [56] layout feature
is used in the SRAF generation domain. In SRAF insertion, the raw training data set
is made up of layout clips, which include a set of target patterns and model-based
SRAFs. Each layout clip is put on a 2-D grid plane with a specific grid size so that
real training samples can be extracted via the CCAS method at each grid. For every
sample, according to the model-based SRAFs, the corresponding label is either
"1" or "0". As Fig. 15.17a illustrates, "1" means inserting an SRAF at this grid
and "0" vice versa. Figure 15.17b shows the feature extraction method in SRAF
generation. However, since adjacent circles contain similar information, the CCAS

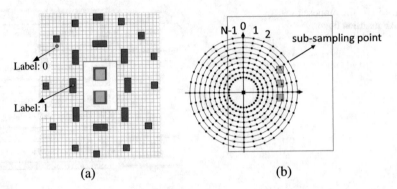

Fig. 15.17 (**a**) SRAF label; (**b**) CCAS feature extraction method in machine learning model-based SRAF generation

feature has much redundancy. In fact, the redundancy will hinder the fitting of a machine learning model.

With the CCAS feature as input, the dictionary learning model is expected to output the discriminative feature of low dimension. In the topic of data representation [57], a self-adaptive dictionary learning model can sparsely and accurately represent data as linear combinations of atoms (i.e., columns) from a dictionary matrix. This model reveals the intrinsic characteristics of raw data. In recent arts, sparse decomposition and dictionary construction are coupled in a self-adaptive dictionary learning framework. As a result, the framework can be modeled as a constrained optimization problem. The joint objective function of a self-adaptive dictionary model for feature revision problem is proposed as Eq. (15.25):

$$\min_{x, D} \frac{1}{N} \sum_{t=1}^{N} \left\{ \frac{1}{2} \left\| y_t - D x_t \right\|_2^2 + \lambda \left\| x_t \right\|_p \right\}, \tag{15.25}$$

where $y_t \in \mathbb{R}^n$ is an input CCAS feature vector and $D = \{d_j\}_{j=1}^{s}, d_j \in \mathbb{R}^n$ refers to the dictionary made up of atoms to encode input features. $x_t \in \mathbb{R}^s$ indicates sparse codes (i.e., sparse decomposition coefficients) with p the type of norm. Meanwhile, N is the total number of training data vectors in memory. The above equation, illustrated in Fig. 15.18, consists of a series of reconstruction error, $\left\| y_t - D x_t \right\|_2^2$, and the regularization term $\left\| x_t \right\|_p$. In Fig. 15.18, every grid represents a numerical value, and dark grid of x_t indicates zero. It can be seen that the motivation of dictionary learning is to sparsely encode input CCAS features over a well-trained dictionary.

However, from Eq. (15.25), it is easy to discover that the main optimization goal is minimizing the reconstruction error in a mean squared sense, which may not be compatible with the goal of classification. Therefore, we try to explore the supervised information and then propose our joint objective function as Eq. (15.26).

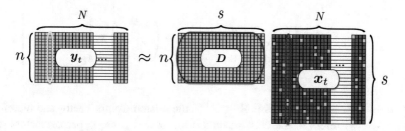

Fig. 15.18 The overview of a dictionary learning model

An assumption has been made in advance that each atom is associated with a particular label, which is true as each atom is selected to represent a subset of the training CCAS features ideally from one class (i.e., occupied with an SRAF or not).

$$\min_{x,D,A} \frac{1}{N} \sum_{t=1}^{N} \left\{ \frac{1}{2} \left\| \left(y_t^\top, \sqrt{\alpha} q_t^\top \right)^\top - \begin{pmatrix} D \\ \sqrt{\alpha} A \end{pmatrix} x_t \right\|_2^2 + \lambda \|x_t\|_p \right\}. \qquad (15.26)$$

In Eq. (15.26), α is a hyperparameter balancing the contribution of each part to reconstruction error. $q_t \in \mathbb{R}^s$ is defined as discriminative sparse code of t-th input feature vector. Hence, $A \in \mathbb{R}^{s \times s}$ transforms original sparse code x_t into discriminative sparse code. In q_t, the non-zero elements indicate that the corresponding atoms share the same label with t-th input. Given dictionary D, it is obvious that q_t has some fixed types.

To illustrate physical meaning of Eq. (15.26) clearly, we can also rewrite it via splitting the reconstruction term into two terms within l_2-norm as (15.27):

$$\min_{x,D,A} \frac{1}{N} \sum_{t=1}^{N} \left\{ \frac{1}{2} \|y_t - Dx_t\|_2^2 + \frac{\alpha}{2} \|q_t - Ax_t\|_2^2 + \lambda \|x_t\|_p \right\}. \qquad (15.27)$$

The first term $\|y_t - Dx_t\|_2^2$ is still the reconstruction error term. The second term $\|q_t - Ax_t\|_2^2$ represents discriminative error, which imposes a constraint on the approximation of q_t. As a result, the input CCAS features from the same class share quite similar representations.

Since the latent supervised information has been used, the label information can also be directly employed. After adding the prediction error term into initial objective function Eq. (15.26), we propose our final joint objective function as Eq. (15.28):

$$\min_{x,D,A,W} \frac{1}{N} \sum_{t=1}^{N} \left\{ \frac{1}{2} \left\| \left(y_t^\top, \sqrt{\alpha} q_t^\top, \sqrt{\beta} h_t \right)^\top - \begin{pmatrix} D \\ \sqrt{\alpha} A \\ \sqrt{\beta} W \end{pmatrix} x_t \right\|_2^2 + \lambda \|x_t\|_p \right\},$$

$$(15.28)$$

where $h_t \in \mathbb{R}$ is the label with $W \in \mathbb{R}^{1 \times s}$ the related weight vector and therefore $\|h_t - W x_t\|_2^2$ refers to the classification error. α and β are hyperparameters that control the contribution of each term to reconstruction error and balance the trade-off. So this formulation can produce a good representation of the original CCAS feature.

Online Algorithm According to our proposed formulation (i.e., Eq. (15.28)), the joint optimization of both dictionary and sparse codes is non-convex, but the sub-problem with one variable fixed is convex. Hence, Eq. (15.28) can be divided into two convex sub-problems. Note that, in a taste of linear algebra, our new input with label information, i.e., $\left(y_t^\top, \sqrt{\alpha} q_t^\top, \sqrt{\beta} h_t \right)^\top$ in Eq. (15.28), can still be regarded as the original y_t in Eq. (15.25). So is the new merged dictionary consisting of D, A, and W. For simplicity of description and derivation, in the following analysis, we will use y_t referring to $\left(y_t^\top, \sqrt{\alpha} q_t^\top, \sqrt{\beta} h_t \right)^\top$ and D standing for merged dictionary with x as the sparse codes.

Two-stage sparse coding and dictionary constructing alternatively perform in iterations. Thus, in the t-th iteration, the algorithm firstly draws the input sample y_t or a mini-batch over the current dictionary D_{t-1} and obtains the corresponding sparse codes x_t. Then it uses two updated auxiliary matrices, B_t and C_t, to help compute D_t.

The objective function for sparse coding is shown in (15.29):

$$x_t \overset{\Delta}{=} \arg\min_x \frac{1}{2} \|y_t - D_{t-1} x\|_2^2 + \lambda \|x\|_1. \tag{15.29}$$

If the regularizer adopts l_0-norm, solving Eq. (15.29) is NP-hard. Therefore, we utilize l_1-norm as a convex replacement of l_0-norm. In fact, Eq. (15.29) is the classic Lasso problem [58], which can be solved by any Lasso solver.

Two auxiliary matrices $B_t \in \mathbb{R}^{(n+s+1) \times s}$ and $C_t \in \mathbb{R}^{s \times s}$ are defined, respectively, in (15.30) and (15.31):

$$B_t \leftarrow \frac{t-1}{t} B_{t-1} + \frac{1}{t} y_t x_t^\top, \tag{15.30}$$

$$C_t \leftarrow \frac{t-1}{t} C_{t-1} + \frac{1}{t} x_t x_t^\top. \tag{15.31}$$

The objective function for dictionary construction is:

$$D_t \stackrel{\triangle}{=} \arg\min_{D} \frac{1}{t} \sum_{i=1}^{t} \left\{ \frac{1}{2} \|y_i - Dx_i\|_2^2 + \lambda \|x_i\|_1 \right\}. \tag{15.32}$$

Algorithm 15.2 Supervised Online Dictionary Learning (SODL)

Require: Input merged features $Y \leftarrow \{y_t\}_{t=1}^N$, $y_t \in \mathbb{R}^{(n+s+1)}$ (including original CCAS features, discriminative sparse code $Q \leftarrow \{q_t\}_{t=1}^N$, $q_t \in \mathbb{R}^s$ and label information $H \leftarrow \{h_t\}_{t=1}^N$, $h_t \in \mathbb{R}$).

Ensure: New features $X \leftarrow \{x_t\}_{t=1}^N$, $x_t \in \mathbb{R}^s$, dictionary $D \leftarrow \{d_j\}_{j=1}^s$, $d_j \in \mathbb{R}^{(n+s+1)}$.

1: **Initialization:** Initial merged dictionary D_0, $d_j \in \mathbb{R}^{(n+s+1)}$ (including initial transformation matrix $A_0 \in \mathbb{R}^{s \times s}$ and initial label weight matrix $W_0 \in \mathbb{R}^{1 \times s}$), $C_0 \in \mathbb{R}^{s \times s} \leftarrow 0$, $B_0 \in \mathbb{R}^{(n+s+1) \times s} \leftarrow 0$;

2: **for** $t \leftarrow 1$ to N **do**

3: Sparse coding y_t and obtaining x_t; ▷ Equation (15.29)

4: Update auxiliary variable B_t; ▷ Equation (15.30)

5: Update auxiliary variable C_t; ▷ Equation (15.31)

6: Update dictionary D_t; ▷ Algorithm 15.3;

7: **end for**

Algorithm 15.2 summarizes the algorithm details of the proposed supervised online dictionary learning (SODL) algorithm. We use coordinate descent algorithm as the solving scheme to Eq. (15.29) (line 3). To accelerate the convergence speed, Eq. (15.32) involves the computations of past signals $y_1, \ldots, y_t$ and the sparse codes $x_1, \ldots, x_t$. One way to efficiently update dictionary is to introduce some sufficient statistics, i.e., $B_t \in \mathbb{R}^{(n+s+1) \times s}$ (line 4) and $C_t \in \mathbb{R}^{s \times s}$ (line 5), into Eq. (15.32) without directly storing the past input data sample y_i and corresponding sparse codes x_i for $i \leq t$. These two auxiliary variables play important roles in updating atoms, which summarize the past information from sparse coefficients and input data. We further exploit block coordinate method with warm start [59] to solve Eq. (15.32) (line 6). As a result, through some gradient calculations, we bridge the gap between Eq. (15.32) and sequentially updating atoms based on Eqs. (15.33) and (15.34).

$$u_j \leftarrow \frac{1}{C[j,j]} (b_j - Dc_j) + d_j. \tag{15.33}$$

$$d_j \leftarrow \frac{1}{\max(\|u_j\|_2, 1)} u_j. \tag{15.34}$$

For each atom d_j, the updating rule is illustrated in Algorithm 15.3. In Eq. (15.33), D_{t-1} is selected as the warm start of D. b_j indicates the j-th column of B_t, while c_j is the j-th column of C_t. $C[j,j]$ denotes the j-th element on diagonal of C_t. Equation (15.34) is an l_2-norm constraint on atoms to prevent atoms from becoming arbitrarily large (which may lead to arbitrarily small sparse codes).

Algorithm 15.3 Rules for Updating Atoms

Require: $D_{t-1} \leftarrow \{d_j\}_{j=1}^s, d_j \in \mathbb{R}^{(n+s+1)},$
$\quad B_t \leftarrow \{b_j\}_{j=1}^s, b_j \in \mathbb{R}^{(n+s+1)},$
$\quad C_t \leftarrow \{c_j\}_{j=1}^s, c_j \in \mathbb{R}^s.$
Ensure: dictionary $D_t \leftarrow \{d_j\}_{j=1}^s, d_j \in \mathbb{R}^{(n+s+1)}.$
1: **for** $j \leftarrow 1$ to s **do**
2: $\quad$ Update the j-th atom d_j; $\quad\quad\quad\quad\quad\quad\quad\quad$ ▷ Equation (15.33) and (15.34)
3: **end for**

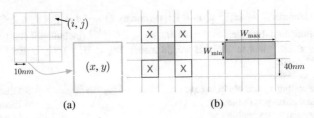

(a)$\quad\quad\quad\quad\quad\quad\quad\quad\quad\quad\quad\quad$ (b)

Fig. 15.19 (a) SRAF grid model construction; (b) SRAF insertion design rule under the grid model

SRAF Insertion via ILP Through SODL model and classifier, the probabilities of each 2-D grid can be obtained. Based on design rules for the machine learning model, the label for a grid with a probability less than the threshold is "0". It means that the grid will be ignored when doing SRAF insertion. However, in [8], the scheme to insert SRAFs is a little naive and greedy. Actually, combined with some relaxed SRAF design rules such as maximum length and width and minimum spacing, the SRAF insertion can be modeled as an integer linear programming (ILP) problem. With the ILP model to formulate SRAF insertion, we will obtain a global view for SRAF generation.

In the objective of the ILP approach, we only consider valid grids whose probabilities are larger than the threshold. The probability of each grid is denoted as $p(i, j)$, where i and j indicate the index of a grid. For simplicity, we merge the current small grids into new bigger grids, as shown in Fig. 15.19a. Then we define $c(x, y)$ as the value of each merged grid, where x, y denote the index of merged grid. The rule to compute $c(x, y)$ is as follows:

$$c(x, y) = \begin{cases} \sum_{(i,j)\in(x,y)} p(i, j), & \text{if } \exists \, p(i, j) \geq \text{threshold}, \\ -1, & \text{if all } p(i, j) < \text{threshold}. \end{cases} \quad (15.35)$$

The motivation behind this approach is twofold. One is to speed up the ILP, where negative decision variables can be determined in advance. The other is to keep the consistency of machine learning prediction.

In ILP for SRAF insertion, our real target is to maximize the total probability of valid grids with feasible SRAF insertion. Accordingly, it manifests to put up with the objective function, which is to maximize the total value of merged grids. The

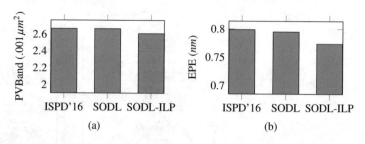

Fig. 15.20 Result comparison with previous work. (**a**) PVBand (**b**) Edge placement error

ILP formulation is shown in Formula (15.36).

$$\max_{a(x,y)} \quad \sum_{x,y} c(x, y) \cdot a(x, y) \tag{15.36a}$$

$$\text{s.t.} \quad a(x, y) + a(x - 1, y - 1) \le 1, \qquad \forall(x, y), \tag{15.36b}$$

$$a(x, y) + a(x - 1, y + 1) \le 1, \qquad \forall(x, y), \tag{15.36c}$$

$$a(x, y) + a(x + 1, y - 1) \le 1, \qquad \forall(x, y), \tag{15.36d}$$

$$a(x, y) + a(x + 1, y + 1) \le 1, \qquad \forall(x, y), \tag{15.36e}$$

$$a(x, y) + a(x, y + 1) + a(x, y + 2)$$
$$+ a(x, y + 3) \le 3, \qquad \forall(x, y), \tag{15.36f}$$

$$a(x, y) + a(x + 1, y) + a(x + 2, y)$$
$$+ a(x + 3, y) \le 3, \qquad \forall(x, y), \tag{15.36g}$$

$$a(x, y) \in \{0, 1\}, \qquad \forall(x, y). \tag{15.36h}$$

Here $a(x, y)$ refers to the insertion situation at the merged grid (x, y). According to the rectangular shape of an SRAF and the spacing rule, the situation of two adjacent SRAFs on the diagonal is forbidden by Constraints (15.36b) to (15.36e); e.g., Constraint (15.36b) requires the $a(x, y)$, and the left upper neighbor $a(x - 1, y - 1)$ cannot be 1 at the same time; otherwise, design rule violations can be generated. Constraints (15.36f) to (15.36g) restrict the maximum length of SRAFs. Figure 15.19b actively illustrates these linear constraints coming from design rules.

Experiments Figure 15.20 compares the results with a state-of-the-art machine learning based SRAF insertion tool "ISPD'16" [8]. It can be seen from the figure that the SODL algorithm outperforms [8] in terms of EPE and PVB by 3%. This indicates the predicted SRAFs by the SODL model match the reference results better than [8]. In other words, the proposed SODL-based feature revision can efficiently improve machine learning model generality.

Fig. 15.21 The proposed
GAN-OPC architecture

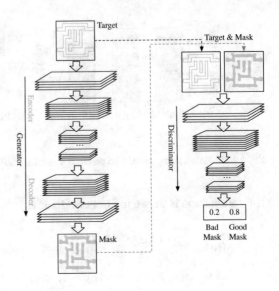

15.4.2.2 GAN-OPC [54]

GAN-OPC [54] is the earliest work bringing GAN to OPC problems, which aims to
train a generative network to provide an initial mask for ILT optimization instantly.
The major contributions include customized conditional GAN design and efficient
training flow.

Conditional GAN As shown in Fig. 15.21, the GAN-OPC framework is back-
boned with conditional GAN structure, which consists of an encode-decoder
generator design and a regular CNN-based discriminator. The objective is therefore
given by,

$$\min_{G}\max_{D}\mathbb{E}_{\mathbf{Z}_t\sim\mathcal{Z}}[1-\log(D(\mathbf{Z}_t,G(\mathbf{Z}_t)))+||\mathbf{M}^*-G(\mathbf{Z}_t)||_n^n]$$

$$+\mathbb{E}_{\mathbf{Z}_t\sim\mathcal{Z}}[\log(D(\mathbf{Z}_t,\mathbf{M}^*))], \tag{15.37}$$

where $\mathcal{Z}$ denotes the distribution of target design layouts, $G(\cdot)$ and $D(\cdot)$ represent
the generator and the discriminator in GAN-OPC architecture, $\mathbf{Z}_t$ denotes the target
design layout, and $\mathbf{M}^*$ is the golden mask for $\mathbf{Z}_t$.

Previous analysis shows that the generator and the discriminator have different
objectives; therefore, the two sub-networks are trained alternatively, as shown in
Algorithm 15.4. In each training iteration, we sample a mini-batch of target images
(line 2); gradients of both the generator and the discriminator are initialized to
zero (line 3); a feed-forward calculation is performed on each sampled instances
(lines 4–5); the ground-truth mask of each sampled target image is obtained from
OPC tools (line 6); we calculate the loss of the generator and the discriminator on
each instance in the mini-batch (lines 7–8); we obtain the accumulated gradient of

losses with respect to neuron parameters (lines 9–10); finally the generator and the discriminator are updated by descending their mini-batch gradients (lines 11–12). Note that in Algorithm 15.4, we convert the min-max problem in Eq. (15.37) into two minimization problems such that gradient ascending operations are no longer required to update neuron weights.

Algorithm 15.4 GAN-OPC Training

1: **for** number of training iterations **do**
2:     Sample m target clips $\mathcal{Z} \leftarrow \{Z_{t,1}, Z_{t,2}, \ldots, Z_{t,m}\}$;
3:     $\Delta W_g \leftarrow \mathbf{0}, \Delta W_d \leftarrow \mathbf{0}$;
4:     **for** each $Z_t \in \mathcal{Z}$ **do**
5:         $M \leftarrow G(Z_t; W_g)$;
6:         $M^* \leftarrow$ Groundtruth mask of Z_t;
7:         $l_g \leftarrow -\log(D(Z_t, M)) + \alpha ||M^* - M||_2^2$;
8:         $l_d \leftarrow \log(D(Z_t, M)) - \log(D(Z_t, M^*))$;
9:         $\Delta W_g \leftarrow \Delta W_g + \dfrac{\partial l_g}{\partial W_g}$; $\Delta W_d \leftarrow \Delta W_d + \dfrac{\partial l_d}{\partial W_g}$;
10:     **end for**
11:     $W_g \leftarrow W_g - \dfrac{\lambda}{m} \Delta W_g$; $W_d \leftarrow W_d - \dfrac{\lambda}{m} \Delta W_d$;
12: **end for**

ILT-guided Pretraining Although with OPC-oriented techniques, GAN is able to obtain a fairly good performance and training behavior, it is still a great challenge to train the complicated GAN model with satisfactory convergence. Observing that ILT and neural network training stage share similar gradient descent techniques, we develop an ILT-guided pre-training method to initialize the generator, after which the alternative mini-batch gradient descent is discussed as a training strategy of GAN optimization. The main objective in ILT is minimizing the lithography error through gradient descent.

$$E = ||Z_t - Z||_2^2, \tag{15.38}$$

where Z_t is the target and Z is the wafer image of a given mask. Because mask and wafer images are regarded as continuously valued matrices in the ILT-based optimization flow, we apply translated sigmoid functions to make the pixel values close to either 0 or 1.

$$Z = \frac{1}{1 + \exp[-\alpha \times (I - I_{th})]}, \tag{15.39}$$

$$M_b = \frac{1}{1 + \exp(-\beta \times M)}, \tag{15.40}$$

where I_{th} is the binarization threshold, M_b is the incompletely binarized mask, and α and β control the steepness of relaxed images.

Considering the lithography behavior we discussed in Sect. 15.2.1.1, we can also derive the gradient representation as follows:

$$\frac{\partial E}{\partial M} = 2\alpha\beta \times M_b \odot (1 - M_b)$$

$$\odot (((Z - Z_t) \odot Z \odot (1 - Z) \odot (M_b \otimes H^*)) \otimes H$$

$$+ ((Z - Z_t) \odot Z \odot (1 - Z) \odot (M_b \otimes H)) \otimes H^*), \quad (15.41)$$

where H^* is the conjugate matrix of the original lithography kernel H and $\odot$ is the operator for element-wise product. In traditional ILT flow, the mask can be optimized by iteratively descending the gradient until E is below a threshold.

The objective of the mask optimization problem indicates the generator is the most critical component in GAN. Observing that both ILT and neural network optimization share similar gradient descent procedures, we propose a jointed training algorithm that takes advantage of the ILT engine, as depicted in Fig. 15.22b. We initialize the generator with lithography-guided pre-training to make it converge well in the GAN optimization flow thereafter. The key step of neural network training is back-propagating the training error from the output layer to the input layer, while neural weights are updated as follows:

$$W_g = W_g - \frac{\lambda}{m} \Delta W_g, \quad (15.42)$$

where ΔW_g is accumulated gradient of a mini-batch of instances and m is the mini-batch instance count. Because Eq. (15.42) is naturally compatible with ILT, if we create a link between the generator and ILT engine, the wafer image error can be back-propagated directly to the generator as presented in Fig. 15.22.

The generator pre-training phase is detailed in Algorithm 15.5. In each pre-training iteration, we sample a mini-batch of target layouts (line 2) and initialize

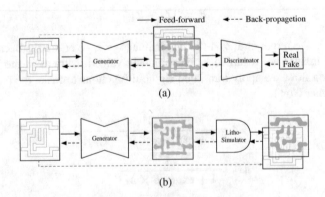

(a)

(b)

Fig. 15.22 (a) GAN-OPC training and (b) ILT-guided pre-training

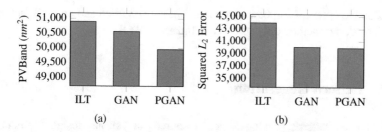

Fig. 15.23 Result comparison with a vanilla ILT engine. (**a**) PVBand (**b**) Squared L_2 Error

the gradients of the generator ΔW_g to zero (line 3). The mini-batch is fed into the generator to obtain generated masks (lines 5). Each generated mask is loaded into the lithography engine to obtain a wafer image (line 6). The quality of wafer image is estimated by Eq. (15.38) (lines 7). We calculate the gradient of lithography error E with respect to the neural networks parameter W_g through the chain rule, i.e., $\dfrac{\partial E}{\partial M}\dfrac{\partial M}{\partial W_g}$ (line 8). Finally, W_g is updated following the gradient descent procedure (line 10).

Algorithm 15.5 ILT-guided Pre-training

1: **for** number of pre-training iterations **do**
2:      Sample m target clips $\mathcal{Z} \leftarrow \{\mathbf{Z}_{t,1}, \mathbf{Z}_{t,2}, \ldots, \mathbf{Z}_{t,m}\}$;
3:      $\Delta W_g \leftarrow 0$;
4:      **for** each $\mathbf{Z}_t \in \mathcal{Z}$ **do**
5:          $M \leftarrow G(\mathbf{Z}_t; W_g)$;
6:          $\mathbf{Z} \leftarrow \texttt{LithoSim}(M)$
7:          $E \leftarrow \|\mathbf{Z} - \mathbf{Z}_t\|_2^2$;
8:          $\Delta W_g \leftarrow \Delta W_g + \dfrac{\partial E}{\partial M}\dfrac{\partial M}{\partial W_g}$;   ▷ Equation (15.41)
9:      **end for**
10:     $W_g \leftarrow W_g - \dfrac{\lambda}{m}\Delta W_g$;   ▷ Equation (15.42)
11: **end for**

Experiments In this experiment, we optimize the ten layout masks in ICCAD 2013 contest benchmark [60] and compare the results with previous work. The quantitative results are illustrated in Fig. 15.23.

Note that all the GAN-OPC and PGAN-OPC results are refined by an ILT engine which generates final masks to obtain wafer images. Column "L_2" is the squared L_2 error between the wafer image and the target image under the nominal condition. Column "PVB" denotes the process variation band under $\pm\ 2\%$ dose error. It is notable that GAN-OPC significantly reduces squared L_2 error of wafer images under the nominal condition by 9%, and with the ILT-guided pre-training, squared L_2 error is slightly improved, and PVB is further reduced by 1%. Because we only focus on the optimization flow under the nominal condition and no PVB factors are considered, our method only achieves comparable PVB areas.

15.5 Pattern Generation

VLSI layout patterns provide critical resources in various design for manufacturability (DFM) researches, from (1) early technology node development to (2) back-end design and sign-off flow [61]. The former includes the perfection of design rules, OPC recipes, lithography models, and so on. The latter covers, but is not limited to, layout hotspot detection and correction [7, 31, 39, 40, 43, 55, 62]. However, layout pattern libraries are sometimes not large and diverse enough for DFM research/solutions due to the long logic-to-chip design cycle. Even some test layouts can be synthesized within a short period; they are usually restricted by certain design rules, and the generated pattern diversity is limited [21, 63].

15.5.1 Machine Learning for Layout Pattern Generation

The development of generative machine learning techniques offers the opportunities to generate desired layouts efficiently. Representative works are [64–66]. DeePattern [64] presents a transforming convolutional auto-encoder (TCAE) architecture for 1D pattern generation. TCAE consists of a recognition unit and a generation unit. The recognition unit is built with stacked convolutional layers that convert layout pattern topologies into latent vectors that allow perturbation for certain transformations. The generation unit is expected to capture layout spatial information (e.g., track and wire direction) well and convert latent vectors back to legal pattern topologies with corresponding deconvolution operations. The work of [65] investigates the possibility of the variational auto-encoder for 2D pattern generation. The techniques in [66] take the VAE in [65] as the backbone and push the technique to advanced technology node with transfer learning and channel-wise attention.

15.5.2 Case Study

15.5.2.1 DeePattern [64]

Squish Pattern Squish pattern is a scan line-based representation that each layout clip is cut into grids aligned at all shape edges, as shown in Fig. 15.24. The squish pattern representation of a given layout clip consists of a topology matrix T and two vectors δ_x and δ_y that contain geometry information in x and y directions. Each entry of T is either zero or one which indicates shape or space, respectively.

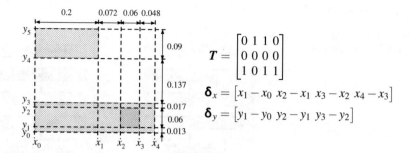

Fig. 15.24 Squish pattern generation

The geometric information describes the size of each grid. For example, the pattern in Fig. 15.24 can be accordingly expressed as in the right side of Fig. 15.24. Here x_is and y_is are the locations of vertical and horizontal scan lines, respectively, and the pattern complexity is accordingly given by $c_x = 4$ and $c_y = 3$. Canonically, (x_0, y_0) is the coordinate of the bottom left corner of the pattern and $x_i < x_{i+1}$, $y_i < y_{i+1}$. Now the problem becomes generating legal topologies and solving associated δ_xs and δ_ys that are much easier than directly generating DRC-clean patterns. The advantages of squish patterns are twofold: (1) Squish patterns are storage-efficient and support neural networks and other machine learning models. (2) Squish patterns are naturally compatible with the simplified pattern generation flow that will be discussed in the following sections.

TCAE for Topology Generation Here the TCAE architecture aims at efficient pattern topology T generation. The TCAE is derived from original transforming auto-encoders (TAEs) [67] which are a group of densely connected auto-encoders and each individual, referred to as a capsule, is targeting certain image-to-image transformations. TAEs cannot be directly applied for pattern generation due to the fact that *transformations are restricted by layout design rules and only very simple pose transformations are supported by original TAEs, which does not satisfy our pattern generation objectives.* Therefore we develop the TCAE architecture for feature learning and pattern reconstruction, as shown in Fig. 15.25. The detection unit in TCAE consists of multiple convolutional layers for hierarchical feature extraction, followed by several densely connected layers as an instantiation of the input pattern in the latent vector space, as in Eq. (15.43).

$$l = f(T; W_f), \tag{15.43}$$

where l is the latent vector, T represents the input topology, and W_f contains all the trainable parameters associated with the recognition unit. The latent vector works similarly as a group of capsule units with each node being a low-level feature

Fig. 15.25 Architecture of transforming convolutional auto-encoder in (a) training phase and (b) testing phase

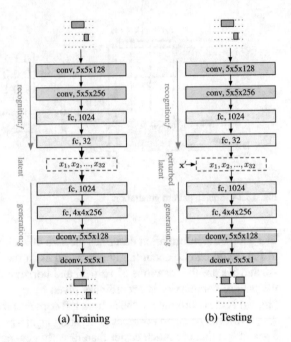

(a) Training   (b) Testing

representation. We will show each latent vector node contributes to pattern shape globally or locally in the experiment section.

The generation unit contains deconvolutional layers [68] that cast the pattern object from the latent vector space back to the original pattern space, as in Eq. (15.44).

$$T' = g(l + \Delta l; W_g), \tag{15.44}$$

where Δl is the perturbation applied on the latent vector that allows inputs to conduct transformations. During training, we force the TCAE to learn an identity mapping with the following objectives.

$$\min_{W_f, W_g} ||T - T'||_2, \text{ s.t. } \Delta l = \mathbf{0}. \tag{15.45}$$

Once the TCAE is trained, we can apply the flow in Fig. 15.25b to generate pattern topologies from perturbed latent vector space of existing layout patterns. During the inference phase, we feed a group of squish topologies into the trained recognition unit that extracts latent vector instantiations of existing topologies. Perturbation on the latent vector space is expected to expand the existing pattern library with legal topologies.

The method introduces random perturbations in the latent vectors. We introduce the concept of feature sensitivity s that statistically defines how easily a legal

Algorithm 15.6 Estimating feature sensitivity. $\mathcal{T} = \{T_1, T_2, \ldots, T_N\}$ is a set of valid pattern topologies, f and g are trained recognition unit and generation unit respectively, t determines the perturbation range, and s is the estimated feature sensitivity

---

Require: $\mathcal{T}, f, g, t.$
Ensure: $s.$
 1: $\mathcal{R}_i \leftarrow \emptyset, \forall i = 1, 2, \ldots, N;$
 2: **for** $i = 1, 2, \ldots, n$ **do**
 3:     **for** $\lambda = -t : t$ **do**
 4:         $\Delta l \leftarrow \mathbf{0}, \Delta l_i \leftarrow \lambda;$
 5:         $T_i \leftarrow g(f(\mathcal{T}) + \Delta l);$
 6:         $\mathcal{R}_i \leftarrow \mathcal{R}_i + T_i;$
 7:     **end for**
 8:     $s_i \leftarrow$ fraction of invalid topologies in $\mathcal{R}_i;$
 9: **end for**
10: **return** $s.$

---

topology can be transformed to illegal when manipulating the latent vector node with everything else unchanged.

Definition 1 (Feature Sensitivity) Let $l = \begin{bmatrix} l_1 & l_2 & \ldots & l_n \end{bmatrix}^\top$ be the output of the layer associated with the latent vector space. The sensitivity s_i of a latent vector node l_i is defined as the probability of reconstructed pattern being invalid when a perturbation $\Delta l_i \in [-t, t]$ is added up on l_i with everything else unchanged.

It can be seen from Definition 1 that a larger s_i indicates the corresponding latent vector node l_i is more likely to create invalid topologies if a large perturbation is applied. We, therefore, avoid manipulating such nodes when sampling random perturbation vectors from a Gaussian distribution. The s_is are estimated following Algorithm 15.6, which requires a set of legal topologies and trained TCAE. The sensitivity of each latent vector node is estimated individually (lines 1–2). We first obtain the latent vectors of all topologies in $\mathcal{T}$ and feed them into the reconstruction unit along with certain perturbations on one latent vector node (lines 3–5). Reconstructed patterns are appended in the corresponding set $\mathcal{R}_i$ (line 6). The sensitivity of the latent vector node i is given by the fraction of invalid topologies in $\mathcal{R}_i$ (line 8).

After we get the estimated sensitivity of all latent vector nodes, we are able to sample perturbation vectors whose elements are sampled independently from $\mathcal{N}(0, \frac{1}{s_i})$. These perturbation vectors will be added up to the latent vectors of existing pattern topologies to formulate perturbed latent vectors which will be fed into the generation unit to construct new topologies. That is, illegal topologies can be filtered out by checking whether shapes appear at any two adjacent tracks.

Experiments In the experiment, we make use of the flow above to augment the pattern space. Pattern library statistics are listed in Table 15.8. Column "Method"

Table 15.8 Statistics of generated patterns

Method	Pattern #	Pattern diversity (H)
Existing design	–	3.101
Industry tool	55408	1.642
DCGAN	1	0
TCAE	**286898**	**3.337**

denotes the approach used to generate layout patterns, column "Pattern #" denotes the number of DRC clean patterns that are different from others, and column "Pattern Diversity" corresponds to the Shannon Entropy of each pattern library in terms of pattern complexity. Row "TCAE" corresponds to the details of 1M patterns generated by perturbing the features of 1000 patterns in existing design with Gaussian noise. "Industry Tool" shows the cataloged results of a test layout generated from a state-of-the-art industry layout generator. The test layout has similar total chip area ($10000 \mu m^2$) as "TCAE" ($14807 \ \mu m^2$). We also implement a DCGAN [24] that has similar number of trainable parameters as the TCAE designed in this framework. 1M patterns are generated by feeding random latent vectors in the trained generator networks. "Existing Design" lists the statistics of a pattern library extracted from an industry layout.

15.6 Conclusion

In this chapter, we survey recent progress and challenges of machine learning solutions on a variety of VLSI DFM problems, ranging from lithography modeling, lithography hotspot detection, mask optimization, to layout generation. These are all critical phases in VLSI design and sign-off flows and seriously affect chip design cycles. We hope the investigation of machine learning techniques in these problems would offer alternate solutions to traditional DFM flows and hence enable faster design closure. Follow-up researches are necessary to prototype these techniques. Future directions will include, but are not limited to, problem-dedicated learning model and algorithm design to meet industrial requirements and constraints, efficient machine learning framework plugins to assist traditional design flows, massive data generation to allow better training convergence and model generality, etc.

References

1. Mack, C.: Fundamental Principles of Optical Lithography: The Science of Microfabrication. Wiley, New York (2008)
2. Greivenkamp, J.E.: Field Guide to Geometrical Optics. SPIE Press, Bellingham (2004)
3. Banerjee, S., Agarwal, K.B., Orshansky, M.: Simultaneous OPC and decomposition for double exposure lithography. In: Proceedings of SPIE, vol. 7973 (2011)

4. Li, X., Luk-Pat, G., Cork, C., Barnes, L., Lucas, K.: Double-patterning-friendly OPC. In: Proceedings of SPIE, vol. 7274 (2009)
5. Gupta, M., Jeong, K., Kahng, A.B.: Timing yield-aware color reassignment and detailed placement perturbation for double patterning lithography. In: IEEE/ACM International Conference on Computer-Aided Design (ICCAD), pp. 607–614 (2009)
6. Kuang, J., Chow, W.K., Young, E.F.Y.: Triple patterning lithography aware optimization for standard cell based design. In: IEEE/ACM International Conference on Computer-Aided Design (ICCAD), pp. 108–115 (2014)
7. Geng, H., Yang, H., Ma, Y., Mitra, J., Yu, B.: SRAF insertion via supervised dictionary learning. In: IEEE/ACM Asia and South Pacific Design Automation Conference (ASPDAC), pp. 406–411 (2019)
8. Xu, X., Matsunawa, T., Nojima, S., Kodama, C., Kotani, T., Pan, D.Z.: A machine learning based framework for sub-resolution assist feature generation. In: ACM International Symposium on Physical Design (ISPD), pp. 161–168 (2016)
9. Lin, T., Robert, F., Borjon, A., Russell, G., Martinelli, C., Moore, A., Rody, Y.: Sraf placement and sizing using inverse lithography technology. In: Optical Microlithography XX, vol. 6520, p. 65202A. International Society for Optics and Photonics, Washington (2007)
10. Cobb, N.B.: Fast optical and process proximity correction algorithms for integrated circuit manufacturing. Ph.D. thesis, University of California, Berkeley (1998)
11. Gao, J.R., Xu, X., Yu, B., Pan, D.Z.: MOSAIC: Mask optimizing solution with process window aware inverse correction. In: ACM/IEEE Design Automation Conference (DAC), pp. 52:1–52:6 (2014)
12. Yang, H., Li, S., Ma, Y., Yu, B., Young, E.F.: GAN-OPC: Mask optimization with lithography-guided generative adversarial nets. In: ACM/IEEE Design Automation Conference (DAC), pp. 131:1–131:6 (2018)
13. Su, Y.H., Huang, Y.C., Tsai, L.C., Chang, Y.W., Banerjee, S.: Fast lithographic mask optimization considering process variation. IEEE Trans. Comput. Aided Des. Integr. Circuits Syst. (TCAD) 35(8), 1345–1357 (2016)
14. Chen, G., Yu, Z., Liu, H., Ma, Y., Yu, B.: DevelSet: Deep neural level set for instant mask optimization. In: IEEE/ACM International Conference on Computer-Aided Design (ICCAD), pp. 1–9. IEEE, New York (2021)
15. Ma, X., Arce, G.R.: Computational lithography, vol. 77. Wiley, New York (2011)
16. Watanabe, Y., Kimura, T., Matsunawa, T., Nojima, S.: Accurate lithography simulation model based on convolutional neural networks. In: Proceedings of SPIE, vol. 10454, p. 104540I (2017)
17. Ye, W., Alawieh, M.B., Lin, Y., Pan, D.Z.: LithoGAN: End-to-end lithography modeling with generative adversarial networks. In: ACM/IEEE Design Automation Conference (DAC), pp. 1–6. IEEE, New York (2019)
18. Chen, G., Chen, W., Sun, Q., Ma, Y., Yang, H., Yu, B.: DAMO: Deep agile mask optimization for full chip scale. IEEE Trans. Comput. Aided Des. Integr. Circuits Syst. (TCAD) vol. 41, no. 9, pp. 3118–3131 (2021)
19. Ye, W., Alawieh, M.B., Watanabe, Y., Nojima, S., Lin, Y., Pan, D.Z.: TEMPO: Fast mask topography effect modeling with deep learning. In: ACM International Symposium on Physical Design (ISPD), pp. 127–134 (2020)
20. Shim, S., Choi, S., Shin, Y.: Machine learning-based 3d resist model. In: Proceedings of SPIE, vol. 10147, p. 101471D. International Society for Optics and Photonics, Washington (2017)
21. Lin, Y., Li, M., Watanabe, Y., Kimura, T., Matsunawa, T., Nojima, S., Pan, D.Z.: Data efficient lithography modeling with transfer learning and active data selection. IEEE Trans. Comput. Aided Des. Integr. Circuits Syst. (TCAD) 38(10), 1900–1913 (2019)
22. Ronneberger, O., Fischer, P., Brox, T.: U-net: Convolutional networks for biomedical image segmentation. In: International Conference on Medical image computing and computer-assisted intervention, pp. 234–241. Springer, Berlin (2015)

23. Zhou, Z., Siddiquee, M.M.R., Tajbakhsh, N., Liang, J.: Unet++: A nested u-net architecture for medical image segmentation. In: Deep learning in Medical Image Analysis and Multimodal Learning for Clinical Decision Support, pp. 3–11. Springer, New York (2018)

24. Radford, A., Metz, L., Chintala, S.: Unsupervised representation learning with deep convolutional generative adversarial networks. In: International Conference on Learning Representations (ICLR) (2016)

25. Wang, T.C., Liu, M.Y., Zhu, J.Y., Tao, A., Kautz, J., Catanzaro, B.: High-resolution image synthesis and semantic manipulation with conditional gans. In: IEEE Conference on Computer Vision and Pattern Recognition (CVPR), pp. 8798–8807 (2018)

26. Johnson, J., Alahi, A., Fei-Fei, L.: Perceptual losses for real-time style transfer and super-resolution. In: European Conference on Computer Vision (2016)

27. Torres, A.J.: ICCAD-2012 CAD contest in fuzzy pattern matching for physical verification and benchmark suite. In: IEEE/ACM International Conference on Computer-Aided Design (ICCAD), pp. 349–350 (2012)

28. Matsunawa, T., Gao, J.R., Yu, B., Pan, D.Z.: A new lithography hotspot detection framework based on AdaBoost classifier and simplified feature extraction. In: Proceedings of SPIE, vol. 9427 (2015)

29. Zhang, H., Yu, B., Young, E.F.Y.: Enabling online learning in lithography hotspot detection with information-theoretic feature optimization. In: IEEE/ACM International Conference on Computer-Aided Design (ICCAD), pp. 47:1–47:8 (2016)

30. Yu, Y.T., Lin, G.H., Jiang, I.H.R., Chiang, C.: Machine-learning-based hotspot detection using topological classification and critical feature extraction. IEEE Trans. Comput. Aided Des. Integr. Circuits Syst. (TCAD) 34(3), 460–470 (2015)

31. Yang, H., Su, J., Zou, Y., Ma, Y., Yu, B., Young, E.F.Y.: Layout hotspot detection with feature tensor generation and deep biased learning. IEEE Trans. Comput. Aided Des. Integr. Circuits Syst. (TCAD) 38(6), 1175–1187 (2019)

32. Jiang, Y., Yang, F., Zhu, H., Yu, B., Zhou, D., Zeng, X.: Efficient layout hotspot detection via binarized residual neural network. In: ACM/IEEE Design Automation Conference (DAC), pp. 1–6. IEEE, New York (2019)

33. Zhu, B., Chen, R., Zhang, X., Yang, F., Zeng, X., Yu, B., Wong, M.D.: Hotspot detection via multi-task learning and transformer encoder. In: IEEE/ACM International Conference on Computer-Aided Design (ICCAD), pp. 1–8 (2021)

34. Geng, H., Yang, H., Zhang, L., Miao, J., Yang, F., Zeng, X., Yu, B.: Hotspot detection via attention-based deep layout metric learning. IEEE Trans. Comput. Aided Des. Integr. Circuits Syst. (TCAD) 41(8), 2685–2698 (2021)

35. Chen, R., Zhong, W., Yang, H., Geng, H., Zeng, X., Yu, B.: Faster region-based hotspot detection. In: ACM/IEEE Design Automation Conference (DAC) (2019)

36. Chen, Y., Lin, Y., Gai, T., Su, Y., Wei, Y., Pan, D.Z.: Semi-supervised hotspot detection with self-paced multi-task learning. IEEE Trans. Comput. Aided Des. Integr. Circuits Syst. (TCAD) 39(7), 1511–1523 (2019)

37. Zhang, H., Zhu, F., Li, H., Young, E.F.Y., Yu, B.: Bilinear lithography hotspot detection. In: ACM International Symposium on Physical Design (ISPD), pp. 7–14 (2017)

38. Yang, F., Sinha, S., Chiang, C.C., Zeng, X., Zhou, D.: Improved tangent space based distance metric for lithographic hotspot classification. IEEE Trans. Comput. Aided Des. Integr. Circuits Syst. (TCAD) 36(9), 1545–1556 (2017)

39. Shin, M., Lee, J.H.: Accurate lithography hotspot detection using deep convolutional neural networks. J. Micro/Nanolithogr. MEMS MOEMS (JM3) 15(4), 043507 (2016)

40. Yang, H., Luo, L., Su, J., Lin, C., Yu, B.: Imbalance aware lithography hotspot detection: a deep learning approach. J. Micro/Nanolithogr. MEMS MOEMS (JM3) 16(3), 033504 (2017)

41. Wallace, G.K.: The JPEG still picture compression standard. IEEE Trans. Consum. Electron. (TCE) 38(1), xviii–xxxiv (1992)

42. Yang, H., Luo, L., Su, J., Lin, C., Yu, B.: Imbalance aware lithography hotspot detection: A deep learning approach. In: SPIE Advanced Lithography, vol. 10148 (2017)

43. Yang, H., Lin, Y., Yu, B., Young, E.F.Y.: Lithography hotspot detection: From shallow to deep learning. In: IEEE International System-on-Chip Conference (SOCC), pp. 233–238 (2017)
44. Szegedy, C., Vanhoucke, V., Ioffe, S., Shlens, J., Wojna, Z.: Rethinking the inception architecture for computer vision. In: IEEE Conference on Computer Vision and Pattern Recognition (CVPR), pp. 2818–2826 (2016)
45. Krizhevsky, A., Sutskever, I., Hinton, G.E.: ImageNet classification with deep convolutional neural networks. In: Conference on Neural Information Processing Systems (NIPS), pp. 1097–1105 (2012)
46. Ren, S., He, K., Girshick, R., Sun, J.: Faster R-CNN: Towards real-time object detection with region proposal networks. In: Conference on Neural Information Processing Systems (NIPS), pp. 91–99 (2015)
47. Liu, W., Anguelov, D., Erhan, D., Szegedy, C., Reed, S., Fu, C.Y., Berg, A.C.: SSD: Single shot multibox detector. In: European Conference on Computer Vision (ECCV), pp. 21–37 (2016)
48. Song, Z., Ma, X., Gao, J., Wang, J., Li, Y., Arce, G.R.: Inverse lithography source optimization via compressive sensing. Opt. Express 22(12), 14180–14198 (2014)
49. Erdmann, A., Fuehner, T., Schnattinger, T., Tollkuehn, B.: Toward automatic mask and source optimization for optical lithography. In: Optical Microlithography XVII, vol. 5377, pp. 646–657. International Society for Optics and Photonics, Washington (2004)
50. Yu, P., Shi, S.X., Pan, D.Z.: Process variation aware OPC with variational lithography modeling. In: ACM/IEEE Design Automation Conference (DAC), pp. 785–790 (2006)
51. Alawieh, M.B., Lin, Y., Zhang, Z., Li, M., Huang, Q., Pan, D.Z.: GAN-SRAF: Sub-resolution assist feature generation using conditional generative adversarial networks. In: ACM/IEEE Design Automation Conference (DAC), pp. 1–6 (2019)
52. Hu, S., Hu, J.: Pattern sensitive placement for manufacturability. In: ACM International Symposium on Physical Design (ISPD), pp. 27–34 (2007)
53. Matsunawa, T., Yu, B., Pan, D.Z.: Optical proximity correction with hierarchical bayes model. J. Micro/Nanolithogr. MEMS MOEMS (JM3) 15(2), 021009 (2016)
54. Yang, H., Li, S., Deng, Z., Ma, Y., Yu, B., Young, E.F.: GAN-OPC: Mask optimization with lithography-guided generative adversarial nets. IEEE Trans. Comput. Aided Des. Integr. Circuits Syst. (TCAD) 39(10), 2822–2834 (2019)
55. Jiang, B., Zhang, H., Yang, J., Young, E.F.: A fast machine learning-based mask printability predictor for OPC acceleration. In: IEEE/ACM Asia and South Pacific Design Automation Conference (ASPDAC), pp. 412–419 (2019)
56. Matsunawa, T., Yu, B., Pan, D.Z.: Optical proximity correction with hierarchical bayes model. In: Proceedings of SPIE, vol. 9426 (2015)
57. Gangeh, M.J., Farahat, A.K., Ghodsi, A., Kamel, M.S.: Supervised dictionary learning and sparse representation-a review. arXiv preprint arXiv:1502.05928 (2015)
58. Tibshirani, R.: Regression shrinkage and selection via the Lasso. J. R. Stat. Soc. Ser. B 58, 267–288 (1996)
59. Friedman, J., Hastie, T., Tibshirani, R.: Regularization paths for generalized linear models via coordinate descent. J. Stat. Softw. 33(1), 1 (2010)
60. Banerjee, S., Li, Z., Nassif, S.R.: ICCAD-2013 CAD contest in mask optimization and benchmark suite. In: IEEE/ACM International Conference on Computer-Aided Design (ICCAD), pp. 271–274 (2013)
61. Tabery, C., Zou, Y., Arnoux, V., Raghavan, P., Kim, R.h., Côté, M., Mattii, L., Lai, Y.C., Hurat, P.: In-design and signoff lithography physical analysis for 7/5nm. In: SPIE Advanced Lithography, vol. 10147 (2017)
62. Yang, H., Pathak, P., Gennari, F., Lai, Y.C., Yu, B.: Detecting multi-layer layout hotspots with adaptive squish patterns. In: IEEE/ACM Asia and South Pacific Design Automation Conference (ASPDAC), pp. 299–304 (2019)
63. Yang, H., Li, S., Tabery, C., Lin, B., Yu, B.: Bridging the gap between layout pattern sampling and hotspot detection via batch active sampling. IEEE Trans. Comput. Aided Des. Integr. Circuits Syst. (TCAD) 40(7), 1464–1475 (2020)

64. Yang, H., Pathak, P., Gennari, F., Lai, Y.C., Yu, B.: Deepattern: Layout pattern generation with transforming convolutional auto-encoder. In: ACM/IEEE Design Automation Conference (DAC), pp. 1–6 (2019)
65. Zhang, X., Shiely, J., Young, E.F.: Layout pattern generation and legalization with generative learning models. In: IEEE/ACM International Conference on Computer-Aided Design (ICCAD), pp. 1–9. IEEE, New York (2020)
66. Zhang, X., Yang, H., Young, E.F.: Attentional transfer is all you need: Technology-aware layout pattern generation. In: ACM/IEEE Design Automation Conference (DAC), pp. 169–174. IEEE, New York (2021)
67. Hinton, G.E., Krizhevsky, A., Wang, S.D.: Transforming auto-encoders. In: International Conference on Artificial Neural Networks (ICANN), pp. 44–51 (2011)
68. Dumoulin, V., Visin, F.: A guide to convolution arithmetic for deep learning. arXiv preprint arXiv:1603.07285 (2018)

Chapter 16
Machine Learning for Agile FPGA Design

Debjit Pal, Chenhui Deng, Ecenur Ustun, Cunxi Yu, and Zhiru Zhang

16.1 Introduction

The end of Dennard scaling [29] and the slowing of Moore's Law made it difficult for homogeneous processors to sustain the performance improvement needed by a diverse range of applications. This motivates a ubiquitous effort across industry and academia to adopt custom hardware accelerators implemented using field-programmable gate arrays (FPGAs) or application-specific integrated circuits (ASICs) [2, 7, 18–20, 31, 40, 41, 45, 55, 70, 89–95, 98, 104, 117, 128, 138, 139].

Modern FPGA fabrics consist of millions of interconnected heterogeneous compute and storage units, e.g., look-up tables (LUTs), flip-flops (FFs), block RAMs (BRAMs), and digital signal processing blocks (DSPs). Collectively, they offer abundant fine-grained parallelism, high on-chip memory bandwidth, and flexible communication schemes. A designer can exploit such architectural features to create efficient application- or domain-specific accelerators that exploit customized memory hierarchy as well as massively parallel and/or deeply pipelined compute engines. As as result, FPGAs have emerged as powerful heterogeneous computing platforms that can support efficient processing of a wide range of applications, such as cryptography [5, 10, 148], communication [13, 61, 78, 140], genomics [14, 35, 37, 38, 56, 102], graph processing [9, 21, 44, 49, 118, 121, 137, 146], ML and neural network (NN) [30, 64, 88, 126, 129, 141, 142], web search [27, 97, 100, 101], and image processing [54, 144].

D. Pal (✉) · C. Deng · E. Ustun · Z. Zhang
Cornell University, Ithaca, NY, USA
e-mail: debjit.pal@cornell.edu; cd574@cornell.edu; eu49@cornell.edu; zhiruz@cornell.edu

C. Yu
University of Utah, Salt Lake City, UT, USA
e-mail: cunxi.yu@utah.edu

© The Author(s), under exclusive license to Springer Nature Switzerland AG 2022
H. Ren, J. Hu (eds.), *Machine Learning Applications in Electronic Design Automation*, https://doi.org/10.1007/978-3-031-13074-8_16

471

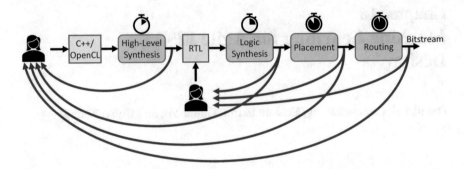

Fig. 16.1 An end-to-end FPGA design flow

As shown in Fig. 16.1, the present-day FPGA design flow starts with an untimed high-level software program, e.g., C/C++/OpenCL, that is automatically synthesized into a cycle-accurate register-transfer level (RTL) model in Verilog or VHDL. High-level synthesis (HLS) generates RTL models and detailed reports summarizing the expected performance and clock frequency, estimated resource usage, timing, etc. For the downstream FPGA design flow, the RTLs then run through logic synthesis, technology mapping, and place-and-route (PnR) to generate a bitstream for the target FPGA device, which is collectively known as the *implementation* process. The FPGA implementation often takes hours or even days to complete for reasonably complex designs [36]. Besides the bitstream, the FPGA CAD tool also generates detailed reports on the quality of results (QoRs) of the implemented design, including post-PnR resource usage and timing. However, such reports are of widely varying fidelity, progressively becoming more accurate from HLS to subsequent steps of the implementation process. Such multi-fidelity reports help the designers to make informed optimization decisions earlier without running the full set of the implementation process.

It is worth noting that the HLS reports represent the crucial, and oftentimes, main evidence based on which the design is iteratively modified to achieve more desirable performance on an actual FPGA device. Despite their importance, many reported values related to performance, resource usage, and clock frequency in the HLS reports are often highly inaccurate when compared to the implementation reports. In particular, final resource usage and timing depend on the cumulative effects of many nontrivial transformations through the series of implementation stages beyond HLS as shown in Fig. 16.1 and are therefore difficult to estimate even by state-of-the-art HLS tools. Post logic synthesis, the area utilization estimates are much more accurate than HLS as the RTL is optimized and mapped into the target FPGA-specific logic networks. However, since the actual physical location of components is still unknown, delay estimates are still inaccurate. To obtain more accurate QoR estimates, designers may have to spend an enormous amount of time iterating the downstream implementation stages for each design point. While HLS and earlier steps of the implementation process are several orders of magnitude faster than the

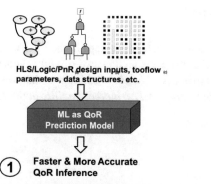
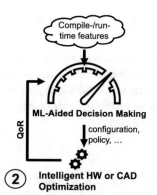

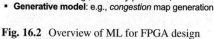

HLS/Logic/PnR design inputs, tooflow parameters, data structures, etc.

ML as QoR Prediction Model

(1) **Faster & More Accurate QoR Inference**

- **Regression:** e.g., *QoR* prediction (*delay, area, power*, etc.)
- **Classification:** e.g., *routability* prediction,
- **Generative model:** e.g., *congestion* map generation

ML-Aided Decision Making

configuration, policy, ...

(2) **Intelligent HW or CAD Optimization**

- **Autotuning for CAD tools**
- **Algorithmic exploration for CAD**
- **Architecture search**

Fig. 16.2 Overview of ML for FPGA design

later steps of the implementation process, inaccuracy of estimated QoRs *creates a strong tension between speed and accuracy* in the entire FPGA design flow.

In recent years, ML has gained considerable momentum for FPGA design flow automation. Based on the recent works, we believe that ML provides a promising way to holistically capture the multitude of factors affecting estimation accuracy. Specifically, the recent progress in leveraging ML techniques to accelerate FPGA design procedure, optimize QoR, and minimize human supervision can be summarized in the following two categories (c.f, Fig. 16.2): (1) developing fast and accurate QoR estimation models using ML—the miscorrelation between multi-fidelity design stages has become the major bottleneck to efficient design closure. ML techniques such as regression, classification, and generative models have demonstrated their capabilities for faster and more accurate estimation to bridge the cross-stage QoR miscorrelation gap; (2) harnessing ML for intelligent decision-making in FPGA tool flow and architecture explorations. With the massive configuration and design space in modern FPGA CAD tools and architectures, the task of searching for the optimal CAD tool configuration and architectures has become more challenging. Recent developments of compile-time intelligent decision-making frameworks in auto-tuning and FPGA architecture search demonstrate successes in design closure via online or reinforcement learning.

In this chapter, we provide a comprehensive review of the application of ML to FPGA design automation problems. We survey prominent ML techniques that accelerate QoR estimation and timing closure. We highlight several key challenges in FPGA design automation that the research community must address for mainstream adoption of FPGA-based computing. By summarizing related studies and challenges, we hope to motivate more ML-based solutions to further streamline fast and accurate FPGA design automation.

16.2   Preliminaries and Background

In this section, we provide a brief overview of each of the steps in the FPGA design
flow. For an in-depth understanding of the entire design flow, we strongly encourage
the reader to consult the papers referred to in Sect. 16.2.1. In addition to that, we also
elaborate on our insights on *how we can leverage ML to increase agility in the entire
FPGA design flow*.

16.2.1   FPGA Design Flow

We show the FPGA design flow in Fig. 16.1. First, HLS compiles the high-level
software program into a control-data flow graph (CDFG), followed by scheduling [24, 28, 136] operations in the CDFG into discrete clock cycles and binding
them to different hardware resources. Such a CDFG is used for RTL generation.
By raising the level of abstraction, HLS reduces design effort while enabling more
productive optimization over a larger space of performance, area, and timing. Due
to these benefits of HLS, there is a growing development and adoption of both
commercial [25, 50, 60] and open-source academic HLS tools [12, 33].

Logic synthesis in FPGA design flow compiles an RTL design into target-
specific logic networks, e.g., LUT-level netlists. Logic synthesis typically consists
of two stages—technology-independent optimization followed by technology mapping. The technology-independent optimization restructures the logic to reduce area
and logic depth while maintaining the correctness of the functionality. Most logic
synthesis algorithms that optimize the multilevel Boolean network circuits are based
on DAG-aware rule-based logic transformations [1, 6, 83, 113, 119, 130, 134] and
formal proof-based exact synthesis [109, 110]. Although DAG-aware synthesis
approaches are effective and versatile, they mostly rely on heuristic rule-based
transformations to provide "good enough" results without optimality guarantees. Technology mapping reconstructs the optimized Boolean networks using
technology-dependent cells. For LUT-based FPGA designs, technology mapping
maps a Boolean network into a LUT network [22, 23, 62, 65, 67, 84]. Like many
other important optimizations in HLS and logic synthesis, the technology mapping
problem is NP-hard.

Placement and Routing (PnR) is the physical design automation process that
transforms a technology-mapped netlist into the layout that describes the cell
locations and interconnection routes between cells. Placement is the process of
assigning each cell in the netlist to a physical location on the device[17, 42, 107].
Commercial tools allow users to floorplan their designs in the interest of reducing
congestion and improving timing results [123]. Floorplanning is implemented in the
form of a set of physical constraints, where each constraint controls the placement
of logic on the device. Finally, given a placed design, the routing problem is to
assign each net to wire segments and switches such that all nets are connected with

user-defined optimization objectives to meet the design specifications. In the modern electronic design flow, routing is particularly slow compared to other design stages and becomes more unpredictable as the technology advances [17, 85, 112, 132].

16.2.2 Motivation of ML for FPGA Design

As discussed earlier, a primary barrier to rapid hardware specialization is the lack of guarantee by existing FPGA's CAD tools on achieving design closure in an out-of-the-box fashion. Using these tools usually requires extensive manual effort to calibrate and configure a large set of design parameters and tool options to achieve a high QoR. Unfortunately, evaluating just one design point can already be painfully time-consuming, as design steps such as PnR usually consume hours to days for large circuits. To enable agile FPGA-based compute acceleration, there is an urgent need to lower the design cost by (1) significantly reducing the time required to obtain accurate QoR estimation and (2) minimizing human supervision in the design tuning process.

Recent years have seen an increasing application of ML to accelerate the FPGA design process and reduce human engineering efforts and are believed to have great potential to address more critical challenges in FPGA design. Specifically, from the modern FPGA design point of view, the motivations can be summarized as follows:

- **Fast and accurate approximation via predictive modeling.** ML can be used as a statistical technique that mines domain knowledge from historical and existing data to forecast future or unseen outcomes with respect to specific algorithmic or mathematical objectives. With the recent progress in advanced ML algorithms and neural architectures, ML can be used to construct generic and accurate approximations for given objectives, which can significantly boost the FPGA design process. For example, it is too expensive to run through a complete FPGA design flow for each design point, while early stage results estimations often lack the accuracy in presenting the right design trade-offs. A well-calibrated ML predictive model can replace such heavy computations with a fast approximation.
- **Flexible and versatile modeling.** Unlike traditional statistical data analysis methods, modern ML techniques provide a wide range of modeling options to cover the complex FPGA design processes. On one hand, ML offers various predictive formulations that are essential to cover a large number of FPGA design challenges, e.g., classification, clustering, regression, generative modeling, etc. On the other hand, modern variants of ML can handle versatile feature representations such as graphs, circuit imaging, functional behaviors, etc. and learn complex behaviors between those features and target metrics.
- **Minimizing human supervision.** Leveraging ML in FPGA design minimizes human supervision in the design process in two directions. First, the traditional CAD tool R&D process heavily relies on expert knowledge in FPGA design and

CAD algorithms, and most heuristics are empirically developed with tremendous experimental efforts. On the other hand, autonomous exploration and learning systems such as reinforcement learning mechanisms can significantly accelerate the exploration efforts with an intelligent self-guided agent.

16.3   Machine Learning Solutions for FPGA

We discuss a set of recently developed ML-based techniques to increase the agility of FPGA development at the various stages in the FPGA design flow. When describing an ML technique, we focus on the design stage of its application, the primary ML algorithm, the input to the ML technique, the learning task, and the improvement it brings to the design stage. Table 16.1 shows an overview of ML-based techniques for FPGA design automation.

16.3.1   ML for Fast and Accurate QoR Estimation

Knowing the QoR, such as timing and resource usage, provides important guidance for improving the solutions in the early stages of the FPGA design flow. However, acquiring the actual QoR involves performing complex optimizations in the physical design stage (e.g., routing), which is very time-consuming and thus slows down the FPGA design closure. To address this issue, ML can be leveraged to obtain an accurate QoR estimation in earlier stages, which avoids the time-consuming post-implementation and therefore accelerates the overall design closure (Part ① of Fig. 16.2).

To bridge the gap between HLS estimation and the actual QoR, Dai et al. first propose to extract input features from the HLS report, which are then fed into the XGBoost model to output a more accurate estimation in terms of resource usage on FPGA [26]. Moreover, Makrani et al. use an ensemble machine learning model to predict the maximum achievable clock frequency based on HLS-reported features [76], followed by Ferianc et al., who exploit the Gaussian process to predict the latency of FPGA-based accelerator for convolutional neural networks [74]. Besides, Ustun et al. propose to leverage graph neural networks (GNN) to incorporate the structural information among operations in a data-flow graph, which is missing in the HLS report [115]. The proposed GNN model drastically improves the accuracy of operation delay estimation in HLS via learning the operation mapping patterns for DSP as well as carry blocks. Furthermore, early and accurate routing congestion estimation is also of great benefit to guide the optimization in HLS and improve the efficiency of implementation. As a representative study, Zhao et al. employ the gradient boosted regression tree (GBRT) to predict the routing congestion during the HLS stage [143]. Specifically, the GBRT model takes as inputs HLS features, including operator interconnection and resource usage, and outputs the estimation

Table 16.1 Overview of ML in FPGA design automation

Design automation task	Task objective	Design stage	ML algorithms	References	Section
Design Closure	Resource usage estimation	HLS	XGBoost	[26]	Sect. 16.3.1
	Max. clock frequency		Ensemble ML model	[76]	
	Accelerator latency		Gaussian process	[74]	
	Operation delay		GNN	[115]	
	Routing congestion	Placement	GBRT	[143]	
			ML model	[75, 86, 87, 99]	
			CGAN	[132]	
			CGAN (HD)	[3]	
	Routability		Three ML models	[111]	
			Reinforcement learning	[32]	
Design Space Exploration	Identify Pareto-optimal HLS designs	HLS	RPCL, Active learning	[69]	Sect. 16.3.2
	Maximize prediction accuracy with fewest samples		Random forest, Gaussian process, Active learning	[68]	
	Predict compiler directives	FPGA CAD flow	ML model	[52]	
			Multi-arm bandit	[127, 131, 135]	
			Sampling-based model search	[77]	
			SA, GA, ACO	[116]	
	Predict compiler directives	HLS	Multi-fidelity model	[72]	
	Estimate system-level metrics	FPGA CAD flow	Hierarchical Gaussian process	[73]	
	Minimize risk of losing Pareto designs		Random forest	[81]	
	Reduce execution latency and resource usage		Bayesian process	[80]	
	Reduce sample complexity/enhance training performance	HLS	Transfer learning	[58, 59]	
	Pareto designs under user-specified constrains/objectives		GNN, Reinforcement learning	[16, 120]	

for the vertical and horizontal congestion metrics. Moreover, the authors further trace the predicted highly congested regions back to the corresponding statement in the source code, which can guide users to resolve congestion issues in HLS without running the time-consuming RTL implementation.

In logic synthesis, structural cut-based techniques play a major role in logic optimization and technology mapping, particularly for LUT-based logic optimization and FPGA technology-mapping [22, 84]. However, the number of available cuts is an exponential relation between the graph size and the number of the cut leaves. Thus, heuristics have been mostly handcrafted through experimental analysis with domain-specific knowledge, e.g., number of leaves, cut level, number of nodes covered by a cut, etc., which is a very time-consuming engineering process. In this context, Lau et al. [86, 87] revisit cut-based technology mapping and pose the question of whether ML techniques can automatically learn novel cut-based heuristics that improve technology mapping results compared to classical methods, as well as improving the runtime complexity. Specifically, Lau et al. [86] propose formulating priority-cuts selection as a multi-class classification problem, which filters out candidate cuts that very likely lead to worse QoRs for technology mapping by embedding the local and global information of cuts and nodes of the Boolean networks. The results demonstrate substantial improvements in both delay and area-delay product improvements compared to exhaustive cut enumeration and state-of-the-art cut-based mapping algorithms in ABC [1].

In addition, several studies are focusing on predicting routing congestion during the placement stage to speed up the physical design closure. The techniques of [99] and [75] extract features such as the wire length and pin count and then apply ML models to predict the routing congestion. Later, the authors in [132] formulate the routing congestion as an image translation (colorization) problem, where the inputs are the post-placement images and the outputs are the congestion heat maps obtained after detailed routing. Consequently, the authors leverage the conditional generative adversarial network (CGAN) to forecast congestion in real time during incremental placement. Similar to [132], the work of [3] adopts a new CGAN model called pix2pixHD that performs high-definition (HD) image translations for large images with high resolutions. Due to the ability to support HD image translation, the proposed approach achieves high congestion estimation accuracy for large FPGA designs. Instead of directly estimating the routing congestion, the authors of [4, 111] propose a method by integrating three unique ML models to predict whether a placement solution is routable, which helps avoid performing further optimization and thus leads to improved run time. Recently, the technique in [32] introduces a way to speed up routing with the aid of reinforcement learning (RL) to minimize the number of routing conflicts. The proposed RL-based routing method achieves a 30% runtime speedup over the existing routing algorithm while achieving a similar QoR.

16.3.2 ML-Aided Decision-Making

A primary obstacle to rapid hardware specialization with FPGAs stems from weak guarantees of existing FPGA tools for achieving high-quality QoR. To meet the diverse requirements of a broad range of application domains, current FPGA tools from academia [12] and industry [25, 50] provide a large set of options across multiple stages of the tool flow in the form of compile-/runtime directives, also known as `pragmas`. Due to the size and complexity of the design space spanned by these options, coupled with the time-consuming evaluation of each design point, deciding an optimal set of tool options, also known as *design space exploration* or DSE, has become remarkably challenging. To tackle this challenge, recently many ML-assisted frameworks have been proposed to automatically and intelligently decide on an optimal set of tool options to accelerate iterative QoR estimation (Part ② of Fig. 16.2). These frameworks utilize design-specific features extracted from the early stages of the design flow to guide the decision process with significant runtime savings.

There are multiple works that apply machine learning to automatically identify an optimal set of compile-time directives for C/C++/OpenCL-based designs at the HLS stage. Liu et al. [69] pose the DSE as a classification problem of identifying beneficial designs for synthesis. It incorporates pruning with an adaptive windowing method to find the candidate Pareto-optimal HLS designs. The adaptive windowing method is derived from the Rival Penalized Competitive Learning (RPCL) model using an important set of features (e.g., estimated area of a register, multiplexer, decoder, number of wires) adjusted on the fly during exploration. Transductive Experimental Design (TED) [68] aims to select a representative and hard-to-predict samples from the design space. The objective is to maximize the accuracy of the predictive model with the fewest possible training samples. TED assumes no a priori knowledge about the learning model and hence can be beneficial to any learning model. Instead of improving the accuracy of the ML model, ATNE (Adaptive Threshold Non-Pareto Elimination) [81] primarily focuses on understanding and estimating the risk of losing good designs due to learning inaccuracy at the system level. Additionally, ATNE provides a Pareto identification threshold by adapting the estimated inaccuracy of the regressor for an efficient DSE. The work of [116] proposes a predictive model-based approach to finding meta-heuristics parameters (hyperparameters) of a multi-heuristic design space explorer consisting of simulated annealing, genetic algorithm, and ant colony optimization. To select an optimal combination of HLS directives, Lo et al. [72] incorporate low-fidelity estimates available from HLS tools in a multi-fidelity model and use a sequential model-based optimization [71] to explore the design space. To further enhance the sequential model, Lo et al. [73] use a hierarchical Gaussian process modeling to combine probabilistic estimates of component designs of a system to obtain exact values of system-level metrics, e.g., area, delay, and latency. Prospector [80] employs Bayesian techniques to optimize HLS synthesis `pragmas` to reduce execution latency and resource usage. Encoding the design space to capture design performance and FPGA costs (e.g., flip-flops, LUTs, BRAMs, DSPs) and sampling a small fraction (typically <1%) of the design space to reveal optimal design are key to Prospector. The authors in [58, 59] propose a transfer learning-based approach to transfer learned design space knowledge from source designs and apply it to a new target design. The key idea is multi-domain transfer learning in which effectively common knowledge between multiple source applications is extracted and is shared with the target applications. The objective is to enhance the training performance and reduce sample complexity. Along the same line of transfer learning, recently, Wu et al. [120] introduce IRONMAN by combining GNN [39] and reinforcement learning to provide either optimal solutions under user-specified constraints or various trade-offs (Pareto solutions) among different objectives (e.g., resources, area, and latency) for an input HLS C/C++ program. IRONMAN exposes concealed optimization opportunities for higher parallelism and shorter latency, accurately predicts the performance of the generated RTL using only the original dataflow graph of the input program consisting of regular and irregular data paths, and explores optimal resource allocation strategies based on user-specified constraints.

In another set of works, researchers apply ML to decide an optimal set of compiler directives for HDL-based designs at logic synthesis or later design stages. Kurek et al. model an objective function with a Gaussian process and use an SVM classifier to estimate whether design constraints are met [57]. To identify a good combination of compiler directives, InTime [52] uses active learning for DSE with ML models to surrogate actual design synthesis during design evaluation. InTime builds a machine learning model from a database of results of preliminary runs of FPGA tools and predicts the next series of FPGA tool options to improve timing results. To further improve the objective, InTime relies on a limited degree of statistical sampling. DATuner [127] is a parallel iterative autotuning framework for FPGA compilation that uses a multi-arm bandit technique to automatically select a set of appropriate compiler directives for a complete FPGA flow. DATuner uses a dynamic solution space partitioning based on information (e.g., runtime, search quality) gained from previous iterations and intelligently allocates computing resources to previously unexplored design subspaces and subspaces containing high-quality solutions. Mametjanov et al. propose a model-based search framework that incorporates sampling-based reduction of compiler directive space and guides the search toward promising directives configurations [77]. LAMDA takes an RTL description as input and automatically configures tool options across logic synthesis, placement, and routing stages [114]. LAMDA avoids iterating over the time-consuming FPGA implementation tool flow, especially in PnR stages, by exploring potential speedups that can be achieved by introducing high-fidelity QoR estimations in early and low-fidelity design stages as an effective way to prune the large and complex search space early in the design flow. Unlike previous approaches, LAMDA tackles the DSE problem from a multi-stage perspective as a way to balance the trade-off between computing effort and estimation accuracy.

16.3.3 Challenges and Strategies of Data Preparation for Learning

As the core of ML, the volume and quality of data are critical to the performance of ML models. Acquiring a large amount of high-quality data for the FPGA-related tasks, however, faces several challenges as follows.

Data Collection To achieve promising results, ML models typically require learning from a large volume of data. On the one hand, however, there are limited open-sourced datasets that can be used for FPGA-related tasks. On the other hand, it can be very time-consuming to build a dataset from scratch. For instance, the routing congestion estimation task requires obtaining the ground truth of routing congestion to guide the model during the training stage, which requires running through the physical design that takes several hours or even a few days per data sample, making it difficult to have enough data (e.g., tens of thousands of data samples) for model training. Besides, lacking the knowledge of technology nodes also hinders the data collection for some tasks (e.g., logic synthesis), since information such as area, delay, and power per logic gate is missing.

Data Transferability Apart from difficulties in data collection, it is also nontrivial to have high-quality data that enable ML models to generalize to unseen designs. Typically, ML models have the assumption that the training and testing data samples are drawn from the same distribution, which may not always be true for the FPGA-related tasks. Data samples are manually generated without any theoretical insights, and there is no standard way to check the quality of the dataset. As a result, the model may overfit the training set and generalize poorly to the unseen testing set.

Several prior arts are addressing some of the above challenges. Specifically, the authors in [26] generate multiple augmented designs per input design by changing the clock period and target device to increase the data volume. Further, the authors obtain the ground truth before the PnR and thus speed up the data collection process. To enable the ML model to predict operation mapping patterns in unseen designs, the work of [115] randomly generates a large number of synthetic designs to cover as many operation mapping patterns in unseen realistic designs as possible.

16.4 ML for FPGA Case Studies

16.4.1 QoR Estimation

16.4.1.1 Resource Usage Estimation in HLS Stage

Accurate QoR estimation at the HLS stage is difficult since the final implemented design reflects the cumulative effects of many nontrivial transformations through the series of implementation stages shown in Fig. 16.1. Moreover, final resource usage and timing depend on constraints imposed by the target FPGA device, such as the number, structure, and interconnection of device resources including LUTs, FFs, DSP blocks (i.e., hardened multipliers), and BRAMs. To enable fast resource and timing estimation, HLS tools pre-characterize different functional units ahead of time and sum up the contributions of instantiated functional units during the synthesis of each design. However, such an additive estimation approach fails to correctly capture the effects of post-HLS optimizations across functional units and neglects to consider limitations imposed by finite on-chip compute and routing resources.

Dai et al. first propose to leverage ML techniques for accurately estimating resource usage in HLS [26]. Specifically, they formulate the QoR estimation problem as a supervised learning task, where samples have features and targets (i.e., QoR) extracted from the HLS reports and implementation reports of HLS tools (e.g., Vivado HLS [25]), respectively. They collect over 1300 samples across 65 individual designs implemented under realistic design constraints from well-known HLS benchmark suites, including CHStone [43], Machsuite [103], S2CBench [106], and Rosetta benchmarks [145]. After extraction, the dataset contains an 87-dimension feature vector $\mathbf{x_i}$ and four-dimension target vector $\mathbf{y_i}$ per design sample in $\{(\mathbf{x_i}, \mathbf{y_i}) \mid i = 1, 2, ..., n\}$, where $n = 1300$ represents the total number of samples. Then, Dai et al. train ML models to estimate the 4 targets (i.e., resource usages): the number of LUTs, FFs, DSPs, and BRAMs. They exploit three different regression models in the following.

Linear Model The classic linear regression model, $\mathbf{y_i} = \mathbf{x_i^T w}$, models the target $\mathbf{y_i}$ as a linear combination of features $\mathbf{x_i}$. Linear regression fits this model onto the training data to determine the $\mathbf{w}$ such that a loss function is minimized, where $\mathbf{w}$ represents the vector of coefficients for the learned model. A Lasso linear model is adopted with a loss function of $||\mathbf{Xw} - \mathbf{y}||_2^2 + \gamma ||\mathbf{w}||_1$ to train a linear model that minimizes the least-square penalty on the training data with L1 regularization, where γ is a hyperparameter to control the magnitude of the regularization term. The L1 regularization term induces sparsity into the trainable weights $\mathbf{w}$ and in turn regulates the complexity of the model to improve its performance on the test set.

Neural Network Unlike linear models, a multi-layer perceptron (MLP) is a feed-forward neural network that can capture nonlinearity in the data, as each layer is followed by a nonlinear activation function [47]. Concretely, suppose there are L layers in the model; then MLP can be denoted by $\mathbf{X}^l = \sigma(\mathbf{X}^{l-1}\mathbf{W}^{l-1})$ for $l = 1, 2, ..., L - 1$ and $\hat{\mathbf{Y}} = \mathbf{X}^{L-1}\mathbf{W}^{L-1}$, where σ is the nonlinear activation function (e.g., ReLU), $\mathbf{X}^0$ is the initial feature matrix, $\mathbf{W}^{l-1}$ is a trainable weight matrix in the l-th layer, and $\hat{\mathbf{Y}}$ contains predicted target values for all samples. The trainable weights are iteratively updated to minimize the difference between predicted $\hat{\mathbf{Y}}$ and the ground truth $\mathbf{Y}$ via backpropagation. Since the number of features is relatively small, and the amount of training data is limited, Dai et al. use MLP with only a few hidden layers to ensure the model converges. Compared to linear models, MLP requires tuning more hyperparameters (e.g., number of layers, neurons per layer) and results in non-convex loss functions that require more effort in training than the linear model above.

Gradient Tree Boosting Based on building a "strong" model by combining a series of "weak" ones (e.g., regression trees), boosting represents another promising nonlinear technique [79]. Among various boosting models, Dai et al. choose to use XGBoost [15], as it has demonstrated accuracy competitive to or even better than neural networks while attaining better efficiency in both training and inference [108]. Specifically, XGBoost models the target as the sum of regression trees, each of which maps the features to a score for the target. Target estimation is determined by accumulating scores across all trees. XGBoost leverages gradient descent to repeatedly identify a new tree pointing to the negative gradient direction and adds the tree to the model to reduce the prediction error.

For evaluating the estimation quality, Dai et al. use the relative root square error (RRSE), defined by

$$\epsilon = \sqrt{\frac{\|\hat{\mathbf{y}} - \mathbf{y}\|_2}{\|\mathbf{y} - \bar{\mathbf{y}}\|_2}},$$

where $\hat{\mathbf{y}}$ is a vector of values predicted by the model for a particular target and $\mathbf{y}$ is a vector of actual ground truth values in the testing set for that target. $\bar{\mathbf{y}}$ denotes the mean value of $\mathbf{y}$. The experimental results show that all ML models can achieve lower RRSE scores than those of commercial HLS tools. In particular, XGBoost stands out as the most competitive model with less than 30% error for LUT and FF. The explanation is that the features extracted from the HLS report are tabular data, comprising samples (rows) with the same set of features (columns), which XGBoost is known to perform quite well [108].

16.4.1.2 Operation Delay Estimation in HLS Stage

Existing academic and commercial HLS tools rely on simple delay models: the delay per operation is pre-characterized in isolation, and the overall delay is calculated additively over individual operations [12, 25]. Nonetheless, operations are clustered into device resources during technology mapping, which invalidates the additive approach. As a result, accurately estimating the operation delay requires the knowledge of operation mapping patterns. There are two typical operation mappings as discussed below.

DSP Mapping Modern heterogeneous FPGA devices include specialized DSP blocks to support high-performance implementations of arithmetic operations, such as FIR filters and FFT. DSP blocks are highly configurable to realize a wide range of functionalities and system requirements. They have pre-fabricated datapaths which can be configured by various control bits. For example, the datapath of the DSP48E2 primitive in Xilinx Ultrascale devices [125] includes a 27-bit pre-adder unit, a 27 × 18-bit multiplier, a 48-bit ALU, registers, wide logic operations, and multiplexers controlled by configuration bits to be able to implement various circuit subgraphs. Depending on the optimizations in logic synthesis and technology mapping stages, subgraphs in a design can be mapped to DSP blocks in various ways.

Adder Cluster Mapping In addition to the specialized DSP blocks, modern heterogeneous FPGA devices also include dedicated carry logic to speed up arithmetic operations (e.g., CARRY8 blocks in Xilinx Virtex UltraScale+ devices [124]). RTL synthesis can take advantage of the carry blocks by mapping two or more adjacent addition/subtraction operations on a single carry chain along with other logic operations (e.g., sign extension, bit truncation, zero extension). Ustun et al. define such patterns of two or more adjacent addition/subtraction operations along with other logic operations as an adder cluster [115]. One such adder cluster is shown in Fig. 16.3, where all the add and sign-extension operations within the blue block are mapped to a single carry chain.

Although the work introduced in Sect. 16.4.1.1 can improve the estimation accuracy of resource usage by learning features from the HLS report, it completely ignores the structural information among operations in the dataflow graph (DFG) and thus fails to capture the operation mapping pattern, which is the key for accurately predicting the operation delay. As indicated in [105], DSP mapping patterns are essentially local structures around multiply operations in DFG. As a consequence, whether an operation is mapped onto a DSP block is fully determined by its local structure and attribute information of its neighbors (e.g., operation type, bit width). Similarly, whether an add operation is clustered with other nearby adjacent add and logic operations (if any) is also determined by its local structure. Since graph neural networks (GNN) can produce node embeddings that incorporate both local structure and node attributes in a graph, Ustun et al. [115] propose to exploit GNN to automatically learn operation mapping patterns in a DFG.

Fig. 16.3 Adder-cluster for an input DFG

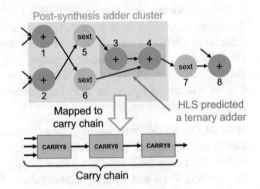

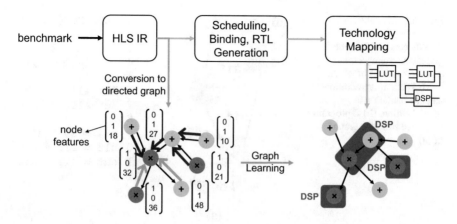

Fig. 16.4 Overall flow—GNN-based learning of operation mapping and clustering in arithmetic-intensive designs

Specifically, they formulate the problem as a supervised edge classification task: the GNN model takes as inputs a DFG G as well as a matrix X that contains the node features (e.g., operation type, bit width). The goal of the GNN model is to predict which edges in DFG will be mapped onto DSP or adder cluster. The proposed approach consists of three steps as shown in Fig. 16.4:

(Step 1) Microbenchmark Generation To learn operation mapping patterns and generalize to unseen designs, Ustun et al. propose to generate synthetic combinational microbenchmarks composed of arithmetic operations. Those microbenchmarks are used to train the GNN model in an inductive manner such that the trained model can directly be exploited to predict mapping patterns of unseen (realistic) designs. To this end, microbenchmarks should contain various subgraph structures that can be mapped to device resources after technology mapping. Specifically, they randomly generate 2000 directed graphs of addition and multiplication operations with different connectivities and set the total number of operations for each microbenchmark to 20 with 4–8 multiplication operations. Moreover, they further vary the number of inputs of each design from 8 to 12 to generate various data-sharing patterns, contributing to a wide spectrum of design samples in the training set. Generated graphs are then converted to C programs to be input into the HLS flow as shown in Fig. 16.5.

(Step 2) Feature and Ground Truth Extraction In the context of arithmetic operation mapping, it is of great importance to capture both structural and contextual information. Structural information stands for connectivity patterns among nodes in DFG and is necessary to identify the neighborhood of an arithmetic operation as well as its mapping patterns. Contextual information, i.e., node features, is supplementary information of the operations, including operation type and bitwidth. Node features are essential for two reasons. First, mapping patterns are local

Fig. 16.5 Microbench
generation procedure—(**a**)
Weakly connected directed
graph comprising a total of
$n = 3$ operations of which
two operations are addition
and one operation is
multiplication. (**b**) C program
generated from the graph
of (**a**)

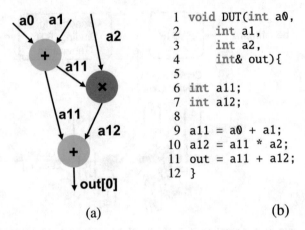

```
1  void DUT(int a0,
2           int a1,
3           int a2,
4           int& out){
5
6  int a11;
7  int a12;
8
9  a11 = a0 + a1;
10 a12 = a11 * a2;
11 out = a11 + a12;
12 }
```

(a)   (b)

structures around arithmetic operations, bringing about the need to distinguish
different operation types. Second, hardened blocks (e.g., DSPs) support functionalities of certain bitwidths, making operation bitwidths an important factor in the
identification of operation clustering. To capture both structural and contextual
information in the dataset, Ustun et al. run microbenchmarks through Vivado HLS
and obtain the HLS intermediate representations (IRs), which are then used to
generate DFGs as the structural feature. Operations in the HLS IR are represented
as nodes, and data dependencies are represented as edges in DFGS. Figure 16.6a
shows a simple DFG, where nodes A, B, and D denote addition operations and
node C represents multiplication operation. Moreover, they also extract operation
type and bitwidth information from the HLS IR as the contextual feature. To train a
GNN model in a supervised manner, they further extract ground truth (edge labels)
by matching operations in HLS IR to device resources in post-mapping netlists.
The ground truth of an edge is **1** if it is mapped onto a DSP or adder cluster and **0**
otherwise.

(**Step 3**) **Learning Operation Mapping via GNN** After obtaining structural and
contextual features as well as ground truth, Ustun et al. adopt the idea of "message
passing" from GraphSAGE [39], a popular and powerful GNN model, for this binary
edge classification task (Fig. 16.6b). Since GraphSAGE is unable to support directed
graphs, it cannot be directly applied to the directed dataflow graph. To overcome
this issue, Ustun et al. propose a customized GNN model called D-SAGE for
this task. Specifically, D-SAGE learns an aggregation function (AGG) that collects
the information from predecessors and successors separately and produces node
embeddings h^k at the k-th layer as follows:

$$AGG^{(k)} \left(\left\{ h_j^{(k)}, j \in \mathcal{N}(i) \right\} \right) = \frac{1}{|\mathcal{N}(i)|} \sum_{j \in \mathcal{N}(i)} h_j^k \tag{16.1}$$

$$h_{i,pr}^{(k+1)} = MLP_{pr}^{(k)} \left(h_i^{(k)} \| AGG_{pr}^{(k)} \left(\left\{ h_j^{(k)}, j \in \mathcal{PR}(i) \right\} \right) \right) \qquad (16.2)$$

$$h_{i,su}^{(k+1)} = MLP_{su}^{(k)} \left(h_i^{(k)} \| AGG_{su}^{(k)} \left(\left\{ h_j^{(k)}, j \in \mathcal{SU}(i) \right\} \right) \right) \qquad (16.3)$$

$$h_i^{(k+1)} = MLP^{(k)} \left(h_{i,pr}^{(k+1)} \| h_{i,su}^{(k+1)} \right) \qquad (16.4)$$

where $\mathcal{N}(i)$ denotes the neighbor set of node i; $MLP_{pr}^{(k)}$, $MLP_{su}^{(k)}$, and $MLP^{(k)}$ are three different multilayer perceptrons at the k-th layer; $\cdot \| \cdot$ is feature-wise concatenation; and $\mathcal{PR}(i)$ and $\mathcal{SU}(i)$ denote the predecessors and successors of node i, respectively. Intuitively, D-SAGE (Fig. 16.6c) works on directed graphs by separately collecting the information from predecessors and successors of each node and then combining the information from both sides to update node embeddings at each layer. Note that $h_i^{(0)} = x_i$, where x_i is the initial feature vector of a node i that contains the operation type and bitwidth information. The edge embedding vector, which is used to determine if the edge is mapped to DSP or adder cluster, is calculated via averaging the embeddings of its two end nodes (Fig. 16.6).

To evaluate the proposed approach, Ustun et al. choose the F1 score as the metric for measuring the edge classification accuracy. Specifically, the F1 score represents a balance between precision and recall, where the precision is the percentage of the correctly classified samples among all samples with the predicted label **1** and the recall is the percentage of the correctly classified samples among all samples with the actual label **1**. F1 scores take values from **0** to **1**, while values closer to

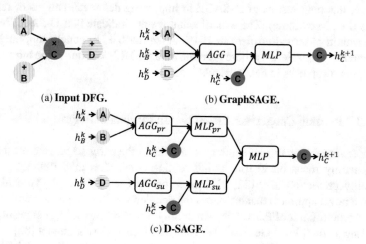

(a) **Input DFG.**   (b) **GraphSAGE.**

(c) **D-SAGE.**

Fig. 16.6 Comparison of GraphSAGE and D-SAGE at the kth layer—(**a**) Input dataflow graph. (**b**) Compute embedding vector of node C (i.e., multiply operation) with GraphSAGE. (**c**) Compute embedding vector of node C with D-SAGE

1 mean higher accuracy. To account for the potential impact of network structure on the learning process, the test set only contains designs with a certain number of multiplication operations and hides these designs from training. For example, the model is trained on designs with 5, 6, 7, and 8 multiplication operations and tested on those with four multiplication operations.

Ustun et al. show that D-SAGE reduces false-positive cases by 62% and false-negative cases by 67% on average for the edge classification, reflected in the significant increase in the F1 score over a commercial HLS tool with a margin up to 22%. The HLS tool fails to capture many clustering schemes of operations into DSP blocks, which has a direct impact on timing. Furthermore, the D-SAGE model trained on microbenchmarks is further directly evaluated on real designs from MachSuite [103]. Testing results show that the D-SAGE model achieves at least a 24% accuracy improvement over the existing HLS tool for predicting DSP mapping patterns in real designs. In addition, the D-SAGE model is evaluated on microbenchmarks for carry chain clustering tasks, which achieves the best F1 score compared to the baseline, confirming its efficacy in learning operation mapping patterns.

In addition, Ustun et al. further integrate the mapping-awareness delay prediction into the HLS tool to correct logic delays based on operation mapping patterns learned by GNN and compare the delay prediction results with the vanilla HLS tool. Concretely, they first evaluate delay estimation results of synthetic designs based on the HLS tool, which shows those designs have up to 143% error in delay estimation. In contrast, the D-SAGE-based method achieves less than 88% error in delay estimation for those synthetic designs. Moreover, they further incorporate operation mapping results of D-SAGE in improving delay estimations of realistic designs (i.e., MachSuite). The experimental results indicate that D-SAGE reduces the maximum observed error from 164% down to 40%. In conclusion, the GNN-based model is considerably more accurate than the HLS tool due to the introduction of mapping awareness in delay calculations.

16.4.1.3 Routing Congestion Estimation in Placement Stage

For improving the quality of routing estimation at the early stages, the most recent works mainly focus on a) forecasting routing congestion map [99, 132] and b) routability prediction [99, 122, 132]. Specifically for FPGA PnR, Yu et.al [132] present a novel approach that estimates the detailed routing congestion with a given placement solution for FPGA PnR, which fully forecasts the routing congestion heat map using a conditional Generative Adversarial Nets (cGANs) model [82].

Generative Adversarial Networks (GANs) are neural network models that are used in unsupervised ML tasks. GANs learn a transformation from random noise vector z to a corresponding mapping g, denoted as $G(z)$, which implements a differentiable function that maps $z \rightarrow y$ [34]. GANs include two multilayer perceptrons, namely, generator G and discriminator D. The goal of discriminator D is to distinguish between samples generated from the generator and samples from the training dataset. The goal of generator G is to generate a mapping of input that cannot be distinguished to be true or false by the discriminator D. The network is trained in two parts and the loss function $L_{(G,D)}$ is shown in Eq. 16.5.

- train D to maximize the probability of assigning the correct label to both training examples and samples from G.
- train G to minimize $log(1 - D(G(z)))$.

$$L_{(G,D)} = \min_{D} \min_{G} (\mathbb{E}_x log D(x) + \mathbb{E}_z log(1 - D(G(z)))) \qquad (16.5)$$

In contrast to GANs, cGANs [82] learn a mapping by observing both input vector x and random noise vector z, denoted as $G(x, z)$, which maps the input x and the noise vector z to g, $(x, z) \rightarrow g$. The main difference compared to GANs is that the generator and discriminator observe the input vector x. Accordingly, the loss function $cL_{(G,D)}$ (Eq. 16.6) and training objectives will be the following:

$$cL_{(G,D)} = \min_{D} \min_{G} \left(\mathbb{E}_{x,g} log D(x, g) + \mathbb{E}_{x,z} log(1 - D(G(x, z)))\right) \qquad (16.6)$$

GANs and cGANs have demonstrated great performance in image synthesis such as image colorization. In particular, Yu et al. [132] formulate the post-placement routing estimation as an *image translation (colorization)* problem. Specifically, the input image is the visualization of the post-placement designs, and the expected output is the congestion heat map obtained after detailed routing. One of the inputs for the physical design process is a technology-mapped netlist, which is represented using directed graphs such that the cells are nodes V and interconnects are edges E. For FPGA placement, it is a packed netlist where each cluster-based logic block (CLB) could contain one or more basic logic elements (BLEs). Let Graph(V, E) be the packed netlist, and all the elements in V are placed within a given floorplan via placement. The nodes and edges in the graphs have a specific 2-D location on the floor plan, denoted as Graph($V, E', grids$), where $grids$ represent the 2-D locations of V. Meanwhile, the edges are updated with locations $E \rightarrow E'$. Finally, routing connects all the elements with respect to Graph($V, E', grids$). The intermediate results, i.e., post-placement and post-routing results, can be visualized as images $\in \mathbb{R}^{w \times w \times 3}$, denoted as img_{place} and img_{route}, respectively. An important observation is that these images are incrementally changed while PnR proceeds, {Graph(V, E), img_{floor}} $\rightarrow img_{place}$, or from post-placement to post-routing, {Graph($V, E', grids$), img_{place}} $\rightarrow img_{route}$. Thus, the problem of forecasting routing heat maps can be mimicked as an *image to the image translation* problem, i.e., colorizing img_{place} into a congestion heatmap annotated img_{route} using conditional generative adversarial networks (cGANs). Specifically, as shown in Fig. 16.7, the generator in the cGAN model will take placement results and netlist connectivity information as inputs and colorizes (estimates) the placement results accordingly. This approach has been evaluated with eight designs obtained from VTR 8.0 [85], where the ground truth img_{route} are generated with VTR as well. The

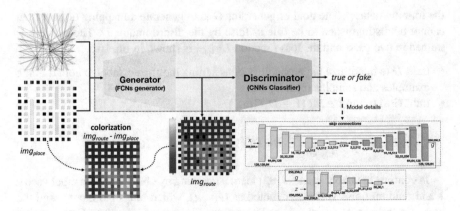

Fig. 16.7 Overview of cGAN-based routing congestion estimation

image generator is implemented in C++ based on VPR [85] with a resolution of $256 \times 256 \times 3$. Yu et.al demonstrate the capability of the cGAN routing congestion approach can effectively prune out the data points that have high routing congestion [132].

16.4.2 Design Space Exploration

To meet the diverse requirements of a broad range of application domains, FPGA CAD tools provide users with an extensive set of configuration options (in form of pragmas) for each design stage as a means of selecting between different heuristics or controlling the behavior of a heuristic. For instance, logic synthesis and PnR options in Intel Quartus Pro translate to a search space of over 1.8×10^{24} design points. These options need to be calibrated with significant manual effort and expert knowledge to achieve desired QoR. However, the size and complexity of the search space make manual exploration extremely inefficient and often infeasible. A similar challenge exists in software compilation, where a number of autotuners have been developed for automatically configuring compiler flags [8]. Such approaches cannot directly be applied to the hardware domain because of the wide gap in runtime between software compilers and hardware compilers. The cost of exploring different configurations in an iterative manner is significantly higher in hardware compilation. Therefore, recent years have seen increasing employment of learning-based techniques to enable rapid auto-configuration of hardware tool flows [51, 63, 68, 96, 149], some of them targeting FPGAs in particular [52, 114, 127, 135].

16.4.2.1 Auto-Configuring CAD Tool Parameters

The greatest limitation of the existing FPGA autotuners is iterating over the time-consuming CAD tool flow, especially PnR stages. LAMDA [114] explores potential speedups that can be achieved by introducing high-fidelity QoR estimations in early and low-fidelity design stages as an effective way to prune the large and complex search space early in the design flow. LAMDA tackles the problem from a multi-stage perspective as a way to balance the trade-off between computing effort and estimation accuracy [72].

The overall autotuning flow of LAMDA is illustrated in Fig. 16.8. It takes an RTL description as input and automatically configures tool options across logic synthesis, placement, and routing stages. Table 16.2 lists a subset of tunable tool options of Intel Quartus Pro. LAMDA leverages multiple fast and low-cost stages of the FPGA design flow to estimate timing both quickly and accurately and use these estimations to effectively prune the design space defined by the tool options. The rest of this section describes the key components of LAMDA in more detail.

Multi-Stage QoR Inference Multi-stage inference model of LAMDA, which uses XGBoost [15], estimates post-routing results based on tool features (i.e., configurations of the tool options) and design-specific netlist features. Collecting early stage features are fast but could lack fidelity. Collecting features from later stages is more informative, yet time-consuming. Therefore, the *fast and low-cost design stages* in Fig. 16.8 need to be carefully selected to balance the trade-off between accuracy and runtime. Analyzing the effects of the features on the estimation error in Table 16.3, one can draw three conclusions. First, design-specific features help estimate QoR more accurately compared to using tool options only (i.e., pre-synthesis). Second, accuracy increases as features from later stages are included in the feature set, bringing about an accuracy-runtime trade-off. Third, although tool estimates are less accurate under tight constraints, design-specific

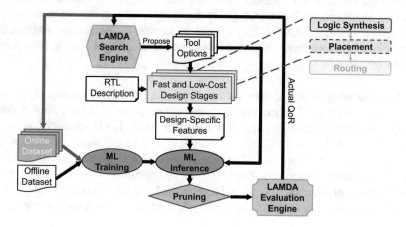

Fig. 16.8 Overview of LAMDA

Table 16.2 A subset of tunable Intel Quartus Pro tool options

Stage	Options	Values
Logic synthesis	auto_dsp_recognition, timing_driven_synthesis	{On, Off}
	disable_register_merging, mux_restructure	{Auto, On, Off}
	optimization_technique	{Area, Speed, Balanced}
	synthesis_effort	{Auto, Fast}
Placement and routing	fitter_effort	{Standard Fit, Auto Fit}
	final_place_optimization, routability_optimization	{Always, Never, Automatically}
	register_packing_effort	{High, Low, Medium}
	allow_register_retiming, auto_delay_chains	{On, Off}
	route_timing_optimization	Normal, Maximum, Minimum

Table 16.3 Effect of features on timing estimation – RMSE on bfly for different feature sets and tight (5.0 ns) and moderate (6.8 ns) constraints, for a training set of size 100

	Target (ns)	
Features	5.0	6.8
Pre-synthesis	0.72	0.46
Pre-place	0.63	0.35
Pre-route	0.60	0.29

Table 16.4 Design-specific features provided by Intel Quartus Pro

Design Stage	Type	Features
Logic synthesis	Resource	#ALM, #LUT, #registers, #DSP,#I/O pins, #fan-out, etc.
	Timing	WS, TNS

features still help improve estimation accuracy compared to using tool options only. To balance the aforesaid trade-off, LAMDA assigns logic synthesis as the *fast and low-cost design stage*. Table 16.4 shows design-specific features for these stages. For certain designs, LAMDA may choose to include the placement stage in the feature set if the logic synthesis stage fails to model the complex timing behavior in the design.

Online Learning Many of the existing autotuning techniques explore supervised learning and train an ML model using a static dataset, also called an offline dataset. Simply relying on offline training is not realistic in the FPGA domain due to the lack of labeled data and the cost of collecting end-to-end data. To balance the data collection efforts and the estimation accuracy, LAMDA uses online learning for its multi-stage ML model. The idea is to first perform offline training using a small number of static data points and then periodically update the ML model (i.e., retraining) with newly collected data points throughout the autotuning process.

Implementation At the first iteration of autotuning, the ML model is trained on the offline dataset. Given an optimization objective, LAMDA picks a set of unseen configurations proposed by the search engine which is built from OpenTuner [8]. Design-specific features are collected after running the tool flow up to logic synthesis. A trained ML model is used to estimate the QoR of these proposals based on tool features and design-specific features from the early stages. Proposed configurations are ranked with respect to these ML inference results. Top-ranking configurations are evaluated through the complete FPGA tool flow to obtain the actual QoRs, whereas the remaining configurations are pruned away. Data obtained from the evaluation is used to update OpenTuner's search engine and the ML model iteratively. The autotuning process terminates when one of the newly proposed configurations meets timing or timeout is reached. Collecting design-specific features and the evaluation process are parallelized to speed up the autotuning process.

Evaluation LAMDA is evaluated for the timing closure of five benchmarks. To demonstrate the effectiveness of multi-stage and online learning separately, different ML-based autotuning modes are compared, namely, *online-multi* (i.e., LAMDA), *online-single*, and *offline-multi*. Here, *single* represents estimating timing based on only tool options; *multi* represents leveraging design-specific features from early stages; *online* represents online learning; and *offline* represents offline learning. It is shown that LAMDA has higher auto-tuning efficiency compared to *online-single* and *offline-multi*. This suggests that both design-specific features and online learning improve estimation accuracy and thereby help prune the design space more effectively.

16.4.2.2 Automatic Synthesis Flow Generation

In addition to end-to-end tool parameter space exploration is exploring the sequence of synthesis transformations (also called synthesis flow) in FPGA design to accelerate the DSE process. Synthesis flows are a set of synthesis transformations that apply iteratively to the Boolean network or circuit design, which are mostly a combination of DAG-aware synthesis algorithms that can benefit from ML techniques. Specifically, the exploration of synthesis flows has been considered a *sequential decision-making* problem, as synthesis flows involve a sequence of synthesis transformations. For example, an iterative training-exploration fashion through deep learning (e.g., CNNs, LSTMs) [133, 135] and reinforcement learning [48, 131, 147] have demonstrated the capability of learning sequential decision-making to achieve a better quality of results [48, 135]. Recently, Yu [131] proposes a novel end-to-end sequential flow tuning platform, *FlowTune*, based on a novel domain-specific multi-arm-bandit (MAB) approach, which offers the best performance in flexibility, theoretical guarantees, and exploration efficiency for Boolean optimization and technology mapping in FPGA designs.

Specifically, FlowTune introduces a generic end-to-end and high-performance domain-specific, multi-stage, multi-armed bandit framework for Boolean logic optimization, utilizing domain-specific knowledge of graph-based synthesis algorithms. The action states are the synthesis flows that are applied in a sequence of actions (synthesis transformations) that are sampled from different arms (partitioned sampling space of sequential decision-making). The reward functions can be flexibly selected from a wide range of logic representations, including both technology-independent (e.g., logic-level and node count) and technology-dependent rewards (e.g., post-LUT-mapped metrics). The partitions of the entire sequential decision space are the most critical to decide at the initialization stage, which are generated with domain knowledge concluded from two important observations: (1) Given a randomly permuted synthesis flow using DAG-aware transformations, earlier graph-aware transformations can effectively optimize the logic network, and (2) the optimization performance of synthesis flow is dominated by the first transformation since the first transformation dominates the "optimizable" nodes for future optimizations.

To gather the domain knowledge of synthesis algorithms summarized from these two observations, Yu [131] proposed a novel MAB environment by re-defining the arms and actions. Let $\mathbb{P}(\mathcal{X})$ be a random permutation function over a set of decisions (synthesis transformations) $\mathcal{X}$. Let $\mathbb{P}(x \| \mathcal{X})$ be a random permutation function over the set $\mathcal{X}$, $x \in \mathcal{X}$, such that $\mathbb{P}(x_i \| \mathcal{X})$ is a random permutation with x_i always being the first element in the permutation, $\mathbb{P}(x_i \| \mathcal{X}) \in \mathbb{P}(\mathcal{X})$. Then, the arms in MAB are defined as $\mathbb{P}(x \| \mathcal{X})$, such that $\mathcal{A} = \{ \mathbb{P}(x_0 \| \mathcal{X}), \mathbb{P}(x_1 \| \mathcal{X}), ..., \mathbb{P}(x_n \| \mathcal{X}) \}$, where n is the number of available decisions in the exploration problem. Specifically, n corresponds to the number of available synthesis transformations. Unlike using traditional MAB algorithms, an action a_t at time t is a sampled permutation from $\mathbb{P}(x_i \| \mathcal{X})$. In other words, a_t is a *multiset* over set $\mathcal{X}$. Let $Q(a_t)$ be the action value that is obtained by applying a_t to the given logic circuit at t time step, and the reward is r_t

$$r_t = Q(a_t) - Q(a_{t-1}) \implies Q(\mathbb{P}(x_i \| \mathcal{X})) - Q(\mathbb{P}(x_j \| \mathcal{X}))$$

where the ith arm is played at t time step and jth arm is played at $t - 1$ time step. Finally, we use upper confidence bound (UCB) bandit algorithm as the agent, such that a_t is chosen with estimated upper bound $U_t(a)$. The upper bound in this work is shown below.

$$a_t = \arg\max_{a \in \mathcal{A}} Q(a) + U_t(a), \, U_t(a) = \sqrt{\frac{\log t}{2N_t(a)}}$$

The performance of the multi-armed bandit algorithm is evaluated in terms of its regret, defined as the gap between the expected payoff of the algorithm and that of an optimal strategy. In this work, the number of regrets equals to the number of synthesis flows that have been evaluated in the synthesis tool. Using the UCB algorithm, an asymptotic logarithmic total regret L_t will be achieved

$$\lim_{t \to \infty} L_t = 2log\ t \sum \Delta_a$$

where Δ_a is the difference between arms in $\mathcal{A}$. As a result, FlowTune integrates with several front-end and back-end tools such as ABC [1], Yosys [130], and VTR [85] and demonstrates the versatility of the domain-specific MAB algorithms on various optimization tasks, e.g., FPGA mapping (LUT) minimization, Boolean SAT formula minimization, Boolean logic minimization, standard-cell technology mapping, etc. More details can be found in [131] and https://github.com/Yu-Utah/FlowTune.

16.5   Future Research Scopes

Data Collection and Transferability There are several possible solutions for the data collection and transferability challenges. For instance, a large volume of training designs can be obtained by augmenting open-source designs by changing their target frequency and device. To speed up the process of ground truth collection in the time-consuming physical design stages, self-supervised learning techniques can be leveraged to reduce the amount of ground truth required for the training process. Moreover, to increase the data transferability, the synthetic benchmark can be generated to cover as many learning patterns as possible by leveraging domain-specific knowledge, such that the ML model trained on the synthetic benchmark can also perform well on unseen (realistic) designs.

Placement and Routing   Placement and routing are time-consuming stages in the FPGA design flow, making FPGA design and optimization highly challenging. There are works on estimating certain placement and routing metrics and using these estimations to guide decision-making so as to speed up design closure. However, ML for placement and routing remains a big challenge due to the amount of training data needed to build accurate estimation models, coupled with the cost of collecting placement and routing data. Recently, VLSI and FPGA placement has been cast as training a neural network, leading to significant speedup with the use of deep learning toolkits [66]. Similar advancements in accelerating FPGA placement and routing that utilize the power of modern computing diagrams such as GPUs, TPUs, and heterogeneous supercomputing infrastructure would greatly benefit design closure efforts.

ML-Assisted Functional Verification of FPGA Designs Despite considerable progress in applying ML to accelerate the FPGA design, little work exists to certify the correctness of the realistic FPGA designs after optimizations are applied. Given that FPGA design is a multi-stage process at varying abstraction, the compiler, the user, or even the optimization engine can introduce functional bugs in the design.

Translation validation [11, 53], and, more recently, Verticert [46], takes a big step toward formally proving equivalence between the input C code and the generated RTL code. However, these methods can only scale to small designs with limited functional constructs and work between the input and output of the HLS stage of Fig. 16.1. To the best of our knowledge, there exists no technique to prove the functional correctness of the FPGA design at logic synthesis or beyond. We envision ML, specifically, the GNN, can be highly effective in providing correctness guarantees between the input and the output of the different stages of the FPGA design flow. Our intuition is that GNN, when trained with appropriate node embeddings capturing domain-specific knowledge, is highly effective in learning graph topology in graph-structured data. Almost all the inputs and the outputs of the different FPGA design stages can be modeled as a directed graph, making those inputs and outputs suitable for graph-based learning. In addition to identifying domain-specific features relevant for the verification task, such a verification solution should also tackle challenges related to modeling FPGA architecture and the compiler intricacies.

Reproducibility Challenges Reproducibility has become a major challenge in many ML applications in various scientific and engineering domains. It shows that results or observations obtained from an experiment or study should be replicated and consistent with the methodologies released publicly, such as presentations and scientific publications. Particularly, due to the high engineering efforts in FPGA design and CAD algorithm development, reproducibility offers high value to any continuous integration or continuous delivery in advancing ML for FPGA design in both industries and academia. In general, reproducibility of applied ML works can be enhanced by (1) code and hyperparameter availability, (2) data accessibility and track of changes, and (3) computing environment. Code and hyperparameter are arguably the biggest boosts to reproducible experiments in ML, so as ML for FPGAs. However, one of the unique challenges in ML for FPGAs is randomness. In addition to the ML setups, there are full of randomizations in existing FPGA CAD algorithms, especially in dealing with large-scale FPGA architecture and designs. Such randomness also leads to the challenge of data accessibility, while most of the FPGA ML datasets are generated with those CAD tools, and increasing randomness could increase the generalizability of the dataset and the ML models. There is a likely trade-off between the dataset's randomness and the performance of the ML models and consistency. Finally, FPGA design tool flows usually involve a very complex compilation system, where the performance highly depends on versions of libraries, compilers, frameworks, etc. Combining with the ML frameworks, the versions of software and used hardware and all other parts of the environment must be carefully recorded and maintained. However, current system maintenance mostly requires a tremendous number of engineering efforts.

16.6 Conclusion

This chapter summarizes the recent advancements in applying ML for accelerating FPGA design procedures. Primarily we focused on two facets of the application of ML in FPGA design—as an estimator to optimize QoR estimation and as a decision-maker to systematically explore design state space to improve iterative QoR estimation. We have also discussed multiple case studies showcasing such ML applications to provide readers with further insights. We hope that this chapter will motivate researchers from academia and industry to develop efficient ML-based solutions to further accelerate FPGA design.

References

1. ABC: A System for Sequential Synthesis and Verification. http://www.eecs.berkeley.edu/alanmi/abc. Accessed December 14, 2022
2. Abts, D., Ross, J., Sparling, J., Wong-VanHaren, M., Baker, M., Hawkins, T., Bell, A., Thompson, J., Kahsai, T., Kimmell, G., Hwang, J., Leslie-Hurd, R., Bye, M., Creswick, E., Boyd, M., Venigalla, M., Laforge, E., Purdy, J., Kamath, P., Maheshwari, D., Beidler, M., Rosseel, G., Ahmad, O., Gagarin, G., Czekalski, R., Rane, A., Parmar, S., Werner, J., Sproch, J., Macias, A., Kurtz, B.: Think fast: a tensor streaming processor (TSP) for accelerating deep learning workloads. In: International Symposium on Computer Architecture (ISCA) (2020)
3. Alawieh, M.B., Li, W., Lin, Y., Singhal, L., Iyer, M.A., Pan, D.Z.: High-definition routing congestion prediction for large-scale FPGAs. In: Asia and South Pacific Design Automation Conference (ASP-DAC) (2020)
4. Al-Hyari, A., Szentimrey, H., Shamli, A., Martin, T., Gréwal, G., Areibi, S.: A deep learning framework to predict routability for FPGA circuit placement. ACM Trans. Reconfig. Technol. Syst. 14(3), (2021)
5. Al-Khaleel, O., Baktır, S., Küpçü, A.: FPGA Implementation of an ECC processor using Edwards curves and DFT modular multiplication. In: International Conference on Information and Communication Systems (ICICS) (2021)
6. Amaru, L., Gaillardon, P.E., De Micheli, G.: Majority-inverter graph: a new paradigm for logic optimization. IEEE Trans. Comput. Aided Des. Integr. Circuits Syst. 35(5), 806–819 (2015)
7. An In-Depth Look at Google's First Tensor Processing Unit. https://cloud.google.com/blog/big-data/2017/05/an-in-depth-look-at-googles-first-tensor-processing-unit-tpu. Accessed: December 14, 2022
8. Ansel, J., Kamil, S., Veeramachaneni, K., Ragan-Kelley, J., Bosboom, J., O'Reilly, U.M., Amarasinghe, S.: OpenTuner: an extensible framework for program autotuning. In: International Conference on Parallel Architectures and Compilation Techniques (PACT) (2014)
9. Asiatici, M., Ienne, P.: Large-scale graph processing on FPGAs with caches for thousands of simultaneous misses. In: International Symposium on Computer Architecture (ISCA) (2021)
10. Balupala, H.K., Rahul, K., Yachareni, S.: Galois field arithmetic operations using Xilinx FPGAs in cryptography. In: International IOT, Electronics and Mechatronics Conference (IEMTRONICS) (2021)
11. Banerjee, K., Karfa, C., Sarkar, D., Mandal, C.: Verification of code motion techniques using value propagation. IEEE Trans. Comput. Aided Design Integ. Circuits Syst. (2014)
12. Canis, A., Choi, J., Aldham, M., Zhang, V., Kammoona, A., Anderson, J.H., Brown, S., Czajkowski, T.: LegUp: high-level synthesis for FPGA-based processor/accelerator systems. In: International Symposium on Field-Programmable Gate Arrays (FPGA) (2011)

13. Capligins, F., Litvinenko, A., Aboltins, A., Kolosovs, D.: FPGA Implementation and study of synchronization of modified Chua's circuit-based chaotic oscillator for high-speed secure communications. In: Workshop on Advances in Information, Electronic and Electrical Engineering (AIEEE) (2021)

14. Castells-Rufas, D., Marco-Sola, S., Moure, J.C., Aguado, Q., Espinosa, A.: FPGA acceleration of pre-alignment filters for short read mapping with HLS. IEEE Access, 10, 22079–22100 (2022)

15. Chen, T., Guestrin, C.: XGBoost: a scalable tree boosting system. In: International Conference on Knowledge Discovery and Data Mining (KDD) (2016)

16. Chen, X., Tian, Y.: Learning to perform local rewriting for combinatorial optimization. In: International Conference on Neural Information Processing Systems (NeurIPS) (2019)

17. Cheng, L., Wong, M.D.: Floorplan design for multimillion gate FPGAs. IEEE Trans. Comput. Aided Design Integr. Circuits Syst. 25(12), 2795–2805 (2006)

18. Chen, T., Du, Z., Sun, N., Wang, J., Wu, C., Chen, Y., Temam, O.: DianNao: a small-footprint high-throughput accelerator for ubiquitous machine-learning. In: International Conference on Architectural Support for Programming Languages and Operating Systems (ASPLOS) (2014)

19. Chen, Y., Luo, T., Liu, S., Zhang, S., He, L., Wang, J., Li, L., Chen, T., Xu, Z., Sun, N., Temam, O.: DaDianNao: a machine-learning supercomputer. IEEE Micro, 609–622 (2014)

20. Chen, Y.H., Emer, J., Sze, V.: Eyeriss: a spatial architecture for energy-efficient dataflow for convolutional neural networks. In: International Symposium on Computer Architecture (ISCA) (2016)

21. Chen, X., Cheng, F., Tan, H., Chen, Y., He, B., Wong, W.F., Chen, D.: ThunderGP: resource-efficient graph processing framework on FPGAs with HLS. ACM Trans. Reconfig. Technol. Syst. (2022)

22. Cong, J., Ding, Y.: FlowMap: an optimal technology mapping algorithm for delay optimization in lookup-table based FPGA designs. IEEE Trans. Comput. Aided Des. Integr. Circuits Syst. (1994)

23. Cong, J., Ding, Y.: On area/depth trade-off in LUT-based FPGA technology mapping. IEEE Trans. Very Large Scale Integr. Syst. 2(2), 137–148 (1994)

24. Cong, J., Zhang, Z.: An efficient and versatile scheduling algorithm based on SDC formulation. In: Design Automation Conference (DAC) (2006)

25. Cong, J., Liu, B., Neuendorffer, S., Noguera, J., Vissers, K., Zhang, Z.: High-level synthesis for FPGAs: from prototyping to deployment. IEEE Trans. Comput. Aided Des. Integr. Circuits Syst. 30(4), 473–491 (2011)

26. Dai, S., Zhou, Y., Zhang, H., Ustun, E., Young, E.F., Zhang, Z.: Fast and accurate estimation of quality of results in high-level synthesis with machine learning. In: IEEE Symposium on Field Programmable Custom Computing Machines (FCCM) (2018)

27. Damiani, A., Fiscaletti, G., Bacis, M., Brondolin, R., Santambrogio, M.D.: BlastFunction: a full-stack framework bringing FPGA hardware acceleration to cloud-native applications. ACM Trans. Reconfig. Technol. Syst. 15(2), 1–27 (2022)

28. De Micheli, G.: Synthesis and Optimization of Digital Circuits. McGraw Hill, New York (1994)

29. Dennard, R., Gaensslen, F., Yu, H.N., Rideout, V., Bassous, E., LeBlanc, A.: Design of ion-implanted MOSFET's with very small physical dimensions. IEEE J. Solid State Circuits, 9(5), 256–268 (1974)

30. Du, Y., Hu, Y., Zhou, Z., Zhang, Z.: High-performance sparse linear algebra on HBM-equipped FPGAs using HLS: a case study on SpMV. In: International Symposium on Field-Programmable Gate Arrays (FPGA) (2022)

31. Du, Z., Fasthuber, R., Chen, T., Ienne, P., Li, L., Luo, T., Feng, X., Chen, Y., Temam, O.: ShiDianNao: shifting vision processing closer to the sensor. In: International Symposium on Computer Architecture (ISCA) (2015)

32. Farooq, U., Hasan, N.U., Baig, I., Zghaibeh, M.: Efficient FPGA routing using reinforcement learning. In: International Conference on Information and Communication Systems (ICICS) (2021)

33. Ferrandi, F., Castellana, V.G., Curzel, S., Fezzardi, P., Fiorito, M., Lattuada, M., Minutoli, M., Pilato, C., Tumeo, A.: Bambu: an open-source research framework for the high-level synthesis of complex applications. Design Automation Conf. (DAC) (2021)
34. Goodfellow, I., Pouget-Abadie, J., Mirza, M., Xu, B., Warde-Farley, D., Ozair, S., Courville, A., Bengio, Y.: Generative adversarial nets. In: International Conference on Neural Information Processing Systems (NeurIPS) (2014)
35. Gudur, V.Y., Maheshwari, S., Acharyya, A., Shafik, R.: An FPGA based energy-efficient read mapper with parallel filtering and in-situ verification. ACM Trans. Comput. Biol. Bioinformat. 1–1 (2021)
36. Guo, L., Maidee, P., Zhou, Y., Lavin, C., Wang, J., Chi, Y., Qiao, W., Kaviani, A., Zhang, Z., Cong, J.: RapidStream: parallel physical implementation of FPGA HLS designs. In: International Symposium on Field-Programmable Gate Arrays (FPGA) (2022)
37. Haghi, A., Marco-Sola, S., Alvarez, L., Diamantopoulos, D., Hagleitner, C., Moreto, M.: An FPGA accelerator of the wavefront algorithm for genomics pairwise alignment. In: International Conference on Field Programmable Logic and Applications (FPL) (2021)
38. Ham, T.J., Lee, Y., Seo, S.H., Song, U.G., Lee, J.W., Bruns-Smith, D., Sweeney, B., Asanovic, K., Oh, Y.H., Wills, L.W.: Accelerating genomic data analytics with composable hardware acceleration framework. IEEE Micro, 41(3), 42–49 (2021)
39. Hamilton, W., Ying, Z., Leskovec, J.: Inductive representation learning on large graphs. In: International Conference on Neural Information Processing Systems (NeurIPS) (2017)
40. Han, S., Liu, X., Mao, H., Pu, J., Pedram, A., Horowitz, M.A., Dally, W.J.: EIE: efficient inference engine on compressed deep neural network. In: International Symposium on Computer Architecture (ISCA) (2016)
41. Han, S., Kang, J., Mao, H., Hu, Y., Li, X., Li, Y., Xie, D., Luo, H., Yao, S., Wang, Y., Yang, H., Dally, W.B.J.: ESE: efficient speech recognition engine with sparse LSTM on FPGA. In: International Symposium on Field-Programmable Gate Arrays (FPGA) (2017)
42. Handa, M., Vemuri, R.: An efficient algorithm for finding empty space for online FPGA placement. In: Design Automation Conference (DAC) (2004)
43. Hara, Y., Tomiyama, H., Honda, S., Takada, H., Ishii, K.: CHStone: a benchmark program suite for practical C-based high-level synthesis. In: International Symposium on Circuits and Systems (ISCAS) (2008)
44. Hassan, M.W., Athanas, P.M., Hanafy, Y.Y.: Domain-specific modeling and optimization for graph processing on FPGAs. In: International Symposium on Applied Reconfigurable Computing. Architectures (ARC) (2021)
45. Hegde, K., Asghari-Moghaddam, H., Pellauer, M., Crago, N., Jaleel, A., Solomonik, E., Emer, J., Fletcher, C.W.: ExTensor: an accelerator for sparse tensor algebra. IEEE Micro, 319–333 (2019)
46. Herklotz, Y., Pollard, J.D., Ramanathan, N., Wickerson, J.: Formal verification of high-level synthesis. In: Intl'l Conference on Object-Oriented Programming, Systems, Languages, and Applications (OOPSLA) (2021)
47. Hornik, K., Stinchcombe, M., White, H.: Multilayer feedforward networks are universal approximators. Neural Netw. 2(5), 359–36 (1989)
48. Hosny, A., Hashemi, S., Shalan, M., Reda, S.: Drills: deep reinforcement learning for logic synthesis. In: Asia and South Pacific Design Automation Conference (ASP-DAC) (2020)
49. Hu, Y., Du, Y., Ustun, E., Zhang, Z.: GraphLily: accelerating graph linear algebra on HBM-equipped FPGAs. In: International Conference on Computer-Aided Design (ICCAD) (2021)
50. Intel HLS Compiler. https://www.intel.com/content/www/us/en/software/programmable/quartus-prime/hls-compiler.html. Accessed: December 14, 2022
51. Jia, W., Shaw, K.A., Martonosi, M.: Stargazer: automated regression-based GPU design space exploration. In: International Symposium on Performance Analysis of Systems and Software (ISPASS) (2012)
52. Kapre, N., Ng, H., Teo, K., Naude, J.: InTime: a machine learning approach for efficient selection of FPGA CAD tool parameters. In: International Symposium on Field-Programmable Gate Arrays (FPGA) (2015)

53. Karfa, C., Mandal, C., Sarkar, D., Pentakota, S.R., Reade, C.: A formal verification method of scheduling in high-level synthesis. In: International Symposium on Quality Electronic Design (ISQED) (2006)
54. Kim, J., Kang, J.K., Kim, Y.: A resource efficient integer-arithmetic-only FPGA-based CNN accelerator for real-time facial emotion recognition. IEEE Access, 9, 104367–104381 (2021)
55. Knag, P., Kim, J.K., Chen, T., Zhang, Z.: A sparse coding neural network ASIC with on-chip learning for feature extraction and encoding. IEEE J. Solid State Circuits, 50(4), 1070–1079 (2015)
56. Knaust, M., Seiler, E., Reinert, K., Steinke, T.: Co-design for energy efficient and fast genomic search: interleaved bloom filter on FPGA. In: International Symposium on Field-Programmable Gate Arrays (FPGA) (2022)
57. Kurek, M., Becker, T., Chau, T.C., Luk, W.: Automating optimization of reconfigurable designs. In: IEEE Symposium on Field Programmable Custom Computing Machines (FCCM) (2014)
58. Kurek, M., Deisenroth, M.P., Luk, W., Todman, T.: Knowledge transfer in automatic optimisation of reconfigurable designs. In: IEEE Symposium on Field Programmable Custom Computing Machines (FCCM) (2016)
59. Kwon, J., Carloni, L.P.: Transfer learning for design-space exploration with high-level synthesis. In: ACM/IEEE Workshop on Machine Learning for CAD (MLCAD) (2020)
60. Lai, Y., Ustun, E., Xiang, S., Fang, Z., Rong, H., Zhang, Z.: Programming and synthesis for software-defined FPGA acceleration: status and future prospects. ACM Trans. Reconfig. Technol. Syst. 14(4), 1–39 (2021)
61. Lee, J., Song, T., He, J., Kandeepan, S., Wang, K.: Recurrent neural network FPGA hardware accelerator for delay-tolerant indoor optical wireless communications. Opt. Express, 29(16), 26165–26182 (2021)
62. Li, H., Katkoori, S., Mak, W.K.: Power minimization algorithms for LUT-based FPGA technology mapping. ACM Trans. Design Automat. Electron. Syst. 9(1), 33–51 (2004)
63. Li, D., Yao, S., Liu, Y.H., Wang, S., Sun, X.H.: Efficient design space exploration via statistical sampling and AdaBoost learning. In: Design Automation Conference (DAC) (2016)
64. Liang, S., Yin, S., Liu, L., Luk, W., Wei, S.: FP-BNN: binarized neural network on FPGA. Neurocomputing, 275(31), 1072–1086 (2018)
65. Lin, J.Y., Jagannathan, A., Cong, J.: Placement-driven technology mapping for LUT-based FPGAs. In: International Symposium on Field-Programmable Gate Arrays (FPGA) (2003)
66. Lin, Y., Jiang, Z., Gu, J., Li, W., Dhar, S., Ren, H., Khailany, B., Pan, D.Z.: DREAMPlace: deep learning toolkit-enabled GPU acceleration for modern VLSI placement. IEEE Trans. Comput Aided Design Integr. Circuits Syst. 40(4), 748–761 (2021)
67. Ling, A., Singh, D.P., Brown, S.D.: FPGA technology mapping: a study of optimality. In: Design Automation Conference (DAC) (2005)
68. Liu, H.Y., Carloni, L.P.: On learning-based methods for design-space exploration with high-level synthesis. In: Design Automation Conference (DAC) (2013)
69. Liu, D., Schafer, B.C.: Efficient and reliable high-level synthesis design space explorer for FPGAs. In: International Conference on Field Programmable Logic and Applications (FPL) (2016)
70. Liu, D., Chen, T., Liu, S., Zhou, J., Zhou, S., Temam, O., Feng, X., Zhou, X., Chen, Y.: PuDianNao: a polyvalent machine learning accelerator. In: International Conference on Architectural Support for Programming Languages and Operating Systems (ASPLOS) (2015)
71. Lo, C., Chow, P.: Model-based optimization of high-level synthesis directives. In: International Conference on Field Programmable Logic and Applications (FPL) (2016)
72. Lo, C., Chow, P.: Multi-fidelity optimization for high-level synthesis directives. In: International Conference on Field Programmable Logic and Applications (FPL) (2018)
73. Lo, C., Chow, P.: Hierarchical modelling of generators in design-space exploration. In: IEEE Symposium on Field Programmable Custom Computing Machines (FCCM) (2020)
74. Luk, W.: Improving performance estimation for FPGA-based accelerators for convolutional neural networks. In: International Symposium on Applied Reconfigurable Computing. Architectures (ARC) (2020)

75. Maarouf, D., Alhyari, A., Abuowaimer, Z., Martin, T., Gunter, A., Grewal, G., Areibi, S., Vannelli, A.: Machine-learning based congestion estimation for modern FPGAs. In: International Conference on Field Programmable Logic and Applications (FPL) (2018)
76. Makrani, H.M., Farahmand, F., Sayadi, H., Bondi, S., Dinakarrao, S.M.P., Homayoun, H., Rafatirad, S.: Pyramid: machine learning framework to estimate the optimal timing and resource usage of a high-level synthesis design. In: International Conference on Field Programmable Logic and Applications (FPL) (2019)
77. Mametjanov, A., Balaprakash, P., Choudary, C., Hovland, P.D., Wild, S.M., Sabin, G.: Autotuning FPGA design parameters for performance and power. In: IEEE Symposium on Field Programmable Custom Computing Machines (FCCM) (2015)
78. Manco, A., Castrillo, V.U.: An FPGA scalable software-defined radio platform for UAS communications research. J. Commun. 16(2), 42–51 (2021)
79. Mason, L., Bartlett, P., Baxter, J., Frean, M.: Boosting algorithms as gradient descent. In: International Conference on Neural Information Processing Systems (NeurIPS) (1999)
80. Mehrabi, A., Manocha, A., Lee, B.C., Sorin, D.J.: Prospector: synthesizing efficient accelerators via statistical learning. In: Design, Automation, and Test in Europe (DATE) (2020)
81. Meng, P., Althoff, A., Gautier, Q., Kastner, R.: Adaptive threshold non-pareto elimination: rethinking machine learning for system-level design space exploration on FPGAs. In: Design, Automation, and Test in Europe (DATE) (2016)
82. Mirza, M., Osindero, S.: Conditional generative adversarial nets (2014). Preprint. arXiv:1411.1784
83. Mishchenko, A., Chatterjee, S., Brayton, R.K.: DAG-aware AIG rewriting a fresh look at combinational logic synthesis. In: Design Automation Conference (DAC) (2006)
84. Mishchenko, A., Chatterjee, S., Brayton, R.K.: Improvements to technology mapping for LUT-based FPGAs. IEEE Trans. Comput. Aided Design Integr. Circuits Syst. (TCAD), 26(2), 240–253 (2007)
85. Murray, K.E., Petelin, O., Zhong, S., Wang, J.M., Eldafrawy, M., Legault, J.P., Sha, E., Graham, A.G., Wu, J., Walker, M.J., et al.: VTR 8: high-performance CAD and customizable FPGA architecture modelling. ACM Trans. Reconfig. Technol. Syst. 13(2), 1–55 (2020)
86. Neto, W.L., Moreira, M.T., Amaru, L., Yu, C.: SLAP: a supervised learning approach for priority cuts technology mapping. In: Design Automation Conference (DAC) (2021)
87. Neto, W.L., Moreira, M.T., Amaru, L., Yu, C., Gaillardon, P.E.: Read your circuit: leveraging word embedding to guide logic optimization. In: Asia and South Pacific Design Automation Conference (ASP-DAC) (2021)
88. Nurvitadhi, E., Sheffield, D., Sim, J., Mishra, A., Venkatesh, G., Marr, D.: Accelerating binarized neural networks: comparison of FPGA, CPU, GPU, and ASIC. In: International Conference on Field Programmable Technology (FPT) (2016)
89. Nurvitadhi, E., Sim, J., Sheffield, D., Mishra, A., Krishnan, S., Marr, D.: Accelerating recurrent neural networks in analytics servers: comparison of FPGA, CPU, GPU, and ASIC. In: International Conference on Field Programmable Logic and Applications (FPL) (2016)
90. Nurvitadhi, E., Cook, J., Mishra, A., Marr, D., Nealis, K., Colangelo, P., Ling, A., Capalija, D., Aydonat, U., Dasu, A., Shumarayev, S.: In-package domain-specific ASICs for Intel Stratix 10 FPGAs: a case study of accelerating deep learning using TensorTile ASIC. Int'l Conf. on Field Programmable Logic and Applications (FPL). (2018)
91. NVIDIA DGX-1. https://www.nvidia.com/content/dam/en-zz/Solutions/Data-Center/dgx-1/dgx-1-ai-supercomputer-datasheet-v4.pdf. Accessed: December 14, 2022
92. NVIDIA Hopper H100. https://nvidianews.nvidia.com/news/nvidia-announces-hopper-architecture-the-next-generation-of-accelerated-computing. Accessed: December 14, 2022
93. NVIDIA PASCAL GP100. https://images.nvidia.com/content/pdf/tesla/whitepaper/pascal-architecture-whitepaper.pdf. Accessed: December 14, 2022
94. NVIDIA Tegra - Parker. https://blogs.nvidia.com/blog/2016/08/22/parker-for-self-driving-cars/. Accessed: December 14, 2022
95. NVIDIA VOLTA GV100. https://devblogs.nvidia.com/parallelforall/inside-volta/. Accessed: December 14, 2022

96. Papamichael, M.K., Milder, P., Hoe, J.C.: Nautilus: fast automated IP design space search using guided genetic algorithms. In: Design Automation Conference (DAC) (2015)
97. Papaphilippou, P., Meng, J., Gebara, N., Luk, W.: Hipernetch: high-performance FPGA network switch. ACM Trans. Reconfig. Technol. Syst. 15(1), 1–31 (2021)
98. Parashar, A., Rhu, M., Mukkara, A., Puglielli, A., Venkatesan, R., Khailany, B., Emer, J.S., Keckler, S.W., Dally, W.J.: SCNN: an accelerator for compressed-sparse convolutional neural networks. In: International Symposium on Computer Architecture (ISCA) (2017)
99. Pui, C.W., Chen, G., Ma, Y., Young, E.F., Yu, B.: Clock-aware ultrascale FPGA placement with machine learning routability prediction. In: International Conference on Computer-Aided Design (ICCAD) (2017)
100. Pundir, N., Rahman, F., Farahmandi, F., Tehranipoor, M.: What is all the FaaS about? – remote exploitation of FPGA-as-a-service platforms. Cryptology ePrint Archive, Report 2021/746 (2021)
101. Rafii, A., Chow, P., Sun, W.: Pharos: a performance monitor for multi-FPGA systems. In: IEEE Symposium on Field Programmable Custom Computing Machines (FCCM) (2021)
102. Ramachandra, C.N., Nag, A., Balasubramonion, R., Kalsi, G., Pillai, K., Subramoney, S.: ONT-X: an FPGA approach to real-time portable genomic analysis. In: IEEE Symposium on Field Programmable Custom Computing Machines (FCCM) (2021)
103. Reagen, B., Adolf, R., Shao, Y.S., Wei, G.Y., Brooks, D.: MachSuite: benchmarks for accelerator design and customized architectures. In: International Symposium on Workload Characterization (IISWC) (2014)
104. Reagen, B., Whatmough, P., Adolf, R., Rama, S., Lee, H., Lee, S.K., Hernández-Lobato, J.M., Wei, G.Y., Brooks, D.: Minerva: enabling low-power, highly-accurate deep neural network accelerators. In: International Symposium on Computer Architecture (ISCA) (2016)
105. Ronak, B., Fahmy, S.A.: Mapping for maximum performance on FPGA DSP blocks. IEEE Trans. Comput. Aided Design Integr. Circuits Syst. 35(4), 573–585 (2016)
106. Schafer, B.C., Mahapatra, A.: S2CBench: synthesizable systemC benchmark suite for high-level synthesis. IEEE Embed. Syst. Lett. 6(3), 53–56 (2014)
107. Sechen, C.: VLSI Placement and Global Routing using Simulated Annealing, vol. 54. Springer Science & Business Media, Berlin (2012)
108. Shwartz-Ziv, R., Armon, A.: Tabular data: deep learning is not all you need. Infor. Fusion, 81, 84–90 (2022)
109. Soeken, M., Amaru, L.G., Gaillardon, P.E., De Micheli, G.: Exact synthesis of majority-inverter graphs and its applications. IEEE Trans. Comput. Aided Design Integr. Circuits Syst. 36(11), 1842–1855 (2017)
110. Soeken, M., Haaswijk, W., Testa, E., Mishchenko, A., Amarù, L.G., Brayton, R.K., De Micheli, G.: Practical exact synthesis. In: Design, Automation, and Test in Europe (DATE) (2018)
111. Szentimrey, H., Al-Hyari, A., Foxcroft, J., Martin, T., Noel, D., Grewal, G., Areibi, S.: Machine learning for congestion management and routability prediction within FPGA placement. ACM Trans. Design Automat. Electron. Syst. (TODAES), 25(5), 1–25 (2020)
112. Tang, X., Giacomin, E., Alacchi, A., Chauviere, B., Gaillardon, P.E.: OpenFPGA: an opensource framework enabling rapid prototyping of customizable FPGAs. In: International Conference on Field Programmable Logic and Applications (FPL) (2019)
113. Testa, E., Soeken, M., Amarù, L., De Micheli, G.: Reducing the multiplicative complexity in logic networks for cryptography and security applications. In: Design Automation Conference (DAC) (2019)
114. Ustun, E., Xiang, S., Gui, J., Yu, C., Zhang, Z.: LAMDA: Learning-assisted multi-stage autotuning for FPGA design closure. In: IEEE Symposium on Field Programmable Custom Computing Machines (FCCM) (2019)
115. Ustun, E., Deng, C., Pal, D., Li, Z., Zhang, Z.: Accurate operation delay prediction for FPGA HLS using graph neural networks. In: International Conference on Computer-Aided Design (ICCAD) (2020)
116. Wang, Z., Schafer, B.C.: Machine learning to set meta-heuristic specific parameters for high-level synthesis design space exploration. In: Design Automation Conference (DAC) (2020)

117. Wang, W., Bolic, M., Parri, J.: pvFPGA: accessing an FPGA-based hardware accelerator in a paravirtualized environment. In: Intl'l Conference on Hardware/Software Codesign and System Synthesis (CODES+ISSS) (2013)

118. Wang, Q., Zheng, L., Huang, Y., Yao, P., Gui, C., Liao, X., Jin, H., Jiang, W., Mao, F.: GraSU: a fast graph update library for FPGA-based dynamic graph processing. In: International Symposium on Field-Programmable Gate Arrays (FPGA) (2021)

119. Wille, R., Soeken, M., Drechsler, R.: Reducing the number of lines in reversible circuits. In: Design Automation Conference (DAC) (2010)

120. Wu, N., Xie, Y., Hao, C.: IronMan: GNN-assisted design space exploration in high-level synthesis via reinforcement learning. In: Great Lakes Symposium on VLSI (2021)

121. Wu, Y., Wang, Q., Zheng, L., Liao, X., Jin, H., Jiang, W., Zheng, R., Hu, K.: FDGLib: a communication library for efficient large-scale graph processing in FPGA-accelerated data centers. J. Comput. Sci. Technol. 36, 1051–1070 (2021)

122. Xie, Z., Huang, Y.H., Fang, G.Q., Ren, H., Fang, S.Y., Chen, Y., Hu, J.: RouteNet: routability prediction for mixed-size designs using convolutional neural network. In: International Conference on Computer-Aided Design (ICCAD) (2018)

123. Xilinx Inc.: Floorplanning Methodology Guide (2013)

124. Xilinx Inc.: UltraScale Architecture Configurable Logic Block (2017)

125. Xilinx Inc.: UltraScale Architecture DSP Slice User Guide (2019)

126. Xin, G., Zhao, Y., Han, J.: A multi-layer parallel hardware architecture for homomorphic computation in machine learning. In: International Symposium on Circuits and Systems (ISCAS) (2021)

127. Xu, C., Liu, G., Zhao, R., Yang, S., Luo, G., Zhang, Z.: A parallel bandit-based approach for autotuning FPGA compilation. In: International Symposium on Field-Programmable Gate Arrays (FPGA) (2017)

128. Xu, P., Zhang, X., Hao, C., Zhao, Y., Zhang, Y., Wang, Y., Li, C., Guan, Z., Chen, D., Lin, Y.: AutoDNNchip: an automated DNN chip predictor and builder for both FPGAs and ASICs. In: International Symposium on Field-Programmable Gate Arrays (FPGA) (2020)

129. Yang, L., He, Z., Fan, D.: A fully onchip binarized convolutional neural network FPGA implementation with accurate inference. In: International Symposium on Low Power Electronics and Design (ISLPED) (2018)

130. Yosys Open Synthesis Suite. https://github.com/YosysHQ/yosys. Accessed: December 14, 2022

131. Yu, C.: FlowTune: practical multi-armed bandits in boolean optimization. In: International Conference on Computer-Aided Design (ICCAD) (2020)

132. Yu, C., Zhang, Z.: Painting on placement: forecasting routing congestion using conditional generative adversarial nets. In: Design Automation Conference (DAC) (2019)

133. Yu, C., Zhou, W.: Decision making in synthesis cross technologies using LSTMs and transfer learning. In: ACM/IEEE Workshop on Machine Learning for CAD (MLCAD) (2020)

134. Yu, C., Choudhury, M., Sullivan, A., Ciesielski, M.J.: Advanced datapath synthesis using graph isomorphism. In: International Conference on Computer-Aided Design (ICCAD) (2017)

135. Yu, C., Xiao, H., De Micheli, G.: Developing synthesis flows without human knowledge. Design Automation Conference (DAC) (2018)

136. Zhang, Z., Liu, B.: SDC-based modulo scheduling for pipeline synthesis. In: International Conference on Computer-Aided Design (ICCAD) (2013)

137. Zeng, H., Prasanna, V.: GraphACT: accelerating GCN training on CPU-FPGA heterogeneous platforms. In: International Symposium on Field-Programmable Gate Arrays (FPGA) (2020)

138. Zhang, S., Du, Z., Zhang, L., Lan, H., Liu, S., Li, L., Guo, Q., Chen, T., Chen, Y.: Cambricon-X: an accelerator for sparse neural networks. IEEE Micro, 1–12 (2016)

139. Zhang, X., Wang, J., Zhu, C., Lin, Y., Xiong, J., Hwu, W.m., Chen, D.: DNNBuilder: an automated tool for building high-performance DNN hardware accelerators for FPGAs. In: International Conference on Computer-Aided Design (ICCAD) (2018)

140. Zhang, C., Hu, H., Cao, S., Jiang, Z.: A novel blind detection method and FPGA imple-
     mentation for energy-efficient sidelink communications. In: Workshop on Signal Processing
     Systems (SiPS) (2021)
141. Zhang, Y., Pan, J., Liu, X., Chen, H., Chen, D., Zhang, Z.: FracBNN: accurate and FPGA-
     efficient binary neural networks with fractional activations. In: International Symposium on
     Field-Programmable Gate Arrays (FPGA) (2021)
142. Zhang, Y., Zhang, Z., Lew, L.: PokeBNN: a binary pursuit of lightweight accuracy. In:
     Conference on Computer Vision and Pattern Recognition (CVPR) (CVPR) (2022)
143. Zhao, J., Liang, T., Sinha, S., Zhang, W.: Machine learning based routing congestion
     prediction in FPGA high-level synthesis. In: Design, Automation, and Test in Europe (DATE)
     (2019)
144. Zhou, Y., Gupta, U., Dai, S., Zhao, R., Srivastava, N., Jin, H., Featherston, J., Lai, Y.H.,
     Liu, G., Velasquez, G.A., Wang, W., Zhang, Z.: Rosetta: a realistic high-level synthesis
     benchmark suite for software programmable FPGAs. In: International Symposium on Field-
     Programmable Gate Arrays (FPGA) (2018)
145. Zhou, Y., Gupta, U., Dai, S., Zhao, R., Srivastava, N., Jin, H., Featherston, J., Lai, Y.H.,
     Liu, G., Velasquez, G.A., et al.: Rosetta: a realistic high-level synthesis benchmark suite for
     software programmable FPGAs. In: International Symposium on Field-Programmable Gate
     Arrays (FPGA) (2018)
146. Zhou, S., Kannan, R., Prasanna, V.K., Seetharaman, G., Wu, Q.: HitGraph: high-throughput
     graph processing framework on FPGA. IEEE Trans. Parallel Distrib. Syst. 30(10), 2249–2264
     (2019)
147. Zhu, K., Liu, M., Chen, H., Zhao, Z., Pan, D.Z.: Exploring logic optimizations with
     reinforcement learning and graph convolutional network. In: ACM/IEEE Workshop on
     Machine Learning for CAD (MLCAD) (2020)
148. Zhu, Y., Zhu, M., Yang, B., Zhu, W., Deng, C., Chen, C., Wei, S., Liu, L.: LWRpro: an energy-
     efficient configurable crypto-processor for Module-LWR. IEEE Trans. Circuits Syst. I, 68(3),
     1146–1159 (2021)
149. Ziegler, M.M., Bertran, R., Buyuktosunoglu, A., Bose, P.: Machine learning techniques for
     taming the complexity of modern hardware design. IBM J. Res. Develop. 61(4/5), 13:1–13:14
     (2017)

Chapter 17
Machine Learning for Analog Layout

Steven M. Burns, Hao Chen, Tonmoy Dhar, Ramesh Harjani, Jiang Hu, Nibedita Karmokar, Kishor Kunal, Yaguang Li, Yishuang Lin, Mingjie Liu, Meghna Madhusudan, Parijat Mukherjee, David Z. Pan, Jitesh Poojary, S. Ramprasath, Sachin S. Sapatnekar, Arvind K. Sharma, Wenbin Xu, Soner Yaldiz, and Keren Zhu

17.1 Introduction

Design automation researchers have addressed the problem of analog layout synthesis for several decades [13, 15, 21, 25, 31, 48, 49, 57, 59, 70]. These techniques were largely based on a toolbox of traditional algorithms such as numerical analysis, graph theory, and optimization techniques. However, these methods were generally unable to compete with a skilled designer in the quality of the solution that was produced. The problem has enjoyed a recent renaissance due to increasing interest in analog design as applications place a greater focus on interactions with the analog real world, and application drivers such as wireless systems and AI hardware make native analog design more attractive.

Several new approaches have been proposed in the recent past to tackle the problem of analog layout synthesis. Frameworks such as BAG [8, 14] have focused

This work was supported in part by the DARPA IDEA program (SPAWAR contracts N660011824048 and N669911824049) and NSF CCF-1704758.

S. M. Burns · P. Mukherjee · S. Yaldiz
Intel Labs, Hillsboro, OR, USA

H. Chen · M. Liu · D. Z. Pan (✉) · K. Zhu
The University of Texas at Austin, Austin, TX, USA
e-mail: sachin@umn.edu

T. Dhar · R. Harjani · N. Karmokar · K. Kunal · M. Madhusudan · J. Poojary · S. Ramprasath · S. S. Sapatnekar (✉) · A. K. Sharma
University of Minnesota, Minneapolis, MN, USA

J. Hu · Y. Li · Y. Lin · W. Xu
Texas A&M University, College Station, TX, USA
e-mail: dpan@utexas.edu

© The Author(s), under exclusive license to Springer Nature Switzerland AG 2022
H. Ren, J. Hu (eds.), *Machine Learning Applications in Electronic Design Automation*, https://doi.org/10.1007/978-3-031-13074-8_17

on procedural design with significant designer input, while approaches such as FASoC [2] leverage digital standard cells and digital layout methodologies for designing "digital analog" circuits. Recent techniques have considered the use of artificial neural networks for the layout of specific topologies [26] or for knowledge migration [27].

A new class of methods in MAGICAL [9, 11, 77] and ALIGN [3, 18, 37] has looked at approaches that could be used with much more limited human intervention, including no-human-in-the-loop automation. These frameworks are facilitated by the advent of machine learning (ML), which has substantially changed the algorithmic landscape. Today, with the emergence of artificial intelligence techniques that can compete with human skill, it is now realistic to think about automated analog layout that delivers a solution that is comparable in quality to manual design.

A typical analog layout generation flow consists of the following steps:

- *Circuit hierarchy specification* takes an input netlist and, through either design annotation or automated recognition techniques, determines the hierarchical blocks in the circuit.
- *Constraint specification* provides guidelines for layout, determining how the blocks in the circuit hierarchy must be arranged. These constraints include (a) *geometric constraints* for symmetrical placement and routing about specified axes, common-centroid, or interdigitated layouts [32] and (b) *electrical constraints* that place limits on allowable interconnect RC parasitics during layout while ensuring that performance constraints are met; typically, this involves trade-offs and complex relationships, i.e., allowing increased RCs on some wires while ensuring lower RCs on others.
- *Cell generation* generates the layout of devices and passives at the lowest level of the hierarchy, parameterized so that cells may be built for a specified set of transistor widths and gate lengths; may have a variety of aspect ratios; may be built with/without body contacts; etc.
- *Placement* is typically performed hierarchically, with the locations of blocks being determined at each step while honoring constraints on symmetry and performance, specified in the constraint generation step.
- *Routing* connects the placed blocks at each hierarchical level, using appropriate wire widths to ensure that performance specifications are met.

When judiciously used, ML methods can be of great help in automating analog design. While some steps can be handled using conventional algorithmic techniques (e.g., cell generation in FinFET technologies, where the limited degrees of freedom render the problem amenable to algorithmic layout generation), ML techniques can help mimic the wisdom of the human designer in other steps. This chapter provides an overview of ML methods for analog layout. A recent paper [1] presents a survey of ML methods in analog design in general, and this chapter complements the overview with an in-depth view of the use of ML in analog layout automation. Section 17.2 presents a variety of approaches for circuit annotation and geometric constraint specification. Next, Sect. 17.3 overviews various ML techniques for well generation, placement, and routing. Finally, the chapter concludes with a view of future directions for ML-assisted analog layout.

17.2 Geometric Constraint Generation

17.2.1 Problem Statement

Geometric constraints help analog designs achieve high performance and high yield by making them resilient to PVT variations. Symmetry considerations are particularly important in specifying constraints to be applied during analog layout synthesis. Analog designs frequently use differential topologies to reject common-mode noise, and layout symmetry helps in reducing mismatch between such devices, which is liable to significantly degrade circuit performance.

The research in extracting geometric constraints automatically has been evolving rapidly in recent years, motivated by advances in ML techniques and the requirements of analog layout automation [83]. This section presents several ML approaches for geometric constraint generation. The approaches in Sects. 17.2.2 and 17.2.3 extract constraints based on the results of circuit annotation using graph convolutional networks and array recognition methods (including approximate graph isomorphism), respectively. The approaches in Sects. 17.2.4 and 17.2.5 directly extract the symmetry constraints. Section 17.2.4 describes a system-level symmetry constraint extraction algorithm leveraging statistical techniques to measure the graph similarity. The circuits with high similarity in their graph structures are identified as symmetry constraints. Section 17.2.5 describes two symmetry constraint extraction techniques using graph neural network, with supervised learning and unsupervised learning.

17.2.2 Subcircuit Annotation Using Graph Convolution Networks

A first step in generating constraints is to find hierarchies within a circuit, identifying specific circuit blocks. If an experienced designer wishes to retain specified hierarchies, this step may be skipped. Otherwise, existing hierarchies may be discarded to find hierarchies that are more amenable to layout and constraint generation. In analog circuits, this is difficult: a large number of circuit variants exist even for a single functionality, e.g., between textbooks [65] and research papers, there are well over 100 widely used operational transconductance amplifier (OTA) topologies of various types (e.g., telescopic, folded cascode, Miller-compensated). Truly generalizable analog automation requires the ability to recognize all variants— including those that have not even been designed to date. Traditional methods have used two approaches: (1) *library-based* [50, 52], matching a circuit to prespecified templates, and requiring an enumeration of possible topologies in an exhaustive database, or (2) *knowledge-based* [30, 71], embedding rules for recognizing circuits; however, the rules must come from an expert designer who may struggle to provide a list (many rules are intuitively ingrained rather than explicitly stated). Moreover, it is difficult to capture rules for all variants. In this section, we describe ML methods for subcircuit recognition, which can be used to derive circuit hierarchies and block-level constraints.

As in [58], a circuit netlist is represented by an undirected bipartite graph $G(V, E)$, where $V = V_e \cup V_n$. The subsets V_e and V_n correspond, respectively, to elements (transistors/passives) in the netlist and the set of nets. The edge set E consists of edges between a vertex in V_e corresponding to an element to the vertices in V_n corresponding to nets connected to its terminals. Similar to [42], each edge connected to a transistor is assigned a three-bit binary label, $l_g l_s l_d$, where $l_g/l_s/l_d$ are set to 1 only if the edge from the transistor vertex connects to the net vertex through its gate/source/drain, respectively. The subcircuit recognition problem is mapped to one of approximate subgraph isomorphism, with approximation allowing for variations around a core structure for a circuit block.

In [38], graph convolutional networks (GCNs) are used for this purpose. General graphs have no unique embedding. A GCN performs convolutions that are independent of the embedding of the graph in the plane. Various types of GCNs have been proposed [16, 28, 29, 34, 80], but they all share a framework that requires three fundamental steps: (i) the application of localized convolutional filters on graphs, (ii) a graph coarsening procedure that groups together similar vertices, and (iii) a graph pooling operation for graph reduction. GCN techniques primarily differ in the nature of the filter that is used for convolution.

A major class of GCNs is based on spectral methods [16, 34], which are independent of graph embedding and have been found to be extremely effective and therefore form the basis of the GCN used in [38]. A spectral representation in the Fourier space of a graph is enabled through the graph Laplacian representation. The Laplacian $L \in R^{n \times n}$ of an unweighted graph $G(V, E)$ with n vertices is often defined as $D - A$ where $A \in R^{n \times n}$ is the adjacency matrix of the graph and $D \in R^{n \times n}$ is a diagonal matrix whose diagonal entry corresponds to the degrees of all vertices, i.e., the row sums of the adjacency matrix. The normalized Laplacian representation is

$$L = I - D^{-1/2} A D^{-1/2} \tag{17.1}$$

The matrix L is symmetric, real, and positive definite: it has real nonnegative eigenvalues that are interpreted as the frequencies of the graph.

The approach proposed by Defferrard et al. [16] creates spectral filters around a vertex, where each filter functions within a region of radius K of (i.e., up to K edges away from) the vertex. A convolution operator on the graph is defined as

$$y = g_\theta(L)x = g_\theta(U \Lambda U^T)x = U g_\theta(\Lambda) U^T x \tag{17.2}$$

where the normalized graph Laplacian is eigendecomposed as $L = U \Lambda U^T$ and g_θ is a filtering operator that acts on an input signal x to produce an output signal y. In this case, the signal corresponds to a region of the graph around a specific vertex.

Next, $g_\theta(L)$ is parameterized as a polynomial Chebyshev expansion, which truncates the filter expansion to order of $K - 1$. Given the K top eigenvalues of L (computed inexpensively using the Lanczos algorithm), for a graph of bounded degree where the number of edges is $O(n)$, this polynomial can be evaluated using K multiplications by a sparse L with a cost of $O(Kn) \ll O(n^2)$.

The GCN topology has two convolutional layers (performing the operations described above) and two pooling layers, which then feed a fully connected (fc) layer, whose outputs provide the classification results. Pooling combines similar vertices in a graph using the greedy Graclus heuristic, built on top of the Metis algorithm [33] for multilevel clustering [20]. The final layer is a fully connected layer of size 512 along with `softmax` function for classification.

Each vertex is associated with 18 features: 12 that annotate the element type, 5 that denote the type of net (input, output, bias, supply, ground), and 1 that describes the label for edges incident on a transistor vertex.

Rather than placing the full burden of recognition on the GCN, a set of simple postprocessing heuristics are used to complete the annotation. *Postprocessing I* involves graph-based heuristics which ensures consistent classification of nodes in the same channel-connected component (CCC) [66]. *Postprocessing II* uses circuit-specific knowledge, e.g., low-noise amplifiers (LNAs) and mixers can be structurally similar, but an LNA has an antenna input, while a mixer has an oscillating input. Such information can be designer-specified or inferred from the testbench in the input netlist.

Testcase 1 is a filter with an OTA and switched capacitors and contains 32 devices and 25 nets. The telescopic OTA subcircuit used in this circuit is not seen by the training set. Using the GCN alone, an accuracy of 56/57 is achieved in identifying OTA and bias circuit nodes. The misclassified vertices belong to the OTA interconnect ports, and all nodes (100%) are correctly classified after Postprocessing I.

Testcase 2 consists of a phased array system [53], illustrated in Fig. 17.1a, containing a mixer (red), LNA (green), BPF (orange), oscillator (gray), VCO buffer (BUF), and inverter-based amplifier (INV) (violet) sub-blocks. The graph for the input netlist has 902 vertices (522 devices + 380 nets). The GCN-based classification identifies nodes belonging to LNA, mixer, and oscillator and passes these results through postprocessing. After Postprocessing I, the BPF is identified as a combination of an oscillator with two input transistors. INV and BUF primitives are identified, and a separate hierarchy is created for them which boosts the accuracy to 87.3%. During Postprocessing II, which uses an antenna label at LNA input and oscillating input for mixer, all nodes are identified correctly. At this point, the classification result after post-processing is shown in Fig. 17.1b: all 522 devices (100%) are classified correctly.

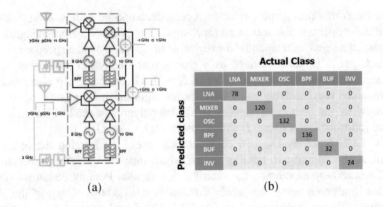

	Actual Class					
	LNA	MIXER	OSC	BPF	BUF	INV
LNA	78	0	0	0	0	0
MIXER	0	120	0	0	0	0
OSC	0	0	132	0	0	0
BPF	0	0	0	136	0	0
BUF	0	0	0	0	32	0
INV	0	0	0	0	0	24

(a)   (b)

Fig. 17.1 (**a**) Phased array system [53] and (**b**) results of GCN after postprocessing, showing the correctness of vertex classification

Once the circuit functionality is determined, graph-based approaches are used to identify primitives (e.g., differential pairs, current mirrors), including annotations for symmetry constraints at the primitive level and at upper levels of hierarchy. The annotation scheme is fast (a few minutes for the phased array on a desktop machine) and is dominated by the runtime of the GCN.

17.2.3 Array-Based Methods for Subcircuit Annotation

The technique above requires the curation of a training set for each type of circuit structure. While this is feasible for standard circuit blocks, an alternative procedure can guide layout by recognizing repeated structures in a circuit and lay them out using array-based methods. Such methods are useful in building structures such as flash ADCs, binary-weighted DACs, R-2R DACs, and equalizers, where the same structure is repeated. For exact replicas, it is possible to use graph-based methods to recognize regularity, but analog designers often witness scenarios where circuit blocks are *nearly* identical and require symmetric/regular layout.

The work in [39] presents an ML method for recognizing approximate matches by error-tolerant matching. The circuit is represented by a bipartite graph as defined earlier. The approach is based on the concept of *graph edit distance* (GED) [82], a measure of similarity between two graphs G_1 and G_2. Given a set of graph edit operations (insertion, deletion, vertex/edge relabeling), the GED is a metric of the number of edit operations required to translate G_1 to G_2.

Let graphs G_1 and G_2 represent, respectively, the CS-LNA and the CG-LNA, as shown in Fig. 17.2, with element vertices at left and net vertices at right. To transform G_1 to G_2, the GED = 4: two edges in G_1 are deleted: (capacitor

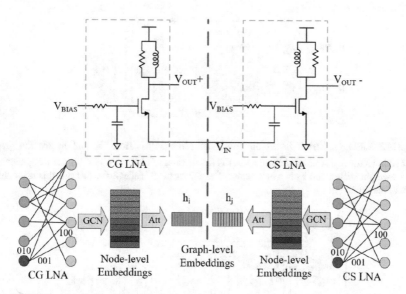

Fig. 17.2 Example showing graph embedding for common gate low noise amplifier (CG LNA) and common source LNA (CS LNA) in noise cancellation LNA

element, ground net) and (transistor element (source label), V_{IN} net); two are added: (capacitor element, V_{IN} net) and (transistor element (source label), ground net).

This work uses a neural network that transforms the original NP-hard problem to a learning problem [4] for computing graph similarity. The method works in four steps. In the first step, each node in the graph is converted to a node-level embedding vector. The second step uses these embedding vectors to create a graph-level embedding of dimension d. The lower half of Fig. 17.2 illustrates these two steps for the graphs for the CG-LNA and CS-LNA. For each subblock in the circuit, these steps need to be carried out once, and the graph embeddings are stored for matching any two pairs of subblocks in later stages. The computational complexity of these two steps is linear in the number of nodes in the graph.

The last two steps are shown in Fig. 17.3. The third step feeds the graph-level embeddings from the second step for two candidate graphs to a trained neural tensor network that generates a similarity matrix between the graphs. The fourth step then processes this matrix using a fully connected neural network to yield a single score. This matching method for two subblocks uses the previously stored graph embeddings instead of the full subblock graphs. The complexity of these two steps is quadratic in d, where d is bounded by a small constant in practice, the procedure is computationally inexpensive as compared to an exact GED computational complexity which is exponential in the number of nodes of the graphs involved.

The optimized model is a three-layer GCN with 128 input channels (the number of channels is halved in each layer), with 8 slices in the neural tensor network

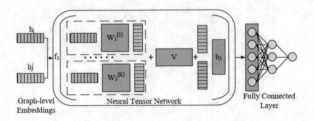

Fig. 17.3 GED prediction based on graph embeddings [4]. Here, h_i and h_j are the graph embeddings for two similarity candidates, f is an activation function, $W_3^{(1\cdots K)}$ is the weight tensor, V is a weight vector, and b_3 is a bias vector. The subscript "3" reflects the fact that this is the third step of the procedure

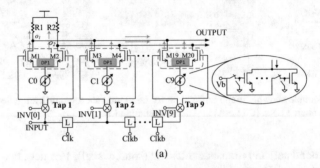

(a)

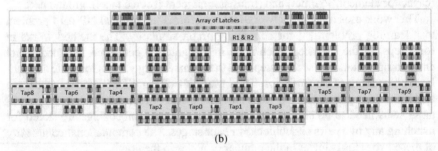

(b)

Fig. 17.4 (a) Schematic and (b) layout of an FIR equalizer [69]

(NTN), and a fully connected network with one hidden layer after the NTN. The trained net uses $d_1 = 64$, $d_2 = 32$, $d_3 = d = 16$.

The method is applied on an FIR Equalizer (Fig. 17.4a) circuit with 10 taps, each containing a differential pair, a current mirror DAC, and an XOR gate. All blocks in each tap share a common symmetry axis for matching. The first four taps use a 7-bit current mirror DAC, and the remaining taps have 5-bit current mirror DAC. To achieve better matching, the first four taps are placed in the center, and the remaining taps are placed around these four, sharing a common symmetry axis. The layout of equalizer, shown in Fig. 17.4b, meets all these requirements. This design demonstrates the detection and use of multiple lines of symmetry in a hierarchical way within primitives, within each tap, and globally at the block level.

17.2.4 System Symmetry Constraint with Graph Similarity

Many studies for symmetry constraint detection focus on generating constraints for building block level circuits such as differential amplifiers and comparators. On the other hand, methods for generating system symmetry constraints have rarely been studied. There still exists a gap in automatic constraint detection and constraint management for system-level analog designs. Previous methods of constraint detection have difficulties in scalability and expressiveness when directly migrating to analog circuit systems.

The system symmetry constraint detection problem for analog circuits can be described as follows. The hierarchical circuit netlist N is given as input. The circuit hierarchy is abstracted into a tree T, with each node representing a subcircuit. For each subcircuit $v \in T$, its children $G = \{g | g \subseteq v\}$ are subcircuits referenced in v. Symmetry constraint pair (g_i, g_j) represents that critical matching is needed between the subcircuits, where $g_i, g_j \in G$. The system symmetry constraint detection problem is to generate constraint pairs for every subcircuit $v \in T$.

A simplified example of system symmetry constraints is shown in Fig. 17.5. The circuit COMP consists of three subcircuits: COMP_CORE, INV1, and INV2. The inverters in COMP need to be matched with symmetry constraint pair (INV1, INV2). System constraints do not consider transistor device symmetry in subcircuits, such as COMP_CORE. If COMP_CORE further contains other subcircuits, the system constraints between these subcircuits should also be considered.

The hierarchies of the netlist is abstracted into a tree representation. Each subcircuit instantiation is abstracted with a node in the tree. The root of the hierarchy tree is the entire analog circuit system. The leaf cell nodes are device-level instances of transistors, resistors, capacitors, and diodes. The primal subcircuits are labeled as

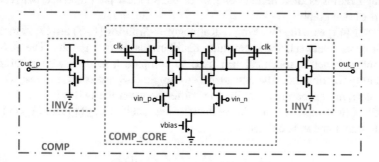

Fig. 17.5 Hierarchical circuit example [43]

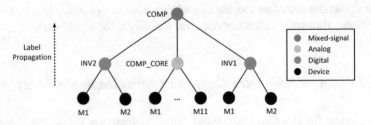

Fig. 17.6 Hierarchy tree of the circuit in Fig. 17.5 [43]

analog or digital in the netlist to improve constraint quality. Figure 17.6 shows an example of the extracted hierarchy tree of the corresponding circuit in Fig. 17.5.

Constraint candidates are classified as valid or invalid during symmetry detection. Symmetry constraints are detected if the circuit graphs are similar. To improve constraint quality and reduce false alarms, the neighboring circuit topologies are extracted on the entire circuit graph. The sizes of extracted subgraphs are determined by graph centrality. The similarities of the extracted subgraphs are measured with a scalable graph similarity metric using spectral graph analysis.

The subgraphs of subcircuits with neighboring circuit topologies are extracted to improve constraint quality and reduce false alarms. Only comparing the subcircuits is not enough to fully characterize symmetry constraints. The extracted neighboring circuit size is critical to the quality of symmetry constraint detection. Extracting small subgraphs would not include enough information. On the other hand, extracted large subgraphs of closely connected subcircuits would fully include both subcircuits and all their neighboring circuit topologies. In this case, the subgraphs would be detected as similar and create unnecessary constrained symmetry. To resolve such issues, graph centrality is used to determine the extracted subgraph radius. In graph theory and network analysis, indicators of centrality assign numbers or rankings to nodes within a graph corresponding to their network position. The subcircuit graph centers and radius are calculated using graph centrality, such as the Jordan Center [51], Eigenvector Centrality [51], or PageRank Center [62]. The extracted subgraph radius is then defined as half of the shortest path distance between the centers of two graphs.

S^3**DET** [43] uses the two-sample Kolmogorov-Smirnov (K-S) test [7] to measure the similarity between the eigenvalue distributions of two graphs. The K-S test is a non-parametric statistical test used to compare a sample with a reference probability distribution. The extended two-sample K-S test is used to test whether the underlying probability distributions differ between the two samples. The K-S statistic of two empirical cumulative distribution function $F_{1,n}(x)$ and $F_{2,m}(x)$ with sample size n and m is defined as:

$$D_n = \sup_x |F_{1,n}(x) - F_{2,m}(x)|. \tag{17.3}$$

It quantifies the difference between the two distributions. In statistics, the p-value is the probability of obtaining results at least as extreme as the observed results of a statistical hypothesis test, assuming that the null hypothesis is correct. The p-value from the K-S test measures how likely these samples comes from the same distribution. A small p-value concludes that the two samples are from different distributions, while a large p-value infers that the distributions match. A scalable graph similarity algorithm is applied using graph spectral analysis. Since the eigenvalue distribution of the graph *Laplacian* is closely linked with the graph structure, the K-S statistic of the eigenvalue distribution can be used as a graph similarity metric, with the following test:

1. The eigenvalues of the two graph *Laplacian* matrices are calculated and sorted.
2. The two-sample K-S test is conducted to test whether the underlying distributions differ for the two sets of eigenvalues.
3. The resulting p-value of the K-S test is used as the graph similarity score.
4. The two graphs are identified as similar if the similarity score is larger than a preset tolerance, *tol*.

17.2.5 Symmetry Constraint with Graph Neural Networks

Recent advances in graph neural networks (GNNs) show great potential toward a more accurate and efficient analog layout constraint annotation. As a circuit netlist can naturally be modeled as a graph, GNNs learn the interconnect structures by mining graph information. In [22], a graph learning framework with path-based features to mimic electric potential in circuit analysis is presented. Leveraging a probability-based filtering technique, the false-positive rates can be reduced.

The method is based on a supervised inductive graph-learning-based methodology for device-level AMS symmetry constraint extraction. Figure 17.7 shows the overall flow, which consists of three main stages: pre-processing, GraphSage-based detection model, and post-processing.

The pre-processing stage takes raw SPICE analog netlists as input and constructs graph representations for each circuit. In this stage, feature vectors are extracted from the type information of the netlists and the structure of the graph. In the detection model stage, symmetry constraint detection problem is mapped into a binary classification problem. GraphSage [28], a general inductive approach that generates node embeddings for graph data, is adapted to measure the similarity between a pair of nodes. The model is modified and trained in a supervised manner. The modified model can produce a set of potential pairs with symmetry constraints. All predicted pairs will go through a rule-based filter and a probability-based filter in the post-processing stage to eliminate most of the false-positive pairs. After filtering, our framework produces the symmetry constraints detected.

The device type information is encoded as part of the node feature. To distinguish device nodes from pin nodes, a two-dimensional vector to indicate whether a node

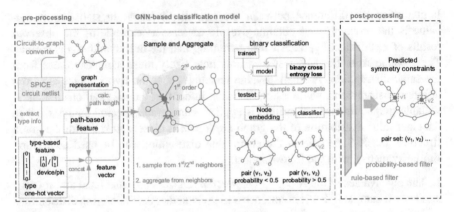

Fig. 17.7 Workflow of the GNN-based framework in [22]

is a device or a pin is used, where $[0, 1], [1, 0]$ stand for a device and a pin, respectively. Then, the device types (i.e., capacitor, resistor, diode, NMOS, PMOS, IO) and pin types (i.e., source, drain, gate, substrate, passive, cathode of a diode, and anode of a diode) are translated into two one-hot vectors. The representation includes power/ground nets and a power node and a ground node as auxiliary nodes. These type-related vectors make up the first two parts of node features.

A novel path-based feature is also proposed inspired by the electric potential in circuit analysis. The feature is not to simulate the circuit but to characterize the "global position" of each node in the graph by VSS/Ground-sourced path lengths. The neighbor structures of the two nodes are not the same, but the electric potential values of their corresponding pins are the same according to DC analysis. The path-based feature intended to capture the intuitive relevance between electric potential and path from ground node by developing a path-based feature.

Aggregator functions are crucial to the sampling and aggregation process. The mean aggregator concatenates the current node representation h_v^{k-1} and the average of aggregated neighbor node representations. $\mathcal{N}(v)$ stands for the sampled neighbor set of node v, and W is a learnable parameter matrix. $\sigma(\cdot)$ denotes an activation function that introduces nonlinearity to our model. ReLU is taken as the activation function $\sigma(\cdot)$ in aggregation process. Equation (17.4) summarizes the mean aggregator,

$$h_v^k = \sigma\left(W \cdot \left\{h_v^{k-1} \oplus MEAN\left(h_u^{k-1}, \forall u \in \mathcal{N}(v)\right)\right\}\right). \tag{17.4}$$

To train the aggregator functions mentioned before, we apply binary cross entropy loss, a classic loss function of binary classification, which facilitates high accuracy in our applications. The loss function is declared as:

$$loss = -\frac{1}{N}\sum_{i=1}^{N} y_i \cdot \log(prob_i) + (1 - y_i) \cdot \log(1 - prob_i), \tag{17.5}$$

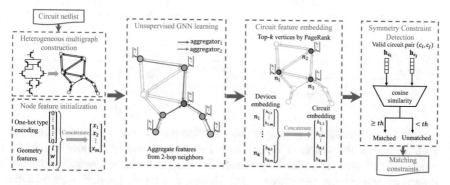

Fig. 17.8 Computation flow of the GNN-based framework in [12]

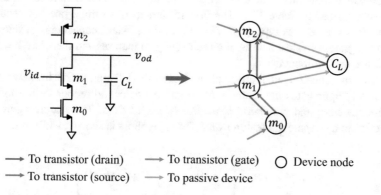

$\longrightarrow$ To transistor (drain) $\longrightarrow$ To transistor (gate)   ○ Device node

$\longrightarrow$ To transistor (source) $\longrightarrow$ To passive device

Fig. 17.9 Example of the heterogeneous multigraph model in [12]

where N denotes the number of node pairs and y_i and $prob_i$ are the ground-truth label and predicted label of the i_{th} pair, respectively.

In [12], an unsupervised inductive graph-learning-based methodology for both system-level and device-level AMS symmetry constraint extraction is proposed. Figure 17.8 shows the overall computation flow. With the unsupervised learning technique, the proposed framework learns a strategy for extracting latent information of matching circuit structures, thus amenable to general AMS designs.

In the framework, a heterogeneous multigraph (i.e., multigraph with various edge types) for circuit netlist representation is first constructed. The proposed heterogeneous multigraph model consists of four different types of edges, representing the connections between different ports of a devices. Figure 17.9 illustrates an example of the heterogeneous multigraph representation. In the figure, the four devices m_0, m_1, m_2, and C_L are mapped to four graph vertices. Consider the connection from the drain of m_1 to the drain of m_2; an edge $e_1 = (m_1, m_2, p_{drain})$ is added. Other edges are added similarly. Note that parallel edges occur if some nets connect to multiple terminals of a transistor.

Table 17.1 Initial vertex features in the heterogeneous multigraph of the input circuit

Feature	Length	Description
Device type	15	The one-hot device type encoding
Geometry	2	The length and width of the device
Layer	1	The number of metal layers

After constructing the heterogeneous multigraph, the initial feature vector for each vertex in the graph is determined. As modern analog devices are much more sophisticated, in order to precisely describe a device's physical structure, dozens of design parameters are needed. However, using all the design parameters might cause overfitting in the learning model and restrict its ability to identify nonidentical matching structures. Therefore, the used features and their dimension are summarized in Table 17.1. The first 15 dimensions form a one-hot vector that represents the devices type (e.g., NMOS, PMOS, MOM capacitor). Besides, [12] also integrates physical details of circuit sizing as features and therefore lowers the false alarms significantly.

With the constructed multigraph and the initial features, the framework then iteratively aggregates the features of neighbor vertices to recognize the localized interconnection and peripheral structures of each vertex. The feature aggregating function to aggregate the features of K-hop neighbors is shown as follows:

$$h_v^{(k)} = GRU\left(h_v^{(k-1)}, \sum_{u \in N_{in}(v)} W_{e_{uv}} h_u^{(k-1)}\right), \tag{17.6}$$

where $h_v^{(k)}$ is the feature vector of vertex v at the kth layers of the GNNs, $GRU(\cdot, \cdot)$ denotes a gated recurrent unit [17], $N_{in}(v)$ is the in-neighbors of v, and $W_{e_{uv}}$ represents the linear transformation matrix with respect to edge e_{uv}. In [12], K is set to 2, and the output dimension of each neural network is set to 18. The GNN model is then trained with an unsupervised loss function

$$\mathcal{L}_{tot} = \sum_{v \in V} \mathcal{L}(z_v),$$

$$\mathcal{L}(z_v) = -\sum_{u \in N_{in}(v)} \log\left(\sigma\left(z_u^{\mathsf{T}} z_v\right)\right) - \sum_{i=1}^{B} \mathbb{E}_{\tilde{u} \sim Neg(v)} \log\left(1 - \sigma\left(z_{\tilde{u}}^{\mathsf{T}} z_v\right)\right),$$

$$\tag{17.7}$$

where $z_v = h_v^{(K)}$ is the final feature representation of a vertex v, $\sigma(x) = 1/(1 + e^{-x})$ is the sigmoid function, $\mathbb{E}$ denotes the expected value, $Neg(v)$ is the negative sampling distribution with respect to v, and B denotes the total number of negative samples. Minimizing the overall loss $\mathcal{L}_{tot}$, the GNN models learns the strategy to improve feature similarity between each vertex and its neighbors while enlarging its discrepancy with the negative samples, and thus implicitly integrates the localized structure information into each vertex.

With the trained feature vectors of each device, the subcircuit feature representation is then determined by concatenating the top-M representative vertices. The PageRank algorithm [62] is utilized to select the top-M representative nodes. The PageRank score of each vertex v can be computed as follows:

$$PR(v) = \frac{1 - \gamma}{|V|} + \gamma \cdot \sum_{u \in \mathcal{N}_{in}(v)} \frac{PR(u)}{|\mathcal{N}_{out}(v)|}, \tag{17.8}$$

where V is the vertex set, γ is the damping factor, $\mathcal{N}_{in}(v)$ is the set of in-neighbors (i.e., $\forall u \in V$ such that there exists a direct edge from u to v) of v, and $\mathcal{N}_{out}(v)$ is the set of out-neighbors (i.e., $\forall u \in V$ such that there exists a direct edge from v to u) of v.

Finally, symmetry constraints are generated by comparing the cosine similarity of the trained features for the subcircuits and devices, where the cosine similarity λ_{sim} of two trained features z_i and z_j is defined as

$$\lambda_{sim} = \frac{z_i \cdot z_j}{\|z_i\| \|z_j\|}. \tag{17.9}$$

Given a similarity threshold λ_{th}, if the cosine similarity of the features of two devices (resp. subcircuits) is larger than λ_{th}, then a device-level (resp. system-level) symmetry constraint between the two objects will be added.

17.3 Constrained Placement and Routing

17.3.1 Placement Quality Prediction

During iterative placement and routing, the ability of a candidate layout to meet performance specifications depends on interconnect RC parasitics. This is illustrated in Table 17.2: post-layout parasitics can cause as much as 22% loss of unity gain frequency from the schematic (pre-layout) values for this testcase.

We overview a set of compact ML-based models for a cost function component that rapidly predicts the quality of a candidate layout. This can be used to reward

Table 17.2 Schematic and post-layout performance of an OTA

Characteristic	Schematic	Layout	Change
DC Gain (dB)	39.30	37.25	−5%
Bandwidth (MHz)	10.64	10.47	−2%
Unity gain frequency (MHz)	440	383	−22%

(or penalize) the optimization cost function when the layout meets (or fails to meet) performance specifications.

17.3.1.1 Applying Standard ML Models

In [41], the performance of a circuit is evaluated by a set of performance functions $z_1, z_2, \ldots$ for an analog block. For each z_i, a *satisfaction function* is defined as

$$
\psi_i(z_i) = \begin{cases} 1 & z_i \geq \theta_i \\ \frac{z_i}{\theta_i} & z_i < \theta_i \end{cases} \tag{17.10}
$$

where θ_i is the design specification for z_i. For K performance functions, if w_i represents the weight factor of function i s.t. $\sum_{i=1}^{K} w_i = 1$, the performance Figure of Merit (FOM) is defined as

$$
\Phi = \sum_{i=1}^{K} w_i \cdot \psi_i(z_i) \tag{17.11}
$$

If all performance specifications are satisfied, $\Phi = 1$. To estimate the probability that a specification is satisfied, a Probability of Demerit (POD) is defined as:

$$
\Delta = \sum_{i=1}^{K} w_i \cdot P(z_i < \theta_i) \tag{17.12}
$$

where P indicates the probability of violating specifications. ML models are built for Δ, and it is incorporated as an additional component of the placement cost function.

In [41], the feature space corresponds to the set of lengths of all interconnects in the circuit. All wires are assumed to have the same width and thickness, and therefore, the RCs of these wires depend on their lengths. Three ML models are evaluated: neural network (NN) [36], random forest (RF) [6], and support vector machine (SVM) [67]. Training is performed using a pre-routing estimate based on the star model. The accuracy of these models on different circuit characteristics of different OTA designs is shown in Fig. 17.10. It is seen that NN outperforms both RF and SVM on every case, while SVM is better than RF on gain, bandwidth, and phase margin.

17.3.1.2 Stratified Sampling with SVM/MLP Models

In [19], a framework was proposed for extracting these relationships by building ML models for each performance constraint, extracting both linear and nonlinear

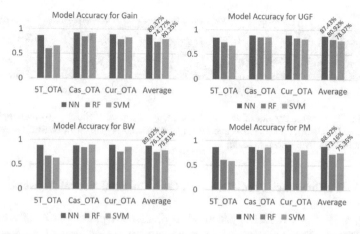

Fig. 17.10 ML model accuracy on gain, unity gain frequency (UGF), bandwidth (BW), and phase margin (PM)

correlations among all the sensitive parasitics over a multidimensional search space of RC parasitics. It consists of the following steps:

(1) Feature space pruning: This step reduces the dimension of the feature space of RC parasitics by (a) identifying variables that the performance constraints are insensitive to and (b) by range reduction, which determines upper bounds on the parasitics.

(2) Sparse sampling and linear SVM classification: Next, a sparse sample set in the updated feature space is generated using stratified sampling based on the Latin hypercube sampling method. These samples are labeled for each performance constraint associated with performance parameter p_k. For each p_k, a support vector machine (SVM) model is built to extract correlations among the features. If the classification error falls below a user-specified threshold, ϵ_0, the SVM provides a good model for p_k.

(3) Dense sampling and classification: If the error exceeds ϵ_0, a larger number of samples is generated to drive higher accuracy, labeling each sample. Next, an SVM is built with the denser samples: if its error is within a user-specified threshold, ϵ_1, the SVM is a good model; else, a multilevel perceptron (MLP) model is built.

The constraint framework is applied to guide the placement of the above OTA and VCO circuits within the placement engine in ALIGN [3, 18, 37], adding a penalty for violating a performance constraint. The performances of the three OTA layouts (Fig. 17.11), extracted from post-layout analyses, are summarized in Table 17.3. All three of the automatically generated layouts maintain the required design constraints ($\checkmark$), but layouts generated without the framework fail one or more constraints ($\boldsymbol{\times}$). For the VCO schematic in Fig. 17.12a, b shows the circuit layout, and it meets all specifications.

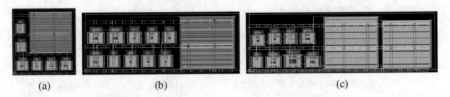

(a)   (b)   (c)

Fig. 17.11 Automated layouts of the (**a**) 5T OTA (9.63μm × 9.60μm), (**b**) telescopic OTA (6.85μm × 18.65μm), and (**c**) two-stage OTA (7.42μm × 24.49μm) with constraints generated by the framework [Not drawn to scale]

17.3.1.3 Performance Prediction with Convolutional Neural Networks

In [46], a 3D CNN model is proposed to predict the placement quality of OTA circuit layouts. The complex and intricate nature of analog circuit behaviors make extracting performance relevant features from placement extremely important. The performance impact of a device placement lies in both the placement location and circuit topology. As an example, the mismatch of differential input pairs has a larger impact toward offset compared with the load. Thus, to ensure a good and generalized model, extracted features must be both easily extendable to different circuit topologies and able to encode effective placement information.

To leverage the success of convolutional neural networks in computer vision tasks, we represent intermediate layout placement results into 2D images. Instead of compacting the entire circuit placement into a single image, we separate devices into different images based on the circuit topology. For OTA circuits, we propose to divide the circuit into subcircuits based on functionality as shown in Fig. 17.13.

The devices are abstracted into rectangles and scaled according to the placement results into a image. The net routing demand is the aggregated pin boundary box for each net. In all our experiment, the image size is selected to be 64*64. Device types are encoded as different intensities in the image. Figure 17.14 shows an OTA layout with the corresponding extracted placement feature images.

Convolutional neural networks have been primarily applied on 2D images as a class of deep models for feature construction. Conventional 2D CNNs extract features from local neighborhoods on feature maps in the previous layer. Formally, given the pixel value at position (x, y) in the jth feature map in the ith layer, the convolutional layer output v_{ij}^{xy} is given by

$$v_{ij}^{xy} = \sigma \left(\sum_m \sum_{p=0}^{P_i-1} \sum_{q=0}^{Q_i-1} w_{ijm}^{pq} v_{(i-1)m}^{(x+p)(y+q)} + b_{ij} \right), \tag{17.13}$$

where $\sigma(\cdot)$ is the activation function, b_{ij} is the bias for feature map, m indexes over the set of feature maps in this layer, and w_{ijm}^{pq} is the value of the weight kernel at the position (p, q) connected to the kth feature map. The output feature is thus the

Table 17.3 Post-layout performance of the OTA test cases

Performance specifications	5T OTA				Telescopic OTA				Two-stage OTA			
	With framework		Without framework		With framework		Without framework		With framework		Without framework	
Gain (dB)	20.57	✓	19.09	✓	42.13	✓	38.12	✗	26.57	✓	24.38	✗
BW (MHz)	103.26	✓	126.20	✓	5.49	✓	7.64	✓	46.84	✓	41.22	✓
UGF (GHz)	1.17	✓	1.14	✓	0.70	✓	0.61	✗	1.00	✓	0.92	✗
PM (°)	110.33	✓	116.77	✓	133.41	✓	106.50	✓	94.43	✓	82.05	✓
CMRR (dB)	52.08	✓	52.92	✓	69.15	✓	62.14	✗	32.71	✓	38.27	✓
PSRR (dB)	21.39	✓	18.47	✗	42.45	✓	53.52	✓	26.94	✓	24.37	✗
SR (V/μS)	156.62	✓	156.63	✓	414.24	✓	424.23	✓	408.19	✓	386.07	✓
ICMR (V)	0.60–0.75	✓	0.60–0.75	✓	0.55–0.85	✓	0.55–0.85	✓	0.60–0.75	✓	0.60–0.75	✓

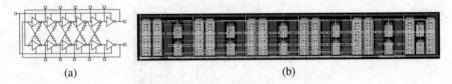

(a)   (b)

Fig. 17.12 VCO (**a**) schematic, (**b**) layout (10.11μm × 72.78μm) using the method

Fig. 17.13 Dividing an OTA
into subcircuits for
performance prediction [46]

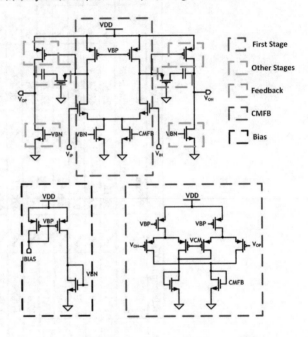

activation output of a weighted sum over all the kernel maps with the previous layer images.

3D convolution layers were first proposed to incorporate both spatial and temporal information for action recognition in videos. In contrast to 2D CNNs where the convolution kernel is a 2D map, 3D convolution is achieved by convolving a 3D kernel to the cube formed by stacking multiple contiguous images together:

$$
v_{ij}^{xyz} = \sigma \left(\sum_m \sum_{p=0}^{P_i-1} \sum_{q=0}^{Q_i-1} \sum_{r=0}^{R_i-1} w_{ijm}^{pqr} v_{(i-1)m}^{(x+p)(y+q)(z+r)} + b_{ij} \right), \tag{17.14}
$$

with r being the value across the third dimension. Images captured across time from videos were stacked to form a 3D input tensor for action recognition.

The use of 3D CNNs is to effectively capture the relative location information between the different placement subcircuits. Figure 17.15 shows the overall model of the 3D CNN network for placement quality prediction. Each extracted placement

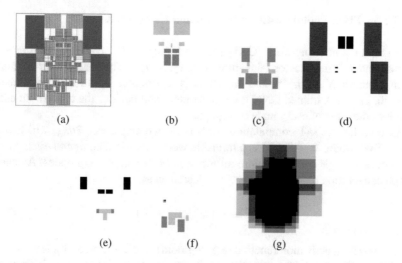

Fig. 17.14 (**a**) is the OTA layout. (**b**)–(**g**) are the extracted image features of the layout with first stage, other stage, feedback, CMFB, bias, and net routing demand, respectively [46]

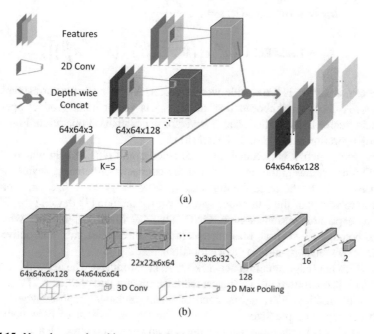

Fig. 17.15 Neural network architecture. (**a**) Initial separate 2D CNN. (**b**) 3D CNN classifier [46]

feature image is augmented into feature sets with coordinate channels. Initial features are then extracted separately for each feature sets with 2D convolutional layers. The outputs are then stacked to form 3D tensors. The 3D tensors are fed to the 3D CNN for placement quality prediction.

17.3.1.4 PEA: Pooling with Edge Attention Using a GAT

In a graph attention network (GAT) [68] on a graph $G(V, E)$, each node $v_i \in V$ is associated with a vector of features $(x_1, x_2, ..., x_d)$. The features for all nodes form a matrix $X \in \mathbb{R}^{n \times d}$, and the set E is represented by an adjacency matrix $A \in \{0, 1\}^{n \times n}$. A trained GNN takes X as input and decides the class of an entire graph or the class of every node in a graph.

In each layer, GAT computation consists of two steps: *weighting*, which computes XW, where $W \in \mathbb{R}^{d \times d}$ is a trainable weight matrix, and *aggregation*, where each node $v_i \in \mathcal{V}$ collects feature information of its neighboring nodes. A generic graph convolution operation in layer l is described as

$$Z^{(l)} = \sigma \left(\Phi_A^{(l)} X^{(l)} W^{(l)} \right) \tag{17.15}$$

where $\sigma(\cdot)$ is an activation function (e.g., sigmoid), and the form of Φ_A is elaborated in [68] and Chapter 4. The **attention** coefficient α_{ij} from node v_j to v_i is given by

$$\alpha_{ij} = \text{softmax}_{row}(\tau_{ij}) = \frac{e^{\tau_{ij}}}{\sum_{k \in \mathcal{N}_i} e^{\tau_{ik}}}$$

$$\tau_{ij} = \text{LeakyReLU} \left(a^{(l)} \cdot \left[\left(W^{(l)\text{T}} X_i^{(l)} \right) \| \left(W^{(l)\text{T}} X_j^{(l)} \right) \right] \right) \tag{17.16}$$

where $a^{(l)} \in \mathbb{R}^{2d_l + 1}$ is a trainable weight vector, $X_i^{(l)}$ is a vector corresponding to node v_i, $\mathcal{N}_i$ is the neighborhood of node v_i, $\cdot$ is vector inner product operation, T means vector transposition, and $\|$ is vector concatenation operation. Finally, the **pooling** operation is based on DiffPool [81].

The circuit netlist is encoded into a directed graph $\mathcal{G}(\mathcal{V}, \mathcal{E})$, in which devices and IO pins are the graph nodes $\mathcal{V}$ and the connections between devices are the graph edges $\mathcal{E}$. In the node feature matrix, $X_i \in \mathbb{R}^d$, $i = 1, 2, ..., n$, represents the feature vector of the i-th node. The d features include (1) device type, PMOS, NMOS, capacitor, current source, GND, etc.; (2) functional module where the device belongs to, such as bias current mirror, differential pair, and active load; (3) device dimension; and (4) device location.

In [40], a new edge attention network, PEA, is introduced. The network, shown in Fig. 17.16, is composed of two stages: feature extractor and predictor. The extractor consists of multiple PEA layers, each of which includes graph convolution and graph pooling. The predictor is an MLP. The key ingredient of PEA network is the integration between edge feature/attention and graph pooling. PEA is composed of four phases: (1) edge-aware attention construction and compression; (2) graph convolution; (3) node pooling; and (4) edge pooling. PEA customizes a standard GAT primarily in phases 1 and 4.

Edge attention was previously proposed in [23] by expanding the attention matrix from 2D to 3D using channels so that $a^{(l)} \in \mathbb{R}^{n_l \times n_l \times p_l}$, but this approach is expensive in runtime and memory use. The edge-aware attention model in PEA overcomes this by defining the raw attention as

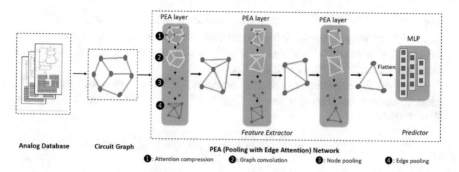

Fig. 17.16 Overview of the PEA network

$$\hat{\alpha}_{ijk}^{(l)} = f^{(l)}\left(X_i^{(l)}, X_j^{(l)}, E_{ijk}^{(l)}\right) = \tau_{ij} E_{ijk}^{(l)} \qquad (17.17)$$

where $E_{ijk}^{(l)}$ corresponds to the edge featue and τ_{ij} is given by Eq. (17.16). The attention matrix is obtained through bidirectional normalization (BN) as

$$\alpha^{(l)} = BN(\hat{\alpha}^{(l)}) = \begin{cases} \tilde{\alpha}_{ijk}^{(l)} = \text{softmax}_{row}\left(\hat{\alpha}_{ijk}^{(l)}\right) \\ \\ \alpha_{ijk}^{(l)} = \sum_{m=1}^{n_l} \dfrac{\tilde{\alpha}_{imk}^{(l)} \tilde{\alpha}_{jmk}^{(l)}}{\sum_{u=1}^{n_l} \tilde{\alpha}_{umk}^{(l)}} \end{cases} \qquad (17.18)$$

where n_l is the number of nodes at layer l. This normalization avoids computing overflow from multiplication and guarantees that in each channel k, the sum in each row and each column of $\alpha^{(l)}$ is 1.

After the attention compression, graph convolution is performed in the same way as the conventional approach, except that the attention is replaced by the compressed version, $g(\alpha^{(l)}; b^{(l)})$, and this is followed by node pooling. This is succeeded by *edge pooling*, an original contribution of [40], consisting of two sub-steps: channel pooling and node-space pooling. Please note the node-space here is for edge features and hence the node-space pooling for edge features is different from conventional node pooling.

The channel pooling operation $h : \mathbb{R}^{p_l} \to \mathbb{R}^{p_{l+1}}$ is performed as follows:

$$Q_{ij}^{(l)} = h\left(\alpha_{ij}^{(l)}; W_{edge}^{(l)}\right)$$

$$Q_{ijk}^{(l)} = \sum_{m=1}^{p_l} \alpha_{ijm}^{(l)} W_{edge}{}_{mk}^{(l)} \qquad (17.19)$$

where $Q^{(l)} \in \mathbb{R}^{n_l \times n_l \times p_{l+1}}$ is edge-feature-encoded attention and $W_{edge}^{(l)} \in \mathbb{R}^{p_l \times p_{l+1}}$ is a trainable weight matrix. This transformation changes the channel dimension from p_l for attention $\alpha^{(l)}$ to p_{l+1} for $Q^{(l)}$. Since attention α incorporates edge feature information in Eq. (17.17), so does Q.

Based on the edge-feature-encoded attention $Q^{(l)}$, the node-space pooling for edge features is designed to be $t : \mathbb{R}^{n_{l+1} \times n_l} \times \mathbb{R}^{n_l \times n_l \times p_{l+1}} \times \mathbb{R}^{n_l \times n_{l+1}} \rightarrow \mathbb{R}^{n_{l+1} \times n_{l+1} \times p_{l+1}}$:

$$
\begin{aligned}
E^{(l+1)} &= t\left(S^{(l)^{\mathrm{T}}}, Q^{(l)}, S^{(l)}\right) \\
&= ||_{k=1}^{p_{l+1}} \left(S^{(l)^{\mathrm{T}}} Q_{\cdot \cdot k}^{(l)} S^{(l)}\right)
\end{aligned}
\tag{17.20}
$$

where $E^{(l+1)} \in \mathbb{R}^{n_{l+1} \times n_{l+1} \times p_{l+1}}$ is the edge feature matrix after the complete pooling. In the pooling step, "$\cdot \cdot k$" is a slicing operation defined by

$$
\left(Q_{\cdot \cdot k}^{(l)}\right)_{ij} = Q_{ijk}^{(l)} \quad i, j \in 1, 2, ..., n_l
\tag{17.21}
$$

where $Q_{\cdot \cdot k}^{(l)}$ is a 2D matrix for channel k, and all channels are concatenated by $||$: $\mathbb{R}^{n_{l+1} \times n_{l+1}} \rightarrow \mathbb{R}^{n_{l+1} \times n_{l+1} \times p_{l+1}}$. This is defined as

$$
U = ||_{k=1}^{p_{l+1}} V_k, \quad U_{ijk} = (V_k)_{ij}
\tag{17.22}
$$

where $V_k \in \mathbb{R}^{n_{l+1} \times n_{l+1}}$, $k \in 1, 2, ..., p_{l+1}$. Edge pooling is illustrated in Fig. 17.17.

A PEA network with L layers can be applied to determine whether performance constraints $y_i \geq \phi_i, i = 1, 2, ..., m$ are satisfied. The overall performance cost can be defined as

$$
Q_I = \sum_{i=1}^{m} w_i \cdot P(y_i < \phi_i)
\tag{17.23}
$$

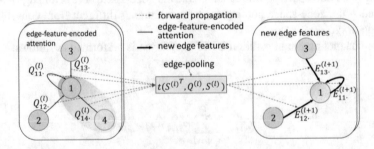

Fig. 17.17 Edge pooling in the PEA network

where $P(y_i < \phi_i)$ is the probability of violating design specification and can be obtained by the softmax output at PEA network. The weighting factors $w_i, i = 1, 2, ..., m$ are decided by users and satisfy $\sum_{i=1}^{m} w_i = 1$. Alternatively, the performance cost can be defined by

$$Q_{II} = P\left(\sum_{i=1}^{m} w_i \cdot \min\left(\frac{y_i}{\phi_i}, 1\right) < \mathcal{T}\right) \tag{17.24}$$

where $\mathcal{T}$ is the specification of overall performance and can be obtained according to legacy designs. The classification on whether $\sum_{i=1}^{m} w_i \cdot \min(\frac{y_i}{\phi_i}, 1) < \mathcal{T}$ can be obtained through PEA network. Cost Q_{II} relies on an additional threshold $\mathcal{T}$ compared to Q_I. However, it requires only one output from PEA, while Q_I needs m outputs from PEA.

A sample result for a cascode OTA is shown in Table 17.4. Here, *self-sustained learning* means the model is trained and applied on the same topology and 80% of the total data is employed for the training. The testing data are the remaining 20% of the entire data. The "Transfer" results are obtained by a major training with 80% of data on S (source) and minor fine-tuning with 10% of data on T (target) and predicting on T. The three PEA-guided results are significantly closer to manual layout than both the previous work [48] and placement guided by the CNN-based model [46].

17.3.2   Analog Placement

The placement stage determines the locations of each module and device. Analog placement constraints are imposed to the placer to mitigate the layout effects on circuit electrical behaviors. A typical analog placer optimizes area and wirelength.

17.3.2.1   Incorporating Extracted Constraints into Analog Placers

Conventionally, analog placement engines take manually labeled constraints as input. Recent fully automated analog layout frameworks, such as MAGICAL [11, 77] and ALIGN [3, 18, 37], have moved toward replacing human-generated constraints with automatic generated constraints. Geometric constraints such as symmetry are extracted from the netlist, as discussed in Sect. 17.2.

To handle performance constraints, the cost function for placement must be augmented using the techniques discussed in Sect. 17.3.1. The stochastic approach is a classical paradigm on solving AMS placement problems, where a placement solution is represented with an intermediate data structure, such as O-tree [64] and segment tree [5]. Then a randomized search scheme, such as a simulated annealing-based optimizer, perturbs the underlying data structure and searches for an optimal

Table 17.4 Results of cascode OTA (SS: Self-Sustained Learning)

| | Schematic | Manual | Conventional | | CNN | | PEA | | | |
			Automatic		SS Q_{11}	Transfer Q_{11}	SS Q_{11}	Transfer Q_{11}	Transfer Q_1
Gain (dB)	37.0	33.0	23.7		27.7	30.1	32.2	32.5	33.1
UGF (MHz)	1522.9	1167.0	947.6		1003.0	617.1	1072.0	948.9	1042.0
BW (MHz)	21.8	26.8	56.0		33.8	17.5	26.9	22.4	24.8
PM (degree)	82.1	80.7	108.5		113.7	104.7	90.8	93.0	85.5
FOM	1.00	0.85	0.71		0.75	0.66	0.82	0.80	0.83
Area (μm²)	–	26.5	24.1		40.4	37.1	34.0	34.4	32.4

solution. Other solution algorithms used for placement include mixed integer linear programming (MILP) [73, 75] and nonlinear programming (NLP) [61, 74, 85], where the input constraints are coded as part of MILP problem and are automatically handled during optimization. However, the scalability of MILP-based placer is limited. NLP, on the other hand, relaxes the hard constraints into penalties in the objective function and is therefore, in general, more efficient in computation compared to MILP. When NLP is used, the final placement must be legalized after the termination of the nonlinear optimization step.

17.3.2.2 Well Generation and Well-Aware Placement

The well layer defines the doping area that acts as the bulks of MOSFETs. For example, a typical P-MOS device needs to be built on an N-well layer and needs contacts to supply the bulk voltage (usually VDD). In a typical digital design methodology, wells are pre-designed within standard cells so that well generation is not needed in layout automation flow. However, analog design methodologies often use customized device layouts, and wells are usually distinctly drawn in manual designs. Manual designs often share wells between transistors to reduce spacing and the number of contacts to optimize area and interconnection. Furthermore, well geometries also impact circuit performance through layout-dependent effects, such as the well proximity effect (WPE). Therefore, inserting wells is an additional task to analog layout flow compared to its digital counterpart.

In [76], a ML-guided well generation framework is proposed. The WellGAN framework generates wells following the guidance from a generative ML model, a generative adversarial network (GAN) [24]. Figure 17.18 shows the overall flow for the WellGAN framework. In the training phase, an AMS circuit layout database is utilized to build a conditional-GAN model [54] for the inference. In the inference phase, the trained GAN model predicts the well region and guides the well generation. Figure 17.19 gives an illustration of the WellGAN framework. It presents the placement features and wells as images. A generator network G produces the well region guidance from the placement features. A discriminator network D is also utilized to assist the training process.

Data Representation The WellGAN framework represents the layout with images. The oxide diffusion (OD) layers are selected to be the input patterns. The first channel of an image represents the OD layers within the wells, such as PMOS devices for N-Well generation task. The second channel extracts the OD layers outside the wells, such as PMOS devices. The third channel is for the targeting well guidance. The ground truth of well regions and the outputs of ML prediction are encoded in this channel.

Data Preprocessing The data pre-processing step extracts the OD layers and well shapes from the layouts. In the training phase, the manual-designed layouts are pre-processed into the three channel images containing OD and well layers. In the

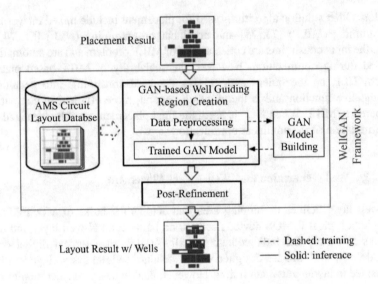

Fig. 17.18 The WellGAN framework [76]

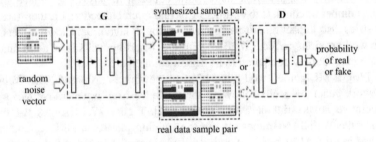

Fig. 17.19 An illustration of the WellGAN framework [76]

inference phase, the OD layers of the placement are extracted into two channels. Then clipping, zero-padding, and scaling are applied to transform the layouts into equally sized image clips to facilitate the modeling.

Well Guidance Generation WellGAN uses a conditional GAN (CGAN) to predict the well regions. It takes the processed data and generates the well region guidance. CGAN simultaneously trains two models: a generative model G and a discriminative model D. The generator G observes input x, the input placement features represented as images, and a random noise vector z. It learns to generate the output y, which includes the placement features and well images. On the other hand, the discriminator D sees y and x and learns to discriminate the "fake" generated y from the ground truth. The CGAN minimizes the loss function in the training process as shown in Eq. (17.25).

$$\min_{G} \max_{D} \mathcal{L}_{CGAN} = \mathbb{E}_{x,y \sim p_d(x,y)}[\log D(x, y)]$$

$$+ \mathbb{E}_{x \sim p_d(x),z \sim p_z}[\log(1 - D(x, G(x, z)))] \qquad (17.25)$$

$$+ \lambda_{L_1} \mathbb{E}_{x,y \sim p_d(x,y),z \sim p_z}[\|y - G(x, z)\|_1],$$

where p_d and p_z denote the probability distributions for the learning targeting and random noise vector z.

The CGAN network is trained to learn the well regions from training data extracted from manual layouts.

Post-Refinement After obtaining the well region guidance in image form from the CGAN model, a post-refinement stage finally generates the well shapes. The image clips are first merged to reconstruct the whole layout. Then the image is transformed into a binary image by applying a threshold the pixel values. The polygons in the binary image is extracted and rectilinearized. The resulting rectangles are then legalized by mapping them to a coarse grid and resolving remaining spacing rule violations. After the post-refinement, the resulting well shapes are inserted into the layout.

Table 17.5 shows the element-wise difference distribution from the manual layouts. The smaller mean error and standard deviation demonstrates the WellGAN framework better mimics the manual layout expertise through ML guidance.

In [84], the WellGAN framework is further extended to integrate with the analog placement engine. Figure 17.20 shows the overall flow of the framework Different from the WellGAN flow where the well generation is an independent step from the placement, in the ML-guided well-aware analog placement engine, placement and well generation are iteratively optimized so that the resulting placement solution is well aware and encourages well sharing to achieve further optimized area.

Table 17.5 Statistics of Manhattan norm of element-wise difference for the test results [76]

Metric	WellGAN	Baseline
Mean	5.67%	12.65%
Standard Deviation	3.58	10.25

Fig. 17.20 The overall flow of the ML-guided well-aware analog placement [84]

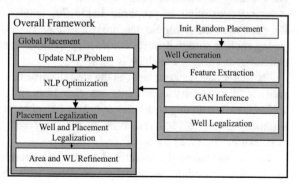

The well-aware global placement is formulated as an NLP optimization problem, where area, wirelength, and constraints are simultaneously considered in the objective function. The well guidance from the GAN model is transformed as a fence region objective and being optimized together with the other terms in the NLP problem. After an iteration of global placement, the current placement is updated, and the GAN model re-predicts the well guidance. This synergistic process seamlessly integrates the placement and ML-guided well generation subroutines.

After the global placement, the well shapes are generated and legalized. The wirelength and area are further optimized with a linear programming-based refinement subroutine. Table 17.6 shows the experimental results of the well-aware placement framework over individual well flow and the WellGAN flow. On the average of ratios, the work reduces area (HPWL) by 82% (26%) over "individual wells" and 74% (46%) over "WellGAN." It demonstrates the effectiveness of the ML-guided well-aware placement algorithm.

17.3.3 *Analog Routing*

The routing stage completes the signal connections using available routing resources to meet various design constraints while meeting certain design objectives.

17.3.3.1 Incorporating Extracted Constraints into Analog Router

In addition to wirelength optimization and design rule handling, which is the main target of conventional digital routing, a comprehensive AMS router should consider more complicated design aspects, including voltage drop, current balancing, parasitics, and signal coupling. For better scalability, these design considerations are usually formulated into geometrical layout constraints. As matching is an essential concept in analog layouts, techniques for various constraints handling such as symmetry, common-centroid, topology-matching, and length-matching have been widely studied. In [63, 72], maze routing algorithms supporting mirror-symmetry constraints are proposed. In [56, 79], length-matching routing approaches for general routing topologies are presented. In [60], an ILP formulation for analog routing simultaneously considering symmetry, common-centroid, topology-matching, and length-matching is presented.

For real-world designs, besides the straightforward mirror symmetry constraints, variants of symmetry constraints are also frequently adopted to describe more sophisticated matching structures of nets. Chen et al. [10] extends the conventional symmetry constraints into four variants, mirror-, cross-, self-, and partial-symmetry, as shown in Fig. 17.21 for a bulk technology node.

The aforementioned geometrical constraints enforced on matching nets during the routing procedure are usually correlated with the placement constraints as placement results determine the overall wiring topologies. As a result, routing

Table 17.6 Comparison of area (μm^2), HPWL(μm), and runtime (RT(s)) [84]

CKTS	Individual wells			WellGAN			Well-aware placement		
	Area	HPWL	RT	Area	HPWL	RT	Area	HPWL	RT
OTA1	360.2	72.3	**1.3**	318.0	68.7	3.2	**290.3**	**60.3**	3.6
OTA2	756.2	234.7	**4.8**	750.7	**203.1**	7.9	**599.0**	205.2	10.6
OTA3	1055.4	586.6	48.9	1325.6	**559.5**	43.2	965.6	651.3	**34.1**
OTA4	3255.2	837.1	**39.7**	3313.6	**799.6**	40.1	3033.7	866	42.6
COMP1	175.1	78.8	**2.0**	144.4	95.1	6.6	**82.2**	**61.8**	3.5
COMP2	192.2	93.1	**3.0**	194.2	105.0	5.6	**84.7**	**48.1**	3.6
BOOTSTRAP	177.9	64.5	**2.0**	130.8	83.4	5.0	97.5	63.2	4.8
RDAC	361.5	209.2	**12.4**	370.4	287.0	30.2	144.3	**137**	23.7
Norm.	1.82	1.26	**0.64**	1.74	1.46	1.33	**1.00**	**1.00**	**1.00**

Best values are indicated in Bold

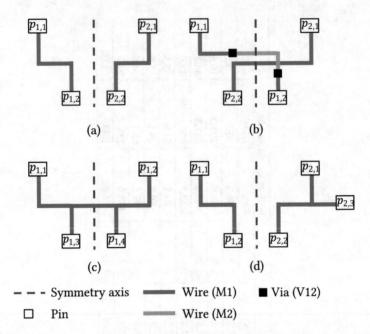

Fig. 17.21 Examples of variants of symmetry constraint (**a**) Mirror-symmetry constraint. (**b**) Cross-symmetry constraint. (**c**) Self-symmetry constraint. (**d**) Partial-symmetry constraint [10]

constraints can also be annotated by the methods such as the graph similarity and graph neural network frameworks described in the previous section.

17.3.3.2 GeniusRoute: ML-Guided Analog Routing

ML methods can also be useful in generating high-quality routing guidances. In [86], a new ML-guided analog routing paradigm, GeniusRoute, is proposed. Existing placement and routing algorithms for AMS circuits usually rely on human-defined heuristics or constraints, such as symmetry and signal flow, to mitigate layout-induced performance issues. In practice, the performance of analog circuits is sensitive to even minor layout changes. The proposed ML-guided analog routing uses a variational autoencoder (VAE) to learn the routing strategies from manual layout database and apply the learned knowledge to guide an automatic routing flow.

Data Representation and Pre-Processing The data in the GeniusRoute are represented as images. Three types of information from a manual layout are extracted to construct a channel in the image. First, the pins of all nets in layout are drawn in the image to give a global view on the placement. Second, the pins of target nets are extracted. Third, the routing region of the nets is extracted to act as ground truth in the training procedure. In the inference flow, only the first two channels are presented.

A training database is constructed by extracting the images from a set of manually routed layouts. To mitigate the shortage of training data, data augmentation is applied by flipping the images.

VAE-Based Routing Guidance Generation For each net type, a VAE model is trained to learn the routing strategy for the given net type. The proposed VAE-based method has two components: an encoder and a decoder. Encoder E_ϕ converts input data x into low-dimensional latent variable vector z, and decoder generates the routing guidance Y from z. VAE uses parametric distribution, usually Gaussian, to model X, Y, and Z, i.e., $P(X|z, \theta) \sim \mathcal{N}(\mu(z), \sigma(z))$ and $z \sim \mathcal{N}(0, I)$, where θ denotes the trainable parameters. The training process maximizes the objective function

$$\log P(Y|z) - \mathcal{D}_{KL}[Q(z|X)||P(z)], \tag{17.26}$$

where $\log P(X|z)$ is a reconstruction log-likelihood of Y from X and $\mathcal{D}_{KL}$ is the Kullback-Leibler (KL) divergence measuring the dissimilarity between the learned distribution Q and training distribution P. The following reparameterization trick is often applied: first sample $\varepsilon \sim \mathcal{N}(0, I)$ and then compute $z = \mu(X) + \sigma^{1/2}(X) * \varepsilon$.

Semi-Supervised Training Algorithm In the GeniusRoute framework, the above VAE training scheme is extended to a semi-supervised approach to achieve better generalization amid limited training data. The proposed training algorithm takes the usage of unlabeled layout data and contains three stages.

In the first stage, the unlabeled training data is used to train the network to reconstruct the layouts. The encoder in this stage can see more training data and learn a more generalized latent space mapping. The objective in this stage is to maximize the standard VAE objective function for reconstruction task:

$$\log P(X|z) - \mathcal{D}_{KL}[Q(z|X)||P(z)]. \tag{17.27}$$

In the second stage, labeled routing region data are used to train the network to learn the routing strategy. The encoder is fixed from the first stage, and only the decoder weights are changed in this stage. The training loss at this stage minimizes the $\mathcal{L}^2$ norm of the distance between ground-truth Y and inferred output $\hat{Y}$:

$$\|Y - \hat{Y}\|_2. \tag{17.28}$$

In the third stage, the GeniusRoute framework fine-tunes the entire model, including both the decoder and encoder. Since the network is already close to a nearly optimal point, the learning rate is set much lower than that in the first two stages. This stage maximizes the objective function shown in Eq. (17.26).

Guided Analog Detailed Routing In the inference phase, GeniusRoute adopts the A^* search algorithm for detailed routing. It leverages the routing guidance generated by machine models to make routing decisions.

The routing guidance is considered as cost functions in A^* search, together with other common routing objective such as wire length and penalty of vias. The cost from the routing guidance is composed of two parts: the penalty of violating the guidance (violating cost), and the cost of routing in the region of other nets demand (competition cost). During the automatic routing process, the nets are encouraged to be routed over the ML-generated guidance.

Figure 17.22 shows output examples of the model inference on testing sets. Figure 17.22a, c is an output of the clock model, Fig. 17.22b, e is from differential nets model, and Fig. 17.22c, f is from the power and ground model. The ML models not only learn well in manual placements but are also capable of generating reasonable outputs for machine-generated placement for clock and differential nets.

Table 17.7 shows the comparison of simulated performance of a comparator. The GeniusRoute result has comparable performance to manual routing and outperforms the router in [72]. Compared to the same router without ML guidance, the proposed method achieves 67% reduction in input-referred offset, with comparable or better results in other metrics.

In the experiments, the GeniusRoute is trained on a dataset consisting of comparators and amplifiers which are representative analog circuit architectures and share similar layout strategies. However, AMS circuits include many different circuit architectures. The layout strategy for those different circuit types can be significantly diverse. How to extend the ML-guided routing to other circuit types is an interesting open question for future research.

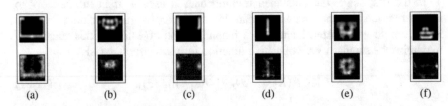

| (a) | (b) | (c) | (d) | (e) | (f) |

Fig. 17.22 Example of inferences of testing set. The upper row shows the ground truth, and the lower row shows the inference [86]

Table 17.7 Comparison of post-layout simulation results for a comparator [86]

	Schematic	Manual	[72]	W/o guide	This work
Offset (μV)	/	480	1230	2530	830
Delay (ps)	102	170	180	164	163
Noise (μV$_{rms}$)	439.8	406.6	437.7	439.7	420.7
Power (μW)	13.45	16.98	17.19	16.82	16.80

17.4 Conclusion and Future Directions

In this chapter, we provide an overview of the current state of the art of applying machine learning to analog layout design automation, from subcircuit recognition and constraint generation to placement and routing. Various machine learning techniques, including supervised, semi-supervised, graph neural network, and reinforcement learning, have been applied to different stages of analog layouts.

The semi- or fully automated open-source analog layout generators ALIGN and MAGICAL have shown great initial success and provided solid foundations for future research and development. An example that incorporates ALIGN into a larger optimization flow is described in [44]: the netlist is sized using a neural network model and iteratively improved with fast layout generation in the inner loop of optimization. A transfer learning method for improving the accuracy of the neural network based on post-silicon measurements is also proposed. The approach is exercised on a VCO circuit, with silicon results. Another example that incorporates MAGICAL into a larger AMS system is described in [45], where an automated end-to-end successive approximation register (SAR) ADC compiler OpenSAR is built. It leverages automated placement and routing in MAGICAL to generate analog building blocks. Meanwhile, capacitor digital-to-analog converter (CDAC) arrays are designed using a template-based layout generator, and digital blocks are done by commercial digital tools. They are then integrated by the hierarchical MAGICAL.

There are still many challenges and research opportunities in applying ML to analog layout.

Integrating with Layout Synthesis Flow The ML-based analog layout automation algorithms have demonstrated many successes in experiments. How to make the whole flow robust and consistent when incorporating ML models is pending a good answer. ML predictions are not always correct and sometimes stochastic. A mature flow might need to compensate the imperfectness of ML-based techniques and leverage their advantages.

In [47] and MAGICAL 1.0 [11], in-loop simulations are involved on building-block-level to guarantee the quality. However, as the simulation costs grow significantly for system-level designs, more scalable and sparser in-loop simulations or better predictive models will be needed.

Analog Layout Dataset and Data-Efficient ML A key challenge for machine learning is always lack of quality training data. Many ML for analog layout techniques require manual labeled data to train the neural network model. However, manually labeling layout data is expensive. How to tackle the limited data is an ongoing open question. There are several possible answers. In [86], a semi-supervised training algorithm is presented to achieve higher data efficiency. More data-efficient ML algorithms might mitigate the issue. In [46], the MAGICAL framework is utilized to generate training data. Machine-generated or automatic labeled data might be a possible solution.

Understanding the Schematics Many existing ML for analog layout techniques use GNN to model the circuit netlist. However, it is still an open question whether GNN is powerful enough to learn the graph topology [78]. Typical convolutional GNN focuses on learning the node-wise features. However, in circuit designs, the topology and interconnections are crucial to circuit functionality. There are several ongoing GNN research trends targeting learning the topology and structures, such as motif-based methods [55] and generative graph models [35].

References

1. Afacan, E., Lourenço, N., Martins, R., Dündar, G.: Review: Machine learning techniques in analog/RF integrated circuit design, synthesis, layout, and test. Integration **77**, 113–130 (2021)
2. Ajayi, T., Cherivirala, Y.K., Kwon, K., Kamineni, S., Saligane, M., Fayazi, M., Gupta, S., Chen, C.H., Sylvester, D., Blaauw, D., Dreslinski, Jr., R., Calhoun, B., Wentzloff, D.D.: Fully autonomous mixed signal SoC design and layout generation platform. In: IEEE Hot Chips Symposium (2020)
3. ALIGN: Analog layout, intelligently generated from netlists (Software repository, accessed November 1, 2021). https://github.com/ALIGN-analoglayout/ALIGN-public
4. Bai, Y., Ding, H., Bian, S., Chen, T., Sun, Y., Wang, W.: SimGNN: A neural network approach to fast graph similarity computation. In: ACM International Conference on Web Search and Data Mining, pp. 384–392 (2019)
5. Balasa, F., Maruvada, S.C., Krishnamoorthy, K.: Efficient solution space exploration based on segment trees in analog placement with symmetry constraints. In: IEEE/ACM International Conference on Computer-Aided Design (ICCAD) (2002)
6. Breiman, L.: Random forests. Mach. Learn. **45**(1), 5–32 (2001)
7. Chakravarti, I.M., Roy, J., Laha, R.G.: Handbook of methods of applied statistics. Wiley, New York (1967)
8. Chang, E., Han, J., Bae, W., Wang, Z., Narevsky, N., Nikolic, B., Alon, E.: BAG2: A process-portable framework for generator-based AMS circuit design. In: IEEE Custom Integrated Circuits Conference (CICC) (2018)
9. Chen, H., Liu, M., Xu, B., Zhu, K., Tang, X., Li, S., Lin, Y., Sun, N., Pan, D.Z.: MAGICAL: An open-source fully automated analog IC layout system from netlist to GDSII. IEEE Des. Test **38**(2), 19–26 (2020)
10. Chen, H., Zhu, K., Liu, M., Tang, X., Sun, N., Pan, D.Z.: Toward silicon-proven detailed routing for analog and mixed signal circuit. In: IEEE/ACM International Conference on Computer-Aided Design (ICCAD) (2020)
11. Chen, H., Liu, M., Tang, X., Zhu, K., Mukherjee, A., Sun, N., Pan, D.Z.: MAGICAL 1.0: An open-source fully-automated AMS layout synthesis framework verified with a 40-nm 1 GS/s $\Delta\Sigma$ ADC. In: IEEE Custom Integrated Circuits Conference (CICC) (2021)
12. Chen, H., Zhu, K., Liu, M., Tang, X., Sun, N., Pan, D.Z.: Universal symmetry constraint extraction for analog and mixed-signal circuits with graph neural networks. In: ACM/IEEE Design Automation Conference (DAC) (2021)
13. Cohn, J., Garrod, D.J., Rutenbar, R.A., Carley, L.R.: KOAN/ANAGRAM II: New tools for device-level analog placement and routing. IEEE J. Solid State Circuits **26**(3), 330–342 (1991)
14. Crossley, J., Puggelli, A., Le, H.P., Yang, B., Nancollas, R., Jung, K., Kong, L., Narevsky, N., Lu, Y., Sutardja, N., An, E.J., Sangiovanni-Vincentelli, A.L., Alon, E.: BAG: A designer-oriented integrated framework for the development of ams circuit generators. In: IEEE/ACM International Conference on Computer-Aided Design (ICCAD), pp. 74–81 (2013)

15. De Ranter, C.R.C., Van der Plas, G., Steyaert, M.S.J., Gielen, G.G.E., Sansen, W.M.C.: CYCLONE: Automated design and layout of RFLC-oscillators. IEEE Trans. Comput. Aided Des. Integr. Circuits Syst. (TCAD) **21**(11), 1161–1170 (2002)

16. Defferrard, M., Bresson, X., Vandergheynst, P.: Convolutional neural networks on graphs with fast localized spectral filtering. In: Neural Information Processing Systems (NeurIPS), pp. 3844–3852 (2016)

17. Dey, R., Salem, F.M.: Gate-variants of gated recurrent unit (GRU) neural networks. In: IEEE International Midwest Symposium on Circuits and Systems (MWSCAS), pp. 1597–1600 (2017)

18. Dhar, T., Kunal, K., Li, Y., Madhusudan, M., Poojary, J., Sharma, A.K., Xu, W., Burns, S.M., Harjani, R., Hu, J., Kirkpatrick, D.A., Mukherjee, P., Sapatnekar, S.S., Yaldiz, S.: ALIGN: A system for automating analog layout. IEEE Des. Test **38**(2), 8–18 (2021)

19. Dhar, T., Poojary, J., Li, Y., Kunal, K., Madhusudan, M., Sharma, A.K., Manasi, S.D., Hu, J., Harjani, R., Sapatnekar, S.S.: Fast and efficient constraint evaluation of analog layout using machine learning models. In: IEEE/ACM Asia and South Pacific Design Automation Conference (ASPDAC), pp. 158–163 (2021)

20. Dhillon, I., Guan, Y., Kulis, B.: Weighted graph cuts without eigenvectors: A multilevel approach. IEEE Trans. Pattern Anal. Mach. Intell. (PAMI) **29**(11), 1944–1957 (2007)

21. Eick, M., Strasser, M., Lu, K., Schlichtmann, U., Graeb, H.E.: Comprehensive generation of hierarchical placement rules for analog integrated circuits. IEEE Trans. Comput. Aided Des. Integr. Circuits Syst. (TCAD) **30**(2), 180–193 (2011)

22. Gao, X., Deng, C., Liu, M., Zhang, Z., Pan, D.Z., Lin, Y.: Layout symmetry annotation for analog circuits with graph neural networks. In: IEEE/ACM Asia and South Pacific Design Automation Conference (ASPDAC) (2021)

23. Gong, L., Cheng, Q.: Exploiting edge features for graph neural networks. In: IEEE Conference on Computer Vision and Pattern Recognition (CVPR), pp. 9211–9219 (2019)

24. Goodfellow, I.J., Pouget-Abadie, J., Mirza, M., Xu, B., Warde-Farley, D., Ozair, S., Courville, A., Bengio, Y.: Generative adversarial nets. In: Neural Information Processing Systems (NeurIPS) (2014)

25. Graeb, H.E. (ed.): Analog Layout Synthesis: A Survey of Topological Approaches. Springer, New York (2010)

26. Guerra, D., Canelas, A., Póvoa, R., Horta, N., Lourenço, N., Martins, R.: Artificial neural networks as an alternative for automatic analog IC placement. In: IEEE International Conference on Synthesis, Modeling, Analysis and Simulation Methods, and Applications to Circuit Design (SMACD) (2019)

27. Gusmão, A., Passos, F., Póvoa, R., Horta, N., Lourenço, N., Martins, R.: Semi-supervised artificial neural networks towards analog ic placement recommender. In: IEEE International Symposium on Circuits and Systems (ISCAS) (2020)

28. Hamilton, W., Ying, Z., Leskovec, J.: Inductive Representation Learning on Large Graphs. In: Neural Information Processing Systems (NeurIPS), pp. 1024–1034 (2017)

29. Hamilton, W.L., Ying, R., Leskovec, J.: Representation learning on graphs: Methods and applications. In: IEEE Data Engineering Bulletin (2017)

30. Harjani, R., Rutenbar, R.A., Carley, L.R.: A prototype framework for knowledge-based analog circuit synthesis. In: ACM/IEEE Design Automation Conference (DAC), pp. 42–49 (1987)

31. Harjani, R., Rutenbar, R.A., Carley, L.R.: OASYS: A framework for analog circuit synthesis. IEEE Trans. Comput. Aided Des. Integr. Circuits Syst. (TCAD) **8**(12), 1247–1266 (1989)

32. Karmokar, N., Madhusudan, M., Sharma, A.K., Harjani, R., Lin, M.P.H., Sapatnekar, S.S.: Common-centroid analog circuit layout. In: IEEE/ACM Asia and South Pacific Design Automation Conference (ASPDAC) (2022)

33. Karypis, G., Kumar, V.: A fast and high quality multilevel scheme for partitioning irregular graphs. SIAM J. Sci. Comput. (SISC) **20**(1), 359–392 (1998)

34. Kipf, T.: Semi-supervised classification with graph convolutional networks. In: International Conference on Learning Representations (ICLR) (2017)

35. Kipf, T.N., Welling, N.: Variational graph auto-encoders. arxiv:1611.07308 (2016)

36. Krizhevsky, A., Sutskever, I., Hinton, G.: ImageNet classification with deep convolutional neural networks. Neural Inf. Proces. Syst. (NeurIPS) **25**(2), 1097–1105 (2012)

37. Kunal, K., Madhusudan, M., Sharma, A.K., Xu, W., Burns, S.M., Harjani, R., Hu, J., Kirkpatrick, D.A., Sapatnekar, S.S.: ALIGN: Open-source analog layout automation from the ground up. In: ACM/IEEE Design Automation Conference (DAC), pp. 77–80 (2019)

38. Kunal, K., Dhar, T., Madhusudan, M., Poojary, J., Sharma, A., Xu, W., Burns, S.M., Hu, J., Harjani, R., Sapatnekar, S.S.: GANA: Graph convolutional network based automated netlist annotation for analog circuits. In: IEEE/ACM Proceedings Design, Automation and Test in Eurpoe (DATE) (2020)

39. Kunal, K., Poojary, J., Dhar, T., Madhusudan, M., Harjani, R., Sapatnekar, S.S.: A general approach for identifying hierarchical symmetry constraints for analog circuit layout. In: IEEE/ACM International Conference on Computer-Aided Design (ICCAD) (2020)

40. Li, Y., Lin, Y., Madhusudan, M., Sharma, A., Xu, W., Sapatnekar, S.S., Harjani, R., Hu, J.: A customized graph neural network model for guiding analog IC placement. In: IEEE/ACM International Conference on Computer-Aided Design (ICCAD) (2020)

41. Li, Y., Lin, Y., Madhusudan, M., Sharma, A., Xu, W., Sapatnekar, S.S., Harjani, R., Hu, J.: Exploring a machine learning approach to performance driven analog IC placement. In: IEEE Annual Symposium on VLSI (ISVLSI), pp. 24–29 (2020)

42. Liou, G.H., Wang, S.H., Su, Y.Y., Lin, M.P.H.: Classifying analog and digital circuits with machine learning techniques toward mixed-signal design automation. In: IEEE International Conference on Synthesis, Modeling, Analysis and Simulation Methods, and Applications to Circuit Design (SMACD), vol. 15, pp. 173–176 (2018)

43. Liu, M., Li, W., Zhu, K., Xu, B., Lin, Y., Shen, L., Tang, X., Sun, N., Pan, D.Z.: S^3DET: Detecting system symmetry constraints for analog circuits with graph similarity. In: IEEE/ACM Asia and South Pacific Design Automation Conference (ASPDAC) (2020)

44. Liu, J., Su, S., Madhusudan, M., Hassanpourghadi, M., Saunders, S., Zhang, Q., Rasul, R., Li, Y., Hu, J., Sharma, A.K., Sapatnekar, S.S., Harjani, R., Levi, A., Gupta, S., Chen, M.S.W.: From specification to silicon: Towards analog/mixed-signal design automation using surrogate NN models with transfer learning. In: IEEE/ACM International Conference on Computer-Aided Design (ICCAD) (2021)

45. Liu, M., Tang, X., Zhu, K., Chen, H., Sun, N., Pan, D.Z.: OpenSAR: An open source automated end-to-end SAR ADC compiler. In: IEEE/ACM International Conference on Computer-Aided Design (ICCAD) (2021)

46. Liu, M., Zhu, K., Gu, J., Shen, L., Tang, X., Sun, N., Pan, D.Z.: Towards decrypting the art of analog layout: Placement quality prediction via transfer learning. In: IEEE/ACM Proceedings Design, Automation and Test in Eurpoe (DATE) (2020)

47. Liu, M., Zhu, K., Tang, X., Xu, B., Shi, W., Sun, N., Pan, D.Z.: Closing the design loop: Bayesian optimization assisted hierarchical analog layout synthesis. In: ACM/IEEE Design Automation Conference (DAC) (2020)

48. Ma, Q., Xiao, L., Tam, Y.C., Young, E.F.Y.: Simultaneous handling of symmetry, common centroid, and general placement constraints. IEEE Trans. Comput. Aided Des. Integr. Circuits Syst. (TCAD) **30**(1), 85–95 (2011)

49. Martins, R.M.F., Lourenço, N.C.C., Horta, N.C.G.: Generating Analog IC Layouts with LAYGEN II. Springer, New York (2010)

50. Massier, T., Graeb, H., Schlichtmann, U.: The sizing rules method for CMOS and bipolar analog integrated circuit synthesis. IEEE Trans. Comput. Aided Des. Integr. Circuits Syst. (TCAD) **27**(12), 2209–2222 (2008)

51. McCulloh, I., Armstrong, H., Johnson, A.N.: Social network analysis with applications. Wiley, New York (2013)

52. Meissner, M., Hedric, L.: FEATS: Framework for explorative analog topology synthesis. IEEE Trans. Comput. Aided Des. Integr. Circuits Syst. (TCAD) **34**(2), 213–226 (2015)

53. Meng, Q., Harjani, R.: A 4GHz instantaneous bandwidth low squint phased array using sub-harmonic ILO based channelization. In: IEEE European Solid-State Circuits Conference (ESSCIRC), pp. 110–113 (2018)

54. Mirza, M., Osindero, S.: Conditional generative adversarial nets. arxiv:1411.1784 (2014)
55. Monti, F., Otness, K., Bronstein, M.M.: MotifNet: A motif-based graph convolutional network for directed graphs. In: IEEE Data Science Workshop (DSW), pp. 225–228 (2018)
56. Mustafa Ozdal, M., Wong, M.D.F.: A length-matching routing algorithm for high-performance printed circuit boards. IEEE Trans. Comput. Aided Des. Integr. Circuits Syst. (TCAD) **25**(12), 2784–2794 (2006)
57. Ochotta, E., Rutenbar, R.A., Carley, L.R.: ASTRX/OBLX: Tools for rapid synthesis of high-performance analog circuits. In: ACM/IEEE Design Automation Conference (DAC), pp. 24–30 (1994)
58. Ohlrich, M., Ebeling, C., Ginting, E., Sather, L.: SubGemini: Identifying subcircuits using a fast subgraph algorithm. In: ACM/IEEE Design Automation Conference (DAC), pp. 31–37 (1993)
59. Ou, H.C., Chien, H.C.C., Chang, Y.W.: Simultaneous analog placement and routing with current flow and current density considerations. In: ACM/IEEE Design Automation Conference (DAC) (2013)
60. Ou, H., Chien, H.C., Chang, Y.: Nonuniform multilevel analog routing with matching constraints. IEEE Trans. Comput. Aided Des. Integr. Circuits Syst. (TCAD) **33**(12), 1942–1954 (2014)
61. Ou, H.C., Tseng, K.H., Liu, J.Y., Wu, I.P., Chang, Y.W.: Layout-dependent effects-aware analytical analog placement. IEEE Trans. Comput. Aided Des. Integr. Circuits Syst. (TCAD) **35**(8), 1243–1254 (2016)
62. Page, L., Brin, S., Motwani, R., Winograd, T.: The PageRank citation ranking: Bringing order to the web. Tech. rep., Stanford InfoLab, Stanford, CA (1999)
63. Pan, P., Chen, H., Cheng, Y., Liu, J., Hu, W.: Configurable analog routing methodology via technology and design constraint unification. In: IEEE/ACM International Conference on Computer-Aided Design (ICCAD), pp. 620–626 (2012)
64. Pang, Y., Balasa, F., Lampaert, K., Cheng, C.K.: Block placement with symmetry constraints based on the o-tree non-slicing representation. In: ACM/IEEE Design Automation Conference (DAC) (2000)
65. Razavi, B.: Design of Analog CMOS Integrated Circuits, 1st edn. McGraw-Hill, Inc., New York (2001)
66. Sapatnekar, S.S.: Timing. Springer, Boston (2004)
67. Vapnik, V.: Statistical learning theory. Wiley, New York (1998)
68. Veličković, P., Cucurull, G., Casanova, A., Romero, A., Lio, P., Bengio, Y.: Graph Attention Networks. arXiv:1710.10903 (2017)
69. Wong, K.-L.J., Yang, C.-K.K.: A serial-link transceiver with transition equalization. In: IEEE International Solid-State Circuits Conference (ISSCC), pp. 223–232 (2006)
70. Wu, C.Y., Graeb, H., Hu, J.: A pre-search assisted ILP approach to analog integrated circuit routing. In: IEEE International Conference on Computer Design (ICCD), pp. 244–250 (2015)
71. Wu, P.H., Lin, M.P.H., Chen, T.C., Yeh, C.F., Li, X., Ho, T.Y.: A novel analog physical synthesis methodology integrating existent design expertise. IEEE Trans. Comput. Aided Des. Integr. Circuits Syst. (TCAD) **34**(2), 199–212 (2015)
72. Xiao, L., Young, E.F.Y., He, X., Pun, K.P.: Practical placement and routing techniques for analog circuit designs. In: IEEE/ACM International Conference on Computer-Aided Design (ICCAD), pp. 675–679 (2010)
73. Xu, B., Basaran, B., Su, M., Pan, D.Z.: Analog placement constraint extraction and exploration with the application to layout retargeting. In: ACM International Symposium on Physical Design (ISPD) (2018)
74. Xu, B., Li, S., Pui, C.W., Liu, D., Shen, L., Lin, Y., Sun, N., Pan, D.Z.: Device layer-aware analytical placement for analog circuits. In: ACM International Symposium on Physical Design (ISPD) (2019)
75. Xu, B., Li, S., Xu, X., Sun, N., Pan, D.Z.: Hierarchical and analytical placement techniques for high-performance analog circuits. In: ACM International Symposium on Physical Design (ISPD) (2017)

76. Xu, B., Lin, Y., Tang, X., Li, S., Shen, L., Sun, N., Pan, D.Z.: WellGAN: Generative-adversarial-network-guided well generation for analog/mixed-signal circuit layout. In: ACM/IEEE Design Automation Conference (DAC) (2019)

77. Xu, B., Zhu, K., Liu, M., Lin, Y., Li, S., Tang, X., Sun, N., Pan, D.Z.: MAGICAL: Toward fully automated analog IC layout leveraging human and machine intelligence. In: IEEE/ACM International Conference on Computer-Aided Design (ICCAD) (2019)

78. Xu, K., Hu, W., Leskovec, J., Jegelka, S.: How powerful are graph neural networks? In: International Conference on Learning Representations (ICLR) (2019)

79. Yan, T., Wong, M.D.F.: BSG-Route: A length-matching router for general topology. In: IEEE/ACM International Conference on Computer-Aided Design (ICCAD), pp. 499–505 (2008)

80. Ying, R., He, R., Chen, K., Eksombatchai, P., Hamilton, W.L., Leskovec, J.: Graph convolutional neural networks for web-scale recommender systems. In: ACM International Conference on Knowledge Discovery and Data Mining (KDD), pp. 974–983 (2018)

81. Ying, Z., You, J., Morris, C., Ren, X., Hamilton, W., Leskovec, J.: Hierarchical graph representation learning with differentiable pooling. In: Neural Information Processing Systems (NeurIPS), pp. 4800–4810 (2018)

82. Zeng, Z., Tung, A.K.H., Wang, J., Feng, J., Zhou, L.: Comparing stars: On approximating graph edit distance. VLDB Endowment 2(1), 25–36 (2009)

83. Zhu, K., Chen, H., Liu, M., Pan, D.Z.: Automating analog constraint extraction: from heuristics to learning. In: IEEE/ACM Asia and South Pacific Design Automation Conference (ASPDAC) (2022)

84. Zhu, K., Chen, H., Liu, M., Tang, X., Shi, W., Sun, N., Pan, D.Z.: Generative-adversarial-network-guided well-aware placement for analog circuits. In: IEEE/ACM Asia and South Pacific Design Automation Conference (ASPDAC) (2022)

85. Zhu, K., Chen, H., Liu, M., Tang, X., Sun, N., Pan, D.Z.: Effective analog/mixed-signal circuit placement considering system signal flow. In: IEEE/ACM International Conference on Computer-Aided Design (ICCAD) (2020)

86. Zhu, K., Liu, M., Lin, Y., Xu, B., Li, S., Tang, X., Sun, N., Pan, D.Z.: GeniusRoute: A new analog routing paradigm using generative neural network guidance. In: IEEE/ACM International Conference on Computer-Aided Design (ICCAD) (2019)

Chapter 18
ML for System-Level Modeling

Erika S. Alcorta, Philip Brisk, and Andreas Gerstlauer

18.1 Introduction

It is expected that future computing architectures will evolve into heterogeneous systems-on-chip (SoCs) that contain tens or even hundreds of CPUs, GPUs, FPGAs, accelerators, and 3D memory stacks in the same package or die. Programmers and architects of these future-generation large-scale heterogeneous systems face a litany of daunting tasks: (1) How many and what mix of heterogeneous cores and components should an architecture contain to best serve emerging workloads? (2) How should applications be rewritten and partitioned to best exploit heterogeneous accelerators, each with different programming models? (3) How can operating and runtime systems effectively manage and orchestrate computations across a heterogeneous sea of components? Answering such questions either manually or, in particular, when automated in the form of design tools requires fast yet accurate models that are at the core of any design flow. In the past, this often involved simulations. However, the time required to perform detailed simulation of heterogeneous platforms running complex applications within a tractable timeframe is well beyond the capabilities of today's most advanced simulators. By contrast, a variety of analytical modeling approaches have been proposed over the years, which can rapidly predict estimates of performance metrics (e.g., IPC) and power consumption, but with much lower accuracy. Bolstered by recent advances in machine learning (ML) at scale, predictive models have emerged, which, in

E. S. Alcorta (✉) · A. Gerstlauer
The University of Texas at Austin, Austin, TX, USA
e-mail: esalcort@utexas.edu; gerstl@ece.utexas.edu

P. Brisk
University of California, Riverside, CA, USA
e-mail: philip@cs.ucr.edu

© The Author(s), under exclusive license to Springer Nature Switzerland AG 2022
H. Ren, J. Hu (eds.), *Machine Learning Applications in Electronic Design Automation*, https://doi.org/10.1007/978-3-031-13074-8_18

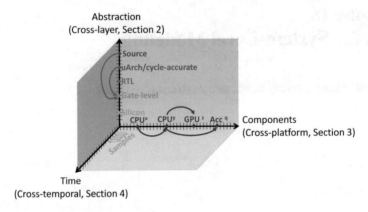

Fig. 18.1 System-level predictive modeling axes

principle, can bridge the gap between traditional simulation-based methodologies and purely analytical or mechanistic models.

A range of work exists in applying ML to system-level modeling (SLM), with a diverse set of prediction targets and purposes. We can categorize predictive modeling approaches along three modeling dimensions, represented by different axes in Fig. 18.1. The first dimension is across abstraction levels, where predictive models learn the relationship between high-level features and low-level implementation metrics. These cross-layer models allow architects to explore design spaces rapidly and evaluate architectures under development for large-scale real-world applications otherwise too complex to be run on accurate design models. They also enable the online estimation of variables whose value is not readily available at execution time due to the lack of sensors, e.g., per-core power consumption. The second SLM dimension is across components, where models learn the relationship of program executions on different platforms. These cross-platform models allow architects, programmers, or runtime systems to profile default CPU code on some host core and use collected statistics to predict expected program behavior on other targets, including GPUs, FPGAs, and dedicated accelerators. Thus, the models will allow making decisions about offloading and refactoring to optimally partition and map applications across heterogeneous architectures. Finally, the third dimension is time, where models learn and characterize past workload patterns to predict future workload behavior. These runtime models allow dynamic management of resources by exploiting fine-grained optimization opportunities that static and application-level runtime management systems would miss. Additionally, they enable systems to behave proactively by anticipating workload changes, adding the capability to adapt the system to upcoming rather than past behaviors.

Vectors in the three-dimensional SLM space constitute different predictive approaches. The starting point of the vectors describes the model inputs or features, while the endpoint represents the model outputs or labels. Much prior work has focused on models that make predictions across a single dimension. Note, however, that these vectors can move in multiple directions simultaneously. While we present a few predictive models that span across more than one dimension, our primary focus in this chapter is models that predict along a single dimension. The set of different combinations is large, and many are yet to be explored. Additionally, we argue that this may be achievable by combining models from separate dimensions.

The rest of this chapter is organized along the three dimensions introduced in Fig. 18.1. In Sects. 18.2, 18.3, and 18.4, we survey predictive cross-layer, cross-platform, and cross-temporal models, respectively, using a representative set of state-of-the-art modeling approaches, and we present selected results that demonstrate their trade-offs. Finally, Sect. 18.5 summarizes the chapter and addresses the challenges and future work in ML for system-level modeling.

18.2 Cross-Layer Prediction

Models of a system and its components at varying levels of abstraction are widely used throughout the design process to evaluate design quality before hardware is available. The level of abstraction and detail represented in a model thereby determines a basic trade-off between simulation accuracy and speed. Traditionally, models at successively higher levels of abstraction are manually constructed by determining the details to include or abstract away. This process has to be repeated, at least partially, for every new architecture to be modeled. Cross-layer prediction is based on the idea that models at higher levels of abstraction can be learned instead of manually constructed. Using observations from a small number of low-level reference simulations and advanced machine learning techniques, models at higher levels of abstraction can be synthesized to predict low-level behavior. In the process, such machine learning-based models can often achieve faster simulation speed and/or accuracy than manual modeling approaches.

Learning-based approaches have been developed to derive both functional and non-functional models of system and their components. On the functional side, so-called surrogate models are learned to predict input-output relationships. Such surrogates are mostly employed in the physical and analog/mixed-signal domains to replace slow high-fidelity simulations. In the digital domain, functional models are readily available as a result of top-down design and synthesis processes. However, non-functional performance, energy, reliability, power, thermal (PERPT), and other models need to be attached to functional simulations to predict corresponding design properties. Some approaches automatically generate high-level PERPT models, but this relies on custom pre-characterization and back-annotation flows [10, 21, 62]. The design of more general learning-based and predictive PERPT modeling techniques is an active area of research [6, 41]. In the remainder of this section, we focus on power models as an example to demonstrate learning-based approaches at different levels of abstraction. Power modeling with any accuracy depends on low-level detail that is challenging to abstract and capture in higher-level models. By contrast, timing and hence performance can more easily be tied to the granularity

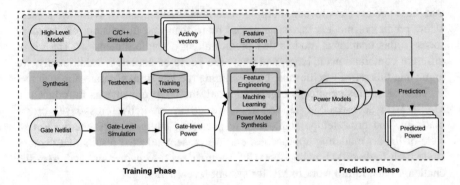

Fig. 18.2 Cross-layer power prediction flow

at which a model is described. However, learning-based performance models have been applied at the highest source- or intermediate-code levels where no structural information is available [39].

Figure 18.2 shows an overview of a learning-based power modeling flow. The flow follows a supervised learning methodology with a training and prediction phase. The primary inputs are the low-level (typically, a gate level) reference model (for training) and a high-level functional simulation model (for training and prediction). During the training phase, simulations are run on both low- and high-level models using the same micro-benchmarks. Reference power traces are generated from the low-level simulations, and activity traces are extracted from the high-level model simulation. In the power model synthesis step, activity features are extracted, and systematic feature selection and decomposition techniques are applied to learn a power model regressor that correlates features with reference power. During prediction, the actual workload to analyze is then simulated in the high-level model, and extracted activity features are fed into the previously trained model to predict low-level power. Such a power modeling flow can be applied at different levels of abstraction. Depending on the level and desired prediction granularity, the high-level simulation model is annotated with the ability to capture relevant activity information. In the following, we describe learning-based power modeling approaches at the micro-architecture and source levels in more detail.

18.2.1 Micro-Architecture Level Prediction

The gold standard for accurate power signoff is gate-level analysis, which comes at the cost of very long simulation times only in very late phases of the design flow. This has traditionally driven the need for power modeling at higher levels of abstraction. At the register-transfer level (RTL), regression-based approaches [5, 54, 57, 59] have been widely adopted including in industrial tools. Early design-space exploration of specifically CPUs is most commonly performed at an abstract micro-architecture level above RTL. Traditionally, generic spreadsheet or library-based analytical power models such as Wattch [7] or McPAT [35] are used to provide power estimates, but they have been shown to be highly inaccurate [58]. Regression and pre-characterization methods have been applied to calibrate existing models [30, 34] or to model power at higher levels [8, 45]. However, all such simple regressions suffer from challenges of modeling nonlinear and inherently activity- and hence data-dependent power characteristics at RTL or higher levels of abstraction. More recently, advances in machine learning have made it possible to accurately capture complex nonlinear regressions with high accuracy and fast prediction. Deep learning approaches specifically have demonstrated the capability for highly accurate RTL power estimation [69], but they require a large amount of training data, where training and inference costs can negate the speed benefits of working at a higher abstraction level. As such, learning formulations that can achieve high accuracy with low complexity are desired.

The effectiveness of supervised learning approaches depends on the selection of features that are uniquely correlated with the values to be predicted as well as appropriate learning models that can capture underlying correlations with low overhead. Deep learning methods have replaced traditional feature engineering with approaches that can learn the feature relevance from an unfiltered union of all possible inputs at the cost of additional training and inference overhead. In case of the power consumption of a circuit, however, it is well known that power of individual combinational and sequential logic blocks is sensitive to switching activity on key input signals. Especially at higher levels of abstraction, instead of costly re-discovering known relationships, we can develop much simpler learning formulations that leverage and encode a priori high-level knowledge. The challenges in power modeling come from the inherently nonlinear relation between activity and power [5] as well as the complexity of larger circuits such complete CPUs consisting of many blocks. To address these challenges, we can combine specialized model decomposition and feature selection techniques with careful selection of the right machine learning models. As such, power modeling provides a case study that highlights the importance of choosing the right approaches when applying machine learning to EDA problems at different abstraction levels.

In [26, 27], we presented an approach that applied a generic and systematic methodology for feature selection and engineering, model decomposition, and model selection to develop learning-based power models for complex CPUs at the micro-architecture level. Our approach applies a hierarchical model composition that assembles a power model for a complete CPU out of individual micro-architecture block models and a separate power model that accounts for glue logic power. Since model complexity grows exponentially with the size of a block and the number of inputs, such a decomposition increases prediction accuracy and speed while allowing for models to be specialized for each block. We further extract features to model different categories of CPU structures using only high-level activity information of each block's inputs available from micro-architectural simulations. Feature selection and engineering are performed based on a categoriza-

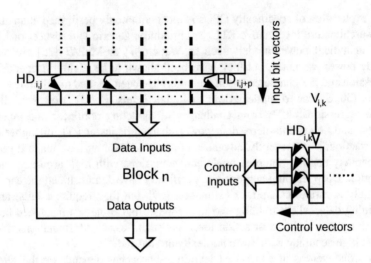

Fig. 18.3 Micro-architecture block modeling and feature selection

tion of common attributes and structures found in typical CPU implementations to efficiently capture their varying power behavior.

Figure 18.3 shows an overview of feature selection and modeling of a micro-architecural CPU block. For each block in a CPU, we develop a model that uses information about data-dependent activity at the block's inputs gathered from a cycle-accurate micro-architecture simulator. We can generally distinguish between data and control signals at block inputs. For datapath blocks, Hamming distance (HD) has been widely used as a feature to capture correlation of power with toggling activity of multi-bit data signals, where larger multi-bit data is decomposed into smaller contiguous bit-groups to account for different bits activating different circuit components. As shown in Fig. 18.3, signal vectors applied to the data inputs of a block are decomposed into bit groups, and Hamming distances $HD_{i,j}$ between bit vectors of group j in cycles i and $i + 1$ are computed and used as features for prediction. To model pipelined blocks, the power consumption depends not only on the current input but also the data stored in internal pipeline registers, which is correlated to the history of the last D inputs $HD_{i,j}, HD_{i-1,j}, \ldots, HD_{i-D-1,j}$, where D is the pipeline depth.

By contrast, control signals determine the mode of operation of different blocks, and both bit-wise values $V_{i,k}$ and Hamming distances $HD_{i,k}$ of the control input k in cycle i can be used to model power variances due to mode switches. Some block inputs possess special characteristics. For example, instruction words at the input of instruction decoders are ideally decomposed at instruction field boundaries, and both bit values and HDs are used to model each sub-field. For buffer blocks, both data and address signals are used to model power variations. To model data and clock gating, corresponding control signals or gated data inputs can be included as features.

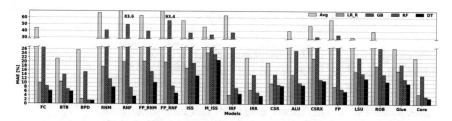

Fig. 18.4 Cross-validation accuracy of BOOM block and whole core models

Finally, a complete CPU model is hierarchically composed from the models of individual blocks and a separate power model for the glue logic that accounts for additional CPU power consumption not captured by the set of blocks and uses the union of all block features as well as external CPU data and control inputs as features. Using selected features, different advanced nonlinear and linear regression formulations can then be explored to learn correlations between features and target power for different micro-architectural blocks with low training overhead and high accuracy. Given cycle-accurate activity information collected from micro-architectural simulations, trained models can provide as fine as cycle-by-cycle power estimates for arbitrary application workloads in a hierarchical fashion from different micro-architecture blocks up to the whole core level.

Figure 18.4 shows the average cycle-by-cycle mean absolute error (MAE) across different explored power models for micro-architecture blocks of a Berkeley Out-of-Order (BOOM) RISC-V core including glue logic as well as a combined model for the whole core. Eight micro-benchmarks from the RISC-V test suite were used for training and tenfold cross-validation. Results show that a model that just predicts average power (*Avg*) performs poorly across the board, demonstrating the importance of capturing data- and control-dependent variations. Among all models, a decision tree (*DT*) performs consistently better than a linear regression (with l2-norm regularization, *LR-R*) as well gradient boosting (*GB*) and random forest (*RF*)-based models of equivalent complexity. As has been observed in other power modeling works, a decision tree-based data representation can efficiently capture the nonlinear but typically discrete power behavior of design blocks, where a single decision tree is capable of better capturing the nonlinear power characteristics of different blocks compared to a forest of shallower (random) trees. Overall, a DT-based power model for the complete BOOM core can predict cycle-accurate power with an MAE of less than 3%. Among different blocks, due to limitations in modeling the queue behaviour of issue and memory issue units, power models of ISS and M_ISS blocks have the highest MAE of 9.05% and 19%, respectively.

Table 18.1 lists the major features selected by the decision tree arranged in ascending order of importance, where "X" denotes the value of signal X in the current cycle, "DEL(X)" denotes the value of signal X in the previous cycle, and "HD(X)" denotes the hamming distance between values in the current and previous cycle of X. The normalized importance of each feature in a decision

Table 18.1 Top decision tree features for different BOOM blocks

Block	Features (Importances)
Fetch controller (FC)	HD(bchecker.io_br_targs_0[7:0]) (0.766)
Branch targ. buf. (BTB)	HD(btb.btb_data_array.RW0_addr[5:0]) (0.57), HD(bim.bim_data_array_1.RW0_addr[8:0]) (0.21), HD(bim.bim_data_array_0.RW0_addr[8:0]) (0.13)
Branch predict. (BPD)	HD(counter_table.d_W0_data_counter[1:0]) (0.87), HD(counter_table.d_W0_data_cfi_idx[1:0]) (0.13)
Decode unit 0 (DEC0)	HD(io_enq_uop_inst[6:0]) (0.8), HD(io_enq_uop_inst[31:25]) (0.18)
Decode unit 1 (DEC1)	HD(io_enq_uop_inst[6:0]) (0.8), HD(io_enq_uop_inst[31:25]) (0.16)
Rename maptbl. (RNM)	DEL(io_ren_br_tags_0_valid) (0.41), io_ren_br_tags_0_valid (0.4)
Rename freelist (RNF)	DEL(io_reqs_0) (0.5), DEL(io_reqs_1) (0.28), DEL(io_ren_br_tags_0_valid) (0.16)
FP maptable (FPM)	DEL(io_ren_br_tags_0_valid) (0.43), io_ren_br_tags_0_valid (0.37)
FP freelist (FPF)	DEL(io_reqs_1) (0.41), DEL(io_ren_br_tags_0_valid) (0.23), DEL(io_ren_br_tags_1_valid) (0.23), DEL(io_reqs_0) (0.12)
Issue unit (ISS)	HD(slots_15.slot_uop_uopc[8:0]) (0.24), HD(slots_15.state[1:0]) (0.19), HD(slots_10.state[1:0]) (0.12)
Mem issue unit (M_ISS)	HD(slots 12.state[1:0]) (0.25), HD(slots 15.state[1:0]) (0.17), HD(slots 11.state[1:0]) (0.12)
Iregister file (IRF)	HD(io_write_ports_0_bits_data[47:40]) (0.78), HD(io_write_ports_0_bits_data[63:56]) (0.11)
Iregister read (IRR)	HD(io rf_read_ports_3_data[7:0]) (0.6), HD(io_bypass_data_0[7:0]) (0.12)
Ctrl./status reg. (CSR)	HD(io_rw_wdata[15:8]) (0.67), HD(io_decode_0_csr[11:0]) (0.15)
Arith./logic unit (ALU)	HD(alu.io_in1[7:0]) (0.72)
CSR exe. unit (CSRX)	HD(alu.io_in2[7:0]) (0.75), HD(alu.io_in2[63:56]) (0.14)
FP Pipeline (FP)	HD(fpu.dfma.fma.io_c[15:8]) (0.83)
Load/store unit (LSU)	HD(io_exe_resp_bits_addr[39:0]) (0.56), io_memresp_valid (0.1)
Reorder buf. (ROB)	io_enq_valids_0 (0.42), HD(io_enq_uops_1_uopc[8:0]) (0.26)

tree-based power model is shown in brackets next to the feature. Such a feature ranking can convey additional information about the power behavior to drive power optimizations.

Figure 18.5 shows a comparison of baseline gate-level power versus power traces predicted by a DT-based model trained on micro-benchmarks running one iteration of the real-world CoreMark benchmark averaged at a granularity of 100 cycles. The zoomed-in subfigure shows cycle-by-cycle power for a region around the peak power cycle. A hierarchically composed core model predicts cycle-by-cycle and average power over the whole execution with 3.6% and 1.2% error, respectively. Furthermore, the model can predict the exact cycle and hence instruction sequence that triggers peak power with 4% error in predicting peak power, which is critical for voltage noise or IR drop analysis but has traditionally not been available at higher abstraction levels.

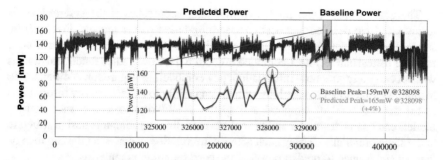

Fig. 18.5 Predicted and baseline BOOM core power traces for CoreMark benchmark

The micro-architecture level power model for the BOOM CPU provides such high-accuracy power estimates when trained with fewer than 30k cycle-level instruction samples. This is in contrast to the 2.2M samples required to train the RTL DNNs in [69]. Furthermore, the hierarchical decomposition of models will require only the models of those blocks that are modified to be re-trained when performing micro-architectural studies. DT model evaluations are fast, performing predictions at a rate of 4 Mcycles/s on average per parallel block. As such, prediction speed is dominated by and almost exclusively depends on micro-architectural simulation times to collect activity information.

18.2.2 Source-Level Prediction

At the system level, co-simulation of different heterogeneous components modeled at RTL or cycle-accurate levels is still too slow to perform rapid design space exploration and optimization. This has driven the need to raise the abstraction level in modeling even further. Virtual platform models capable of simulating whole systems typically incroporate a purely functional source-level modeling of hardware and software behavior [21]. However, the modeling gap between fast, purely functional models for integration into virtual platforms and their corresponding low-level hardware implementations makes accurate PERPT modeling challenging. As regards power modeling, existing approaches require the structure of the design to be modeled and activity information to be collected at the same granularity at which power is estimated, potentially using learning-based methods as described in the previous section. At the system level, this either limits power estimation to coarse-grained mode-based models driven by fast functional C/C++ simulations or relies on slower cycle-based simulations at the intermediate representation (IR) level similar to micro-architectural approaches. Having to simulate all component functionality at the same level of structural detail as activity collection and power estimation is performed fundamentally limits the speed/accuracy trade-off that can be achieved.

The goal of cross-layer predictive modeling approaches at the system level is to instead drive detailed and accurate learning-based power models directly from source-level simulations. This allows functionality to be modeled in source code form for fastest natively host-compiled execution. However, since functional models will not carry any structural information, power modeling requires an additional step to annotate functional simulations with the capability to collect activity information at the desired level of temporal and spatial granularity. Crucially, such activity models can be much simpler and faster than modeling the complete design structure at the same granularity. Using such activity information, a power model can then be learned and trained to predict power at the corresponding granularity.

In the following, we further detail an approach in which we extended source-level functional simulations with the ability to predict power at the cycle, basic block, or whole function invocation level [31–33]. Depending on the observability of hardware internals and their mapping to high-level constructs, we annotate source-level code with activity models to capture data-dependent resource, basic block, or external input and output (I/O) activity in a fully automated flow. Extracted activity data is then used to drive corresponding cycle-, block-, or invocation-level power models, where we synthesize data-dependent, activity-based power models from a given gate-level implementation using machine learning approaches. In the process, specialized model decomposition and feature selection approaches can be applied to improve model accuracy and speed.

Predicting power at the most detailed cycle level requires capturing cycle-accurate activity of datapath resources by back-annotating source code with information about the mapping of source-level operations, such as additions and multiplications into control states and micro-architecture resources. Such mapping information can be either manually provided or automatically extracted from compilers or high-level synthesis (HLS) tools for hardware processors that are software programmed or synthesized by HLS. In practice, we perform such annotation at the IR level to reflect front-end optimizations performed by typical compiler or synthesis tools. Using back-annotated mapping information, IR operand values are traced and re-arranged to collect cycle-by-cycle Hamming distances (HDs) at the inputs and outputs of each resource. This activity is in turn used to drive a power model specific to each control state. Instead of a decomposition into sub-blocks similar to the micro-architecture level (see Sect. 18.2.1), a separate and independent power model for each control state of the hardware is learned. In the process, mapping information is further used to identify and remove signals corresponding to unused resources and thus masked activity in a particular state.

Figure 18.6 shows a hardware micro-architecture and mapped IR operation graph with three resources (MUL0, MUL1, and ADD) executing five operations in three cycles and control states. The power consumption of the complete hardware processor can in general be estimated using a single cycle-level model that uses HD activity for all resources as features. By contrast, a decomposed power model for a given control state can utilize a much smaller subset of signals connecting only to the resources active in the given state. For example, the power consumption of state S3 can be estimated with three signals instead of all nine HD vectors. As such, a

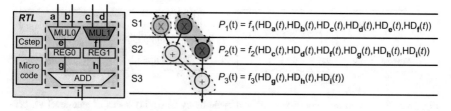

Fig. 18.6   Example of cycle-level power model decomposition

power model decomposition based on structural control state information is able to reduce the complexity of the model with little to no information loss. At the same time, it also allows for state-dependent variations in models that can account for differences in power consumption of resources and other shared logic. In addition, an automatic feature selection can be applied to further specialize each model to the most relevant activity in the corresponding state.

Power modeling at the granularity of complete multi-cycle code blocks can in turn improve prediction speed while maintaining high accuracy compared to cycle-level models. It requires less switching activity collection and fewer invocations of the power model. A basic block-level power model is synthesized by tracing only the HD activity of architectural resources representing data and control signals flowing into and out of each basic source code block. Similar to modeling of pipelined blocks at the micro-architecture level in Sect. 18.2.1, this leverages the fact that hardware-internal switching activity throughout the execution of a multi-stage block is correlated to the history of activity of the block's inputs and outputs. A power model specific to each basic code block is then learned, where each model predicts the average power over the corresponding blocks' execution. To account for control path-specific variations in pipelined block executions, we further extract possible execution paths and build different power models for each possible path through each block.

Finally, an invocation-level power model realizes a black-box approach that does not require any information about the internal hardware architecture and only depends on external I/O activity captured for each function invocation. In turn, it can only predict average power at the granularity of function invocations. Source code is annotated to trace HD activity of transitions on all external data and control ports during function execution. An invocation power model is then synthesized as a decomposed ensemble of specialized models for each cycle of function execution. Decomposed cycle models use the complete transaction activity in an invocation as features, where automatically applied feature selection determines relevant I/O activity for each cycle. Models are further parameterized for different control modes and execution latencies of the function.

Results of applying this flow to generate models for different industrial-strength fixed-function hardware accelerators have shown that source-level power models can predict cycle-, basic block-, and invocation-level power consumption within 90–97% accuracy compared to a commercial gate-level power estimation tool all

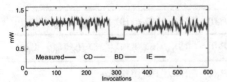 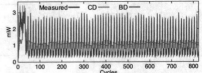

Fig. 18.7 Invocation-by-invocation (left) and cycle-by-cycle (right) traces of gate-level vs. predicted power using cycle-level decomposed (CD), block-level decomposed (BD), and invocation-level ensemble (IE) models for a HDR accelerator

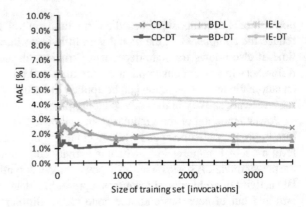

Fig. 18.8 Training curves of cycle-level decomposed (CD), block-level decomposed (BD), and invocation-level ensemble (IE) models for a pipelined DCT accelerator using decision tree (DT) or linear (L) regression

while running at several orders of magnitude faster speeds of 1–10Mcycles/s. Figure 18.7 shows cycle-by-cycle and invocation-by-invocation traces of estimated versus measured power waveforms for a high dynamic range (HDR) weight computation accelerator with 10 basic blocks and 20 control states using 104 traced IR operands, 53 block I/O, or 12 external I/O signals as features. Note that the cycle-level trace of the block-level model shows the averaged power at block granularity. As the profiles show, predictive models can accurately track power behavior within each invocation, as well as data-dependent effects across different invocations of the same design.

Similar to Sect. 18.2.1, decision tree (DT)-based models performed best among different linear and nonlinear regressors that were explored. Figure 18.8 shows the training curves of cycle-, block- and invocation-level models for a DCT accelerator using either a DT-based or a least squares linear regressor. Models with linear regression suffer from overfitting, which again indicates that hardware power behavior is inherently nonlinear. Depending on the granularity and complexity of the model, DT-based models reach stable accuracy after training with less than 3000 and as little as 300 cycle or invocation vectors. For the same training size, models based on more detailed and fine-grain estimation generally show better accuracy than more coarse-grained ones. Combined with opposing trends in estimation speed, this demonstrates a trade-off between accuracy, training efficiency, and speed based on modeling granularity.

18.3 Cross-Platform Prediction

In contrast to cross-layer approaches, which predict across abstraction levels for the same platform or component, cross-platform predictive modeling intuits the existence of a latent relationship between the execution of a program on different platforms. Ongoing work on cross-platform predictive modeling, summarized in this section, has shown that machine learning processes can yield sufficient insight to the nature of hidden correlations between the two scenarios and that machine learning frameworks can automatically construct proxy models to exploit the correlations and accurately predict the performance and/or power consumption of an application target purely from observations on a completely different host. In other words, an application can be analyzed or run on a commercially available machine at fast, native speeds to observe its characteristics and then predict its execution behavior on a completely different platform. Furthermore, to train such models, it is possible to run a few small micro-benchmarks on a much slower simulator or other reference model that characterizes the chosen target processor or core.

Note that cross-platform prediction and cross-layer prediction at the highest source level of abstraction share similar problem statements. In both cases, a source application is executed on a simulation host to predict low-level metrics for a modeled target. However, in contrast to cross-layer prediction, which relies on target-specific activity features collected by annotating the source code, cross-platform prediction executes original (potentially binary) application code and instead performs predictions based on host-specific activity features collected typically using hardware counter support on the host.

Figure 18.9 depicts an overview of the cross-platform prediction concept, which is based on supervised learning comprising a training and a prediction stage. During training, a set of sample programs are profiled on a host to obtain machine-dependent or machine-independent code characteristics; they are also executed on the target platform to obtain measurements of performance and power consumption. Host and target can be two different physical or simulated machines. Host profiling can be done using static analysis or symbolic or native execution. Typically, the host is an existing machine, where feature data is collected using hardware counter support while running an application natively. The prediction target is usually a different physical machine or a simulation model. On the target, execution characteristics such as application performance or power are obtained using the simulator or direct execution on commercially available hardware, such as a development board. The training phase learns the correlation between host features obtained from profiling and execution characteristics obtained from target execution. Specifically, correlation extraction is formulated as a statistical learning problem, which produces the models that are then used by the prediction stage. During prediction, a previously unseen workload is profiled on the host to obtain its features; these features are then input to the model, which produces an estimate of target performance and/or power consumption.

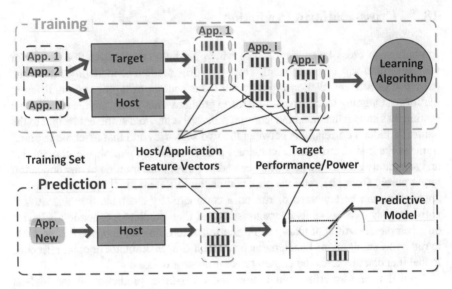

Fig. 18.9 Cross-platform prediction flow

18.3.1 *CPU to CPU*

Some of the earliest work applied the concept of cross-platform prediction to develop approaches for accurate power and performance prediction across CPUs with potentially vastly different micro-architectures and instruction set architectures (ISAs). In [63–66], we presented several such approaches that use hardware performance counters on the host to predict the performance and power consumption of single-threaded programs executing on a single target core both at whole program and program phase granularity (defined as a fixed number of basic blocks).

This cross-platform prediction problem can be formulated as follows: for each sample t, let $x_t \in \mathbb{R}^d$ denote the hardware counter feature vector obtained from host execution and $y_t \in \mathbb{R}$ the corresponding performance (delay) or power measurement or estimate obtained from the target. The predictive model generation process extrapolates a mapping $\mathfrak{F} : \mathbb{R}^d \to \mathbb{R}$ such that for all t, $\mathfrak{F}(x_t) \approx y_t$. A wide variety of models have been proposed in the literature, including but not limited to various forms of linear regression, random forests, and artificial neural networks, among others.

Linear models tend to be easier to generate and interpret but may suffer from high error rates due to the fact that the mapping $\mathfrak{F}$ between sampled features and performance and power consumption of industry-scale applications is inherently nonlinear [38, 63]. In addition to the choice of a nonlinear model, it is also possible to approximate $\mathfrak{F}$ by a piecewise linear relationship specific to each prediction period. We demonstrate the formulation of the cross-platform prediction problem as a *Constrained Locally Sparse Linear Regression (CLSLR)* if $\mathfrak{F}$ is differentiable

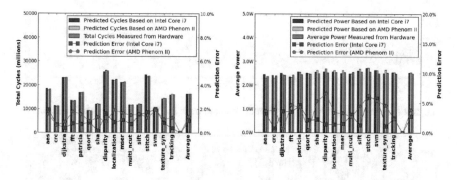

Fig. 18.10 Accuracy of predicting SPEC benchmarks on ARM from executions on x86 hosts

within its domain. A CLSLR applies a local variant of LASSO linear regression at each prediction point.

Given a feature vector $\hat{x}_{\hat{t}}$ at sample $\hat{t}$ to be predicted, let $\{(x_l, y_l), l = 1 \ldots q\}$ be the set of q feature vector and reference performance or power pairs in the training set that are within Euclidean distance $\|\hat{x}_{\hat{t}} - x_l\|_2 \leq \mu$, where μ determines the local neighborhood of interest. Let $X \in \mathbb{R}^{q \times d}$ be the matrix that contains all the x_l^T as its row vectors and $Y \in \mathbb{R}^q$ be the column vector that contains all the y_l as its elements. The model is constructed by solving a quadratic programming problem that minimizes the error between predicted and observed results obtained from the training data; the objective to minimize contains an l_1-regularization term that promotes small and sparse solutions:

$$\underset{w_{\hat{t}}}{\text{minimize}} \quad \frac{1}{2m}\|Xw_{\hat{t}} - Y\|_2^2 + v\|w_{\hat{t}}\|_1 \quad \text{subject to} \quad w_{\hat{t}} \geq 0. \tag{18.1}$$

The solution $w_{\hat{t}}$ is then used for prediction as the parameter for the local linear approximation $\widehat{\mathfrak{F}}$ at $\hat{x}_{\hat{t}}$ (i.e, $\mathfrak{F}(\hat{x}_{\hat{t}}) \approx \widehat{\mathfrak{F}}(\hat{x}_{\hat{t}}) = w_{\hat{t}}^T \hat{x}_{\hat{t}}$). Typically, this optimization problem does not have an analytical solution but can be computed via gradient descent methods. Tuning parameters μ and v can be determined empirically using standard cross-validation techniques.

Results of applying this approach to prediction of ARM behavior from executions on two different x86 hosts showed that time-varying power and performance traces can be predicted at fine-grain temporal granularity with more than 95% accuracy (Figs. 18.10, 18.11, and 18.12) [64]. Similar results were obtained in predicting the performance and power consumption when predicting between the two different x86 host variants [65]. In all cases, predictors were trained on small micro-benchmarks that each require less than a second to execute. Table 18.2 shows the hardware counters that were collected as features for predictions on each of the hosts. Note that features differ due to the underlying differences in hardware counter availability on the two hosts.

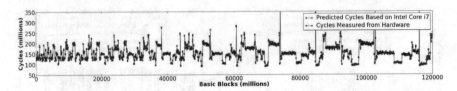

Fig. 18.11 Fine-grain prediction of ARM cycle count trace for SPEC2006 `dealII` benchmark

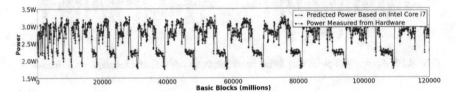

Fig. 18.12 Fine-grain prediction of ARM power trace for SPEC2006 `dealII` benchmark

Table 18.2 Hardware performance counter features profiled on the hosts

Intel Core i7-920	AMD Phenom II X6
L1 Total Cache Misses	L1 Total Cache Misses
L2 Total Cache Misses	L2 Total Cache Misses
L3 Total Cache Misses	Branch Misses
TLB Loads	Instructions
Unconditional Branches	Cycle Stalled
Conditional Branches	Cycles
Branch Misses	L1 Total Cache Accesses
Instructions	Floating Point Operations
Cycle Stalled	
Cycles	
L1 Total Cache Accesses	
L2 Total Cache Accesses	
L3 Total Cache Accesses	
Floating Point Operations	

A basic cross-platform prediction approach requires expensive code instrumentation to collect counters on the host at regular intervals, which incurs runtime overhead and restricts prediction to applications for which source code is available. Instead, a source-oblivious, binary-only, sampling-based approach [66] can support fully autonomous background prediction based on timer interrupts, similar to sampling-based profilers, such as gprof. This allows predictions for arbitrary application, library, or runtime system code with reduced instrumentation overhead. The main challenge, however, is to align samples obtained during training on hosts and targets that progress at different rates. This alignment problem can be solved by interpreting numerical measurements as stochastic signals and using a dynamic coupling heuristic to align them such that their cross-covariance is maximized [66]. The resulting sampling-based prediction models can run at speeds of over 3GIPS,

Fig. 18.13 Sampling-based cross-platform prediction accuracy vs. speed

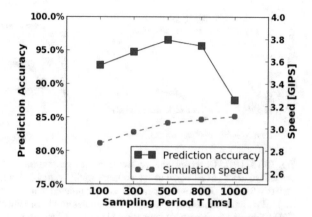

approximately 6x faster than instrumentation. Due to the speed difference between x86 and ARM CPUs, this allows predicting ARM performance at twice the speed of running natively on ARM hardware. Given host limitations on simultaneously collectable hardware counters in a single run, the sampling-based model only uses six counters as features (instructions, cycles, total cache misses and references, and total branches and branch misses).

The granularity at which prediction is performed determines a trade-off between accuracy and speed (Fig. 18.13). Accuracy initially improves with decreasing granularity due to more training data and program phases becoming more similar and thus easier to predict until alignment errors start dominating at very fine granularities ($T < 500$ ms) [65]. At the same time, prediction also becomes slower with smaller granularity, requiring an optimization problem to be constructed and solved at each prediction point. This incurs significant overhead, which may require further investigation of rapidly converging learning methods that sacrifice accuracy for efficiency in system deployment scenarios where prediction speed is a major concern.

18.3.2 GPU to GPU

Cross-platform predictive models using machine learning and hardware performance counter measurements have similarly been constructed for integrated graphics accelerators. In [44], we presented such an approach for Hardware-Assisted Light Weight Performance Estimation (HALWPE) of GPUs. HALWPE runs on commodity graphics accelerator hardware and predicts the performance of a simulator configured to model next-generation graphics accelerators under industrial development, where replacing slow-running simulations with predictive models can significantly reduce the cost of early-stage design space exploration. HALWPE employs a modeling framework consisting of 12 linear and one nonlinear random

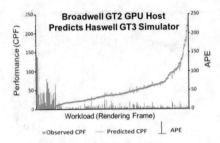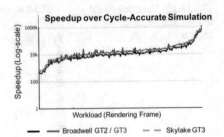

Fig. 18.14 Left: Observed and predicted cycles per frame (CPF) and per-workload absolute percentage error (APE) when using a commercially available Broadwell GT2 GPU to predict the CPF of a Skylake GT3 simulator. Workloads are ordered in non-decreasing order of observed CPF. Right: Predictive modeling speedups over simulators for Broadwell GT2/GT3 and Skylake GT3 GPUs. Frames are ordered by increasing Skylake GT3 speedup

forest (RF) regression models. For any host-target prediction scenario, all 13 models are trained, and results for the best-performing model, i.e., the model that minimizes out-of-sample error, are reported.

HALWPE collects hardware performance counter and DirectX program metrics by running workloads on an Intel Haswell GT2 host, which features 20 Execution Units (EUs). To evaluate HALWPE, models were then trained to predict the performance of an Intel-internal GPU simulator configured as Broadwell GT2 (24 EUs), Broadwell GT3 (48 EUs), and Skylake GT3 (48 EUs).

HALWPE achieved out-of-sample errors of 7.47%, 7.47%, and 10.28% when predicting the performance of a Haswell G2, Haswell GT3, and Broadwell GT3 GPU, respectively, well-within Intel's target. For example, Fig. 18.14(left) reports the observed and predicted CPF (cycles per frame) and Absolute Percentage Error (APE) for the Broadwell GT3 case, across a variety of rendering workloads (taken from commercial 3D games). The average speedups compared to Intel's internal cycle-accurate simulator were 29,481x for Broadwell GT2, 43,643x for Broadwell GT3, and 44,214x for Skylake GT3, as shown in Fig. 18.14(right).

HALWPE established the feasibility of using commercially available hardware to predict the performance of next-generation products presently under development at the pre-silicon stage. HALWPE's models can then be used for early-stage design space exploration, allowing the evaluation of many more workloads for each explored point in the design space than would be feasible using Intel's internal cycle-accurate simulator. Along similar lines, obtaining performance metrics from a coarse-grained functional simulator can predict the performance of a detailed cycle-accurate simulator to better understand the implications for changes to the software stack [42]. In this case, RF was the best-performing model, achieving an out-of-sample error of 14% while running 327x faster, on average, than Intel's simulator. Figure 18.15(left) reports the predicted cumulative probability function (CPF) and average performance errors (APEs) observed for each workload, while Fig. 18.15(right) reports the runtime of the two simulators and the speedups achieved.

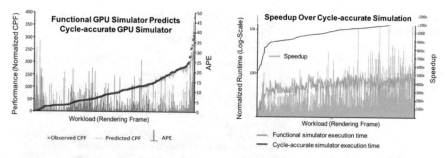

Fig. 18.15 Left: Observed and predicted CPF and per-workload APE when using a functional GPU simulator to predict the CPF of a cycle-accurate GPU simulator. Workloads are ordered from left to right in non-decreasing order of observed CPF. Right: Speedup of functional simulation + predictive modeling over cycle-accurate simulation. Frames are ordered by increasing cycle-accurate simulation time

Haswell GT2 Host Predicts Broadwell GT2 GPU Simulator Configuration			Haswell GT2 Host Predicts Broadwell GT3 GPU Simulator Configuration		
Feature Name (Description)	Feature Category	P-value	Feature Name (Description)	Feature Category	P-value
EU Busy Time (Number of cycles EUs actively execute instructions)	Hardware Counter	1.78 E-125	**Device Context Clear Unordered Access View Float** (Cycles spent accessing memory via a resource)	DirectX Metric	4.95 E-172
EU Stall Time (Number of cycles EUs are inactive)	Hardware Counter	2.35 E-83	**GPU Busy** (Number of cycles the render engine was not idle)	Hardware Counter	2.48 E-91
Domain Shader Thread Busy Time (Number of cycles Domain Shader was active)	Hardware Counter	2.02 E-49	**Draw Indexed Primitive** (Cycles spent rendering a primitive into a vertex array)	DirectX Metric	7.62 E-46
Compute Shader EU Stall Time (Number of cycles Compute Shader stalled)	Hardware Counter	2.75 E-49	**Compute Shader Thread Busy Time** (Cycles the Compute Shader was active)	Hardware Counter	1.25 E-31
Domain Shader EU Stall Time (Number of cycles Domain Shader stalled)	Hardware Counter	6.22 E-39	**Geometry Shader Thread Busy Time** (Cycles Geometry Shader was active)	Hardware Counter	2.68 E-29
Device9 Draw Primitive Up (Number of pointer-referenced primitives rendered)	DirectX Metric	4.26 E-21	**Device9 Draw Primitive Up** (Number of pointer-referenced primitives rendered)	DirectX Metric	8.21 E-23
Query9 Occlusion Query Get Data (Number of queries for occlusion data)	DirectX Metric	3.04 E-18	**Device Context Draw Instance** (Number of non-indexed, instanced primitives drawn)	DirectX Metric	9.17 E-19
Device Context Draw Instance (Number of non-indexed, instanced primitives drawn)	DirectX Metric	1.72 E-16	**Domain Shader EU Stall Time** (Number of cycles Domain Shader stalled)	Hardware Counter	1.57 E-18
OMZ Test Fail (Number of z-test fails after Pixel Shader execution)	Hardware Counter	9.81 E-16	**Query9 Occlusion Query Get Data** (Number of queries for occlusion data)	DirectX Metric	1.79 E-18
Device Context Execute Command List (Number of queue commands issued onto a device)	DirectX Metric	4.56 E-15	**Domain Shader Thread Busy Time** (Number of cycles Domain Shader was active)	Hardware Counter	3.68 E-18

Fig. 18.16 Features ranked by p-values for the best-performing linear regression models that use a Haswell GT2 host to predict a GPU simulator configured as a Broadwell GT2 (Left) and Broadwell GT3 (Right)

HALWPE also demonstrated that machine learning-based prediction models have potentially high *interpretability*, allowing features to be ranked based on how much they influence the model. For linear regression models, feature ranking can be computed using p-values; for random forests (RF), feature ranking is determined by residual sum-of-squares (RSS) error. High-ranking features can suggest potential micro-architecture performance bottlenecks, which can guide architectural design space exploration. Figure 18.16 ranks the top 10 most influential features targeting the Broadwell GT2 and GT3. On the left, the top 5 highest ranked features for the Broadwell GT2 provide information about execution unit (EU) activity relating to front-end render pipeline units, noting that the limited parallelism in the GT2 (24 EUs) can impede performance. The 3rd through 5th highest ranking features report compute and domain shader EU activity, which suggests that interactions between the two shaders and their EU occupations should be analyzed. Notably, the

Fig. 18.17 Features ranked by RSS error (RF model) when using a functional GPU simulator to predict the performance of a cycle-accurate GPU simulator

Feature Name (Description)	Feature Category	RSS Ranking
Pixel Written (The number of pixels written to render target)	Pixel Backend	20.39
Sub-spans Written (The number of Sub-spans written to render target)	Pixel Backend	19.73
Samples Written (Number of samples written to render target)	Pixel Backend	19.5
Passed Early Z Test (The number of passed Early Z tests in output merger stage of render pipeline)	Vertex Testing	15.73
Stencil Tested Subspans (The number of stencils tests performed on sub-spans)	Vertex Testing	14.61
Early Z Tests (The total number of early Z tests performed)	Vertex Testing	14.50
Passed Early Stencil Test (The number of early stencil tests that passed in output merger stage)	Vertex Testing	13.17
Early Stencil Tests (The total number of early stencil tests performed)	Vertex Testing	12.55
Passed Depth Tests (The Number of pixels passing depth tests)	Pixel Testing	12.01
Sub-span Z Tests (The number of Z tests performed on sub-spans)	Vertex Testing	11.87

presence of *Compute Shader EU Stall Time* suggests that limiting the amount of time that the compute shader stalls the EU array could potentially improve overall GPU throughput. On the right, EU busy/stall cycles no longer fall within the top 10 ranked counters, which reflect the increase in parallelism provided by GT3 GPUs (48 EUs). The two counters representing the domain shader are ranked 8th and 10th, suggesting that they still influence performance but that other subsystems with higher-ranking features have emerged as bottlenecks in the presence of greater EU parallelism. Again, the compute shader is within the top 5 ranked features, which suggests that the process by which the compute shader stalls the EU array still influences performance significantly. Similarly, the geometry shader now appears in the top 10 highest ranked features, which suggests that additional front-end units start to influence performance as EU parallelism increases. This points to possible GPU micro-architecture optimization opportunities.

Figure 18.17 similarly reports the ten highest performance counters, ranked in decreasing order of RSS error, for a RF model [42] that predicts performance from a functional to cycle-accurate GPU simulator. The most important performance counters for this prediction scenario were chosen from all portions of the render pipeline, and many of them measure the number of vertices passing front-end depth and stencil tests. This indicates that corresponding subsystems have the greatest impact on GPU performance and could be slated for further study and optimization by architects.

18.3.3   CPU to GPU

In addition to CPU-to-CPU and GPU-to-GPU prediction, several approaches have attempted prediction across both classes of devices. XAPP [3] predicts the performance that would result from porting a C/C++ workload to CUDA and running it on a GPU. XAPP employs instrumented program binaries to collect features that correlate to GPU execution behavior. These features are then used to train a machine learning model to predict GPU performance; however, XAPP's accuracy was rather low, ranging from 64% to 73% for different GPU targets. A follow-up project, Static XAPP [4], collects features using the compiler's intermediate representation and predicts whether or not the speedup will be above or below a user-specified threshold.

CGPredict [56] solves a similar problem to XAPP but employs an analytical model, rather than machine learning. Analytical models are generally easier to interpret compared to machine learning models and are better suited to help with performance tuning or architectural bottleneck analysis. The limitation of analytical models is that they must be designed and validated by architecture experts and must be redesigned or recalibrated for each new target.

18.3.4   CPU to FPGA

The starting point for many application acceleration projects is reference code, often single-threaded, written in a high-level language such as C++. A first step in what may be a much longer application accelerator design project is to port the application to run on an FPGA. With the emergence and maturation of High-Level Synthesis (HLS) tools in recent years [9, 16], the engineering cost of doing so has been reduced in comparison to earlier practice, which was to rewrite the application at the Register Transfer Level (RTL). Nonetheless, the application developer must still appropriately set design pragmas to control program and data transformations such as pipelining, array partitioning, and loop unrolling, to optimize performance and throughput. Finding the right HLS parameter settings is a complex problem and entails some mechanism to evaluate the most promising design points that have been uncovered. Although synthesizing, placing, and routing each design point and then characterizing its performance by direct execution would be ideal, the amount of time required to do so is untenable.

Before embarking on such an arduous task, a typical designer would like to be able to quantify, as accurately as possible, the gains in performance and/or power consumption that would be accrued from the ultimate engineering effort. If the anticipated gains are insufficient, then pursuing acceleration may not be prudent use of engineering resources. Cross-platform predictive modeling from the sequential CPU reference code onto an FPGA, while accounting for the impact of source code transformation, could help guide the designer toward making prudent engineering decisions at early stages of design.

Several existing approaches employ predictive modeling to cover regions of the HLS design space without needing to explicitly explore them. MPSeeker [68] employs an analytical performance model [67] to perform design space exploration using high-level synthesis (HLS) simulation in conjunction with instrumented C/C++ source code. The objective of this work is to synergistically explore fine- and coarse-grained parallelism without repeatedly invoking slow-running HLS tools to obtain empirical results. Eliminating the need to repeatedly invoke HLS tools reduces the runtime of design space exploration from hours/days to minutes. Subsequent work has shown that transfer learning can reduce the number of HLS runs required to train predictive models without compromising quality of results [28].

A more fundamental question is whether performance metrics obtained from running an application on a CPU can provide insight into that application's performance when ported to an FPGA. HLScope [13, 14] uses a combination of representative HLS synthesis reports in conjunction with analytical performance models to perform CPU to FPGA performance prediction. By contrast, in HLSPredict [43], we obtain performance counter measurements from direct execution on a CPU and feed them into machine learning models that predict FPGA performance and power consumption. Similar to MPSeeker, all these methods run orders of magnitude faster than generating FPGA bitstreams to measure performance directly.

Here, we describe HLSPredict in greater detail. HLSPredict creates models for latency (cycle count) and power consumption (static, dynamic, and total) targeting Xilinx FPGAas with and without optimization obtained by using the loop unrolling, array partitioning, and pipelining directives in Vivado HLS; optimization yielded an average speedup of 14.49x. HLSPredict ensures time synchronicity between the host CPU and the target FPGA accelerator by partitioning application execution into subtraces (epochs of workload execution time) for model training; each subtrace entails performance counter measurements obtained from the host CPU as well as FPGA cycle counts for an identical sequence of assembly instructions (CPU) and operations (FPGA). Subtraces are not needed for power prediction, as power-per-epoch is not a meaningful measure; full workload execution suffices to predict power.

The first step is to instrument workloads to produce subtraces. Instrumented host workloads are executed on the host CPU, and performance counter measurements are collected. The next step is to synthesize each workload on the FPGA using HLS and execute it directly to obtain performance and power measurements for each subtrace. Once subtrace data is collected, a model is trained to predict FPGA cycle counts consumption as depicted in Fig. 18.9. Once a model is trained, the user can then execute previously unseen workloads on the host CPU to obtain performance counter readings and then apply the model to rapidly predict pre-RTL estimates of FPGA accelerator cycle counts.

An architectural template specifically designed for the HLSPredict study was developed to allow collection of FPGA performance and power metrics on the

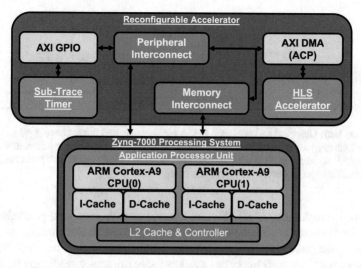

Fig. 18.18 HLSPredict architectural template: each workload encompasses a unique HLS accelerator and its internal trace generation logic

granularity of subtraces (Fig. 18.18). The template uses the Advanced eXtensible Interface (AXI) stream protocol for communication, Direct Memory Access (DMA) for memory transactions, and the DMA's Accelerator Coherency Port (ACP) to interface with the CPU's cache hierarchy. Each workload is specified in synthesizable C and verified via C/RTL co-simulation. This template represents a simple and effective System-on-a-Chip (SoC) and assumes that subtrace input sizes are small enough to fit into the FPGA's internal BRAMs.

Subtraces for CPU workloads consist of performance counter measurements for each epoch; for FPGAs they consist of a one-time reading of the FPGA cycle count, which resets when the next subtrace initiates. On the CPU side, subtrace execution times are dominated by API initialization, file I/O overhead, and repeated workload executions that are necessary to collect all counter measurements, yielding a slowdown of more than 12,000x compared to direct execution of the subtrace once without instrumentation; this is still markedly faster than FPGA synthesis times. Subtraces for FPGA workloads encompass active compute time exclusively; they do not account for FPGA streaming, communication, or data initialization time.

Figure 18.19 reports the observed and predicted FPGA execution times along with per-subtrace Absolute Percentage Error (APE) without (Left) and with (Right) optimization. The out-of-sample error when targeting unoptimized FPGA accelerators was 9.08%, noting that the subtraces with the largest errors tend to be those with the smallest FPGA cycle count, where even small absolute deviates can lead to larger APEs. The out-of-sample error when targeting optimized FPGA accelerators was slightly higher, 9.79%; once again, most of the outliers were present in the smallest subtraces with one notable exception in the middle of Fig. 18.19(Right).

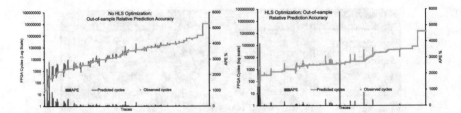

Fig. 18.19 Left: Observed and predicted FPGA cycle counts and per subtrace APE when using HLS without pragma optimization. Right: Observed and predicted FPGA cycle counts and per subtrace APE when using HLS with pragma optimization. In both charts, subtraces are listed in non-decreasing order of observed FPGA cycle count

The out-of-sample error of HLSPredict's power consumption total predictions was 7.84% for the unoptimized and 4.48% for the optimized HLS scenarios.

HLSPredict was also able to rank the importance of features used for each prediction task. The most important features for performance prediction for unoptimized HLS workloads track micro-op execution in the CPU front-end which decode program code into operations such as register arithmetic and data transfer between CPU busses. The most important features for performance prediction of pragma-optimized HLS workloads focused on stall cycles that occurred during the CPU pipeline's requirement stage and communication between the L2 and L3 caches, as well as micro-op execution. The most important features for power prediction for unoptimized HLS workloads correspond to cache hierarhcy performance; for example, the highest ranked feature counts CPU stalls due to cache hierarchy traffic. The model that predicted power for optimized HLS workloads was so simple that only two performance counters were needed: the duration of a page table walk which occurs in response to a TLB miss and the number of cycles required to recover from tasks such as memory disambiguation and SSE exceptions. From these results, we can see that optimizing workloads via HLS directives simplifies prediction but also changes which performance counters are the most impactful features.

18.3.5 FPGA to FPGA

FPGA-to-FPGA prediction is emerging as a new direction for predictive modeling in system-level design. Here, the starting point is an application deployed on a relatively low-cost, low-capacity FPGA. The objective is to predict improvements in performance that could be achieved by porting the design to a larger device in a similar family, which may often entail a substantial reengineering effort.

XPPE [36], for example, is based on the premise that resource utilization correlates to performance when porting to a target FPGA with radically different capacity than the host. Rather than utilizing performance metrics as features, XPPE employs the HLS utilization report, the number of available resources on the target FPGA, and application properties. XPPE is particularly useful when the performance of data-intensive kernels is bandwidth-limited, rather than compute-limited. Recent work [52] has suggested that transfer learning can be applied in this domain and could reduce the amount of data needed to train new cross-platform models.

18.4 Cross-Temporal Prediction

The third axis in the three-dimensional SLM space is time, which accounts for the dynamic behaviors of program applications throughout their execution. This section presents models that learn workload behaviors over time and anticipate workload changes. The primary purpose of these models is to aid in casting proactive runtime management actions such as frequency scaling and task migrations. While much prior work in runtime models has studied the impact of management actions, e.g., [18, 48], such models fail to acknowledge that workload behaviors may change in the upcoming future. In other words, they are not making predictions across the time axis but runtime decisions by reacting to changes. By contrast, the rest of this section focuses on proactive runtime modeling, which is aimed at forecasting workload behaviors across time.

Due to the presence of loops in program applications, their execution history is a good estimator for future workload behaviors. Thus, proactive runtime models learn from the past to predict the future. They look at an execution history of features that characterize workload behaviors—for instance, instruction streams, hardware counters, memory accesses, and performance metrics. We observe two general flavors in proactive runtime models. The first, *workload prediction*, explicitly predicts workload descriptors that a runtime management system will then use for reconfiguration. For example, M. Moghaddam et al. [40] proposed an LSTM model which predicts the CPI and number of instructions for the upcoming execution interval, and they are then used by a separate entity to select an optimized voltage-frequency (VF) state. The second flavor, *action prediction*, embeds the runtime decision into the model with the objective to predict optimized actions for a given policy. For example, R. Jain et al. [22] used reinforcement learning to predict the best cache configuration and a number of active cores. Action prediction models require an objective function that optimizes a specific policy, e.g., minimizing energy-delay products (EDP) or maximizing performance under power constraints. Therefore, these models need separate training for each policy. On the other hand, workload prediction models learn and predict properties that are independent of the target policy, allowing the reuse of the same model across multiple policies. We focus the rest of this section on workload prediction.

Workload prediction can be cast as a time series forecasting problem, where regression models are a typical solution. These models excel at predicting short-term upcoming workload behaviors; however, they struggle to predict abrupt long-term workload changes [2]. Such changes occur because workloads go through phases,

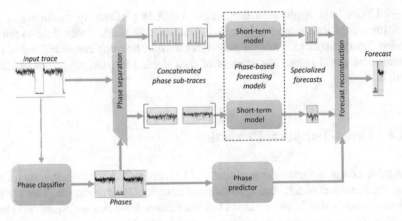

Fig. 18.20 Example of phase-aware forecasting

where phase classification and prediction solutions have been studied by prior work. Predicting phases is a cross-temporal modeling approach that characterizes and forecasts long-term workload behaviors; however, it overlooks sample-by-sample variability. In our recent work [2], we proposed a phase-aware mechanism that combines long-term and short-term predictions to provide more accurate forecasting results. Figure 18.20 depicts the intuition behind our approach. The left side of the figure shows a snippet of a workload going through two different phases, highlighted as red and green at the output of the phase classifier. This approach partitions the trace based on phases, concatenates the subtraces, and trains one short-term prediction model per phase. Therefore, the forecasts belonging to a phase are only dependent on the history of that phase, and phase-specific predictors are specialized to a single phase to increase accuracy. Finally, with knowledge of the future phase behavior from the phase predictor (i.e., long-term predictor), phase-specific forecasts are concatenated and assembled to reconstruct the forecast for the overall time series. Our work showed the potential of this approach by evaluating an oracle long-term predictor. In the following subsections, we survey and discuss different techniques utilized for short-term and long-term workload forecasting as well as present selected results.

18.4.1 *Short-Term Workload Forecasting*

Short-term forecasting models expect to find workload changes on a sample-by-sample basis. As such, new predictions are made for each new sample. Early work investigated simple methods such as exponential averaging and history tables [20]. Later studies proposed more advanced approaches, ranging from linear predictors [50] to, more recently, recurrent neural networks (RNNs) [37]. Their objective is to minimize the forecasting error of periodically measured workload metrics, such

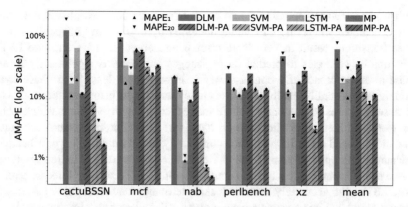

Fig. 18.21 Accuracy of basic versus phase-aware (PA) short-term workload forecasting models

as CPI [2, 40], cache miss rates [20, 37, 50, 60], temperature [15, 17, 24, 47, 60], power consumption [37, 49, 60], video frames [19], etc.

Formally, runtime workload behavior forecasting is a time series forecasting problem, which motivated multiple studies to research the application of traditional auto-regressive models to runtime workload forecasting [17, 19, 50]. However, such models assume linearity in the workload patterns and are limited in their input history lengths [49]. Instead, this modeling task has been addressed as a supervised learning problem where advanced ML techniques have proven to be very accurate [2, 37, 47, 49]. The models can be formalized as follows:

$$(\hat{y}_{t+1}, ..., \hat{y}_{t+k}) = m_w\big((U_{t-h+1}, ..., U_t)\big), \tag{18.2}$$

where the input length is h, U is a multivariate time series, y is the variable of interest, k is the forecast horizon, $\hat{y_{t+i}}$ represents a prediction of y_{t+i} made at time t, and m_w is the forecasting model and its set of trainable parameters w. Note that for the phase-aware approach depicted in Fig. 18.20, the set of trainable parameters w is different for each phase-specific forecasting model.

Basic forecasting techniques show low prediction accuracy for workloads that exhibit distinct long-term phase behavior, even when those phases repeat over time [2]. This is shown in Fig. 18.21 where we compare two learning-based models, a support vector machine (SVM) and a long-short term memory (LSTM) RNN, and two time series analysis techniques, a traditional auto-regressive dynamic linear model (DLM) and a recent matrix profile (MP) approach, using a basic and a phase-aware (PA) setup for different SPEC 2017 benchmarks [2]. When comparing basic models with phase-aware approaches, results show that using phase-specific models consistently decreases the forecasting error of all techniques for all benchmarks except *perlbench*. The workload traces of *perlbench* do not go through phases, and there is no room for improvement for phase-aware forecasting. We observe that *cactuBSSN* exhibits the most impact in forecasting error reduction with a phase-aware approach. Some of its transitions between phases have very abrupt changes where the CPI value increases by 500%. Any mispredictions of these transitions result in very large error penalization. A phase-unaware LSTM in particular struggles to predict those changes and benefits significantly from a phase-aware approach. By contrast, *mcf* is impacted the least from phase-aware models and generally exhibits poor accuracy and larger error variations. This is because its phases continuously change and reduce in length over time, which makes the workload hard to predict overall. Note that matrix profile cannot accurately predict this trend since its predictions are purely based on recalling past behavior unchanged. As opposed to *mcf*, the phases of *nab* have repetitive uniform patterns, where phase-aware models have a significant impact in decreasing forecast error. A basic LSTM can accurately learn both short-term and long-term phase patterns for this workload, but its phase-aware counterpart still had room for improvement.

For the phase-aware results presented in Fig.18.21, we used an oracle phase classifier and predictor to model long-term phase behavior. In the following subsection, we introduce phase classification and prediction as a long-term workload forecasting problem and describe different techniques used in selected previous works.

18.4.2 Long-Term Workload Forecasting

Long-term workload forecasting models focus on coarse-grained patterns and overlook short-term workload variations. These long-term patterns are known as phases. In general, a phase is defined as a group of execution intervals with similar behaviors. Phases typically show repetitive patterns, and predicting them requires two main tasks: (1) identifying the phases, i.e., phase classification, and (2) learning their patterns to anticipate changes, i.e., phase prediction.

Phase classification requires selecting a set of inputs that can characterize phases and an unsupervised learning model that clusters similar inputs together. The input selection in some studies aims at finding similar blocks of code, for instance, based on program counters (PCs) [11, 29, 51, 61] or instruction types [23], while others look at performance metrics, e.g., hardware counters [1, 15, 24, 25, 55]. The collection of PC can be either general or focused on branches. Additionally, it can be dense or sparse. For example, [61] is a dense method focused on branches because it collects the PC of all branches, while [51] is sparse because it samples the PC every certain fixed number of branches. A dense collection typically requires augmenting the hardware or instrumenting the source code. Sparse implementations can be less intrusive, but they rely on the availability of precise event-based sampling [51]. Collecting hardware counters is a more transparent approach, which also incurs less overhead [55]. A feature selection process is necessary since hundreds of hardware counters are available in modern processors. While some studies rely on domain knowledge [1, 25], others use dimensionality reduction [15, 24] or an ML-based feature selection approach [55].

Classification studies have proposed both offline and online model training. When the phases are defined offline, there is an implicit assumption that phases will not change during runtime and that there is data available at design time for training. Examples of offline models are k-means [15, 24, 25] and Gaussian mixture models clustering (GMM) [12]. By contrast, online methods can adapt to different workloads and learn new phases at runtime, where the leader-follower (LF) algorithm [23, 51, 55, 61] has been very popular for phase classification. It holds a table of known phases that is updated as new samples arrive based on a definition of distance and a threshold. While it does not require knowing the number of phases a priori, the definition of phases strongly depends on the order in which samples arrive and the given distance threshold. Additionally, finding an optimized threshold may require offline exploration.

Recent studies have acknowledged the hierarchical nature of phases. In [61], two independent LF classifiers are used with different granularities to define fine and coarse phases simultaneously. In [25], k-means is used to generate one sub-phase per sample, and the sub-phases are then grouped into fixed-size windows to define phases with a second instance of k-means. While fixed-size windows lose precision when detecting phase changes, they generate more stable phases.

Classified phases are used by phase prediction methods to learn patterns between them and foretell future phase behavior. Early work proposed predicting the phase ID of the following sample interval. Later studies realized that many phases last for multiple consecutive intervals and proposed other prediction strategies. We refer to these strategies as *window-based prediction* and *phase change prediction*. *Window-based prediction* aims to accurately determine the phase ID of the immediately upcoming sample window of size k. The inputs to these predictors are typically pre-classified data, and the outputs are phase IDs [1, 12]. Therefore, the models are expected to find the relationship between the raw execution history and the upcoming phases. *Phase change predictions* use run-length encoding to compress phase information and predict both the duration of a phase and the ID of the next phase, where early work proposed using table-based predictors such as global history table (GHT) for this purpose [11, 29, 61]. Predicting the exact duration of a phase is a challenging task. Chang et al. [11] approximated the duration of phases into ranges, limiting the precision. More recent work has treated this as a regression problem [1, 53].

To compare combinations of classifiers and predictors, we introduced error-frequency product (EFP) [1] as a metric that penalizes both prediction inaccuracies and classification overfitting, which leads to very short phases with high phase transition frequencies:

$$EFP = 100 \cdot \frac{\left| \frac{1}{N} \sum_{t=1}^{N} (y_t - \hat{y}_t) \right|}{\bar{d}_j}, \tag{18.3}$$

where N is the number of predicted samples and 100 is a scale factor. We select CPI as our variable of interest, y. To determine the error, we use a vector $(v_1, ..., v_c)$,

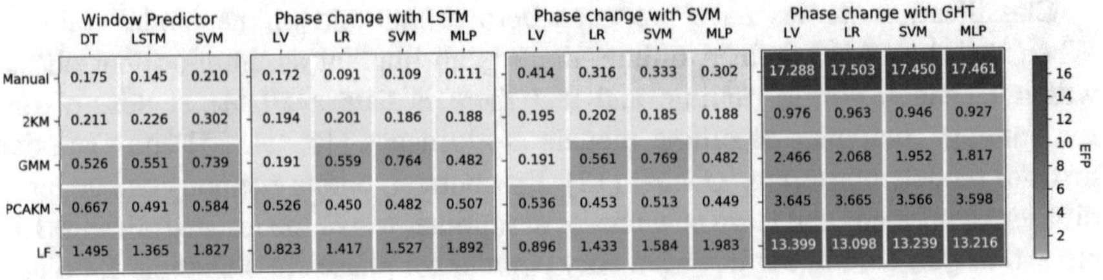

	Window Predictor			Phase change with LSTM				Phase change with SVM				Phase change with GHT			
	DT	LSTM	SVM	LV	LR	SVM	MLP	LV	LR	SVM	MLP	LV	LR	SVM	MLP
Manual	0.175	0.145	0.210	0.172	0.091	0.109	0.111	0.414	0.316	0.333	0.302	17.288	17.503	17.450	17.461
2KM	0.211	0.226	0.302	0.194	0.201	0.186	0.188	0.195	0.202	0.185	0.188	0.976	0.963	0.946	0.927
GMM	0.526	0.551	0.739	0.191	0.559	0.764	0.482	0.191	0.561	0.769	0.482	2.466	2.068	1.952	1.817
PCAKM	0.667	0.491	0.584	0.526	0.450	0.482	0.507	0.536	0.453	0.513	0.449	3.645	3.665	3.566	3.598
LF	1.495	1.365	1.827	0.823	1.417	1.527	1.892	0.896	1.433	1.584	1.983	13.399	13.098	13.239	13.216

Fig. 18.22 Error-frequency product (EFP) of all classifier and predictor combinations

where each element v_i represents the average CPI that characterizes phase i and c is the corresponding total number of phases. At each timestep t, we transform the prediction of phase $\hat{\alpha}_t$ to a prediction $\hat{y}_t = v_{\hat{\alpha}_t}$ and measure the error $y_t - \hat{y}_t$. We normalize average errors by the average phase duration $\bar{d}_j$. We define EFP as the product of the average error and the average transition frequency $1/\bar{d}_j$.

The average EFP results across several benchmarks are shown as a heatmap in Fig. 18.22 [1], where each row is a phase classifier and each column a phase predictor. It includes one oracle classifier (*Manual*), one online leader-follower classifier (*LF*), a two-level k-means classifier (*2KM*), and two offline classifiers, *GMM* and *PCAKM*, where *PCAKM* applies PCA for dimensionality reduction to a k-means input. Our evaluation of phase predictors includes both window-based prediction and phase change prediction. For window predictors, the columns in the heatmap correspond to the different ML models. We evaluate and compare decision trees (DTs) against LSTMs and SVMs. For phase change predictors, we show multiple combinations of duration and next phase prediction. The individual column labels correspond to the duration predictor, and the shared titles correspond to the next phase predictor. We evaluate and compare a last-value (LV), linear regression (LR), SVM, and LSTM duration predictors and three phase change predictors, LSTM, SVM, and GHT.

Without considering the *manual* classifier, the best combination of classification and prediction that reduces the EFP is *2KM* with an SVM for next phase and duration prediction. *2KM* resulted as the best classifier for 13 out of 15 predictors. The most remarkable difference between *2KM* and the other classifiers is its two-level clustering approach which considers a window of samples instead of a single sample to define phases. Out of all the classifiers, we observe that *LF* tends to have the worst EFP values. When comparing window and phase change predictors, a phase change predictor is always best across all classifiers. For the manual classifier, the LSTM model for next phase prediction shows strong superiority over the SVM model. However, such a clear trend is not exhibited by other classifiers. Clearly, GHT shows the highest EFP values for any classifier and duration prediction. This shows that learning-based predictors are more suitable for next phase prediction.

The study aims at finding stable, long-term, predictable phases. A next step for future work is to find phase classifiers and predictors that maximize the accuracy of the phase-aware approach presented in Fig. 18.20 [2]. Therefore, the best combination of short-term and long-term predictions for a phase-aware approach is yet to be found.

18.5 Summary and Conclusions

Machine learning for system-level modeling has emerged to guide rapid and accurate design-time and runtime optimizations. We have surveyed a representative set of studies and presented selected results in three learning dimensions: abstraction, components, and time. These dimensions form a three-dimensional space where vectors represent supervised learning directions with starting and ending points comprising features and labels, respectively. Models that learn across the abstraction layer dimension find cross-layer relationships that allow predicting low-level metrics with high-level features, e.g., collect features from a functional simulator to predict gate-level power values. In the components dimension, cross-platform predictive models learn the relationship between executions of programs in different platforms, enabling predictions of power and performance of applications without running them on the target platforms. Additionally, researchers have discovered that interpretable models, such as random forests can provide insights into architecture bottlenecks. Finally, models that learn across time anticipate dynamic workload changes, enabling proactive runtime optimizations. We have further categorized these models into long-term and short-term forecasting, depending on the granularity of the workload behaviors that the models are aiming to predict.

In all cases, studies presented in these chapters highlight the importance of choosing the right approaches when applying machine learning to EDA problems. EDA problems such as modeling come with specific characteristics and needs, such as limited amounts of training data or availability of a priori knowledge, where just blindly applying deep learning approaches rarely provides an optimal solution. Instead, this calls for specialized learning formulations that are tailored to the problem at hand.

Furthermore, while the majority of the models presented in this chapter learn in one dimension, predictions can target multiple dimensions. For example, cross-platform predictors have been deployed in operating systems to make mapping and scheduling decisions in heterogeneous platforms [46]. However, pure cross-platform models can only predict whether a different core would have been a better target after having run the workload on some core for some time. Combining such cross-platform approaches with cross-temporal workload forecasting models would instead allow proactively predicting the outcome of migrating an upcoming workload phase to a different platform in the system. Such models remain unexplored and represent and interesting direction for future work.

References

1. Alcorta, E.S., Gerstlauer, A.: Learning-based workload phase classification and prediction using performance monitoring counters. In: Workshop on Machine Learning for CAD (MLCAD) (2021)
2. Alcorta, E.S., Rama, P., Ramachandran, A., Gerstlauer, A.: Phase-aware CPU workload forecasting. In: International Conference on Embedded Computer Systems: Architectures, Modeling and Simulation (SAMOS) (2021)
3. Ardalani, N., Lestourgeon, C., Sankaralingam, K., Zhu, X.: Cross-architecture performance prediction (XAPP) using CPU code to predict GPU performance. In: International Symposium on Microarchitecture (MICRO) (2015).
4. Ardalani, N., Thakker, U., Albarghouthi, A., Sankaralingam, K.: A static analysis-based cross-architecture performance prediction using machine learning. Computing Research Repository (CoRR) (2019)
5. Bogliolo, A., Benini, L., De Micheli, G.: Regression-based RTL power modeling. ACM Trans. Des. Autom. Electron. Syst. (TODAES) 5(3), 337–372 (2000)
6. Bridges, R.A., Imam, N., Mintz, T.M.: Understanding GPU power: A survey of profiling, modeling, and simulation methods. ACM Comput. Surv. 49(3), 41:1–41:27 (2016).
7. Brooks, D., Tiwari, V., Martonosi, M.: Wattch: A framework for architectural-level power analysis and optimizations. In: International Symposium on Computer Architecture (ISCA) (2000)
8. Brooks, D., Bose, P., Srinivasan, V., Gschwind, M.K., Emma, P.G., Rosenfield, M.G.: New methodology for early-stage, microarchitecture-level power-performance analysis of microprocessors. IBM J. Res. Dev. 47(5.6), 653–670 (2003)
9. Canis, A., Choi, J., Aldham, M., Zhang, V., Kammoona, A., Czajkowski, T.S., Brown, S.D., Anderson, J.H.: Legup: An open-source high-level synthesis tool for FPGA-based processor/accelerator systems. ACM Trans. Embed. Comput. Syst. 13(2), 24:1–24:27 (2013)
10. Chakravarty, S., Zhao, Z., Gerstlauer, A.: Automated, retargetable back-annotation for host compiled performance and power modeling. In: International Conference on Hardware/Software Codesign and System Synthesis (CODES+ISSS) (2013)
11. Chang, C.H., Liu, P., Wu, J.J.: Sampling-based phase classification and prediction for multithreaded program execution on multi-core architectures. In: International Conference on Parallel Processing (ICPP) (2013)
12. Chiu, M.C., Moss, E.: Run-time program-specific phase prediction for python programs. In: International Conference on Managed Languages & Runtimes (ManLang) (2018)
13. Choi, Y., Zhang, P., Li, P., Cong, J.: Hlscope+,: Fast and accurate performance estimation for FPGA HLS. In: International Conference on Computer-Aided Design (ICCAD) (2017)
14. Choi, Y.K., Cong, J.: Hlscope: High-level performance debugging for FPGA designs. In: International Symposium on Field-Programmable Custom Computing Machines (FCCM) (2017)
15. Cochran, R., Reda, S.: Consistent runtime thermal prediction and control through workload phase detection. In: Design Automation Conference (DAC) (2010)
16. Cong, J., Liu, B., Neuendorffer, S., Noguera, J., Vissers, K.A., Zhang, Z.: High-level synthesis for FPGAs: From prototyping to deployment. IEEE Trans. Comput. Aided Des. Integr. Circuits Syst. 30(4), 473–491 (2011)
17. Coşkun, A.K., Šimunić Rosing, T., Gross, K.C.: Utilizing predictors for efficient thermal management in multiprocessor SoCs. IEEE Trans. Comput. Aided Des. Integr. Circuits Syst. 28(10), 1503–1516 (2009)
18. Curtis-Maury, M., Blagojevic, F., Antonopoulos, C.D., Nikolopoulos, D.S.: Prediction-based power-performance adaptation of multithreaded scientific codes. IEEE Trans. Parallel Distrib. Syst. 19(10), 1396–1410 (2008)
19. Dietrich, B., Goswami, D., Chakraborty, S., Guha, A., Gries, M.: Time series characterization of gaming workload for runtime power management. IEEE Trans. Comput. 64(1), 260–273 (2015)

20. Duesterwald, E., Caşcaval, C., Dwarkadas, S.: Characterizing and predicting program behavior and its variability. In: Parallel Architectures and Compilation Techniques (PACT) (2003)
21. Gerstlauer, A., Chakravarty, S., Kathuria, M., Razaghi, P.: Abstract system-level models for early performance and power exploration. In: Asia and South Pacific Design Automation Conference (ASPDAC) (2012)
22. Jain, R., Panda, P.R., Subramoney, S.: Machine learned machines: Adaptive co-optimization of caches, cores, and on-chip network. In: Design, Automation and Test in Europe (DATE) (2016)
23. Khan, O., Kundu, S.: Microvisor: A runtime architecture for thermal management in chip multiprocessors. In: Lecture Notes in Computer Science (LNCS) (2011)
24. Khanna, R., John, J., Rangarajan, T.: Phase-aware predictive thermal modeling for proactive load-balancing of compute clusters. In: International Conference on Energy Aware Computing (ICEAC) (2012)
25. Khoshbakht, S., Dimopoulos, N.: A new approach to detecting execution phases using performance monitoring counters. In: Lecture Notes in Computer Science (LNCS) (2017)
26. Kumar, A.K.A., Alsalamin, S., Amrouch, H., Gerstlauer, A.: Machine learning-based microarchitecture-level power modeling of CPUs. IEEE Trans. Comput. (TC) (submitted)
27. Kumar, A.K.A., Gerstlauer, A.: Learning-based CPU power modeling. In: Workshop on Machine Learning for CAD (MLCAD) (2019)
28. Kwon, J., Carloni, L.P.: Transfer learning for design-space exploration with high-level synthesis. In: Workshop on Machine Learning for CAD (MLCAD) (2020)
29. Lau, J., Schoenmackers, S., Calder, B.: Transition phase classification and prediction. In: International Symposium on High-Performance Computer Architecture (HPCA) (2005)
30. LeBeane, M., Ryoo, J.H., Panda, R., John, L.K.: Watt Watcher: Fine-grained power estimation for emerging workloads. In: International Symposium on Computer Architecture and High Performance Computing (SBAC-PAD) (2015)
31. Lee, D., Gerstlauer, A.: Learning-based, fine-grain power modeling of system-level hardware IPs. ACM Trans. Des. Autom. Electron. Syst. (TODAES) 23(3), 1–25 (2018)
32. Lee, D., John, L.K., Gerstlauer, A.: Dynamic power and performance back-annotation for fast and accurate functional hardware simulation. In: Design, Automation and Test in Europe (DATE) Conference (2015)
33. Lee, D., Kim, T., Han, K., Hoskote, Y., John, L.K., Gerstlauer, A.: Learning-based power modeling of system-level black-box IPs. In: International Conference on Computer-Aided Design (ICCAD) (2015)
34. Lee, W., Kim, Y., Ryoo, J.H., Sunwoo, D., Gerstlauer, A., John, L.K.: PowerTrain: A learning-based calibration of McPAT power models. In: International Symposium on Low Power Electronics and Design (ISLPED) (2015)
35. Li, S., Ahn, J.H., Strong, R.D., Brockman, J.B., Tullsen, D.M., Jouppi, N.P.: The McPAT framework for multicore and manycore architectures: Simultaneously modeling power, area, and timing. ACM Transaction on Architecture Code Optimization (TACO) 10(1), 1–29 (2013)
36. Makrani, H.M., Sayadi, H., Mohsenin, T., Rafatirad, S., Sasan, A., Homayoun, H.: XPPE: cross-platform performance estimation of hardware accelerators using machine learning. In: Asia and South Pacific Design Automation Conference, (ASPDAC) (2019)
37. Masouros, D., Xydis, S., Soudris, D.: Rusty: Runtime interference-aware predictive monitoring for modern multi-tenant systems. IEEE Trans. Parallel Distrib. Syst. 32(1), 184–198 (2021)
38. McCullough, J.C., Agarwal, Y., Chandrashekar, J., Kuppuswamy, S., Snoeren, A.C., Gupta, R.K.: Evaluating the effectiveness of model-based power characterization. In: USENIX Annual Technical Conference (2011)
39. Mendis, C., Renda, A., Amarasinghe, S., Carbin, M.: Ithemal: Accurate, portable and fast basic block throughput estimation using deep neural networks. In: International Conference on Machine Learning (ICML) (2019)
40. Moghaddam, M.G., Guan, W., Ababei, C.: Investigation of LSTM based prediction for dynamic energy management in chip multiprocessors. In: International Green and Sustainable Computing Conference (IGSC) (2018)

41. O'Neal, K., Brisk, P.: Predictive modeling for CPU, GPU, and FPGA performance and power consumption: A survey. In: Symposium on VLSI (ISVLSI) (2018)

42. O'Neal, K., Brisk, P., Abousamra, A., Waters, Z., Shriver, E.: GPU performance estimation using software rasterization and machine learning. ACM Trans. Embed. Comput. Syst. **16**(5s), 148:1–148:21 (2017)

43. O'Neal, K., Liu, M., Tang, H., Kalantar, A., DeRenard, K., Brisk, P.: Hlspredict: cross platform performance prediction for FPGA high-level synthesis. In: International Conference on Computer-Aided Design (ICCAD) (2018)

44. O'Neal, K., Brisk, P., Shriver, E., Kishinevsky, M.: Hardware-assisted cross-generation prediction of GPUs under design. IEEE Trans. Comput. Aided Des. Integr. Circuits Syst. **38**(6), 1133–1146 (2019)

45. Park, Y.H., Pasricha, S., Kurdahi, F.J., Dutt, N.: A multi-granularity power modeling methodology for embedded processors. IEEE Transactions on Very Large Scale Integration Systems (TVLSI) **19**(4), 668–681 (2011)

46. Prodromou, A., Venkat, A., Tullsen, D.M.: Deciphering predictive schedulers for heterogeneous-ISA multicore architectures. In: International Workshop on Programming Models and Applications for Multicores and Manycores (PMAM) (2019)

47. Rapp, M., Elfatairy, O., Wolf, M., Henkel, J., Amrouch, H.: Towards nn-based online estimation of the full-chip temperature and the rate of temperature change. In: Workshop on Machine Learning for CAD (MLCAD) (2020)

48. Rapp, M., Pathania, A., Mitra, T., Henkel, J.: Neural network-based performance prediction for task migration on s-nuca many-cores. IEEE Trans. Comput. **70**(10), 1691–1704 (2021)

49. Sagi, M., Rapp, M., Khdr, H., Zhang, Y., Fasfous, N., Doan, N.A.V., Wild, T., Henkel, J., Herkersdorf, A.: Long short-term memory neural network-based power forecasting of multi-core processors. In: Design, Automation and Test in Europe (DATE) (2021)

50. Sarikaya, R., Buyuktosunoglu, A.: Predicting program behavior based on objective function minimization. In: International Symposium on Workload Characterization (IISWC) (2007)

51. Sembrant, A., Eklov, D., Hagersten, E.: Efficient software-based online phase classification. In: International Symposium on Workload Characterization (IISWC) (2011)

52. Singh, G., Diamantopoulos, D., Gómez-Luna, J., Stuijk, S., Mutlu, O., Corporaal, H.: Modeling FPGA-based systems via few-shot learning. In: International Symposium on Field Programmable Gate Arrays (2021)

53. Srinivasan, S., Kumar, R., Kundu, S.: Program phase duration prediction and its application to fine-grain power management. In: Symposium on VLSI (ISVLSI) (2013)

54. Sunwoo, D., Wu, G.Y., Patil, N.A., Chiou, D.: PrEsto: An FPGA-accelerated power estimation methodology for complex systems. In: International Conference on Field Programmable Logic and Applications (FPL) (2010)

55. Taht, K., Greensky, J., Balasubramanian, R.: The pop detector: A lightweight online program phase detection framework. In: International Symposium on Performance Analysis of Systems and Software (ISPASS) (2019)

56. Wang, S., Zhong, G., Mitra, T.: Cgpredict: Embedded GPU performance estimation from single-threaded applications. ACM Trans. Embed. Comput. Syst. **16**(5s), 146:1–146:22 (2017)

57. Wu, G., Greathouse, J.L., Lyashevsky, A., Jayasena, N., Chiou, D.: GPGPU performance and power estimation using machine learning. In: International Symposium on High Performance Computer Architecture (HPCA) (2015)

58. Xi, S.L., Jacobson, H., Bose, P., Wei, G.Y., Brooks, D.: Quantifying sources of error in McPAT and potential impacts on architectural studies. In: International Symposium on High Performance Computer Architecture (HPCA) (2015)

59. Yang, J., Ma, L., Zhao, K., Cai, Y., Ngai, T.F.: Early stage real-time SoC power estimation using RTL instrumentation. In: Asia and South Pacific Design Automation Conference (ASPDAC) (2015)

60. Zaman, M., Ahmadi, A., Makris, Y.: Workload characterization and prediction: A pathway to reliable multi-core systems. In: International On-Line Testing Symposium (IOLTS) (2015)

61. Zhang, W., Li, J., Li, Y., Chen, H.: Multilevel phase analysis. ACM Trans. Embed. Comput. Syst. **14**(2), 31:1–31:29 (2015)
62. Zhao, Z., Gerstlauer, A., John, L.K.: Source-level performance, energy, reliability, power and thermal (PERPT) simulation. IEEE Trans. Comput. Aided Des. Integr. Circuits Syst. (TCAD) **36**(2), 299–312 (2016)
63. Zheng, X., Ravikumar, P., John, L., Gerstlauer, A.: Learning-based analytical cross-platform performance prediction. In: International Conference on Embedded Computer Systems: Architectures, Modeling, and Simulation (SAMOS) (2015)
64. Zheng, X., John, L.K., Gerstlauer, A.: Accurate phase-level cross-platform power and performance estimation. In: Design Automation Conference (DAC) (2016)
65. Zheng, X., John, L.K., Gerstlauer, A.: LACross: Learning-based analytical cross-platform performance and power prediction. Int. J. Parallel Prog. (IJPP) **45**(6), 1488–1514 (2017)
66. Zheng, X., Vikalo, H., Song, S., John, L.K., Gerstlauer, A.: Sampling-based binary-level cross-platform performance estimation. In: Design, Automation and Test in Europe (DATE) (2017)
67. Zhong, G., Prakash, A., Liang, Y., Mitra, T., Niar, S.: Lin-analyzer: a high-level performance analysis tool for FPGA-based accelerators. In: Design Automation Conference (DAC) (2016)
68. Zhong, G., Prakash, A., Wang, S., Liang, Y., Mitra, T., Niar, S.: Design space exploration of FPGA-based accelerators with multi-level parallelism. In: Design, Automation and Test in Europe (DATE) (2017)
69. Zhou, Y., Ren, H., Zhang, Y., Keller, B., Khailany, B., Zhang, Z.: PRIMAL: Power inference using machine learning. In: Design Automation Conference (DAC) (2019)

Index

© The Author(s), under exclusive license to Springer Nature Switzerland AG 2022
H. Ren, J. Hu (eds.), *Machine Learning Applications in Electronic Design Automation*, https://doi.org/10.1007/978-3-031-13074-8

Printed in the United States
by Baker & Taylor Publisher Services